Plant Drug Analysis

Second Edition

Hildebert Wagner • Sabine Bladt

Plant Drug Analysis

A Thin Layer Chromatography Atlas

Second Edition

Photographs by Veronika Rickl

Professor Dr. h.c. Hildebert Wagner
Universität München
Institut für Pharmazeutische Biologie
Karlsstraße 29
80333 München
Germany

Dr. rer. nat. Sabine Bladt
Universität München
Institut für Pharmazeutische Biologie
Karlsstraße 29
80333 München
Germany

current address:
Professor Dr. h.c. Hildebert Wagner
Department of Pharmacy
- Center for Pharma Research -
Ludwig-Maximilians-University of Munich
Butenandtstr. 5
Building B
81377 Munich
Germany
H.Wagner@cup.uni-muenchen.de

ISBN 978-3-540-58676-0 (hardcover)
ISBN 978-3-642-00573-2 (softcover)
e-ISBN 978-3-642-00574-9
DOI 10.1007/978-3-642-00574-9
Springer Dordrecht Heidelberg London New York

Library of Congress Control Number: 2009925996

Cover design: WMXdesign GmbH, Heidelberg, Germany

Printed on acid-free paper

Springer is part of Springer Science+Business Media (www.springer.com)

Preface to the Second Edition

More than 12 years have passed since the first and very successful attempt was made to reproduce the thin layer chromatography (TLC) separation of 170 medicinal plant drugs in the form of color TLC fingerprints in a book. The reproduction of natural color photographs in UV 365 nm was a difficult undertaking at that time due to the relatively unsophisticated film and filter technology. The first German edition of this book with its appended English translation met with worldwide acceptance in the field of natural product chemistry and has remained an indispensable aid in the laboratory analysis of medicinal drugs.

Due to the higher demands now placed on plant drug quality, the introduction of herbal preparations with medicinal significance, and the increasing number of phytochemical preparations, the analytical and standardization procedures of the plants have gained even greater importance. We have tried to do justice to this development in this second edition.

This TLC atlas now includes about 230 medicinal plants of worldwide interest. The photographs of the TLC fingerprints and the descriptions of the characteristic compounds of each plant extract are a quick and reliable source for the identification and purity check of plant material and phytopreparations.

Most of the TLC systems are standard systems and have been optimized when necessary. In spite of other available analytic techniques, such as gas chromatography and high performance liquid chromatography, TLC still remains a most useful, quick, effective, and low-cost method for the separation and identification of complex mixtures of herbal drug preparations and plant constituents.

The authors are most grateful to Ms. Ute Redl for her comprehensive technical assistance. We also thank Ms. Veronika Rickl not only for the excellent quality of the photographs, but also for the layout of the TLC fingerprint pages in the book and for the drawing of the chemical formulae.

Munich, March 1996

Sabine Bladt
Hildebert Wagner

Contents

Introduction

Thin-Layer Chromatographic Analysis of Drugs

Of the many chromatographic methods presently available, thin-layer chromatography (TLC) is widely used for the rapid analysis of drugs and drug preparations. There are several reasons for this:

- The time required for the demonstration of most of the characteristic constituents of a drug by TLC is very short.
- In addition to qualitative detection, TLC also provides semi-quantitative information on the major active constituents of a drug or drug preparation, thus enabling an assessment of drug quality.
- TLC provides a chromatographic drug fingerprint. It is therefore suitable for monitoring the identity and purity of drugs and for detecting adulterations and substitutions.
- With the aid of appropriate separation procedures, TLC can be used to analyze drug combinations and phytochemical preparations.

Photographic Record of Thin-Layer Separations of Drug Extracts (A Photographic TLC Drug Atlas)

- A photographic TLC atlas fulfils the same function and purpose as a catalogue of spectra. The identity or non-identity of an official drug can be established by comparison with the chromatogram of the "standard drug".
- Unknown commercial drugs can be classified by comparison with the visual record in the TLC atlas.
- The photographic drug atlas is an aid to the routine identification and purity testing of drugs in control laboratories, and it can be used without previous pharmacognostic training.
- Photographic reproduction of thin-layer separations has a large didactic advantage over mere graphic representation. The TLC photo-drug atlas has an immediate clarity of representation that facilitates the learning of TLC drug analysis for the student.

Compilation of a TLC Drug Atlas

Compilation of a TLC drug atlas was governed by certain preconditions, related to the source of the drugs, the TLC technique in general and the photographic reproduction of the thin-layer chromatograms.

Source of the drugs The drugs used in the compilation of a drug atlas must meet the standards of the official pharmacopoeia, and they must originate from a clearly identified botanical source.

Slight variations in the chromatographic picture, due to botanical varieties or differences in cultivation, climatic conditions, time of harvesting and drying and extraction methods are normal.

Extraction conditions The chosen extraction procedures should be fast, but efficient, according to present scientific knowledge. They have often been adopted from the pharmacopoeias and modified when new drug substances or separation problems have been encountered.

TLC Reproducible TLC separations can be guaranteed only if standardized adsorption layers are used. Commercially available TLC plates were therefore used (Silica gel 60 F_{254}-precoated TLC plates; Merck, Germany). Silica gel is an efficient adsorbent for the TLC separation of most of the drug extracts. In specific cases aluminium oxide- or cellulose-precoated plates (Merck, Germany) have been used.

Since special chromatography rooms are not always available, all TLC separations were performed at room temperature, i.e. 18°–22°C. Details of the TLC technique can be found in pharmacopoeias and books on methodology (see Standard Literature and Pharmacopoeias). Generally a distance of 15 cm is used for the development of a chromatogram.

Chromatography solvents In choosing suitable solvent systems, preference has been given to those which are not too complicated in their composition, which possess minimal temperature sensitivity and which give exact and sufficient separation of constituents, enough for a significant characterization of the drug.

Concentration of substances for TLC In order to obtain sharply resolved zones, the quantity of material applied to the chromatogram should be as small as possible. Rather large sample volumes are, however, often necessary for the detection (by colour reactions) of substances that are present in low concentration; this inevitably results in broadening and overlapping of zones.

Detection methods For the detection of the main, characteristic compounds of a drug, methods were chosen that give the most striking colours.

The active principles of a group of drugs may be very similar (e.g. drugs from Solanaceae or saponin drugs), so that differentiation and identification are difficult or impossible on the basis of the active principles alone. In such cases, other classes of compounds have been exploited for the purposes of differentiation.

For drugs with unknown or incompletely known active principles, identification has been based on other non-active, but easily detectable constituents that can be regarded as "guide substances".

Photography The developed chromatograms were photographed on Kodak Gold 100 (Negativfilm) or Kodak EPY (diapositive film). To achieve authentic colour reproduction, different commercially available yellow and ultraviolet (UV) filters (e.g. B+W 409) are used. Photography in UV-365 nm needs a specific technique of exposure, individual times for each type of fluorescent compound and, last but not least, a great deal of experience. Further information on photography is given in the publication by E. HAHN-DEINSTROP (Chromatographie, GIT Suppl. 3/1989, pp. 29–31).

1 Alkaloid Drugs

Most plant alkaloids are derivatives of tertiary amines, while others contain primary, secondary or quarternary nitrogen. The basicity of individual alkaloids varies considerably, depending on which of the four types is represented. The pK_B values (dissociation constants) lie in the range of 10–12 for very weak bases (e.g. purines), of 7–10 for weak bases (e.g. Cinchona alkaloids) and of 3–7 for medium-strength bases (e.g. Opium alkaloids).

1.1 Preparation of Extracts

Alkaloid drugs with medium to high alkaloid contents (≥1%)

General method, extraction method A

Powdered drug (1g) is mixed thoroughly with 1ml 10% ammonia solution or 10% Na_2CO_3 solution and then extracted for 10min with 5ml methanol under reflux. The filtrate is then concentrated according to the total alkaloids of the specific drug, so that 100µl contains 50–100µg total alkaloids (see drug list, section 1.4).

Exception

Harmalae semen: Powdered drug (1g) is extracted with 10ml methanol for 30min under reflux. The filtrate is diluted 1:10 with methanol and 20µl is used for TLC.

Strychni semen: Powdered seeds (1g) are defatted with 20ml n-hexane for 30min under reflux. The defatted seeds are then extracted with 10ml methanol for 10min under reflux. A total of 30µl of the filtrate is used for TLC.

Colchici semen: Powdered seeds (1g) are defatted with 20ml n-hexane for 30min under reflux. The defatted seeds are then extracted for 15min with 10ml chloroform. After this, 0.4ml 10% NH_3 is added to the mixture, shaken vigorously and allowed to stand for about 30min before filtration. The filtrate is evaporated to dryness and the residue solved in 1ml ethanol; 20µl is used for TLC investigation.

Alkaloid drugs with low total alkaloids (<1%)

Enrichment method, extraction method B

Powdered drug (2g) is ground in a mortar for about 1min with 2ml 10% ammonia solution and then thoroughly mixed with 7g basic aluminium oxide (activity grade I). This mixture is then packed loosely into a glass column (diameter, 1.5cm; length, 20cm) and 10ml $CHCl_3$ is added. Alkaloid bases are eluted with about 5ml $CHCl_3$ and the eluate is collected, evaporated to 1ml and used for TLC.
This method is suitable for the Solanaceae drugs, e.g. Belladonnae or Scopoliae radix and Stramonii semen, which should be defatted first by extraction with n-hexane or light petroleum. Leaf extracts contain chlorophylls, which can interfere with the TLC separation. In such cases extraction with sulphuric acid (described below) is recommended.

Sulphuric acid extraction method C

Powdered drug (0.4–2 g) is shaken for 15 min with 15 ml 0.1 N sulphuric acid and then filtered. The filter is washed with 0.1 N sulphuric acid to a volume of 20 ml filtrate; 1 ml concentrated ammonia is then added. The mixture is shaken with two portions of 10 ml diethyl ether. The ether is dried over anhydrous sodium sulphate, filtered and evaporated to dryness and the resulting residue dissolved in 0.5 ml methanol.
This is the preferred method for leaf drugs, e.g. Belladonnae folium (0.6 g), Stramonii folium (0.4 g), Hyoscyami folium (2 g) or Fumariae herba (1 g).

1.2 Thin-Layer Chromatography

Drug extracts

The samples applied to the TLC plate should contain between 50 and 100 µg total alkaloids, which have to be calculated according to the average alkaloid content of the specific drug (see 1.4 Drug List).

Example: Powdered drug (1 g) with a total alkaloid content of 0.3 %, extracted with 5 ml methanol by the general method described above will yield 3 mg in 5 ml methanolic solution, containing approximately 60 µg total alkaloids per 100 µl.

Reference compounds

- Commercially available compounds are usually prepared in 1% alcoholic solution and 10 µl is applied for TLC, e.g. atropine, brucine, strychnine, berberine, codeine.
- Rauvolfia alkaloids are prepared in 0.5% alcoholic solution, and 10 µl is applied for TLC, e.g. reserpine, rescinnamine, rauwolscine, ajmaline, serpentine.
- Colchicine is prepared as a 0.5% solution in 70% ethanol, and 10 µl is applied for TLC.

Alkaloid references can also be obtained from pharmaceutical products by a simple methanol extraction. The sample solution used for TLC should contain between 50 and 100 µg alkaloid.

- Alkaloid content 10–250 mg per tablet or dragée:
  One powdered tablet or dragée is mixed with 1 ml methanol per 10 mg alkaloid and shaken for about 5 min at 60°C. After filtration or centrifugation, the extract is applied directly; 10 µl then corresponds to 100 µg alkaloid.
- Alkaloid content 0.075–1.0 mg per tablet or dragée:
  Ten powdered tablets or dragées are mixed with 10 ml methanol, shaken for about 5 min at 60°C and filtered and the filtrate evaporated to dryness. The residue is dissolved in 1 ml methanol and, if necessary, the solution cleared by centrifugation; 10 µl of this solution contains 100 µg alkaloid (1.0 mg/tablet), or 100 µl contains 75 µg alkaloid (0.075 mg/tablet).

Test mixtures

- Cinchona alkaloids test mixture for Cinchonae (Chinae) cortex (DAB 10)
  A mixture of 17.5 mg quinine, 0.5 mg quinidine, 10 mg cinchonine and 10 mg cinchonidine is dissolved in 5 ml ethanol, and 2 µl of this solution is applied for TLC.
- Test mixture for Solanaceae drugs (DAB 10)
  A total of 50 mg hyoscyamine sulphate is dissolved in 9 ml methanol and 15 mg scopolamine hydrobromide in 10 ml methanol.

  For Belladonnae folium (T1): 1.8 ml scopolamine hydrobromide solution is added to 8 ml hyoscyamine sulphate solution; 20 µl is used for TLC.

For Hyoscyami folium (T2): 4.2 ml scopolamine hydrobromide solution is added to 3.8 ml hyoscyamine sulphate solution; 20 μl is used for TLC.
For Stramonii folium (T3): 4.2 ml scopolamine hydrobromide solution is added to 3.8 ml hyoscyamine sulphate solution; 20 μl is used for TLC.

Adsorbent

Silica gel 60 F_{254}-precoated TLC plates (Merck, Darmstadt, Germany)
▶ The principal alkaloids of the most common alkaloid drugs can be identified.

Aluminium oxide-precoated TLC plates (Merck, Darmstadt, Germany)
▶ More suitable for the separation of berberine, columbamine and jatrorrhizine.

Chromatography solvents

Solvent system	Drug, alkaloids
Toluene–ethyl acetate–diethylamine (70:20:10)	**Screening system**, suitable for the major alkaloids of most drugs
Chloroform–diethylamine (90:10)	Chinae cortex; Cinchona alkaloids
Acetone–light petroleum–diethylamine (20:70:10)	Gelsemii radix
Cyclohexane–ethanol–diethylamine (80:10:10) Cyclohexane–chloroform–diethylamine (50:40:10)	Aconiti tuber
Chloroform–acetone–diethylamine (50:40:10) Chloroform–methanol–ammonia 10% (80:40:15)	Harmalae semen
Ethyl acetate–isopropanol–ammonia 25% (100:2:1)	Uncariae cortex
Dioxane–ammonia 25% (90:10)	Adhatodae folium
Ethyl acetate–cyclohexane–methanol–ammonia 25% (70:15:10:5)	Ephedrae herba
Ethyl acetate–methanol–water (100:13.5:10)	**Screening system**, suitable e.g. for xanthine derivatives, Colchicum and Rauvolfia alkaloids
Ethyl acetate–methanol (90:10)	Vincae herba
Ethyl acetate–methanol (60:20)	Catharanthi folium
Toluene–chloroform–ethanol (28.5:57:14.5)	Secale alkaloids Ephedrae herba
n-Propanol–formic acid–water (90:1:9)	Berberidis cortex, Hydrastis rhizoma, Colombo radix, Chelidonii herba
n-Butanol–ethyl acetate–formic acid–water (30:50:10:10)	Mahoniae radices cortex

Solvent system	Drug, alkaloids
Ethyl acetate–ethylmethyl ketone–formic acid–water (50:30:10:10)	Fumariae herba, Corydalidis rhizoma
Cyclohexane–chloroform–glacial acetic acid (45:45:10)	Berberine- and protoberberine-type alkaloids
Chloroform–methanol–glacial acetic acid (47.5:47.5:5)	Genistae herba, Sarothamni herba, Spartii scop. flos
n-Butanol–glacial acetic acid–water (40:40:10)	Catharanthus alkaloids

1.3 Detection

- UV-254 nm Pronounced quenching of some alkaloid types such as indoles, quinolines, isoquinolines, purines; weak quenching of e.g. tropine alkaloids
- UV-365 nm Blue, blue-green or violet fluorescence of alkaloids, e.g. Rauvolfiae radix, Chinae cortex, Ipecacuanhae radix, Boldo folium. Yellow fluorescence, e.g. colchicine, sanguinarine, berberine
- Spray reagents (see Appendix A)

- Dragendorff reagent (DRG No.13)
  The alkaloids appear as brown or orange-brown (vis.) zones immediately on spraying. The colour is fairly stable. Some types such as purines or ephedrine need special detection. The colour of alkaloid zones can be intensified or stabilized by spraying first with Dragendorff reagent and then with 10% sodium nitrite solution or 10% ethanolic sulphuric acid.
- Iodoplatinate reagent (IP No.21)
  Directly after spraying, alkaloids appear as brown, blue or whitish zones (vis.) on the blue-grey background of the TLC plate.
- Special detection

Iodine-potassium iodide-HCl reagent (No.20)	→ purines
Iodine $CHCl_3$ reagent (No.19)	→ emetine, cephaeline
Marquis reagent (No.26)	→ opium alkaloids
van Urk reagent (No.43)	→ secale alkaloids
Ninhydrine reagent (No.29)	→ ephedrine
10% ethanolic H_2SO_4 (No.37)	→ china alkaloids

1.4 Drug List

The chromatograms of the specific alkaloid drugs are reproduced according to their alkaloid types (Fig. 1–30).

Drug/plant source Family/pharmacopoeia	Total alkaloids Major alkaloids (for formulae see 1.5 Formulae)	
Indole Alkaloids		**Fig. 3–10**
Rauvolfiae radix Rauvolfia, snake root Rauvolfia serpentina (L.) BENTH ex KURZ. Rauvolfia vomitoria AFZEL Apocynaceae DAB 10, USP XXII, MD	0.6%–2.4% total alkaloids (R. serpentina) 1.3%–3% total alkaloids (R. vomitoria) >50 alkaloids, yohimbane derivatives: Reserpine (0.14%), rescinnamine (0.01%), epi-rauwolscine (0.08%), serpetine (0.08%), serpentinine (0.13%), ajmaline (0.1%), ajmalicine (=raubasine 0.02%), raupine (0.02%)	**Fig. 3**
Yohimbe cortex Yohimbe bark Pausinystalia johimbe PIERRE Rubiaceae	2.3%–3.9% total alkaloids Yohimbine and ten minor alkaloids, e.g. pseudoyohimbine and coryantheine	**Fig. 4**
Quebracho cortex Aspidosperma bark Aspidosperma quebracho-blanco SCHLECHT Apocynaceae DAC 86	0.3%–1.5% total alkaloids (>30) Yohimbine, pseudoyohimbine, aspidospermine, aspidospermatine, quebrachamine, hypoquebrachamine, quebrachocidine	**Fig. 4**
Catharanthi folium Catharanthus leaves Catharanthus roseus (L.) G. DON. (syn. Vinca rosea L.) Apocynaceae MD	0.15%–0.25% total alkaloids Vinblastine (0.01%), vincristine, vindoline, catharanthine, Root: <0.74% total alkaloids	**Fig. 4**
Vincae herba Common periwinkle Vinca minor L. Apocynaceae MD	0.15%–1% total alkaloids Vincamine (0.05%–0.1%), vincaminine, vincamajine, vincine, minovincine, reserpinine	**Fig. 5**
Strychni semen Poison nuts, Nux vomica seeds Strychnos nux-vomica L. Loganiaceae ÖAB, Helv. VII, MD, Japan	2%–3% total alkaloids Strychnine (>1%) and brucine (>1.5%), α- and β-colubrine, vomicine; psendostrychnine, psendobrucine	**Fig. 6**
Ignatii semen St. Ignaz beans Strychnos ignatii BERG Loganiaceae	2.5%–3% total alkaloids Strychnine (45%–50%), brucine, 12-hydroxy strychnine, α-colubrine, vomicine	**Fig. 6**

	Drug/plant source Family/pharmacopoeia	Total alkaloids Major alkaloids (for formulae see 1.5 Formulae)
Fig. 7	**Secale cornutum** Ergot Claviceps purpurea (FRIES) TULASNE Clavicipitaceae (Ascomycetes) ÖAB, MD	0.2%–1% total alkaloids Ergot alkaloids, lysergic acid alkaloids; amide alkaloids (ergometrine), peptide alkaloids (ergotamine), ergotoxin group (ergocristine)
Fig. 8	**Gelsemii radix** Yellow jasmine, wild woodbine Gelsemium sempervirens (L.) AIT. Loganiaceae MD	0.25%–0.7% total alkaloids Gelsemine, sempervirine, (isogelsemine, gelsemicine)
Fig. 9	**Harmalae semen** Syrian (wild) rue Peganum harmala L. Zygophyllaceae	2.5%–4% total alkaloids Carbolinderivatives: harmaline (>60%), harmine, harmalol, harmidine Quinazoline alkaloids: (−)-vasicine (= (−) peganine), vasicinone
Fig. 10A	**Justiciae-adhatodae-folium** Malabarnut leaves Justicia adhatoda L. (syn. Adhatoda vasica NEES.) Acanthaceae MD	0.5%–2% quinazoline alkaloids Vasicine (45–95%), vasicinine Vasicinone, oxyvasicinine (oxidation products, artefacts)
Fig. 10B	**Uncariae radix** Uncaria ("una de gato") Uncaria tomentosa WILLD. Rubiaceae	>0.9% tetracyclic and pentacyclic oxindoles Rhychnophylline, isorhychnophylline, mitraphylline, isomitraphylline, pteropodine, isopteropodine, uncarine A, F
Fig. 11–16	**Quinoline and isoquinoline alkaloids alkaloids of the morphinane type (phenanthrene type)**	
Fig. 11	**Ipecacuanhae radix** Ipecacuanhae root Cephaelis ipecacuanha (BORT.) RICH. (Rio and Matto-Grosso)	1.8%–6% total alkaloids Emetine and cephaeline (>95%), o-methylpsychotrine and psychotrine (corresponding dehydro compounds) 1:1 → 3:1 ratio of emetine to cephaeline
	Cephaelis acuminata KARSTEN (Cartagena, Panama drugs) Rubiaceae DAB 10, Ph. Eur. I, ÖAB, Helv. VII, BP 88, USP XXII, MD, DAC 86	1.7%–3.5% total alkaloids cephaeline (>50%), emetine; o-methylpsychotrine, psychotrine (0.05%)

Drug/plant source Family/pharmacopoeia	Total alkaloids Major alkaloids (for formulae see 1.5)	
Chinae cortex **Cinchonae cortex** Red Cinchona bark Cinchona pubescens VAHL (syn. C. succirubra PAVON) DAB 10, ÖAB, Helv. VII, MD DAC 86 (tinct.)	4%–12% total alkaloids: approximately 20 alkaloids; diastereomeres Quinine/quinidine and cinchonine/cinchonidine quinine (0.8%–4%), quinidine (0.02%–0.4%), cinchonine (1.5%–3%), cinchonidine (1.5%–5%)	**Fig. 12**
Cinchona calisaya WEDDEL Yellow Cinchona bark Rubiaceae USP XI	Yellow Cinchona bark contains up to 90% quinine	
Opium Opium Papaver somniferum L. subsp. somniferum and varieties Papaveraceae DAB 10, ÖAB, Helv. VII, BP'88, MD, Japan (pulv.), USP XXII (tinct.)	20%–29% total alkaloids raw opium: 30 alkaloids Phenanthrene type: morphine (3%–23%), codeine (0.3%–3%), thebaine (0.1%–3%) Benzylisoquinoline type: papaverine (0.1%–2%), noscapine (narcotine; 2%–12%), narceine (0.1%–2%)	**Fig. 13,14**
Corydalidis rhizoma Hollowroot-birthwort Corydalis cava (L.) SCHWEIGG et KOERTE Papaveraceae, Fumariaceae China, Japan	3–5% total alkaloids Berberine type; corydaline, coptisine tetrahydropalmatine, canadine Aporphine type: bulbocapnine (0.2%–0.3%) (+) corytuberine, corydine Protopine	**Fig. 15**
Fumariae herba Fumitory herb Fumaria officinalis L. Papareraceae (Fumariaceae)	0.5%–1% total alkaloids Protoberberine type (0.2%–0.4%) protopine ► 0.5% flavonoids and phenol carboxylic acids, fumaric acid	**Fig. 16**
Miscellaneous classes of alkaloids		**Fig. 17-26**
Sarothamni (Cytisi) herba Scotch broom tops Cytisus scoparius (L.) LINK (syn. Sarothamnus scoparia (L.)) Fabaceae MD, DAC 86	0.3%–1.5% quinolizidine alkaloids >20 alkaloids. (−)-Sparteine (85%–90%), 17-oxo-α-isosparteine, lupanine, 4- and 13-hydroxylupanine ► 0.2%–0.6% flavonoids: spiraeoside, isoquercitrine, scoparoside, ► coumarins; caffeic acid derivatives	**Fig. 17**

	Drug/plant source Family/pharmacopoeia	Total alkaloids Major alkaloids (for formulae see 1.5)
Fig. 17	**Spartii flos** Spartii juncei flos Broomflowers Spartium junceum L. Fabaceae (Leguminosae)	0.3%–0.4% quinolizidine alkaloids Cytisine (40%) N-methylcytisine (45%) anagyrine ► Flavonoids: isoquercitrine, luteolin-4'-O-glucoside
Fig. 18	**Genistae herba** Dyer's weed, Dyer's broom Genista tinctoria L. Fabaceae	0.3%–0.8% quinolizidine alkaloids *N*-methylcytisine, anagyrine, isosparteine, lupanine ► 0.5%–3% flavonoids: luteolin glycosides Isoflavones: genistein, genistin

Note: The trivial name genistein is used for the isoflavone and the alkaloid (α-isosparteine).

Fig. 19	**Chelidonii herba** Tetterwort, greater celandine Chelidonium majus L. Papaveraceae DAB 10	0.35%–1.30% total alkaloids (>20) Benzophenanthridine type: chelidonine (>0.07%), chelerythrine (>0.04%) and sanguinarine (>0.01%) Protoberberine type: coptisin (>1.07%), berberine (0.11%). Protopine
	► Chelidonii radix/rhizoma	2.4%–3.4% total alkaloids: chelidonin (1.2%), and chelerythrine (1%)
Fig. 20	**Colchici semen** Meadow saffron seeds Colchicum autumnale L. Liliaceae DAC 86, MD	0.5%–1% total alkaloids: >20 alkaloids Colchicine (65%), colchicoside (30%), demecolcine, lumialkaloids (artefacts)
Fig. 21	**Berberidis radicis cortex** Barberry root bark Berberis vulgaris L. Berberidaceae MD	>13% total alkaloids Berberine, protoberberine (6%), jateorrhizine (jatrorrhizine), palmatine <5% bisbenzylisoquinolines e.g. oxyacanthine. Magniflorine
Fig. 21	**Hydrastis rhizoma** Golden seal root Hydrastis canadensis L. Ranunculaceae MD	2.5%–6% total alkaloids Berberine (2%–4.5%), tetrahydroberberine (0.5%–1%) (canadine), hydrastine (3.2%–4%; phthalide-isoquinoline alkaloid)
Fig. 21	**Colombo radix** Calumba root Jateorhiza palmata (LAM) MIERS Menispermaceae MD Japan (J. columba MIERS)	2%–3% total alkaloids Palmatine, jatrorrhizine, bisjatrorrhizine, columbamine (protoberberine type) ► Furanoditerpenoid bitter principles (palmarin, columbin)

Drug/plant source Family/pharmacopoeia	Total alkaloids Major alkaloids (for formulae see 1.5)	
Mahoniae radicis cortex Mahonia bark, grape root Mahonia aquifolium (PURSH) NUTT (syn. Berberis aquif.) Berberidaceae	1.8%–2.2% total alkaloids Jatrorrhizine, berberine, palmatine, columbamine (protoberberines); magnoflorine, corytuberine (aporphines); oxyacanthine, berbamine, (bisbenzyl-isoquinolines)	**Fig. 22**
Boldo folium Boldo leaves Peumus boldus J.I.MOLINA Monimiaceae DAC 86, Helv. VII, MD	0.2%–0.5% total alkaloids Aporphine alkaloid boldine ▶ 2%–3% essential oils: p-cymol, cineole, ascaridole (40%–50%) ▶ 1% flavonoids	**Fig. 23**
Nicotianae folium Tobacco leaves Nicotiana tabacum L., N. rustica L. and other varieties Solanaceae	0.06%–10% total alkaloids L-Nicotine, nornicotine, anabasine, nicotyrine	**Fig. 24**
Aconiti tuber Aconite root Aconitum napellus L. Ranunculaceae MD	0.3%–1.5% total alkaloids: 15 ester alkaloids Aconitine, mesaconitine, hypaconitine (benzoylaconine and aconine: hydrolytic cleavage products)	**Fig. 25**
Lobeliae herba Lobelia, Indian tobacco Lobelia inflata L. Campanulaceae (Lobeliaceae) ÖAB, BP 88, MD	0.2%–0.6% total alkaloids Lobeline (piperidine ring system) Isolobinine (dehydro, piperidine ring) DL-lobelidine, lobelanine	**Fig. 26**
Sabadillae semen Caustic barley, Cevadilla seed Schoenocaulon officinale A. GRAY	3%–6% steroid alkaloids (C-nor-C-homo-cholestanes)	**Fig. 26**
Liliaceae MD	"veratrine" = mixture of cevadine, veratridine, devadilline, sabadine, cevine)	
Ephedrae herba Desert tea (Ma-huang) Ephedra sinica STAPF Ephedra shennungiana TANG E. distachya L. or other species Gnetaceae (Ephedraceae) DAB 10, MD, Japan, China	2.5%–3% total alkaloids L-Ephedrine (0.75%–1%), norephedrine (+)-Pseudoephedrine and norpseudoephedrine	**Fig. 26B**

Drug/plant source Family/pharmacopoeia	Total alkaloids Major alkaloids (for formulae see 1.5)

Fig. 27–28 **Tropine alkaloids**

Fig. 27,28	**Belladonnae folium** Belladonna leaves Solanaceae DAB 10, Ph.Eur.I, ÖAB, Helv. VII, BP 88, USP XXII	0.2%–0.5% total alkaloids (−)-Hyoscyamine/atropine (~87%) scopolamine, apoatropine ▶ Flavonoids: quercetin glycosides
Fig. 27,28	**Belladonnae radix** Belladonna root Atropa belladonna L. Solanaceae DAC 86, ÖAB, MD, Japan	0.3%–0.8% total alkaloids (−)-Hyoscyamine and scopolamine Minor alkaloids apoatropine, belladonnine, cuskhygrine, ▶ Coumarins: scopoletin, −7-O-glucoside (see Chap. 5, Fig. 5)
Fig. 27,28	**Scopoliae radix** Scopolia root Scopolia carniolica JACQ. Solanaceae Japan (e.g. Scopolia japonica)	0.4%–0.95% total alkaloids (−)-Hyoscyamine and scopolamine ▶ Coumarins: scopoletin, −7-O-glucoside (see Chap. 5, Fig. 5)
Fig. 27,28	**Hyoscyami folium** Henbane leaves Hyoscyamus niger L. var. niger Solanaceae DAB 10, PhEur. I, ÖAB, Helv. VII, MD	0.04%–0.17% total alkaloids (−)-Hyoscyamine/atropine (60%) scopolamine, belladonine, apoatropine ▶ Flavonoid glycosides
Fig. 27,28	**Hyoscyami mutici folium** Hyoscyamus muticus L. Solanaceae MD	0.8%–1.4% total alkaloids (−)-Hyoscyamine/atropine (90%) scopolamine, apoatropine, belladonnine
Fig. 27,28	**Stramonii folium** Thornapple leaves Datura stramonium L. Solanaceae DAB 10, PhEur. I, ÖAB, Helv. VII, MD	0.1%–0.6% total alkaloids (−)-Hyoscyamine/atropine and scopolamine in ratio of approximately 2:1; belladonnine ▶ Flavonoid glycosides

Drug/plant source Family/pharmacopoeia	Total alkaloids Major alkaloids (for formulae see 1.5)	
Purines		**Fig. 29–30**
Cacao semen Cacao beans Theobroma cacao L. Sterculiaceae MD	0.2%–0.5% caffeine 1%–2% theobromine	**Fig. 29,30**
Coffeae semen Coffee beans Coffea arabica L., other species Rubiaceae MD, DAB 10 (caffeine)	0.3%–2.5% caffeine theophylline (traces) ► Chlorogenic acid	**Fig. 29,30**
Mate folium Mate, Jesuit's tea Ilex paraguariensis St.HIL. Aquifoliaceae DAC 86, MD	0.3%–1.7% caffeine 0.03%–0.05% theophylline 0.2%–0.45% theobromine ► 10% chlorogenic-, iso- and neochlorogenic acid, isoquercitrin ► Triterpene saponines: ursolic and oleanolic acid derivatives	**Fig. 29,30**
Theae folium Tea Camellia sinensis (L.) KUNTZE Theaceae MD	2.5%–4.5% caffeine 0.02%–0.05% theophylline 0.05% theobromine ► Polyphenols; tannins: catechin type (10%–20%), dimeric theaflavins, oligomeric procyanidins; flavonoid glycosides	**Fig. 29,30**

Note: Colae semen contains 0.6%–3% caffeine (Cola nidita, C. acuminata SCHOTT et ENDL, Sterculiaceae)

1.5 Formulae

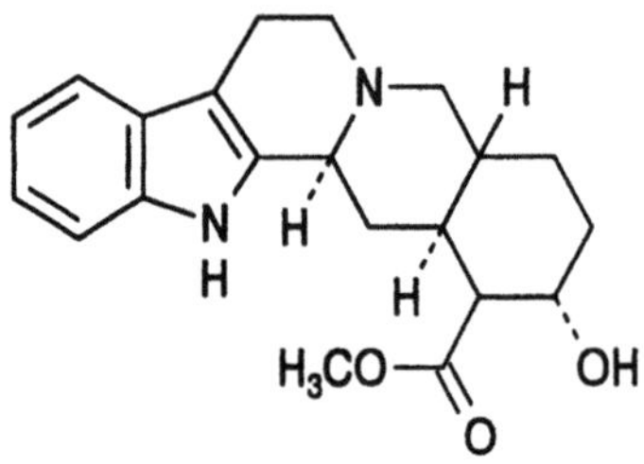

Yohimbine

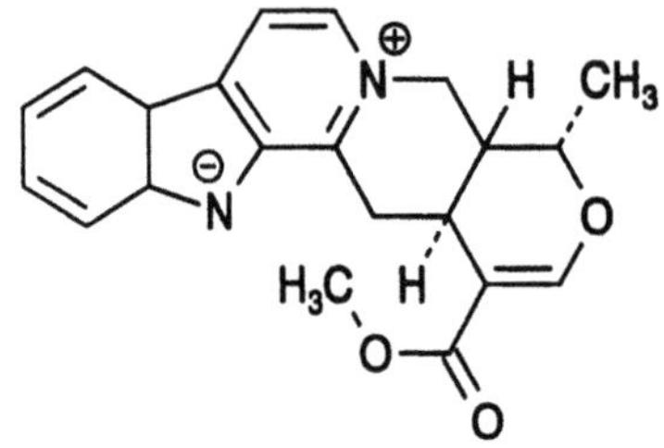

Serpentine

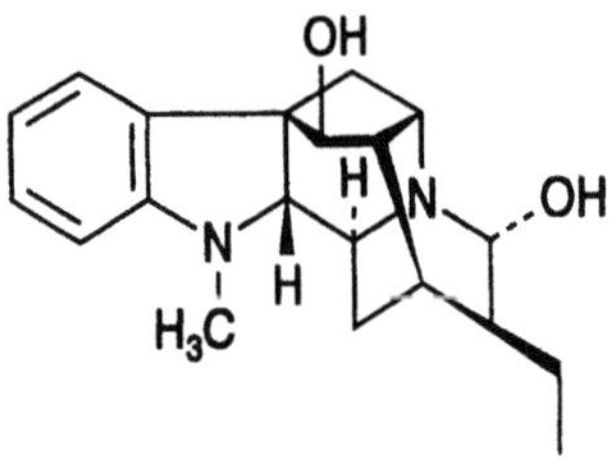

Ajmaline

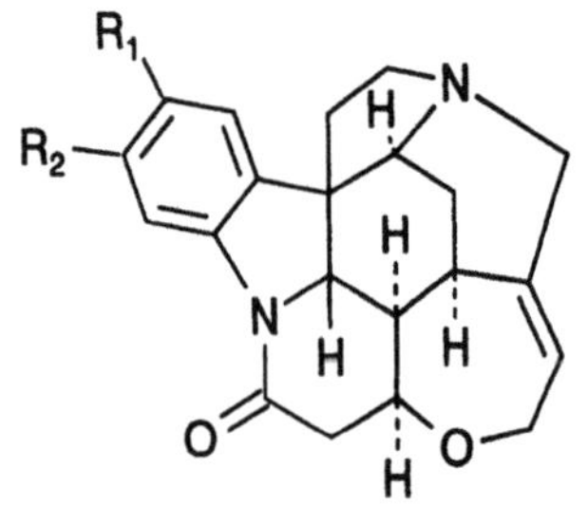

$R_1 = R_2 = H$	Strychnine
$R_1 = R_2 = OCH_3$	Brucine

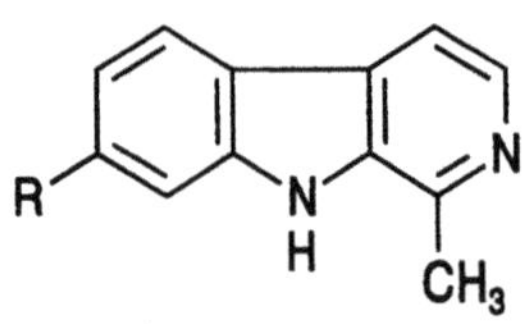

Harman	R = H
Harmine	$R = OCH_3$
Harmol	R = OH

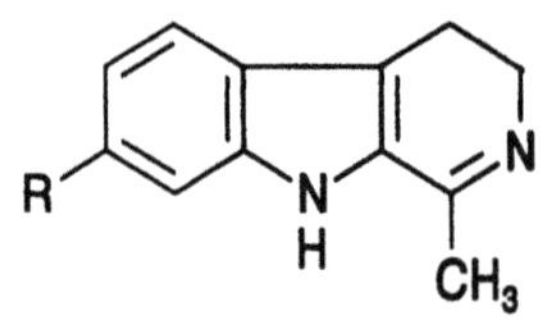

Harmalol	R = OH
Harmaline	$R = OCH_3$

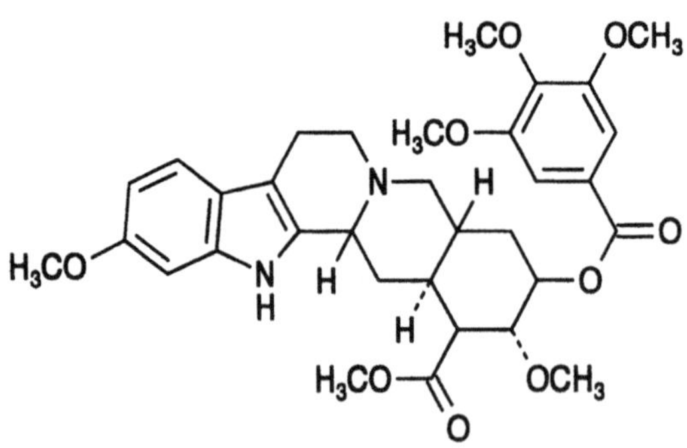

Reserpine

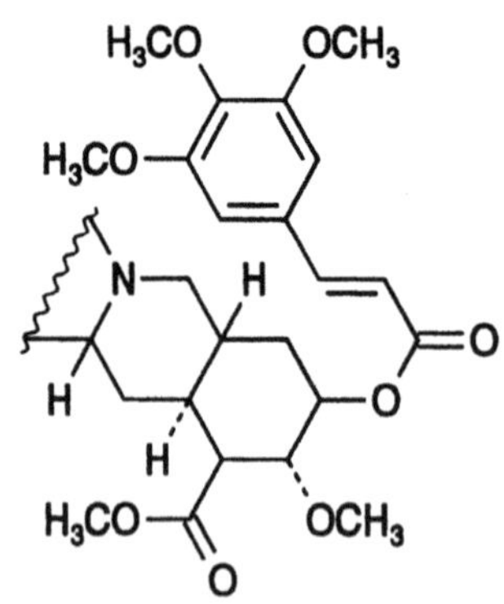

Rescinnamine

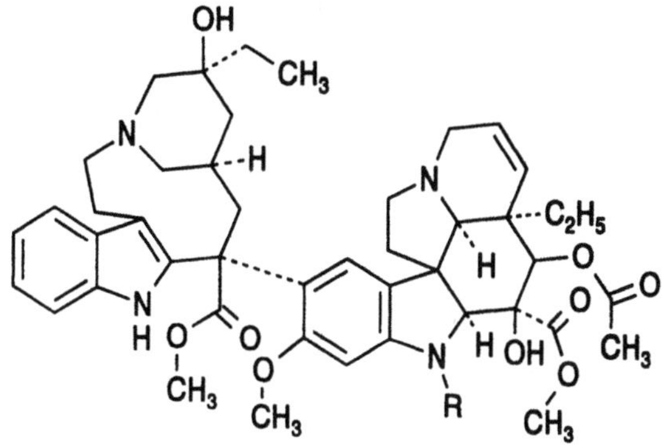

Vincaleucoblastine R = CH_3

Leurocristine R = 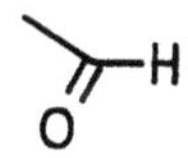

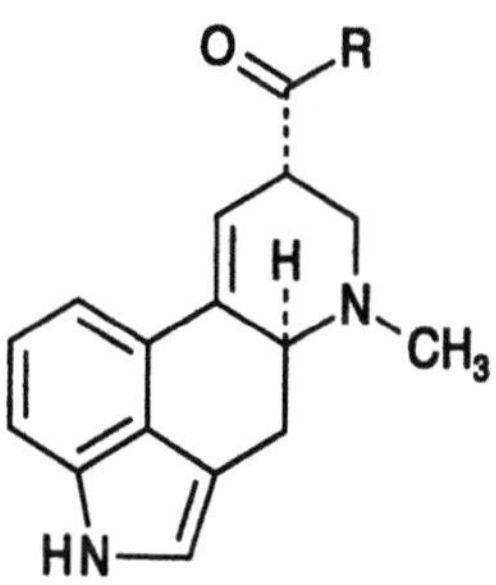

(Pyrrolindol)

Ergotamine R = 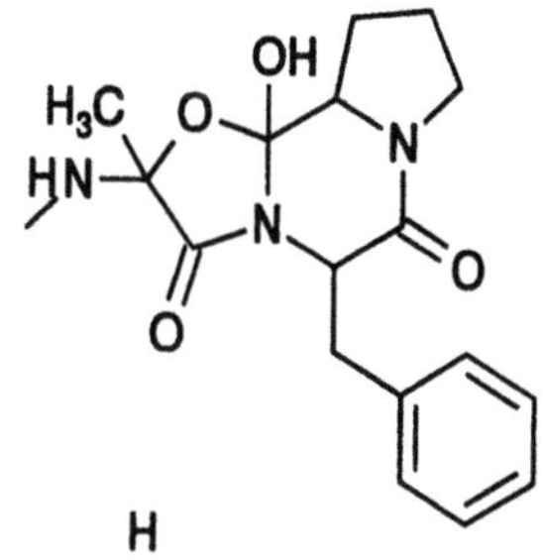

Ergometrine R = 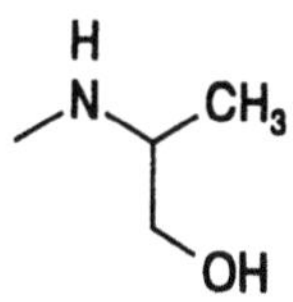

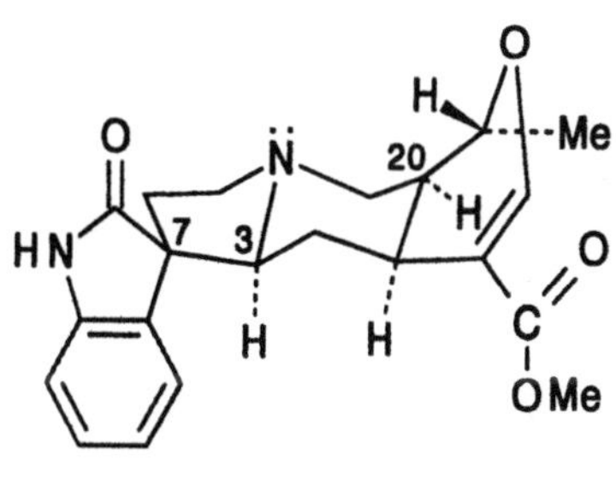

Pteropodine

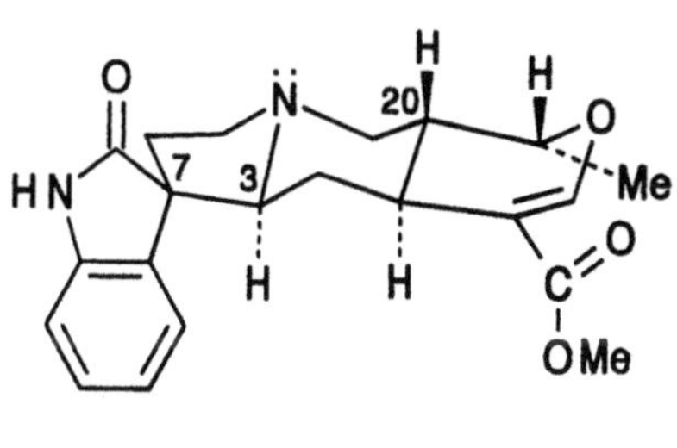

Mitraphylline

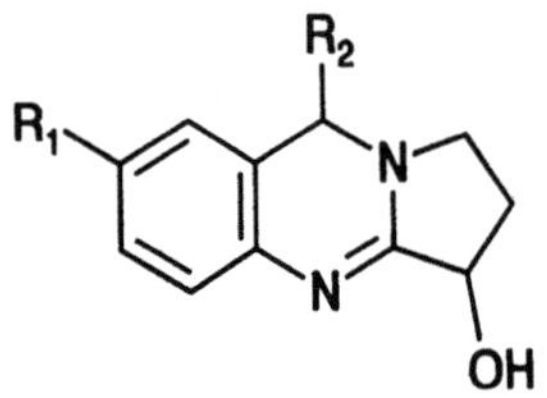

	R_1	R_2
Vasicine	$-H$	$-H_2$
Vasicinone	$-H$	$=O$
Oxyvasicine	$-OH$	$-H_2$

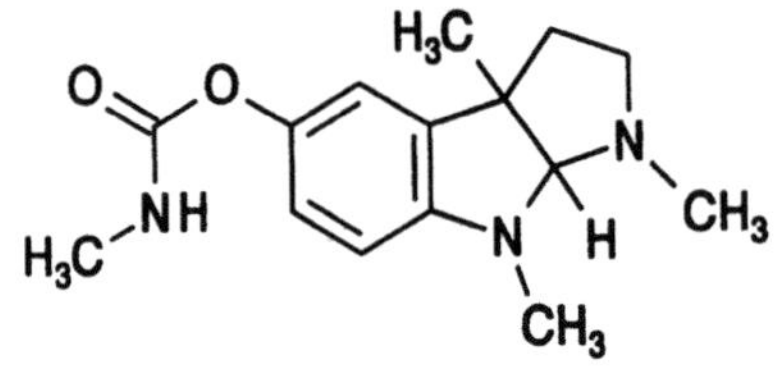

Physostigmine

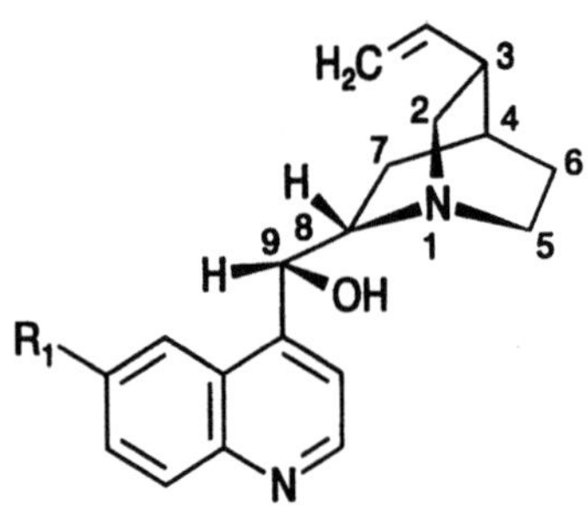

Cinchonidine: R = H

Quinine: R = OCH_3

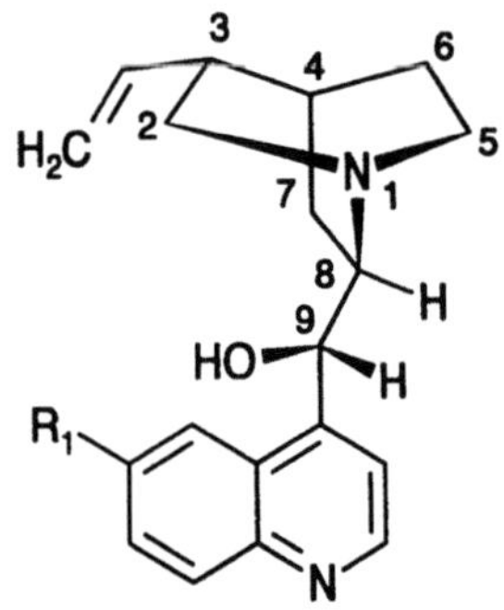

Cinchonine: R = H

Quinidine: R = OCH_3

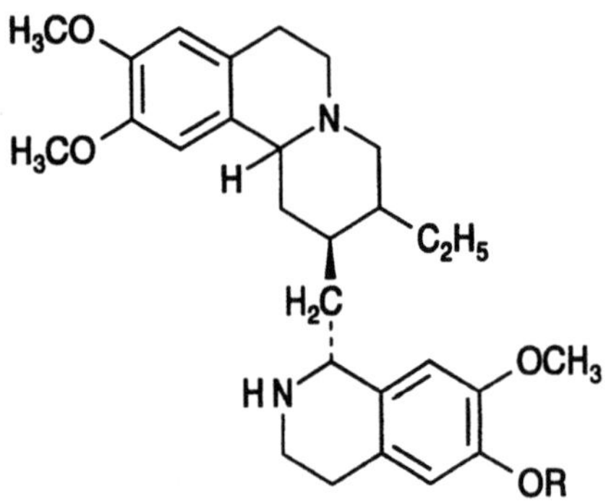

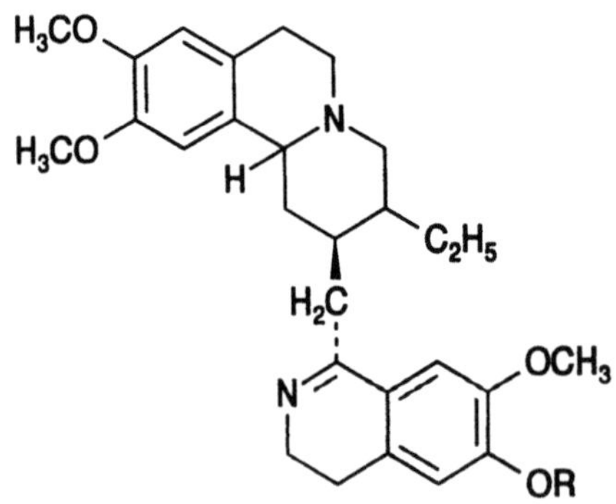

(-) Emetine R = CH_3 $\xrightarrow{-2H}$ O-Methylpsychotrine

(-) Cephaeline R = H $\longrightarrow$ Psychotrine

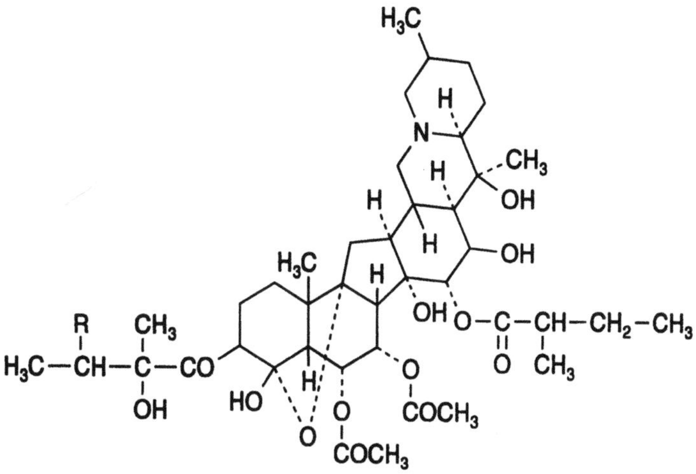

Protoveratrine A : R = H
Protoveratrine B : R = OH

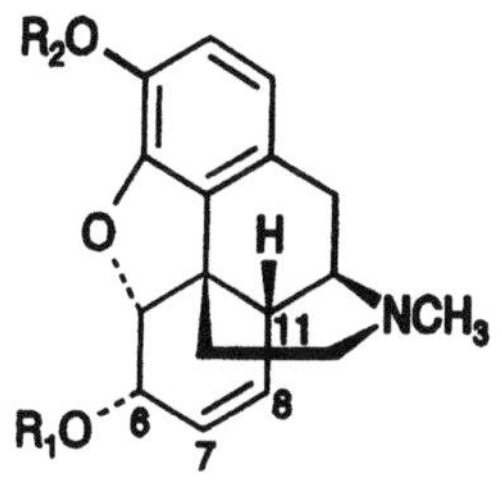

Morphine $R_1 = R_2 = H$
Codeine $R_1 = H$; $R_2 = CH_3$
Thebaine $R_1 = R_2 = CH_3$
(double bond C 6/7 and C 8/11)

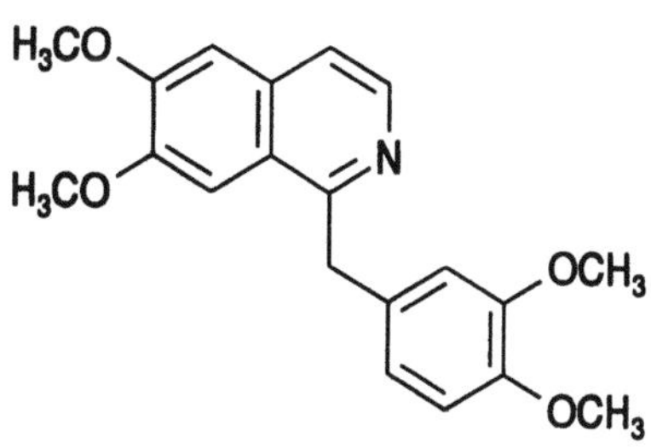

Papaverine

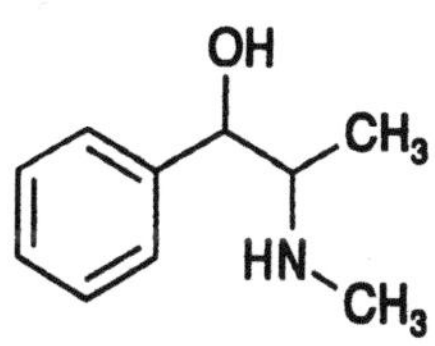

Ephedrine

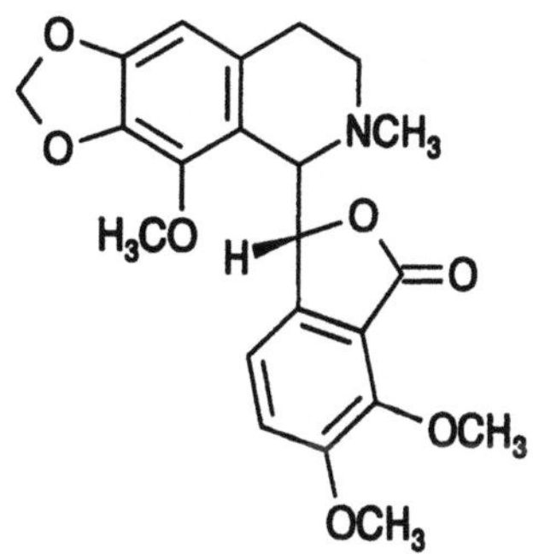

Noscapine

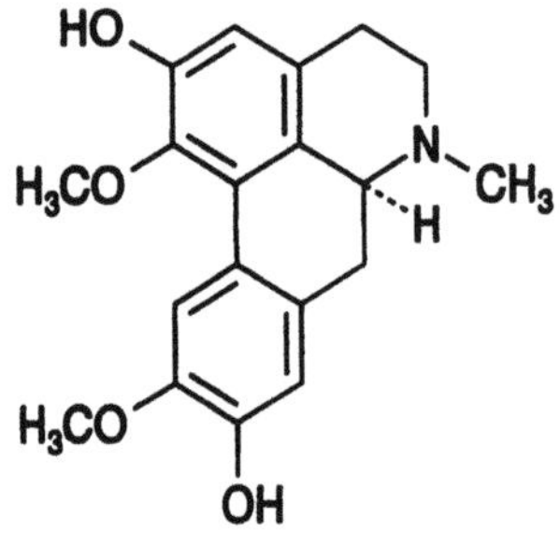

(S)-Boldine

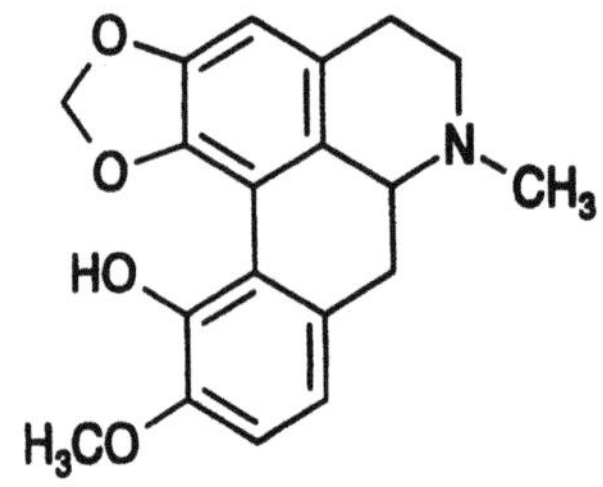

(S)-Bulbocapnine

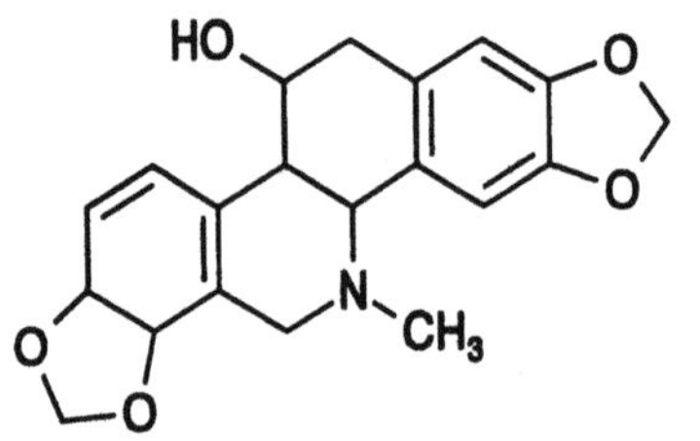

Chelidonine

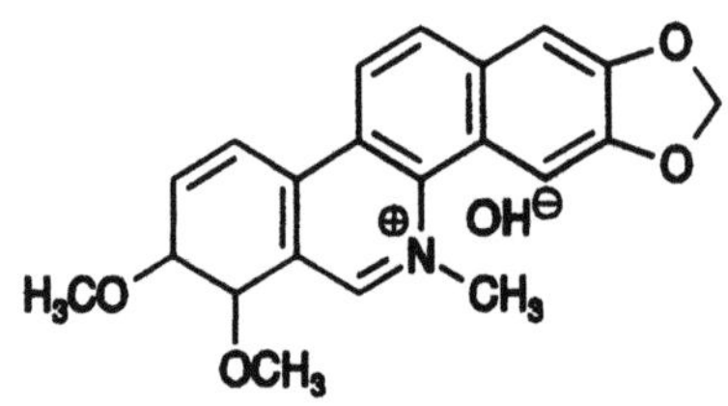

Chelerythrine

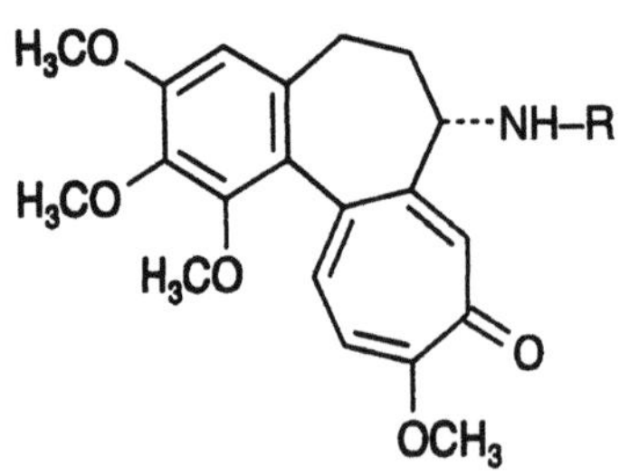

Colchicine R = 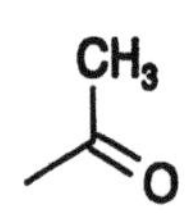

Demecolcine R = CH_3

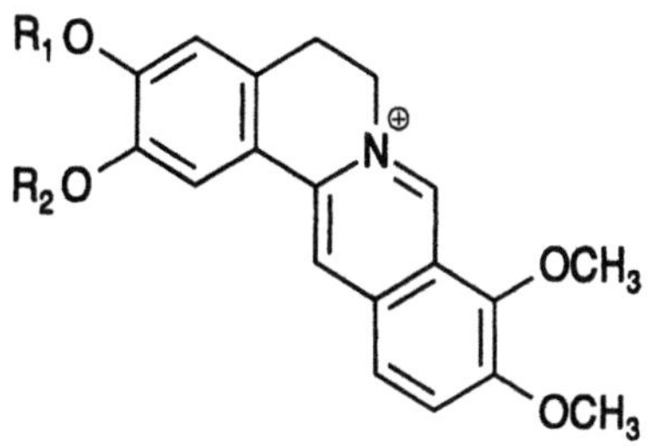

R_1	R_2	Compound
H	CH_3	Jatrorrhizine
CH_3	H	Columbamine
CH_3	CH_3	Palmatine
$-CH_2-$		Berberine

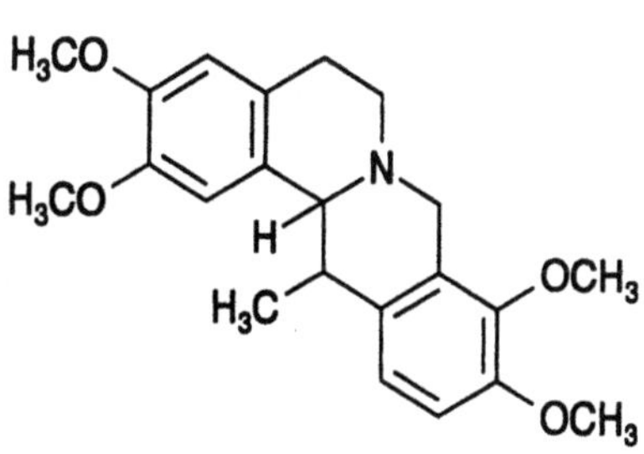

(-)-Corydaline

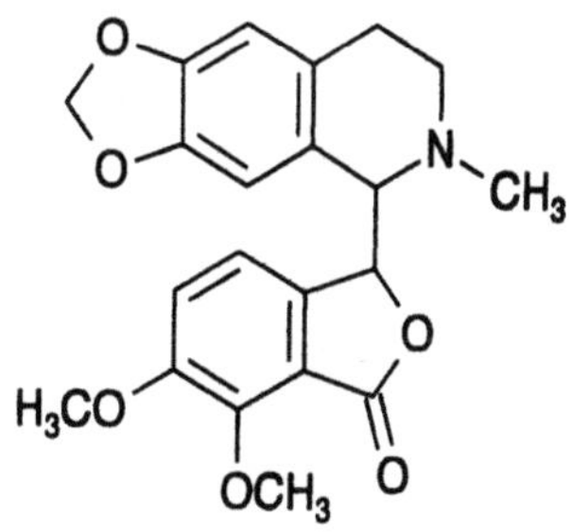

Hydrastine

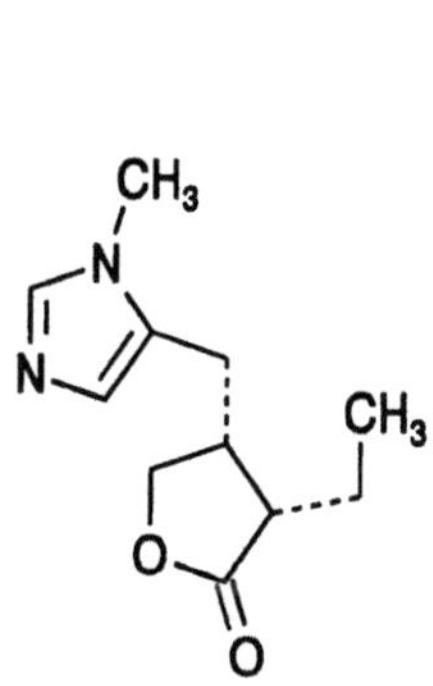

Pilocarpine

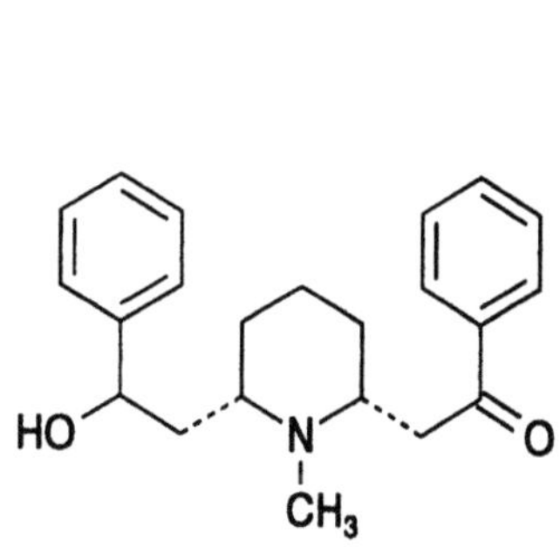

Lobeline

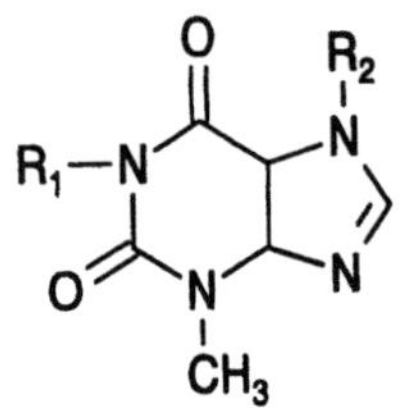

	R_1	R_2
Caffeine	CH_3	CH_3
Theobromine	H	CH_3
Theophylline	CH_3	H

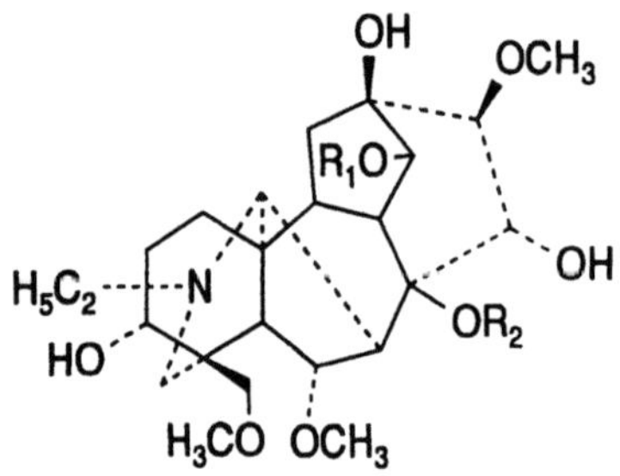

R_1	R_2	
COC_6H_5	$COCH_3$	Aconitine
COC_6H_5	H	Benzoylaconin
H	H	Aconin

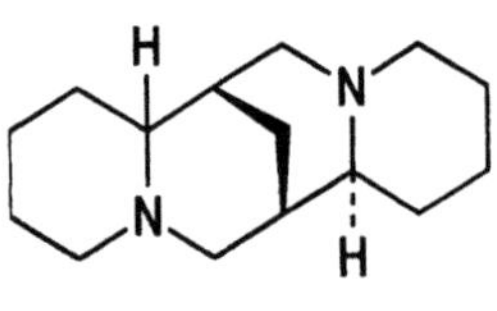

Sparteine

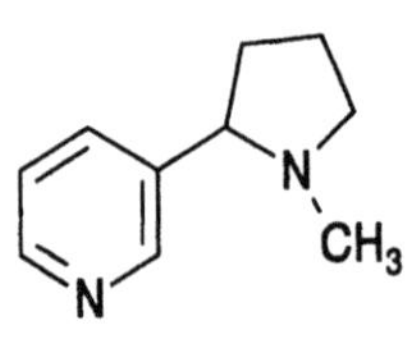

Nicotine

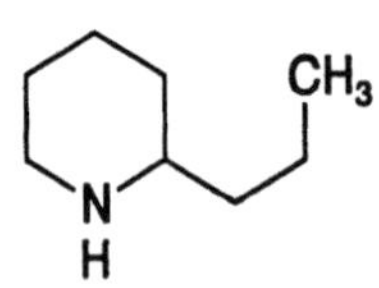

Coniine

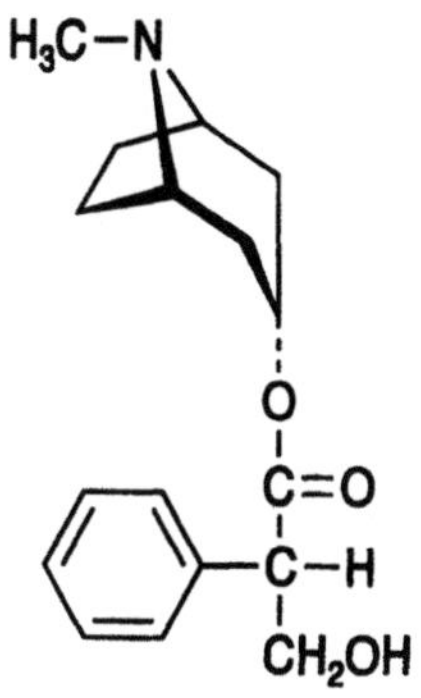

L-Hyoscyamine

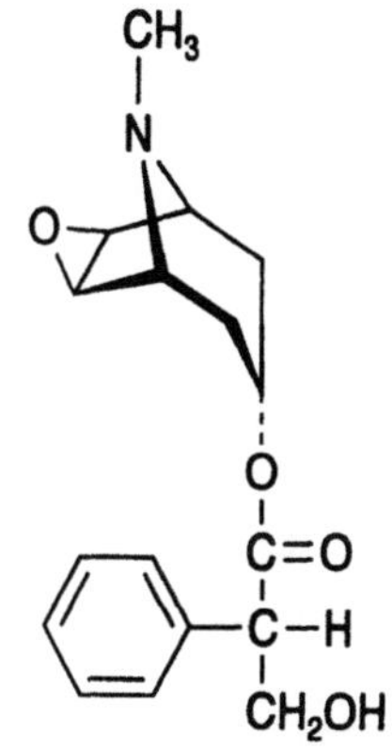

L-Scopolamine

1.6 TLC Synopsis of Important Alkaloids

Alkaloids I Reference compounds detected with Dragendorff reagent

1 colchicine	9 atropine	16 nicotine
2 boldine	10 codeine	17 veratrine
3 morphine	11 cinchonine	18 emetine
4 pilocarpine	12 scopolamine	19 papaverine
5 quinine	13 strychnine	20 lobeline
6 brucine	14 yohimbine	21 mesaconitine ►aconitine
7 cephaeline	15 physostigmine	22 noscapine (=narcotine)
8 quinidine		

Solvent system Fig. 1 toluene–ethyl acetate–diethylamine (70:20:10)

Detection A Dragendorff reagent (No. 13A) → vis
B Dragendorff reagent followed by sodium nitrite (No. 13B) → vis

Fig. 1 With Dragendorff reagent alkaloids spontaneously give orange-brown, usually stable colours in the visible. With some alkaloids, e.g. boldine (2), morphine (3) and nicotine (16), the colour fades rapidly and can be intensified by additional spraying with sodium nitrite reagent. The zones then appear dark brown (e.g. morphine, 3) or violet-brown (e.g. atropine, 9). The colours of pilocarpine (4) and nicotine (16) are still unstable.

Alkaloids II Reference compounds that fluoresce in UV-365 nm

23 serpentine	27 cinchonidine	31 noscapine
24 quinine	28 cephaeline	32 hydrastine
25 cinchonine	29 emetine	33 berberine
26 quinidine	30 yohimbine	34 sanguinarine

Solvent system Fig. 2 toluene–ethyl acetate–diethylamine (70:20:10)

Detection A Dragendorff reagent (No. 13A) → vis
B Sulphuric acid reagent (10%- No. 37A) → UV-365 nm

Fig. 2 The fluorescence of these alkaloids, predominantly light blue, can be intensified by treatment with 10% ethanolic sulphuric acid.
In the case of the quinine alkaloids, the initial light blue fluorescence of quinine and quinidine becomes a radiant blue (this appears white in the photo), while cinchonine and cinchonidine show a deep violet fluorescence (hardly visible in the photo).
Berberine (33) and sanguinarine (34) are exceptions in showing a bright yellow fluorescence.
Colchicine shows a yellow-green fluorescence (see Fig. 20, Alkaloid Drugs).

Remarks: The commercial alkaloid reference compounds (e.g. hydrastine (32)) frequently show additional zones of minor alkaloids or degradation products.

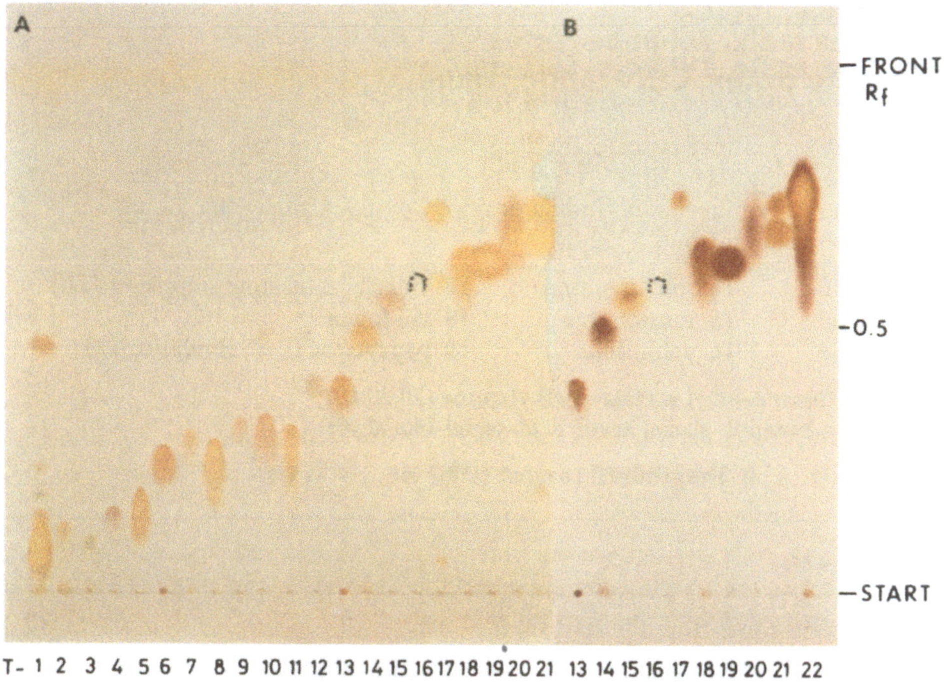
A
B
FRONT
Rf
0.5
START
T- 1 2 3 4 5 6 7 8 9 10 11 12 13 14 15 16 17 18 19 20 21 13 14 15 16 17 18 19 20 21 22

Fig. 1

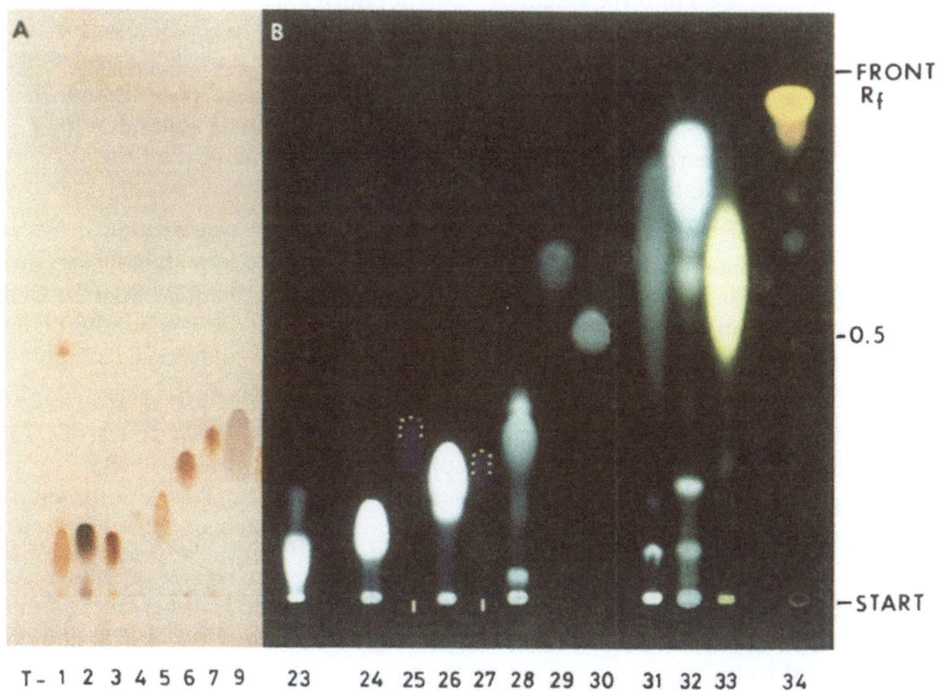
A
B
FRONT
Rf
0.5
START
T- 1 2 3 4 5 6 7 9 23 24 25 26 27 28 29 30 31 32 33 34

Fig. 2

1.7 Chromatograms

Rauvolfiae radix, Yohimbe cortex, Quebracho cortex, Catharanthi folium

Drug sample
1 Rauvolfiae serpentinae radix (Siam drug)
2 Rauvolfiae vomitoriae radix
3 Rauvolfiae serpentinae radix (Indian drug)
(alkaloid extraction method A, 30 μl)
4 Yohimbe cortex
5 Quebracho cortex
6,7 Catharanthi folium

Reference compound
T1 serpentine
T2 ajmaline
T3 reserpine
T4 rescinnamine
T5 rauwolscine
T6 yohimbine
T7 vincaleucoblastine sulphate (VLB)
T8 vindoline
T9 papaverine (→R_f similar to T8)

Solvent system
Fig. 3,4 A toluene–ethyl acetate–diethylamine (70:20:10)
Fig. 4 B n-butanol–glacial acetic acid–water (40:10:10)

Detection
A UV-365 nm B Dragendorff reagent (DRG No. 13) →vis

Fig. 3 Rauvolfiae radix

A The drug extracts 1–3 are generally characterized in UV-365 nm by seven to ten intense blue fluorescent zones from the start till $R_f \sim 0.8$:

$R_f \sim 0.05$ (T1)	Serpentine
0.15–0.25	Two to three alkaloids, not identified
0.30 (T2)	Ajmaline[a]
0.40 (T5)	Rauwolscine[b]
0.45 (T3, T4)	Reserpine/rescinnamine[b]
0.6–0.8	Two to three alkaloids, e.g. raubasine

[a] Ajmaline shows a prominent quenching in UV-254 nm and only develops a dark blue fluorescence when exposed to UV-365 nm for 40 min.

[b] Rescinnamine and rauwolscine show three to four zones due to artefacts formed in solution and on silica gel.

Rauvolfiae serpentinae radix (1,3) show varying contents of the major alkaloids according to drug origin. The Indian drug mostly has a higher serpentine content than the Siam drug. **Rauvolfiae vomitoriae radix** (2) differs from (1) and (3) by a generally higher content of reserpine, rescinnamine and ajmaline and by the additional compound rauwolscine.

B All Rauwolfia alkaloids give with Dragendorff reagent orange-brown zones (T2/T1). *Note*: Ajmaline immediately turns red when sprayed with concentrated HNO_3.

Fig. 4A Yohimbe and Quebracho cortex (4,5)

Both drug extracts are characterized in UV-365 nm by the blue fluorescent zone of yohimbine at $R_f \sim 0.45$ (T6). A variety of additional alkaloids are seen as ten blue zones in the lower R_f range (e.g. quebrachamine, aspidospermine in 5), whereas Yohimbe cortex (4) has two prominent alkaloid zones in the upper R_f range (R_f 0.7–0.75) and one near the solvent front.

B **Catharanthi folium** (6,7)

After treatment with the DRG reagent the extracts reveal five to seven alkaloid zones mainly in the R_f range 0.05–0.75. Two prominent brown zones with vindoline at $R_f \sim 0.7$ (T8) dominate the upper R_f range. Slight differences are noticed in the lower R_f range between the fresh leaf sample (6) and the stored material (7). Vincaleucoblastine (T7) migrates to $R_f \sim 0.2$. It is present at very low concentration in the plant ($<0.002\%$) and therefore not detectable in these drug extracts without prior enrichment.

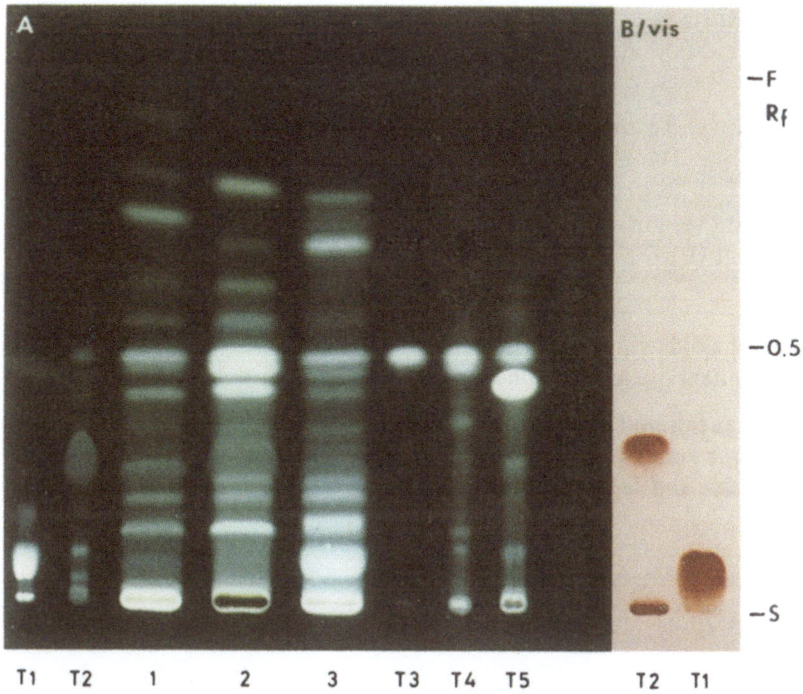

Fig. 3

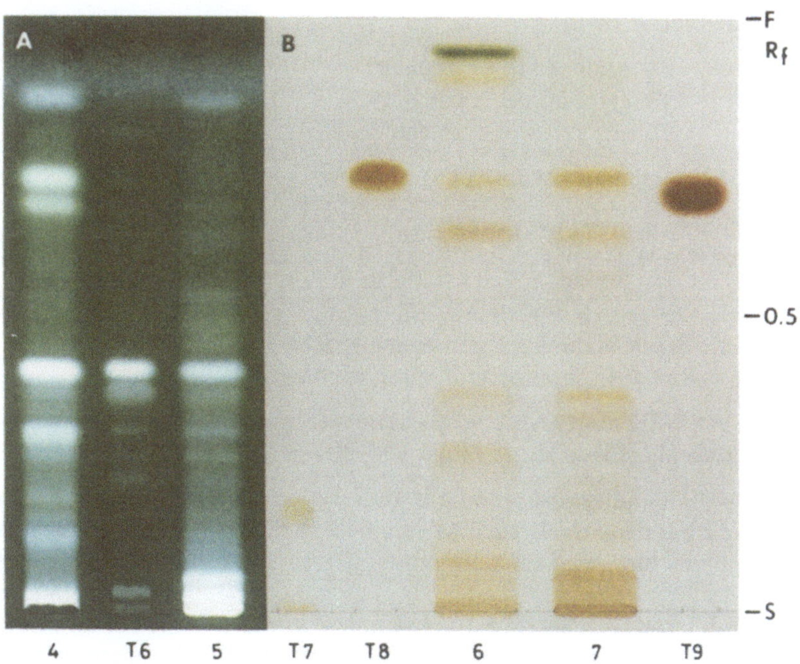

Fig. 4

Vincae minoris folium

Drug sample 1 Vinca minor (fresh leafs), (alkaloid extraction method C, 40 µl)

Reference compound
T1 vincamine
T2 vincaminine
T3 vincine
T4 vincamajine
T5 minovincine
T6 reserpinine

Solvent system Fig. 5 ethyl acetate–methanol (90:10)

Detection
A UV-254 nm (without chemical treatment)
B Dragendorff reagent (DRG No. 13B) → vis

Fig. 5A The four principal alkaloids vincamine, vincaminine, vincine and vincamajine (T1-T4) are detected as prominent quenching zones in the R_f range 0.25–0.4.

B The alkaloids of **Vincae folium** (1) show four weak brown zones in the R_f range 0.15–0.45 (T1-T4) and two major zones at $R_f \sim$ 0.8–0.85 (T5-T6). The colour obtained with the DRG reagent is unstable and fades easily in vis.

Secale cornutum

Drug sample
1 Secale cornutum (freshly prepared alkaloid fraction)
2 Secale cornutum (stored alkaloid fraction)
(alkaloid extraction method A, 30 µl)

Reference compound
T1 ergocristine
T2 ergotamine
T3 ergometrine
T4 egometrine + artefact▸)
T5 ergotamine + artefact▸)
T6 ergocristine + artefact▸)

Solvent system Fig. 6 toluene–chloroform–ethanol (28.5:57:14.5)

Detection
A UV-254 nm (without chemical treatment)
B, C van URK reagent (No. 43) → vis

Fig. 6A The three characteristic **Secale** alkaloids ergometrine at $R_f \sim$ 0.05, ergotamine at $R_f \sim$ 0.25 and ergocristine at $R_f \sim$ 0.45 show prominent quenching in UV-254 nm.

B After treatment with van URK reagent, the Secale extract (1) generates three blue zones of the principal alkaloids (T1-T3) in the R_f range 0.05–0.4.

C Secale alkaloids in solution and exposure to light undergo easy epimerization and also form lumi-compounds. Secale extracts such as sample 2 then show artefacts, such as isolysergic acid derivatives, lumi- and aci-compounds seen as additional, usually weaker zones with *higher* R_f values.
The artefacts (>) are detectable in Secale extract sample 2 as well as in solutions of the reference compounds T4-T6. They also form blue zones with van URK reagent (vis).

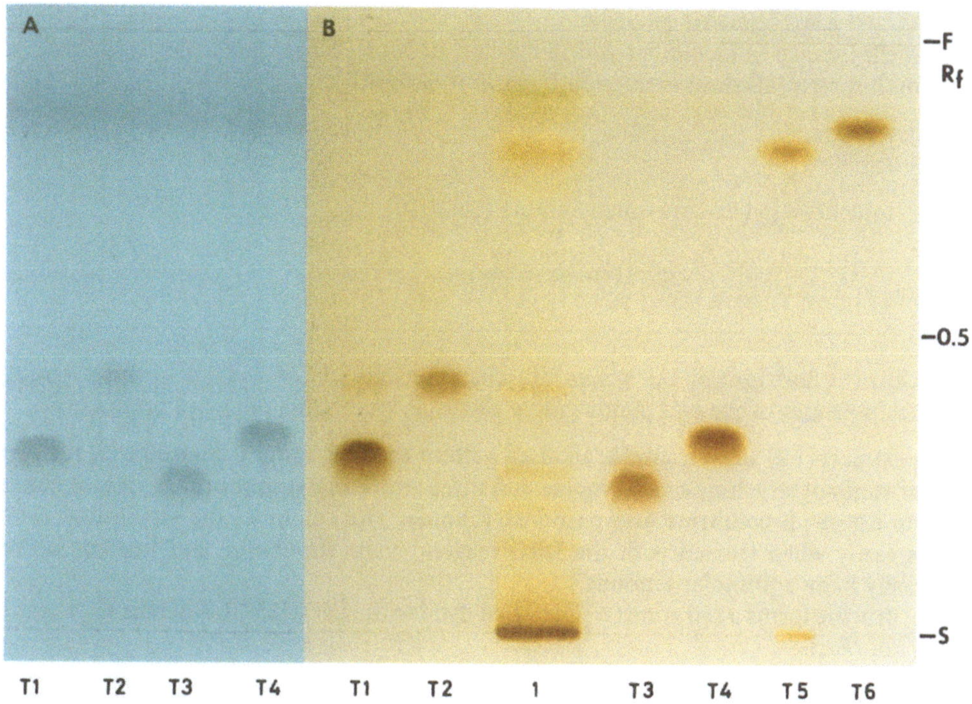
A
B
–F
Rf
–0.5
–S
T1
T2
T3
T4
T1
T2
1
T3
T4
T5
T6

Fig. 5

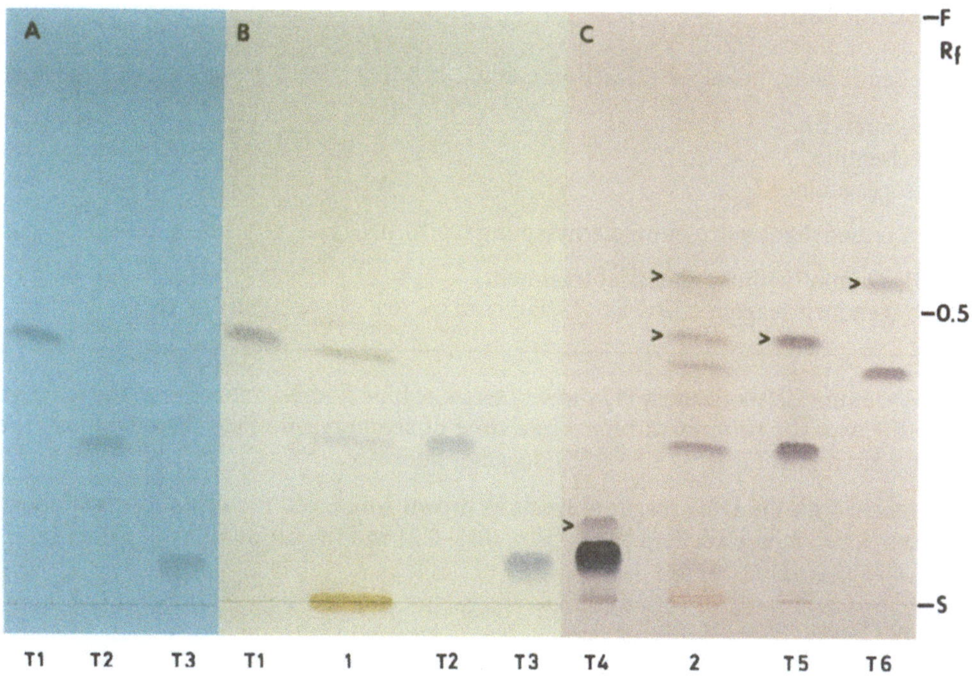
A
B
C
–F
Rf
–0.5
–S
T1
T2
T3
T1
1
T2
T3
T4
2
T5
T6

Fig. 6

Strychni and Ignatii semen

Drug sample 1 Strychni semen (alkaloid extraction method A, 30µl)
2 Ignatii semen (alkaloid extraction method A, 30µl)

Reference compound T1 strychnine
T2 brucine

Solvent system Fig. 7 toluene–ethyl acetate–diethylamine (70:20:10)

Detection A UV-254m (without chemical treatment)
B Dragendorff reagent (DRG No. 13) → vis

Fig. 7A **Strychni** (1) and **Ignatii** (2) **semen** are characterized in UV-254nm by their strong quenching zones of the two major indole alkaloids strychnine (T1) and brucine (T2).

B Both extracts (1,2) show a similar alkaloid pattern in the R_f range 0.25–0.55 with the two major zones of strychnine and brucine and three additional minor orange-brown zones due to e.g. α-, β-colubrine and pseudostrychnine. The colour of the strychnine zone fades easily when treated with the DRG reagent (vis). Strychnine and brucine occur normally in an equimolar amount.
Note: Brucine forms a red zone (visible when dyed with HNO_3 (25%), whereas strychnine does not react.

Gelsemii radix

Drug sample 1 Gelsemii radix, (alkaloid extraction method B, 40µl)

Reference compound T1 sempervirine
T2 gelsemine
T3 isogelsemine

Solvent system Fig. 8 aetone–light petroleum–diethylamine (20:70:10)

Detection A UV-365nm (without chemical treatment)
B Dragendorff reagent (DRG No. 13/followed by 10% $NaNO_2$/13B) → vis

---

Fig. 8A In UV-365nm **Gelsemii radix** (1) shows a series of blue fluorescent zones in the R_f range 0.05–0.7 with the prominent blue white zone of sempervirine (T1) directly above the start. Gelsemine (T2/→ B: $R_f \sim 0.35$) does not fluoresce.

B Treatment with the DRG reagent reveals as brown zones: sempervirine (directly above the start), two minor alkaloid zones ($R_f \sim 0.15$–0.2) and the major alkaloid gelsemine at $R_f \sim 0.35$ (T2; vis.).

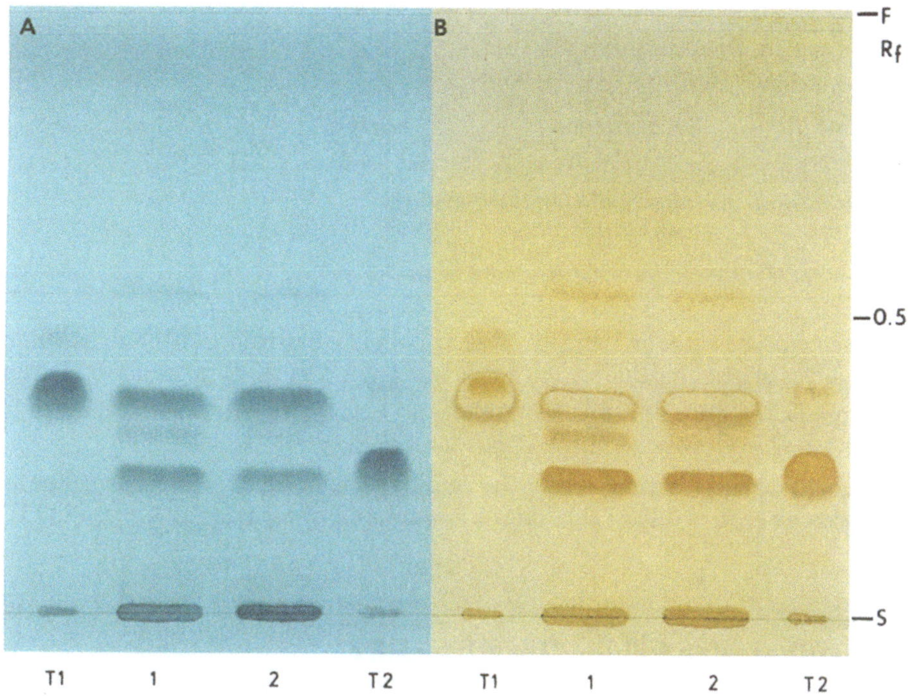

Fig. 7

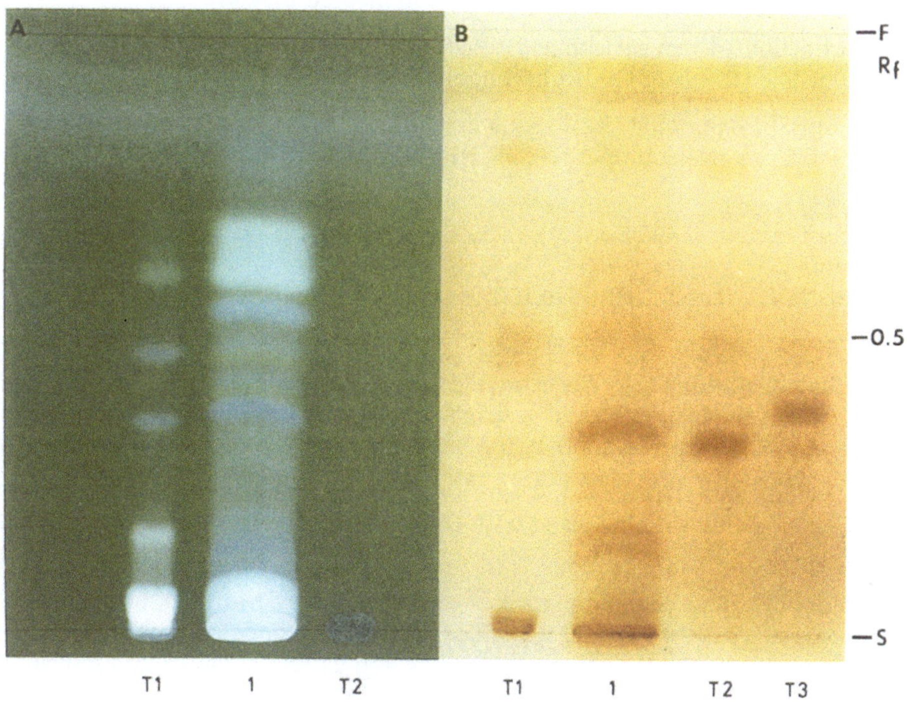

Fig. 8

Harmalae semen

Drug sample 1 Harmalae semen, (methanol extract, 30 µl)

Reference compound T1 harmalol; T2 harmaline; T3 harmane; T4 harmine; T5 harmol

Solvent system Fig. 9A chloroform–methanol–10% NH_3 (80:40:1.5)
B chloroform–acetone–diethylamine (50:40:10)

Detection A, B UV-365 nm (without chemical treatment)

Fig. 9A **Harmalae semen.** The carbolin derivatives harmalol (T1), harmaline (T2) and harmine (T4) are found as bright blue fluorescent zones in solvent A in the R_f range 0.1–0.75. The Harmalae semen sample 1 shows as major alkaloids harmalol and harmaline in the low R_f range 0.05–0.25 and harmine in the upper R_f range 0.75.

B Development in solvent system B reveals the zone of harmalol at $R_f \sim 0.05$, harmaline at $R_f \sim 0.4$, harmine at $R_f \sim 0.45$ (T2) besides a low amount of harmane at $R_f \sim 0.55$ (T3).

Justiciae-adhatodae folium, Uncariae radix

Drug sample 1 Adhatodae folium, (alkaloid extraction method B, 30 µl)
2 Uncariae tomentosae cortex, (alkaloid extraction method B, 40 µl)

Reference compound T1 alkaloid fraction/vasicin enrichment/Adhatodae folium
T2 rychnophylline ($R_f \sim 0.35$) + isorhychnophylline ($R_f \sim 0.75$)

Solvent system Fig. 10A,B dioxane–ammonia (90:10) → Adhatoda
C,D ethyl acetate–isopropanol–conc.NH_3 (100:2:1) → Uncaria

Detection A UV-254 nm
B Dragendorff reagent (DRG No. 13) → vis.
C UV-254 nm
D DRG/10% $NaNO_2$ reagent (DRG No 13B) → vis

Fig. 10A **Justiciae-adhatodae-folium** (1). The extract (1) and the alkaloid fraction (T1) show the quenching zone of the major alkaloid vasicine at $R_f \sim 0.55$; vasicinone at $R_f \sim 0.6$ and some other alkaloids (e.g. vasicinol) in the lower R_f range 0.2–0.25. Vasicinone is an artefact due to oxydative processes during extraction.

B From the alkaloids only vasicine reacts with Dragendorff reagent as an orange-brown zone in vis.

C **Uncariae radix** (2). This alkaloid extract is characterized by two pairs of quenching zones in the R_f ranges 0.7–0.8 and 0.25–0.3. The pentacyclic oxindoles, such as isomitraphylline, isopteropodine and uncarine A + F, as well as tetracyclic oxindols such as isorhychnophylline are found in the R_f range 0.7–0.8. The pentacylic mitraphylline and the tetracyclic rhychnophylline give prominent zones in the R_f range 0.25–0.3. The alkaloid distribution is subject to change. The alkaloid pattern of an individual plant changes over the year.

D All alkaloid zones turn orange-brown with Dragendorff/$NaNO_2$ reagent (vis.).

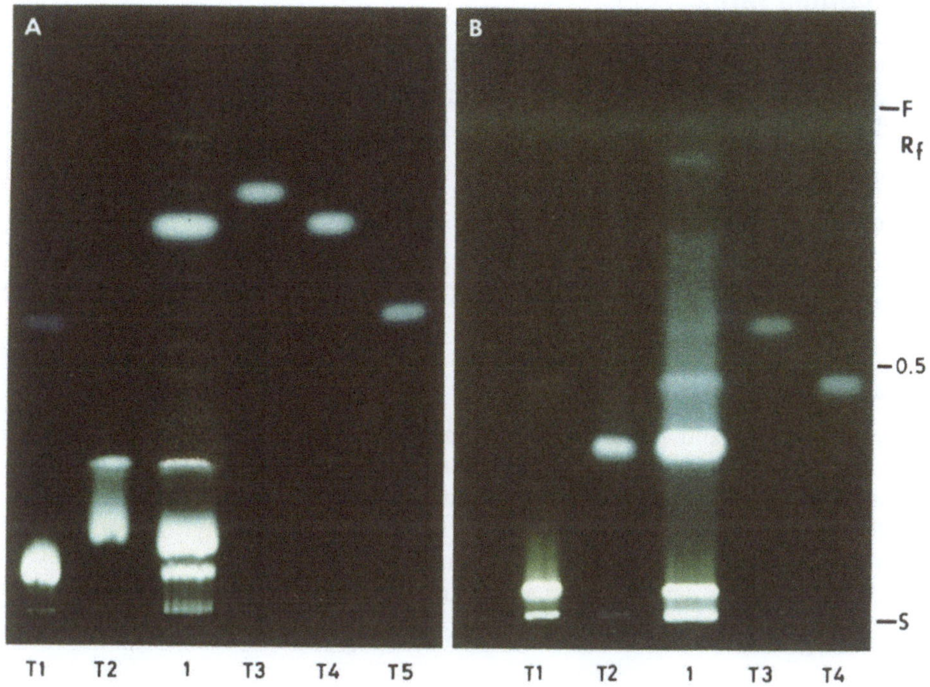
A
B
—F
Rf
—0.5
—S
T1
T2
1
T3
T4
T5
T1
T2
1
T3
T4

Fig. 9

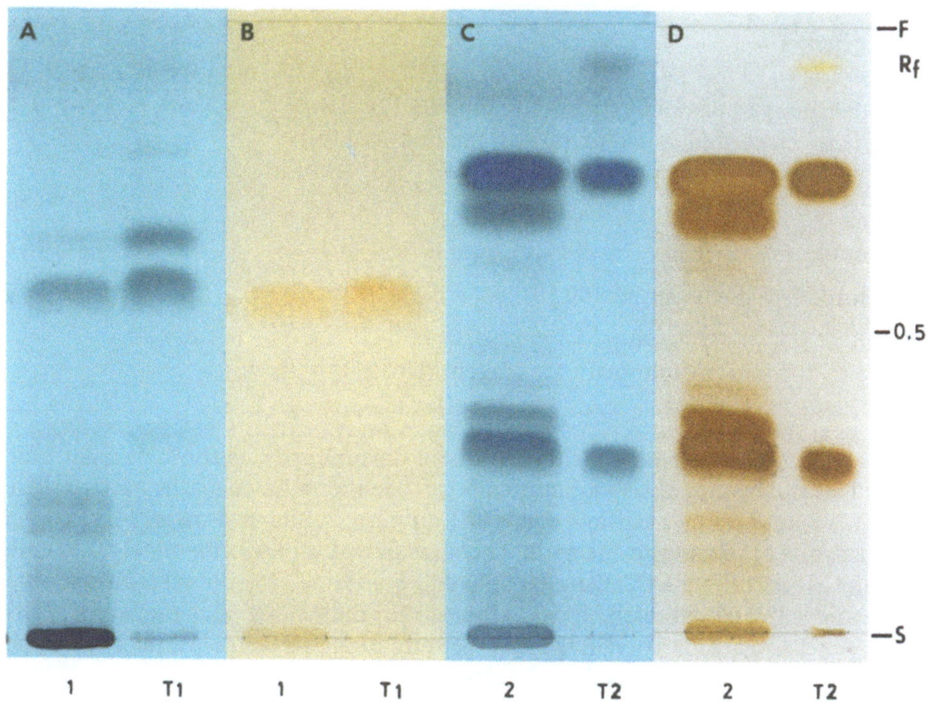
A
B
C
D
—F
Rf
—0.5
—S
1
T1
1
T1
2
T2
2
T2

Fig. 10

Ipecacuanhae radix

Drug sample 1 Cephaelis acuminata "Cartagena/Panama drug"
2 Cephaelis ipecacuanha "Rio/Matto-Grosso drug"
(alkaloid extraction method A, 30 μl)

Reference T1 cephaeline ($R_f \sim 0.2$) ► emetine ($R_f \sim 0.4$)

Solvent system Fig. 11 toluene–ethyl acetate–diethylamine (70:20:10)

Detection A, B Iodine/$CHCl_3$ reagent (No. 19) A → UV-365 nm; B → vis
C Dragendorff reagent (DRG No. 13A) → vis

Fig. 11 **Ipecacuanhae radix** (1,2)

A,B Cephaeline ($R_f \sim 0.2$) and emetine ($R_f \sim 0.4$) are the major alkaloids, which fluoresce light blue in UV-365 nm without chemical treatment. With iodine reagent cephaeline fluoresces bright blue and emetine yellow-white in UV-365 nm and they turn red and weak yellow, respectively, in vis. (→ B). Minor alkaloids, e.g. *O*-methylpsychotrine, are found in R_f range of emetine, or psychotrine in the R_f range of cephaeline.
The yellow fluorescence develops after approximately 30 min.

C With DRG reagent the major alkaloids are seen as orange-brown zones (vis).

Chinae cortex

Drug sample 1 Cinchona calisaya (alkaloid extraction method A, 20 μl)
2 Cinchona succirubra (alkaloid extraction method A, 20 μl)

Reference compound TC China alkaloid mixture (T1-T4 see section 1.2)
T1 quinine — T3 quinidine
T2 cinchonidine — T4 cinchonine

Solvent system Fig. 12 chloroform–diethylamine (90:10)

Detection A 10% eth. H_2SO_4 → UV-365 nm
B 10% H_2SO_4 followed by iodoplatinate reagent (No. 21) → vis

Fig. 12A In the R_f range 0.05–0.25 both **Cinchona (Chinae Cortex)** extracts show six light blue fluorescent alkaloid zones in UV-365 nm. They can be differentiated on the basis of their quinine (T1) content. In C. calisaya cortex (1) quinine counts as a major alkaloid. C. succirubrae cortex (2) contains the main cinchona alkaloids in approximately the same proportions as test mixture TC. Quinine (T1) and quinidine (T3) fluoresce bright blue after spraying with 10% ethanolic H_2SO_4, while cinchonidine (T2) and cinchonine (T4) turn dark violet and are hardly visible in UV-365 nm. In the extracts (1) and (2) the zone of cinchonidine (T2) is overlapped by the strong blue fluorescence of quinidine (T1).

B Treatment with iodoplatinate reagent results in eight mostly red-violet zones in the R_f range 0.05–0.65 (vis). The violet-brown zone of quinine is followed by the grey-violet zone of cinchonidine, a weak red-violet zone of quinidine and the more prominent brown-red cinchonine (TC). Three additional red-violet zones are found in the R_f range 0.4–0.6.

Remark: The slight variation in R_f values of the cinchona alkaloids (→ A:B) are due to the great sensitivity of the chloroform–diethylamine solvent system to temperature.

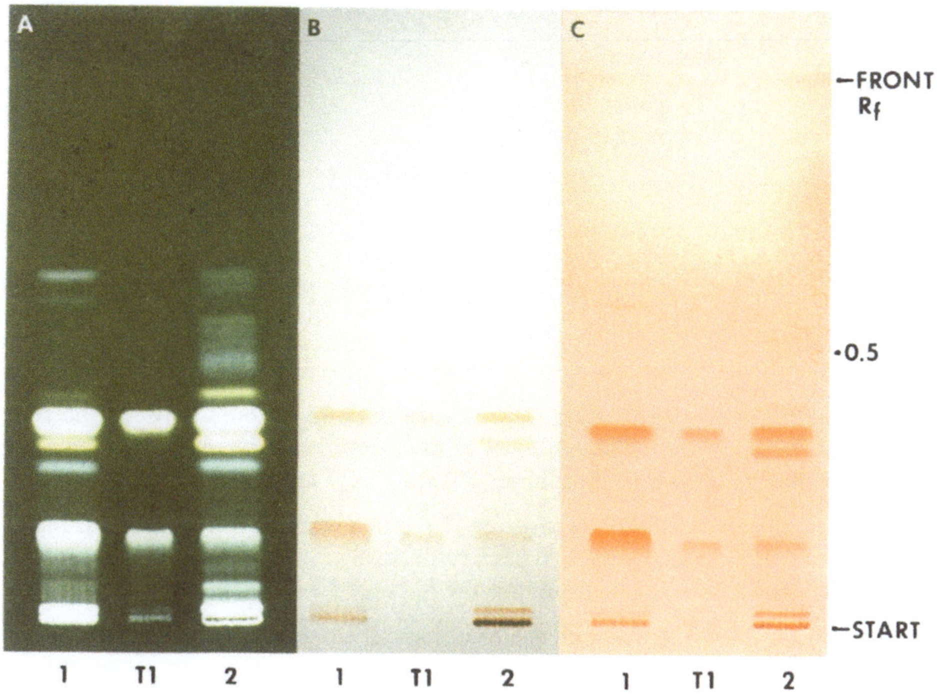
A
B
C
–FRONT
Rf
•0.5
–START
1 T1 2
1 T1 2
1 T1 2

Fig. 11

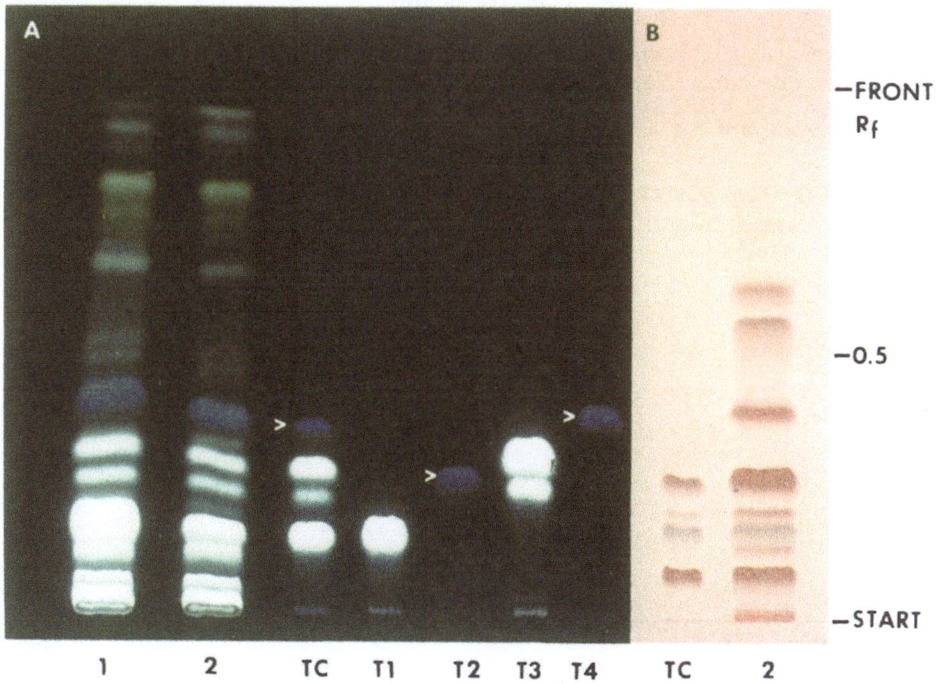
A
B
–FRONT
Rf
–0.5
–START
1 2 TC T1 T2 T3 T4
TC 2

Fig. 12

Opium

Drug sample 1 Opium extract (5% total alkaloids, 5 µl)

Reference compound		
	T1 morphine	T3 papaverine
	T2 codeine	T4 noscapine

Solvent system Figs. 13, 14 toluene–ethyl acetate–diethylamine (70:20:10)

Detection
A UV-254nm (without chemical treatment)
B Dragendorff reagent (DRG No. 13A followed by $NaNO_2$; No. 13B) → vis
C Natural products, polyethylene glycol reagent (NP/PEG No. 28) → UV-365nm
D Marquis reagent (No. 26) → vis

Figs. 13A **Opium extract** (1) shows six to eight fluorescence-quenching zones between the start and $R_f \sim 0.85$ in UV-254nm. The alkaloids of the morphinane/phenanthrene type are found in the lower R_f range with morphine (T1) at $R_f \sim 0.1$ and codeine (T2) at $R_f \sim 0.2$.

The benzyl isoquinoline alkaloids papaverine (T3) and noscapine (T4) are seen as major quenching zones at $R_f \sim 0.65$ and $R_f \sim 0.85$, respectively.
Thebaine and minor alkaloids migrate into the R_f range 0.3–0.5.

B With Dragendorff-$NaNO_2$ reagent all major opium alkaloids turn orange-brown (vis). Narceine remains at the start.

Fig. 14C Treatment with the NP/PEG reagent reveals a sequence of blue fluorescent zones at the beginning of the R_f range up to $R_f \sim 0.9$ (UV-365nm).

Except codeine (T2), which does not fluoresce, the main alkaloids morphine (T1), papaverine (T3) and noscapine (T4) give a blue fluorescence in UV-365nm.

D With Marquis reagent the alkaloids morphine and codeine are immediately stained typically violet.
A nonspecific reaction is given by papaverine, with a weak violet, and by noscapine, with a weak yellow-brown colour.

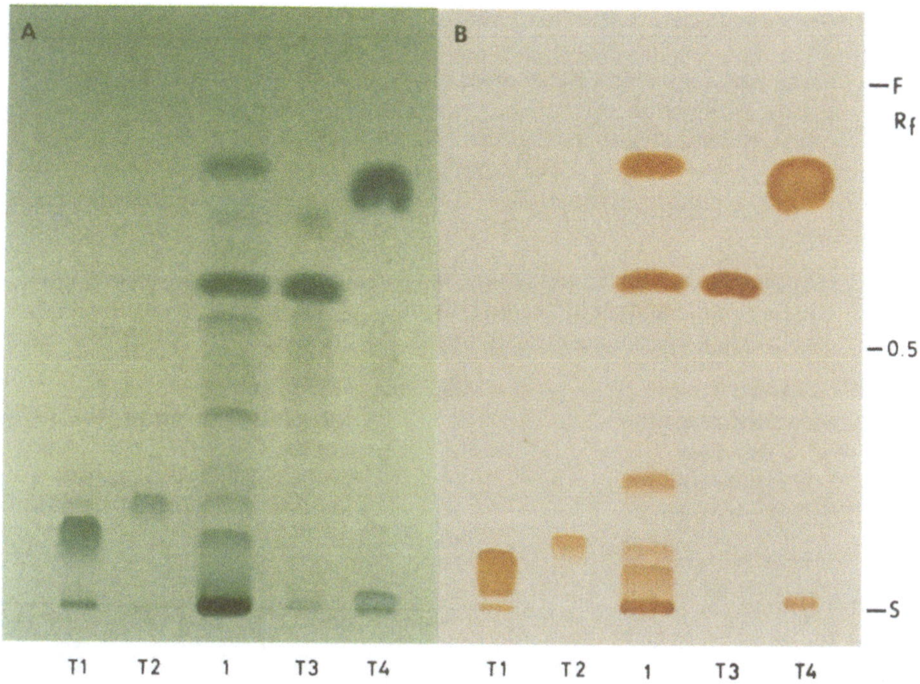
A
B
F
Rf
0.5
S
T1
T2
1
T3
T4
T1
T2
1
T3
T4

Fig. 13

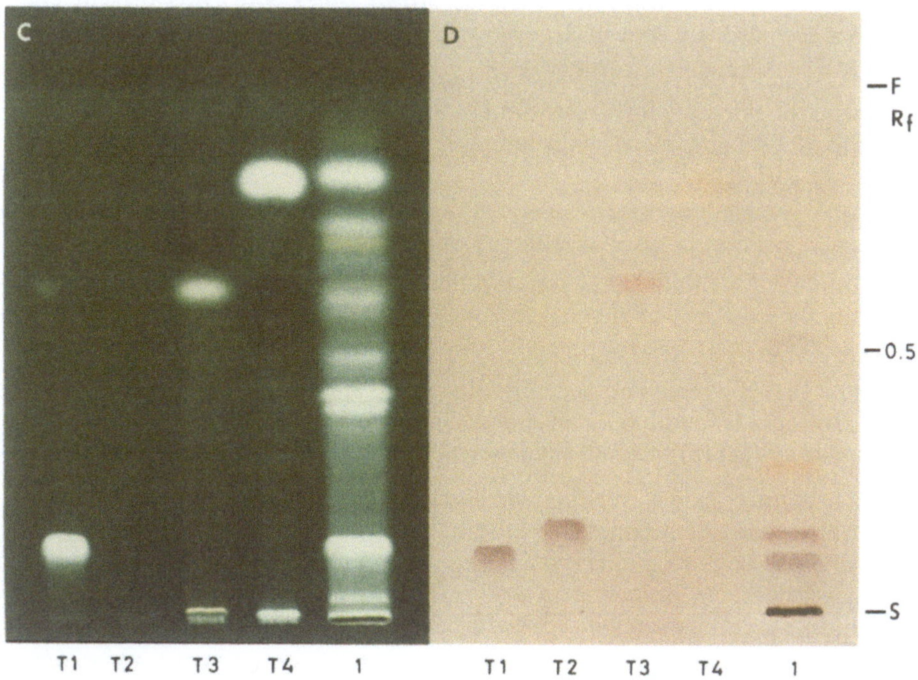
C
D
F
Rf
0.5
S
T1
T2
T3
T4
1
T1
T2
T3
T4
1

Fig. 14

Corydalidis rhizoma, Fumariae herba

Drug sample
1 Corydalidis rhizoma (alkaloid extraction method A, 30 µl)
2 Fumariae herba (methanolic extract 1 g/10 ml, 10 µl)
3 Fumariae herba (alkaloid extraction method C, 30 µl)

Reference Compound
T1 corytuberine
T2 corydaline
T3 rutin ($R_f \sim 0.35$) ▶ chlorogenic acid ($R_f \sim 0.4$) ▶ hyperoside ($R_f \sim 0.55$) = Flavonoid test mixture

Solvent system
Fig. 15A–C ethyl acetate–methylethyl ketone–formic acid–water (50:30:10:10) system 1
Fig. 16D,E ethyl acetate–methylethyl ketone–formic acid–water (50:30:10:10) system 1
F ethyl acetate–glacial acetic acid–formic acid–water (100:11:11:26) system 2

Detection
Fig. 15A UV-254 nm
B Dragendorff reagent (No. 13 B) → vis.
C UV-365 nm (without chemical treatment)

Fig. 16D UV-365 nm
E Dragendorff reagent (No. 13 B) → vis.
F Natural products reagent (NP/PEG No. 28) – UV-365 nm

Fig. 15A **Corydalidis rhizoma** (1)
The extract shows seven to eight quenching zones distributed up to R_f 0.75. The prominent zones at $R_f \sim 0.35$ can be identified as corytuberine (T1) and at $R_f \sim 0.7$ as corydaline (T2).

B Most of the major quenching zones react as brown zones with DRG reagent (vis). Corydaline is seen as main zone at $R_f \sim 0.7$, while bulbocapnine and corytuberine (T1) are found at $R_f \sim 0.45$ and 0.35 respectively.

C Direct viewing of extract 1 in UV-365 nm shows a series of predominantly blue (e.g. corydaline at $R_f \sim 0.7$) or yellow-white fluorescent zones (e.g. berberine-type alkaloids) in the R_f range 0.05–0.7.

Fig. 16D **Fumariae herba** (2,3)
A methanolic extraction of the drug (2) and an alkaloid enrichment (3) show in UV-365 nm 4–6 blue fluorescent zones in the R_f range 0.25–0.55 with an additional yellow-white zone at $R_f \sim 0.55$ (phenol carboxylic acids, sanguinarine, protoberberines) in sample 2.

E With DRG reagent two main and one minor brown alkaloid zone (vis) are detectable in sample 3. Protropin is found at $R_f \sim 0.6$ and allocryptopine in the lower $R_f \sim$ range. In the methanolic extract (2) these alkaloids are present in low concentration only.

F Separation of extract (2) in solvent system 2 and spraying with NP/PEG reagent reveals a series of blue fluorescent zones from the start till the solvent front, mostly due to phenol carboxylic acids (e.g. chlorogenic acid at $R_f \sim 0.45$) and a yellow fluorescent flavonoid glycosides, e.g. isoquercitrin at $R_f \sim 0.6$, as well as minor compounds in the lower R_f range (e.g. rutin, quercetin-3,7-diglucosido-3-arabinoside) and the aglycones at the solvent front.

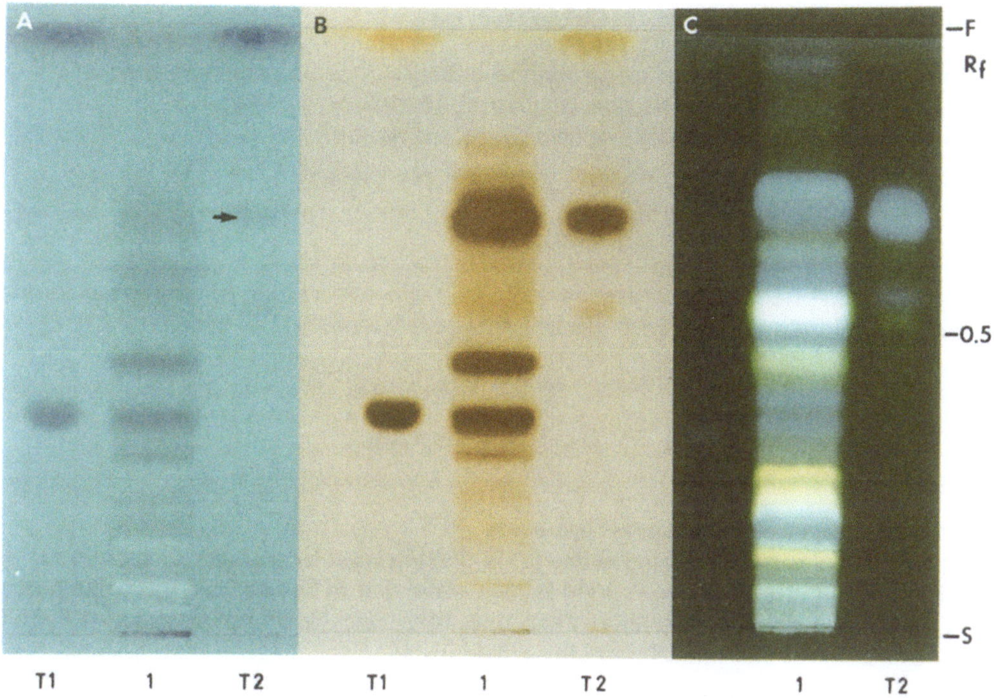

Fig. 15

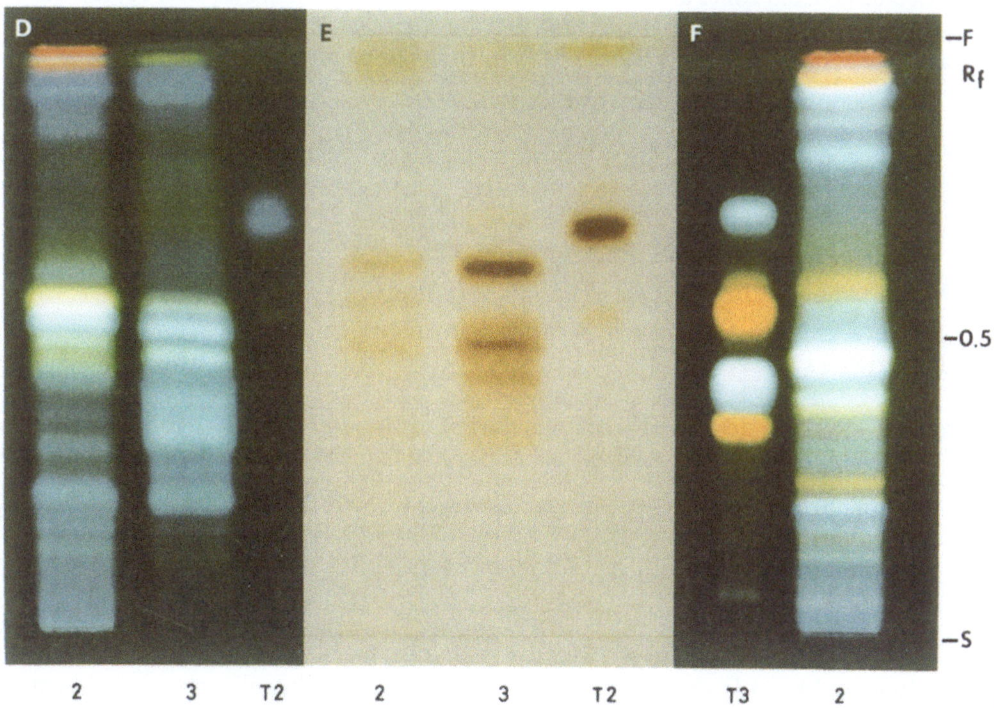

Fig. 16

Spartii flos, Sarothamni (Cytisi) herba

Drug sample
1 Spartii flos (MeOH extract 1 g/10 ml, 10 µl)
1a Spartii flos (alkaloid extraction method A, 50 µl)
2 Sarothamni herba (MeOH extract 1g/10 ml, 10 µl)
2a Sarothamni herba (alkaloid extraction method A, 30 µl)

Reference compound
T1 rutin ($R_f \sim 0.45$) ▶ chlorogenic acid ($R_f \sim 0.5$) ▶ hyperoside ($R_f \sim 0.6$) ▶ isochlorogenic acid = Flavonoid test mixture
T2 sparteine sulphate

Solvent system
Fig. 17A ethyl acetate–glacial acetic acid–formic acid–water (100:11:11:26) → flavonoids
B chloroform–methanol–glacial acetic acid (47.5:47.5:5) → alkaloids

Detection
A Natural products–polyethylene glycol reagent (NP/PG No. 28) → UV-365 nm ▶ flavonoids
B Iodoplatinate reagent (IP No. 21) → vis ▶ alkaloids

Fig. 17A NST/PEG reagent UV-365 nm ▶ Flavonoids
The methanolic extract of **Spartii flos** (1) is characterized by a major orange zone at R_f 0.65 (isoquercitrin, luteolin-4'-O-glucoside), while that of **Sarothamni scopariae herba** (2) shows two yellow-green fluorescent zones of spiraeoside and scoparoside at R_f 0.6–0.7 as well as the aglycone close to the solvent front.

B Iodoplatinate reagent vis. ▶ Alkaloids
Dark blue alkaloid zones are developed with IP reagent. Sparteine (R_f 0.25/T2) is a major alkaloid in Sarothamni scop. herba (2a). Besides sparteine sample 2a shows an additional dark blue zone at R_f 0.15. Cytisine and N-methylcytisine are present in Spartii flos (1a).

Genistae herba

Drug sample
3 Genistae herba (MeOH extract 1 g/10 ml/10 µl)
3a Genistae herba (alkaloid extraction method A, 30 µl)

Reference compound
T1 rutin ▶ chlorogenic acid ▶ hyperoside ▶ isochlorogenic acid
T2 sparteine sulfate

Solvent system
Fig. 18A ethyl acetate–glacial acetic acid–formic acid–water (100:11:11:26) → flavonoids
B chloroform–methanol–glacial acetic acid (47.5:47.5:5) → alkaloids

Detection
A Natural products–polyethylene glycol reagent (NP/PEG No. 28) → UV-365 nm ▶ flavonoids
B Dragendorff reagent (DRG No. 13) followed by $NaNO_2$ (No 13 B) → vis ▶ alkaloids

---

Fig. 18A NST/PG reagent, UV-365 nm ▶ Flavonoids
Genistae herba (3) is characterized by a high amount of luteolin glycosides, seen as bright yellow fluorescent zones in the R_f range 0.55–0.8, the aglycone at the front and blue fluorescent isoflavones (e.g. genistin) and phenol carboxylic acids (e.g. chlorogenic acid) at R_f 0.5.

B DRG/$NaNO_2$, vis ▶ Alkaloids
Two brown alkaloid zones in the R_f range 0.1–0.2 of (3a) are due to sparteine type alkaloids such as *N*-methylcytisine, anagyrine and cytisine.

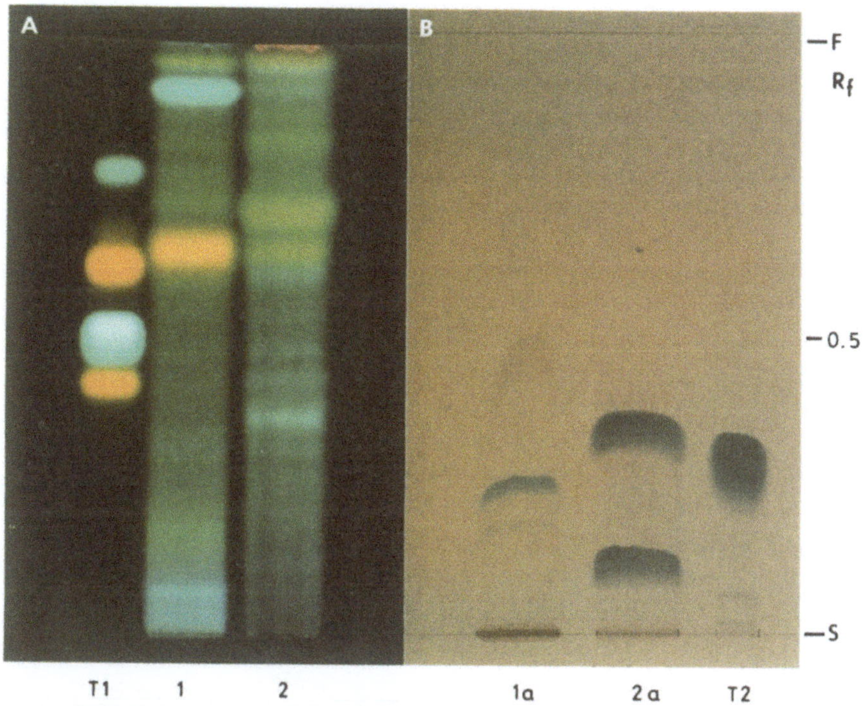

Fig. 17

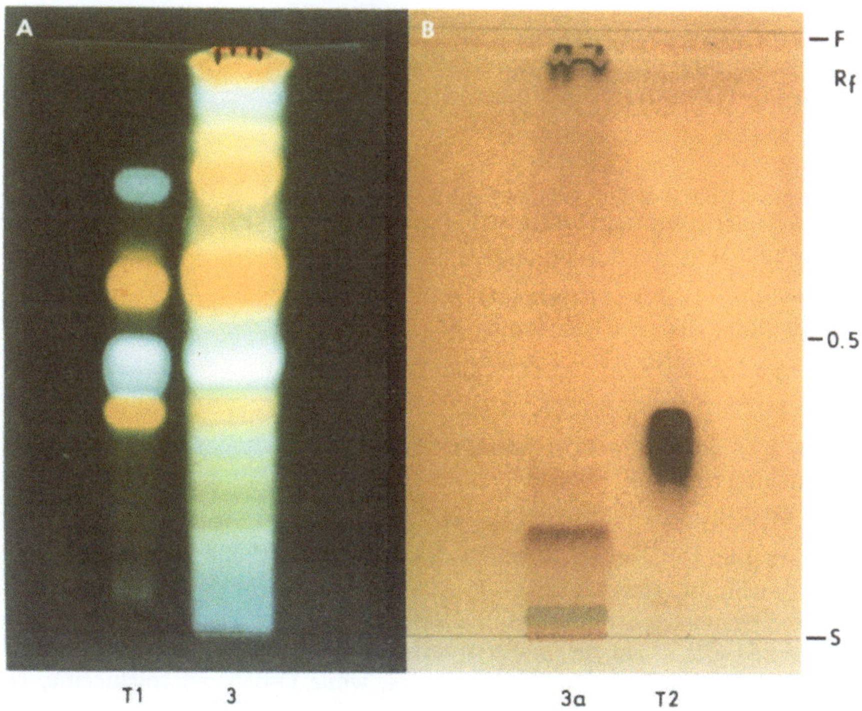

Fig. 18

Chelidonii herba

Drug sample 1–3 Chelidonii herba different trade samples (alkaloid extraction method A, 40 µl)

Reference compound T1 sanguinarine
T2 papaverine
T3 methyl red

Solvent system Fig. 19 1-propanol–water–formic acid (90:9:1)

Detection A UV-365 nm (without chemical treatment)
B Dragendorff reagent [DRG reagent No. 13A] → vis

Fig. 19A **Chelidonii herba** (1–3). The extracts of the samples 1–3 are characterized in UV-365 nm by bright yellow fluorescent zones: the major alkaloid coptisin at $R_f \sim 0.15$, followed by minor alkaloids berberine and chelerythrine directly above and sanguinarine (T1) as a broad yellow band in the R_f range 0.3–0.4. In the R_f range 0.75–0.85 weak yellow-green (e.g. chelidonine) and blue-violet zones are found.

B The fluorescent alkaloid zones in the R_f range 0.15–0.85 respond to DRG reagent with brown, rapidly fading colours (vis.). Papaverine (T2) can serve as reference compound for sanguinarine ($R_f \sim 0.4$), and methyl red (T3) for the alkaloidal zones at $R_f \sim 0.8$.

Colchici semen

Drug sample 1 Colchici semen (alkaloid extraction method A, 30 µl)
2 Colchici semen (MeOH extract 3 g/10 ml, 10 µl)

Reference compound T1 colchicine
T2 colchicoside

Solvent system A ethyl acetate-glacial acetic acid formic acid–water (100:11:11:26)
B ethyl acetate–methanol–water (100:13.5:10)

Detection A UV-254 nm (without chemical treatment)
B UV-365 nm (without chemical treatment)
C Dragendorff reagent/$NaNO_2$ (DRG No. 13 B) → vis.

Fig. 20A **Colchici semen** (1,2). Both extracts are characterized by colchicine, which is seen as a prominent quenching zone at $R_f \sim 0.6$ (T1), while colchicoside ($R_f \sim 0.15$/T2) is found in the methanolic extract (2) only.

B In the alkaloid fraction (1) a series of seven to nine prominent blue and yellow-white fluorescent zones from the start till $R_f \sim 0.35$, six weaker blue zones at R_f 0.4–0.85 and two zones at the solvent front are detected in UV-365 nm. Besides colchicine at $R_f \sim 0.25$ (T1) minor alkaloids such as colchiceine, *N*-acetyl demecolcine and 1-ethyl-2-demethyl colchiceine also show a yellow-white fluorescence, while O-benzoyl colchiceine, *N*-formyl-deacetyl colchiceine and *N*-methyl demecolcine fluoresce blue.

C Colchicine and minor alkaloids react as brown zones with DRG reagent (vis). Artefacts of colchicine ($R_f \sim 0.6$) appear as a blue zone at $R_f \sim 0.5$ (vis)

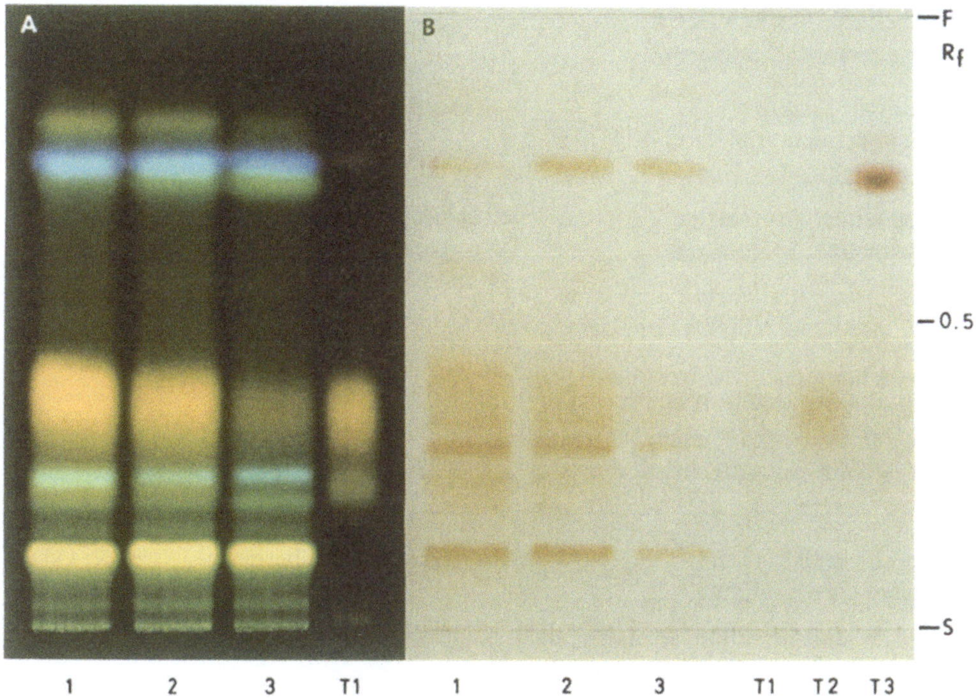

Fig. 19

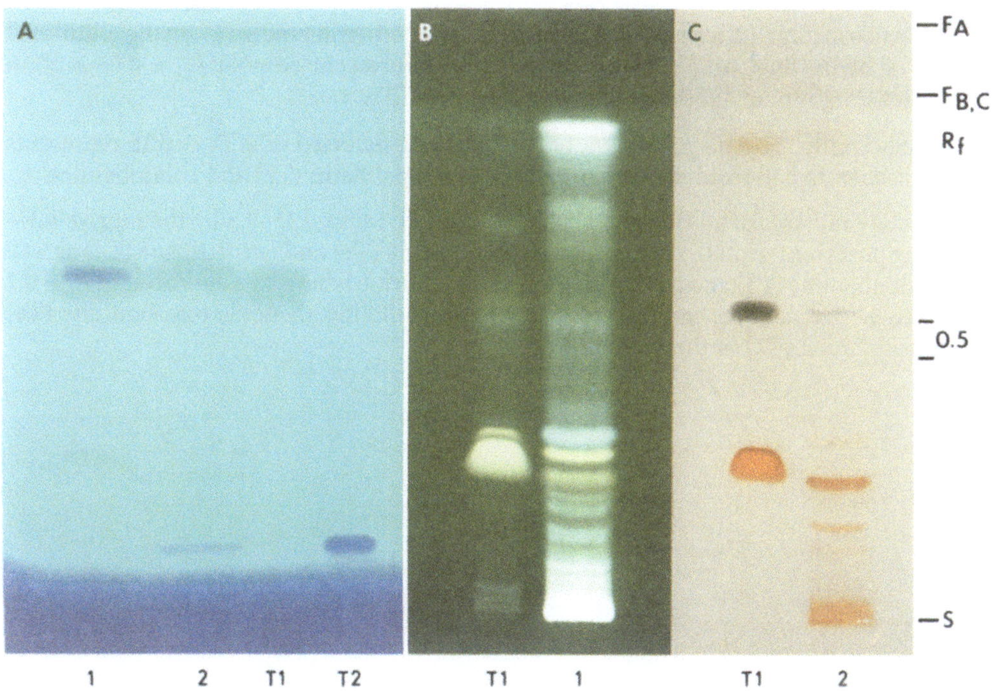

Fig. 20

Berberidis cortex, Colombo radix, Hydrastis rhizoma, Mahoniae radix/cortex

Drug sample
1 Berberidis radix
2 Hydrastis rhizoma
3 Colombo radix
4 Mahoniae radix/cortex
(alkaloid extraction method A, 30 μl)

Reference compound
T1 berberine
T2 palmatine/jatrorrhizine
T3 hydrastine
T4 jatrorrhizine
T5 columbamine
T6 oxyacanthine
T7 berbamine
T8 palmatine

Solvent system
Fig. 21 n-propanol–formic acid–water (90:1:9)
Fig. 22 n-butanol–ethyl acetate–formic acid–water (30:50:10:10)

Detection
A vis (without chemical treatment)
B Dragendorff reagent [DRG No. 13A] → vis
C UV-365 nm (without chemical treatment)
D UV-365 nm (without chemical treatment)

Fig. 21A **Berberidis radixs** (1) shows the characteristic yellow zone of berberine ($R_f \sim 0.2$/T1) on untreated chromatogram (vis.).

B Berberine and the minor alkaloids, such as jatrorrhizine and palmatine, react with a brown-red colour with DRG reagent (vis.).

C Extracts of **Berberidis radix** (1) and **Hydrastis rhizoma** (2) both show the major alkaloid berberine as a prominent lemon-yellow fluorescent zone at $R_f \sim 0.25$.

Hydrastis rhizoma (2) can be differentiated from Berberidis radix (1) by the additional zone of hydrastine, which forms a blue-white fluorescent zone at $R_f \sim 0.03$ and an additional light blue fluorescent zone at $R_f \sim 0.9$ (T3).

Colombo radix (3). The yellow-white alkaloid zone detected in at $R_f \sim 0.15$ represents the unseparated alkaloid mixture of jatrorrhizine, palmatin (T2) and columbamine.

Fig. 22D **Mahoniae radix/cortex** (4) is characterized in the R_f range 0.45–0.5 by the four yellow-green fluorescent protoberberine alkaloids berberine (T1) and jatrorrhizine (T4) as well as columbamine (T5) and palmatine (T8). Magnoflorine is seen as a dark zone at $R_f \sim 0.2$ directly above the blue fluorescent bisbenzylisoquinoline alkaloids oxyacanthine (T6) and berbamine (T7) in the R_f range 0.05–0.1.

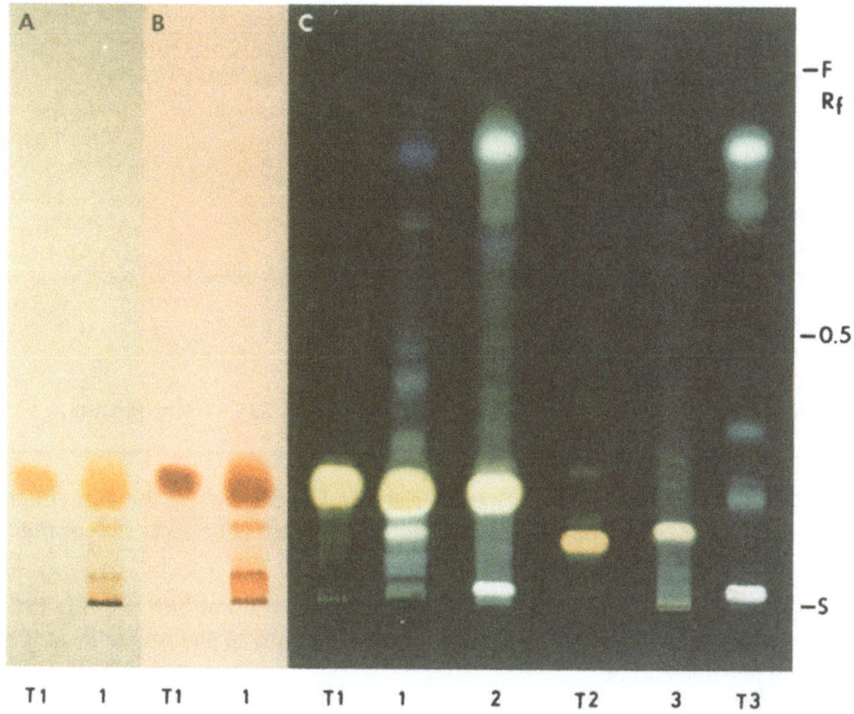

Fig. 21

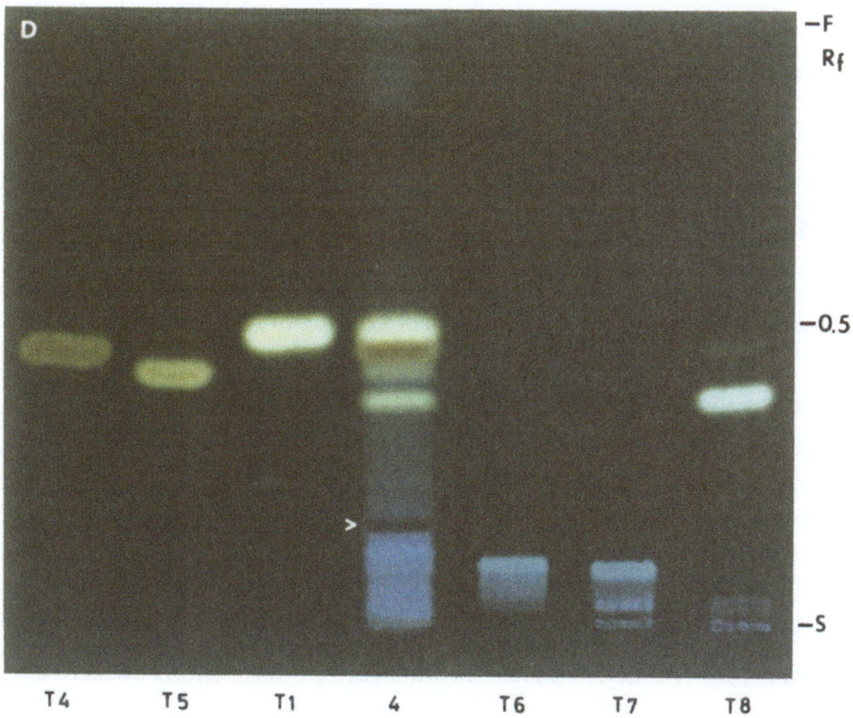

Fig. 22

Boldo folium

Drug samples	1 alkaloid extract (method A, 30 µl) 3 methanol extract (1 g/10 ml, 10 µl) 2 essential oil (TAS method, 100 mg)
Reference compound	T1 boldine T2 rutin (R_f 0.4) ▶ chlorogenic acid (R_f 0.5) ▶ hyperoside (R_f 0.65) favonoid test
Solvent system	Fig. 23 A,B toluene–ethyl acetate–diethylamine (70:20:10) C toluene–ethyl acetate (93:7) D ethyl acetate–formic acid–glacial acetic acid–water (100:11:11:26)
Detection	A UV-365 nm (without chemical treatment) B Dragendorff reagent (DRG No. 13B) → vis C Vanillin-H_2SO_4 reagent (VS No. 42) → vis D Natural products-polyethylene glycol reagent (NP/PEG No. 28) → UV-365 nm

Fig. 23A **Boldo folium.** The alkaloid extract (1) is characterized in UV-365 nm by the two violet fluorescent zones in the R_f range of the boldine test T1, as well as various red-orange fluorescent chlorophyll zones in the upper R_f range.

B With DRG reagent two dark brown zones in the R_f range of the boldine test T1, two minor alkaloid zones above the start and greenish-brown zones in the upper R_f range due to chlorophyll are detectable.

C The volatile oil compounds (2) yield ten grey or blue zones between the start and R_f 0.85 with 1,4-cineole ($R_f \sim 0.4$) and ascaridole ($R_f \sim 0.8$) as major terpenoides.

D The methanolic extract (3) is characterized by its high amount and variety of flavonol glycosides. Five almost equally concentrated yellow-green fluorescent zones appear in the R_f range 0.4–0.65 (rutin ▶ hyperoside/T2) accompanied by two prominent zones at R_f 0.75–0.8 and three minor zones in the lower R_f range.

Nicotianae folium

Drug samples	1 alkaloid extract (method A, 40 µl) 2 commercial cigarette (method A, 40 µl) 1a methanol extract (1 g/10 ml, 10 µl) 2a methanol extract of (2) (1 g/10 ml, 10 µl)
Reference compound	T1 nicotine T2 rutin (R_f 0.4) ▶ chlorogenic acid (R_f 0.5) ▶ hyperoside (R_f 0.6) favonoid test
Solvent system	Fig. 24 A,B toluene–ethyl acetate–diethyl amine (70:20:10) C ethyl acetate–formic acid–glacial acetic acid–water (100:11:11:26)
Detection	A UV-254 nm (without chemical treatment) B Dragendorff reagent (DRG No. 13B) → vis. C Natural products-polyethylene glycol reagent (NP/PEG No. 28) → UV-365 nm

Fig. 24A **Nicotianae folium** (1,2). The major alkaloid nicotine (T1/$R_f \sim 0.75$) shows quenching in UV-254 nm.

B The alkaloid extracts of sample (1) and (2) both contain nicotine and two additional alkaloids at R_f 0.35–0.4 (e.g. nornicotine, anabasine) which turn orange-brown with DRG reagent (vis.).

C The methanolic extracts (1a) and (2a) show, in addition to the alkaloids, the flavonol glycoside rutin and the chlorogenic acid (T2), more highly concentrated in 1a.

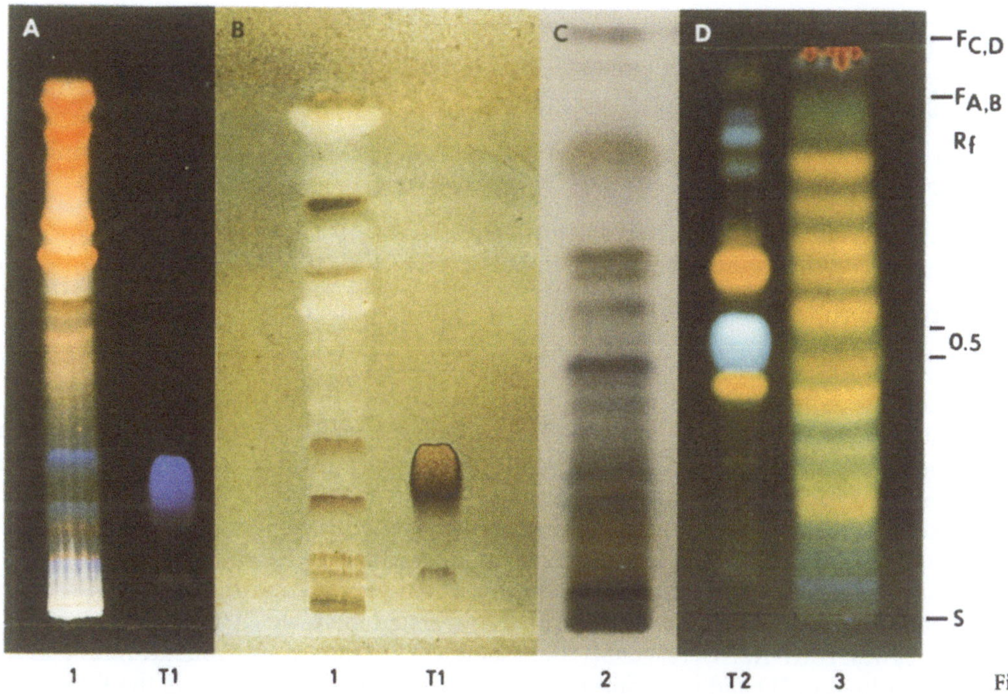

Fig. 23

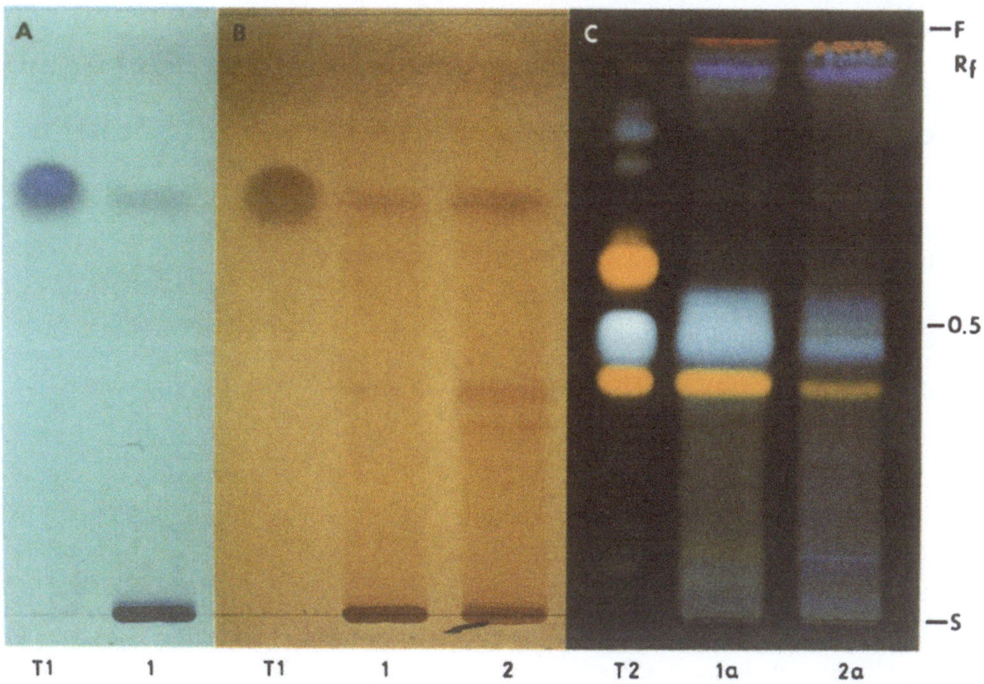

Fig. 24

Aconiti tuber

Drug sample	1 trade sample (1992) 2 A. napellus L. ssp. napellus (alkaloid extraction method A, 30–40 µl)	3 trade sample (1984) 4 A. paniculatum ssp. paniculatum	
Reference compound	T1 aconitine/mesaconitine T2 aconitine	T3 deoxyaconitine T4 hypaconitine	T5 benzoylaconine T6 aconine
Solvent system	Fig. 25 A toluene-ethyl acetate-diethylamine (70:20:10) B cyclohexane-ethanol-diethyamine (80:10:10)		
Detection	Dragendorff reagent (DRG No. 13A) → vis	DRG/$NaNO_2$ reagent (No. 13B) → vis	

Fig. 25 The European **Aconitum napellus** group comprises three species: A. napellus, A. pentheri and A. angustifolium. The TLC pattern of their alkaloid distribution varies: a dominating aconitine amount, aconitine and mesaconitine as prominent zones or mainly mesaconitine and/or hypaconitine.

A Extract (1) contains aconitine and mesaconitine (T1) which appear in system A at R_f 0.6–0.75 as brown, fast-fading zones after treatment with DRG reagent (vis).

B The alkaloids deoxyaconitine (T3) and hypaconitine (T4) and the cleavage products benzoylaconine (T5) and aconine (T6) are separated in system B and show fast-fading zones with DRG-$NaNO_2$ reagent (vis). In samples (1,2) the aconitine/mesaconitine zones at R_f 0.35–0.4 (T1) and in sample (3) various, additional brown zones in the R_f range of benzoylaconine (T5) and aconine (T6) are found. A. paniculatum extract (4) has an obviously different TLC pattern with a main zone in the R_f range of hypaconitine (T4) and at $R_f \sim 0.55$.

Aconiti tuber, Sabadillae semen, Lobeliae herba, Ephedrae herba

Drug sample	1 Aconiti tubera (trade sample) 2 Sabadillae semen (alkaloid extraction method A, 30 µl)	3 Lobeliae herba 4 Ephedrae herba
Reference compound	T1 aconitine/mesaconitine T2 veratrine (alkaloid-mixture)	T3 lobeline T4 ephedrine
Solvent system	Fig. 26 A toluene-ethyl acetate-diethylamine (70:20:10) B ethyl acetate-cyclohexane-methanol-ammonia (70:15:10:5) C toluene-chloroform-ethanol (28.5:57:14.5)	
Detection	A Iodoplatinate reagent (IP No. 21) → vis B Ninhydrine reagent (NIH No. 29) → vis C Dragendorff reagent (DRG No. 13A) → vis	

Fig. 26A **Aconiti tuber** (1), **Sabadillae semen** (2), **Lobeliae herba** (3). Their major alkaloids are found in the R_f range 0.6–0.65 as white zones against a grey-blue background.
Aconiti tuber (1): aconitine/mesaconitine (T1) and six minor zones (R_f range 0.25–0.7)
Sabadillae semen (2): veratrine (T2) and eight minor zones (R_f 0.5–0.55/0.8).
Lobeliae herba (3): one prominent zone of lobeline (R_f 0.65/ref T3).

B, C **Ephedrae herba** (4): ephedrine is detected as a violet-red band (R_f 0.4–0.5) with ninhydrine, or with DRG reagent as a brown zone at $R_f \sim 0.2$ in solvent system C.

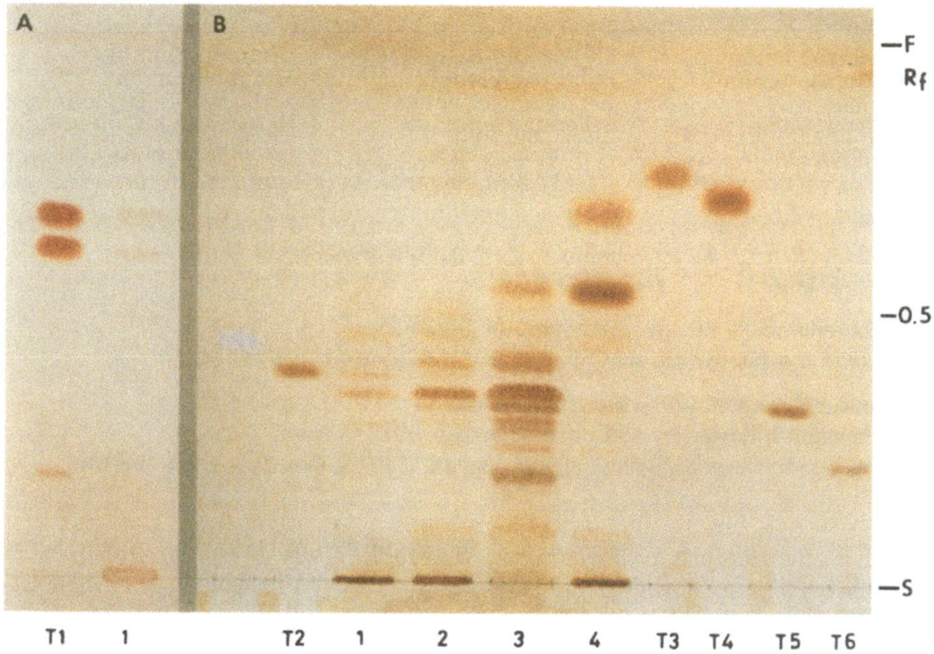

Fig. 25

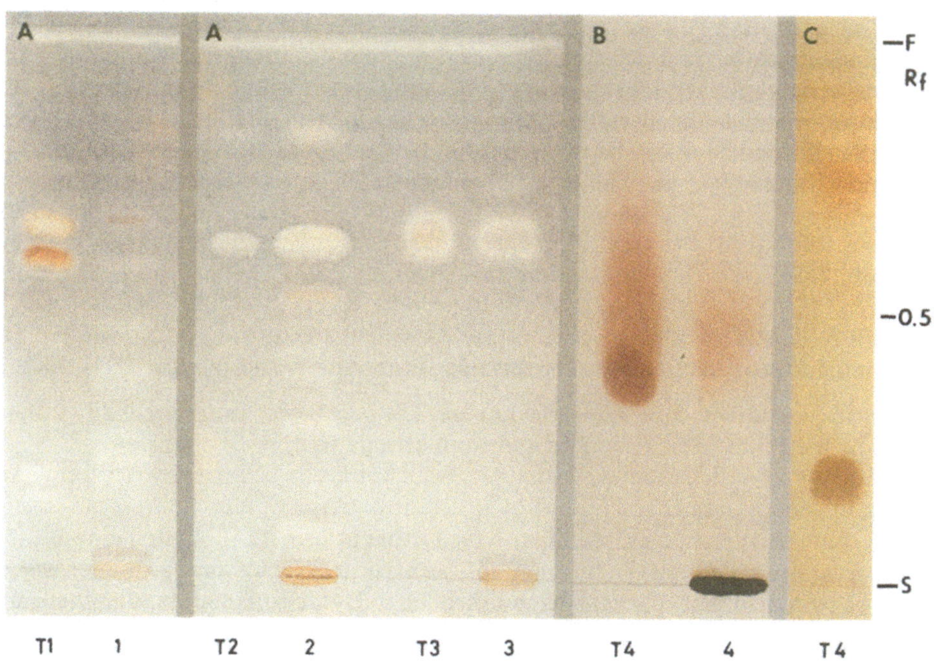

Fig. 26

Solanaceae drugs

Alkaloid extract	1 Belladonnae folium	2 Hyoscyami folium	3 Stramonii folium
Methanol extract	4 Scopoliae radix	6 Belladonnae folium	8 Hyoscyami nigri folium
	5 Belladonnae radix	7 Stramonii folium	9 Hyoscyami mutici folium

(alkaloid extraction method C: (1)–(3) 30 µl, flavonoids (1g/10ml MeOH): (4)–(9) 20 µl)

Reference compound
T1–T3 alkaloid test: hyoscyamine ▶ scopolamine mixture (defined ratio see sect. 1.2)
T4 rutin (R_f 0.35) ▶ chlorogenic acid (R_f 0.45) ▶ hyperoside (R_f 0.6)
T5 scopoletin T6 caffeic acid

Solvent system
Fig. 27 toluene–ethyl acetate–diethylamine (70:20:10)
Fig. 28 ethyl acetate–formic acid–glacial acetic acid–water (100:11:11:26)

Detection
A Dragendorff reagent (DRG No. 13A) → vis
B DRG reagent followed by sodium nitrite (No. 13B) → vis
C Natural products–polyethylene glycol reagent (NP/PG No. 28) → UV 365 nm

---

Fig. 27A,B **Alkaloids in Belladonnae, Hyoscyami and Stramonii folium** (1–3). The tropane alkaloids (-)-hyoscyamine (during extraction procedures partly changed into (±) atropine) and scopolamine as major compounds of the alkaloidal fraction of Solanaceae drugs respond to Dragendorff reagent with orange, unstable colour. Treatment with $NaNO_2$ increases the colour stability of the hyoscyamine zones.

A TLC differentiation of the three drugs is based on the hyoscyamine to scopolamine ratio and, to a limited extent, on the contents of the minor alkaloids belladonnine, atropamine and cuskhygrine.

For drug identification and determination of the alkaloid content, DAB 10 describes a TLC comparison with alkaloid mixtures containing defined ratios of atropine-SO_4 to scopolamine-HBr (T1–T3). Identification of the drug is then based on the similarity of colour intensity and zone size between the standard solutions and drug extracts.

Belladonnae folium (1): the ratio of hyoscyamine (R_f 0.25) to scopolamine (R_f ~ 0.4) corresponds to that of T1 at about 3:1. Both alkaloids are also present in the roots and seeds.

Hyoscyami folium (2): the ratio of the two main alkaloids is about 1.2:1. The total alkaloid content is less than the standard solution T2.

Stramonii folium (3): a higher scopolamine content than in (1) and (2). The typical hyoscyamine to scopolamine ratio for this drug is about 2:1.

Fig. 28 **Caffeic acid derivatives, coumarins, flavonoids.** The Solanaceae drugs are easily differentiated by their individual flavonoid and coumarin pattern.

Scopoliae- (4) and **Belladonnae radix** (5), which have a similar hyoscyamine to scopolamin content, are characterized by different patterns of blue fluorescent caffeic acid and coumarin derivatives (see Chap. 5 for further information). In **Belladonnae** (6) and **Hyoscyami nigri folium** (8), the main zones are rutin (R_f ~ 0.4; orange fluorescence) and chlorogenic acid (R_f ~ 0.45; blue fluorescence). In Hyoscyami nigri folium, these are the only two detectable zones, whereas Belladonnae folium shows additional blue, yellow-green and orange fluorescent zones in the R_f range 0.05–0.1 (7-glucosyl-3-rhamnogalactosides of kaempferol and quercetin).

Stramonii folium (7) is characterized by five orange fluorescent quercetin glycosides in the R_f range 0.03–0.25. The absence of rutin and chlorogenic acid clearly distinguishes the drug from Belladonnae and Hyoscyami folium. Hyoscyami mutici folium (9) has only a very low flavonoid content.

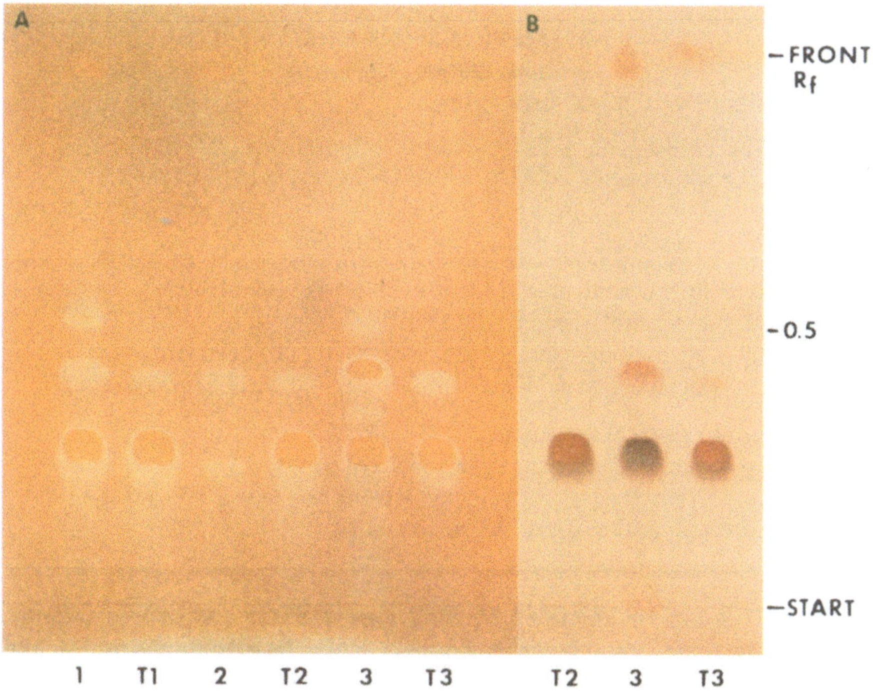
A
B
– FRONT
Rf
- 0.5
–START
1 T1 2 T2 3 T3 T2 3 T3

Fig. 27

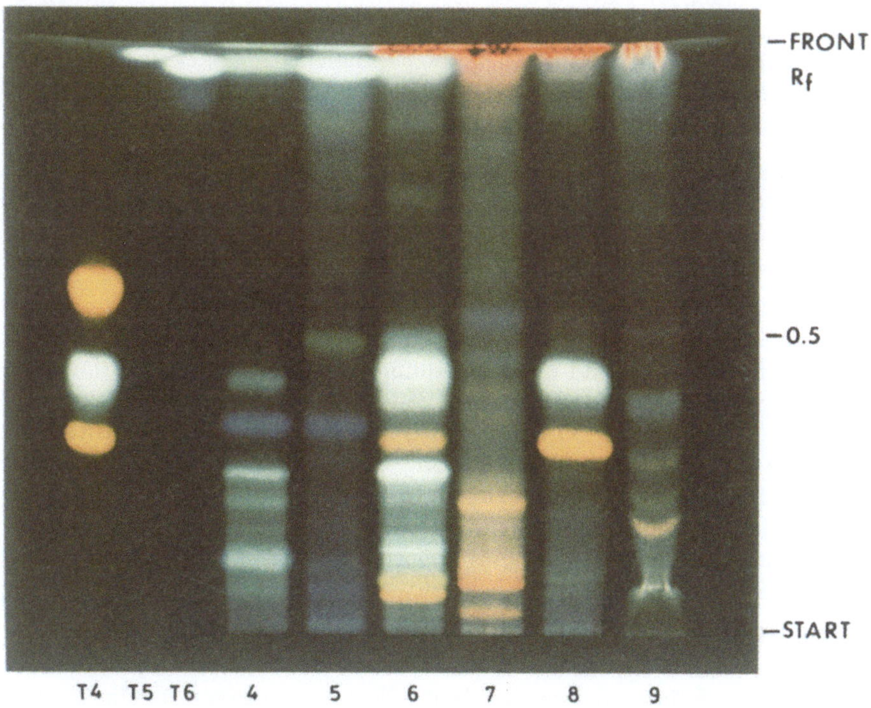
–FRONT
Rf
–0.5
–START
T4 T5 T6 4 5 6 7 8 9

Fig. 28

Purine drugs

Drug sample 1 Coffeae semen; 2 Mate folium; 3 Theae folium (black tea); 4 Cacao semen
(methanolic extraction, 1 g/10 ml, 30 μl)

Reference compound
T1 rutin (R_f ~ 0.35) ► chlorogenic acid (R_f ~ 0.45) ► hyperoside (R_f ~ 0.6)
T2 caffeine
T3 theobromine
T4 aescin (R_f ~ 0.25) + aescinols (R_f ~ 0.45) = saponin test

Solvent system
Fig. 29 A ethyl acetate–formic acid–glacial acetic acid–water (100:11:11:26) → system A
B ethyl acetate–methanol – water (100:13.5:10) → system B
Fig. 30 C ethyl acetate–formic acid–glacial acetic acid–water (100:11:11:26) → system A
D chloroform–glacial acetic acid–methanol–water (60:32:12:8) → system D

Detection
A UV-254nm (without chemical treatment)
B Iodine-potassium iodide-HCl reagent (I/HCl No. 20) → vis
C Natural products-polyethylene glycol reagent (NP/PG No. 28) → UV-365nm
D Anisaldehyde-sulphuric acid reagent (AS No. 3) → vis

The Purine drugs 1–4 can be identified by their characteristic contents of caffeine, theobromine, theophylline, various caffeoylquinic acids, flavonoid glycosides and saponines.

Fig. 29A **Puridnerivatives.** (System A). Extracts of Coffeae semen (1), Mate folium (2) and Theae folium (3) show one to four prominent fluorescence-quenching zones in the R_f range 0.4–0.6 with caffeine as the main zone at R_f ~ 0.60. Caffeine migrates in this solvent system directly above the hyperoside (T1/R_f ~ 0.6). → For detection of caffeoyl quinic acids and flavonoids see reagent C.

B (System B) Treatment with I/HCl reagent generates a dark-brown zone of caffeine at R_f ~ 0.45 (T2) in extracts (1) and (3), less concentrated in (2) and (4). Theobromine at R_f ~ 0.4 (T3) is detected as a grey, fast-fading zone in Mate folium (2). The concentration of theobromine in Cacao semen (4) is low, the amount of theophylline (R_f ~ 0.6) in the extracts 1–4 is not sufficient for detection.

Fig. 30C **Phenol carboxylic acids, flavonoids and saponines.** (System A) Treatment with NP/PEG reagent reveals caffeoyl (CQA) and dicaffeoyl quinic acids as blue and the flavonoid glycosides as orange-yellow or green fluorescent zones in UV-365nm. Coffeae semen (1) and Mate folium (2): the blue 5-CQA, 3-CQA (R_f 0.45–0.5) and additional dicaffeoyl quinic acids in the upper R_f range are characteristic. One additional orange-yellow zone of rutin at R_f ~ 0.4 (T1) is found in Mate folium (2) only.
Theae folium (3): four mainly yellow fluorescent flavonoid glycosides in the R_f range of hyperoside and rutin (T1) and two flavonoid glycoside zones at R_f 0.25–0.3 with yellow and green fluorescence, respectively.

D (System D) **Saponines** (aescin T4) respond as blue-violet zones to AS reagent (vis). In Mate folium (2) the main triterpene saponins are seen as six blue-violet zones in the R_f range 0.4–0.8. In Theae folium (3) broad bands of yellow-brown zones from the start till R_f ~ 0.4 (“thea flavines”) dominate in the lower R_f range.

Note: Caffeine migrates in solvent system A up to the solvent front.

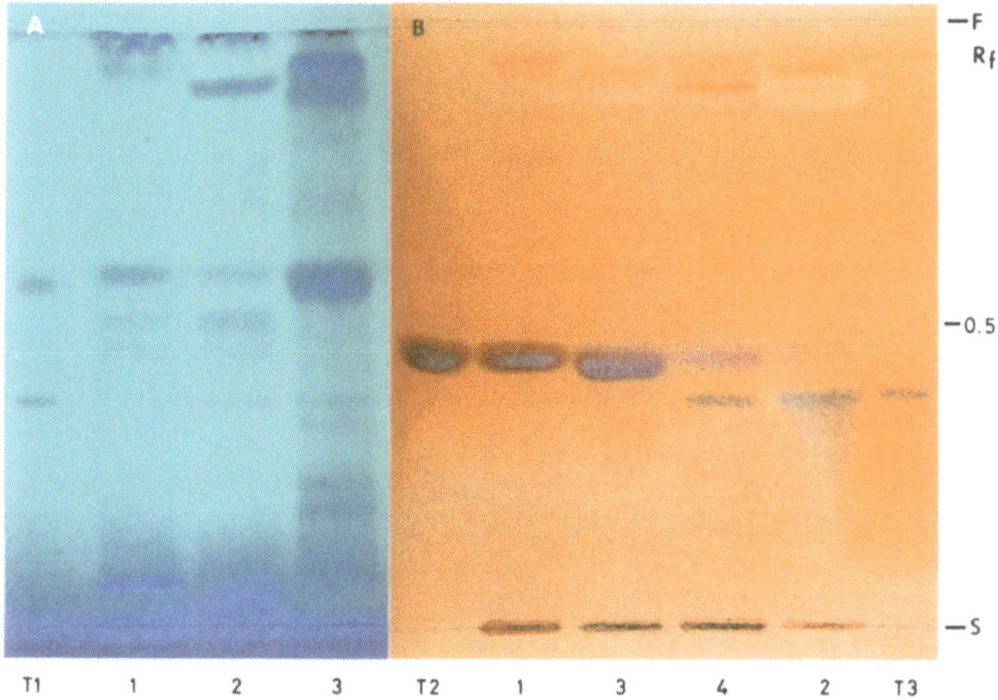

Fig. 29

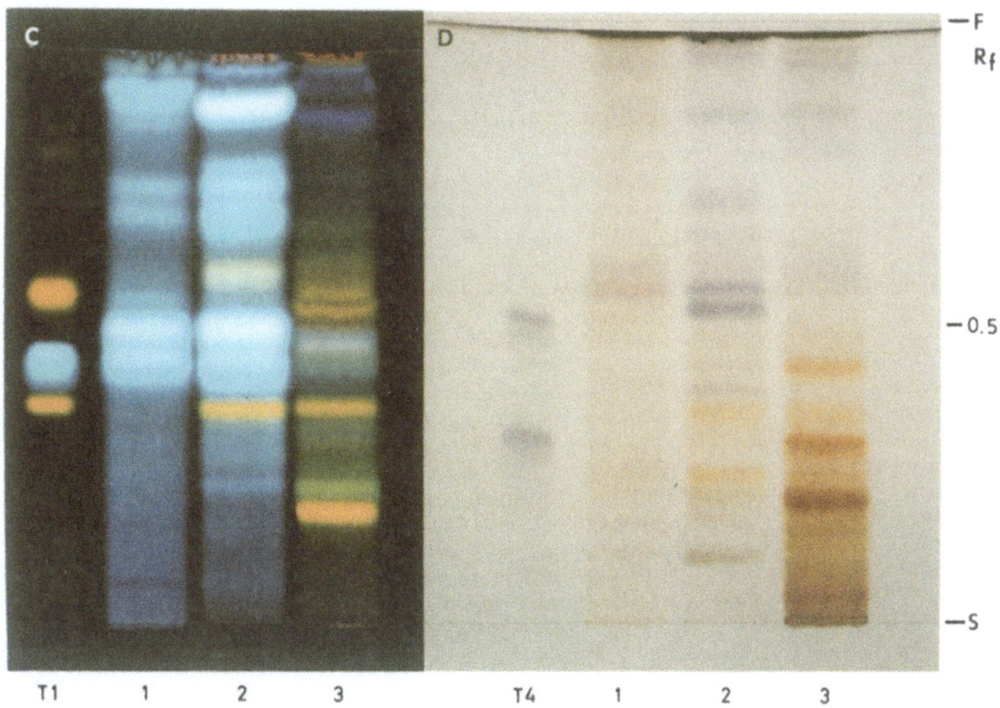

Fig. 30

2 Drugs Containing Anthracene Derivatives

The characteristic constituents of this drug group are anthraquinones, oxanthrones, anthranols and anthrones with laxative properties. The anthraquinones possess phenolic groups on C-1 and C-8 and keto groups on C-9 and C-10. In the anthrones and anthranols, only C-9 carries an oxygen function. In addition, a methyl, oxymethyl or carboxyl group may be present on C-3, and a hydroxy or methoxy group on C-6. Most compounds in this group are present in the plant as O-glycosides. The glycoside linkage is usually at C-1, C-8 or C-6-OH. C-Glycosides occur as anthrones only, with the C–C bond always at C-10. In the O- and C-glycosides, the only sugars found so far are glucose, rhamnose and apiose.

2.1 Preparation of Extracts

General method, methanolic extract

Powdered drug (0.5 g) is extracted for 5 min on a water bath with 5 ml methanol. The filtrate is used for TLC: 5 µl (Aloe) and 20 µl (Rheum, Frangula, Cascara).

Senna

Sennae folium or fructus are extracted with 50% methanol; 20 µl is used for TLC.

Hydrolysis of anthraquinone glycosides

Powdered drug (0.5 g) and 25 ml 7.5% hydrochloric acid are heated under reflux for 15 min. After cooling, the mixture is extracted by shaking with 20 ml ether. The ether phase is concentrated to about 1 ml, and 10 µl is used for TLC (e.g. Rhei radix).

2.2 Thin-Layer Chromatography

Reference solutions

Aloin, frangulin A/B, glucofrangulin A/B, rhein, aloe-emodin and rhaponticoside (stilbene glucoside) are applied as 0.1% methanolic solutions.
Sennosides A and B are prepared as a 0.1% solution in methanol-water (1:1).
A total of 10 µl of each reference solution is used for TLC.

Adsorbent

Chromatography is performed on silica gel 60F_{254} precoated plates (Merck, Germany).

Chromatography solvents

- Ethyl acetate-methanol-water (100:13.5:10)
  With the exception of Senna preparations, the solvent system is suitable for the chromatography of all anthracene drug extracts.
- n-propanol-ethyl acetate-water-glacial acetic acid (40:40:29:1) ► Senna
- light petroleum-ethyl acetate-formic acid (75:25:1) ► anthraquinone aglycones

- toluene-ethyl formiate-formic acid (50:40:10) or (50:20:10) ▶ for the non-laxative dehydrodianthrones of Hyperici herba

2.3 Detection

- UV 254 nm All anthracene derivatives quench fluorescence
- UV 365 nm All anthracene derivatives give yellow or red-brown fluorescence

- Spray reagents (See Appendix A)

  - Potassium hydroxide (KOH No. 35; → Bornträger reaction)
    After spraying with 5% or 10% ethanolic KOH, anthraquinones appear red in the visible and show red fluorescence in UV-365 nm.
    Anthrones and anthranols: yellow (vis.), bright yellow fluorescence (UV-365 nm). Dianthrones do not react.

  - Natural products-polyethylene glycol reagent (NP/PEG No.28)
    Anthrones and anthranols: intense yellow fluorescence (UV-365 nm).

  - Sennoside detection
    The TLC plate is sprayed with concentrated HNO_3 and then heated for 10 min at 120°C. It is then sprayed with 5% ethanolic KOH. After further heating, sennosides appear as brown-red zones in UV-365 nm and brown zones in the visible.
    Sennosides can also be detected with a 1% solution of sodium metaperiodate in 10% ethanolic KOH. After spraying and heating (approximately 5 min), yellow-brown zones are obtained in UV-365 nm.

  - Rhaponticoside detection
    Phosphomolybdate-H_2SO_4 reagent (PMA-H_2SO_4 No.36)
    Rhaponticoside gives dark blue zones in the visible

  - Hypericin detection
    A 10% solution of pyridine in acetone intensifies the red fluorescence of hypericin in UV-365 nm.

  - "Isobarbaloin" test for the differentiation of Aloe capensis and Aloe barbadensis. One drop of saturated $CuSO_4$ solution, 1 g NaCl and 10 ml 90% ethanol are added to 20 ml of an aqueous solution of Aloe barbadensis (Curacao aloe, 1:200). A wine-red colour is produced, which is stable for at least 12 h. Solutions of Aloe capensis fade rapidly to yellow.

2.4 Circular TLC in Addition to the Ascending TLC

This method is generally useful for the separation of drug extracts containing a high proportion of ballast substances, e.g. mucilages from Sennae folium.

Application Two diagonal pencil lines are drawn from the corners of the TLC plate. The centre point of the plate is marked and a circle is drawn around it with a diameter of approximately 2 cm. The circle is thus divided into four segments by the diagonals. The perimeter of each segment serves for the application of drug extracts or reference solutions (see figure below).

Procedure 100 ml of solvent are placed in a round, straight-sided chamber (glass trough, 10 cm high, 20 cm in diameter). A glass funnel is loosely packed with cotton, which extends as a wick through the tube of the funnel. The funnel is placed in the solvent system, so that the solvent soaks into the cotton. With the loaded side facing downwards, the TLC plate is placed over the top of the trough, so that the cotton makes contact exactly at the marked centre.
The solvent migrates circularly from the point of application. The zones of the separating substances form single arcs, which increase in length from the starting point to the periphery of the spreading solvent.
The same adsorbent (silica gel 60 F_{254} precoated plates, 20 × 20 cm; Merck, Darmstadt), the solvent systems and detection methods can be used as described for ascending TLC. Good separations are obtained by solvent migrations of 6 cm only.

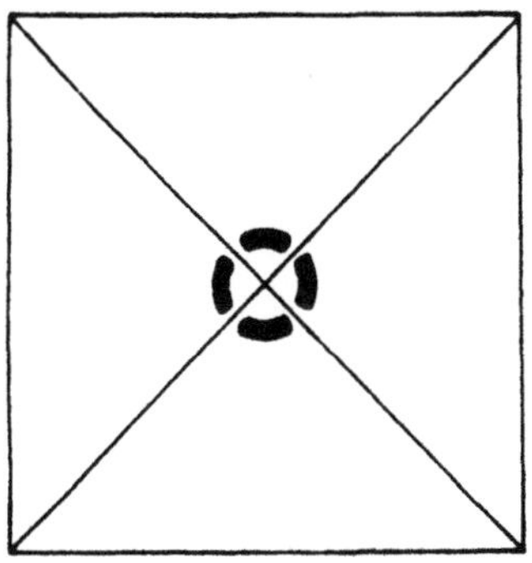

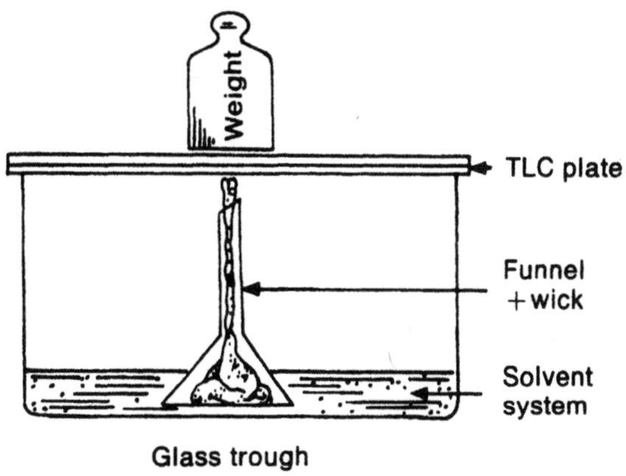

2.5 Drug List

	Drug/plant source Family/pharmacopoeia	Main constituents Hydroxyanthracenes
Fig. 1,2	**Aloes** Various types such as: Cape and Curacao aloes Socotrine aloes DAB 10, Helv VII, USP XXII, MD Uganda, Kenya aloe, Indian aloe Asphodelaceae (Liliaceae)	Dried juice of aloe leaves. Aloin A, B (10-C-β-D glucopyranoside of aloe emodin-anthrone), α- and β-stereoisomers Aloinoside A and B (stereoisomers of aloin-11-α-L-rhamnoside), aloe-emodin (aglyone) Aloeresins (non-laxative compounds): aloesin A (chromone-C-glucoside), aloesin B (p-coumaric acid ester of aloeresin A), aloesin C (glucoside of aloesin B)
Fig. 1,2	**Aloe capensis** Cape aloes Aloe ferox MILLER and hybrids DAB 10, BHP 90, ÖAB 90, USP XXII, Helv VII, Jap XI	Not less than 18% hydroxyanthracenes calculated as aloin (e.g. DAB 10) Aloin A/B, aloeresins A/B (type I) Aloin A/B, aloinosides A/B, aloesin A/B (type II), 5-hydroxyaloin A/B, aloe-emodin (<1%)
Fig. 1,2	**Aloe barbadensis** Curacao aloes, Aloe vera Aloe barbadensis MILL. DAB 10, BHP 90, Helv VII, ÖAB 90, USP XXII, MD	Not less than 28% hydroxyanthracenes calculated as aloin (DAB 10) Aloin A/B, 7-hydroxyaloin A/B (3%) 8-Methyl-7-hydroxyaloin A/B, aloesin B/D
Fig. 2	**Aloe perryi** Socotrine aloes Aloe perryi BAK. MC	Up to 14% hydroxyanthracene derivatives calculated as aloin Aloin A/B, aloinosides A/B, aloeresins A/B
Fig. 4	**Rhamni purshiani cortex** **Cascarae sagradae cortex** Cascara sagrada bark Sacred bark, chitten bark Rhamnus purshianus D.C. Rhamnaceae DAB 10, PhEur II, ÖAB 90, Helv VII, MD USP XXII (extract)	Not less than 8% hydroxyanthracenes with at least 60% cascarosides calc. as cascaroside A (DAB 10) Cascarosides A and B (diastereoisomers of aloin-8-O-β-D-glucoside); cascarosides C and D (diastereoisomers of deoxyaloin-8-O-β-D-glucoside); Aloin, deoxyaloin (10%–20%), small amounts of emodine; frangula-emodin-O-glycosides (10%–20%)
Fig. 3	**Frangulae cortex** Rhamni frangulae cortex Alder buckthorn bark Rhamnus frangula L. Rhamnaceae DAB 10, PhEur II, Helv VII, MD	Not less than 6% anthraquinone glycosides Glucofrangulin A and B (emodin-6-O-α-L-rhamnosyl-8-O-β-D-glucoside and -6-O-α-L-apiosyl-8-O-β-D-glucoside). Frangulin A and B (emodin-6-O-α-L-rhamnoside and emodin-6-O-α-L-apioside). Emodin-8-β-O-glucoside, -diglucoside Physcion, chrysophanol glycosides

Drug/plant source Family/pharmacopoeia	Main constituents Hydroxyanthracenes	
Frangulae fructus Alder buckthorn fruits Rhamnus frangula L. Rhamnaceae ÖAB	Low concentrations of anthraquinone aglycones and traces of anthraquinone glycosides.	Fig. 3,4
Oreoherzogiae cortex Rhamni fallaci cortex Rhamnus alpinus L. ssp. fallax (BOISS.) PETITMAIRE Rhamnaceae	1%–3% Hydroxyanthracene derivatives Emodin-glucoside, physcion-rutinoside Flavonoids: e.g. xanthorhamnin ► adulterant of Frangulae cortex	Fig. 3
Rhamni cathartici fructus Buckthorn berries Rhamnus catharticus L. MD	Low contents of anthraquinones in fruit flesh, 0.7%–1.4% hydroxyanthracenes in semen: frangulaemodin, -emodinanthrons Flavonol glycosides >1%: xanthorhamnines = triglycosides of rhamnocitrin (7-methyl-kaempferol and 7-methyl-quercetin) Catharticin (rhamnocitrin-3-O-β-rhamnoside)	Fig. 3,4
Rhei radix Rhubarb rhizome Rheum officinale BAILLON Rheum palmatum L. and hybrids Polygonaceae DAB 10, ÖAB. MD, Japan, China	1%–6% Hydroxyanthracenes (not less than 2.5%): 60%–80% of mono- and diglucosides of physcion, chrysophanol and rhein (e.g. physcion-8-O-gentiobioside); rhein, physcion, chrysophanol, emodin, aloe-emodin; bianthronglycosides: rheidin A–C, sennidin C,D, galloyl-β-D-glucose	Fig. 5,6
Rhei rhapontici radix Garden rhubarb Rheum rhaponticum L. Polygonaceae	0.3%–0.5% anthraquinone aglycones and glucosides, 7%–10% stilbene derivatives: rhaponticoside 5%, desoxyrhaponticoside, Adulterant of Rhei radix	
Sennae folium Senna leaves Cassia senna L. (Alexandrian senna) Cassia angustifolia VAHL (Tinnevelly senna) Caesalpiniaceae DAB 10, ÖAB 90, Helv VII, Jap XI, MD	2%–3.5% dianthrone glycosides (not less than 2.5%). calc. as sennoside B for Alexandrian and Tinevelly senna (e.g. DAB 10). As principal active compounds: sennoside A and B as 8,8′-diglucosides of sennidin A/B (= stereoisomeric 10-10′-dimers of rhein anthrone) Sennoside A (dextrorotary), sennoside A_1 (optical isomer), sennoside B (optically inactive mesoform) low amounts of Sennoside C and D (=heterodianthrons), rhein, emodin and their mono- and diglycosides	Fig. 7,8

Drug/plant source Family/pharmacopoeia	Main constituents Hydroxyanthracenes

Fig. 7,8

Sennae fructus Senna pods Cassia senna L. (Alexandrian senna) Cassia angustifolia VAHL (Tinevelly senna) Caesalpiniaceae DAB 10, PhEur I, ÖAB, Helv VII, MD, USP XXII	2.2%–3.4% dianthrone glycosides Alexandrian senna pods > 3.4% (DAB 10) Tinnevelly senna pods > 2.2% (DAB 10) Sennoside A,B besides C,D; rhein, mono- and diglycosides of emodin and rhein Naphthalenes: 6-hydroxy musizin glucoside (C. senna); tinevellin-glucoside (C. angustifolia)

Fig. 9,10

Hyperici herba St. John's wort Hypericum perforatum L. Hypericaceae (Glusiaceae) DAC 86, Helv VII, MD	0.05–0.6% dehydrodianthrones Hypericin, pseudohypericin, protohypericin Flavonol glycosides: rutin, hyperoside, quercitrin, isoquercitrin; quercetin; biapigenin Chlorogenic acid. Hyperforin (fresh plant)

2.6 Formulae

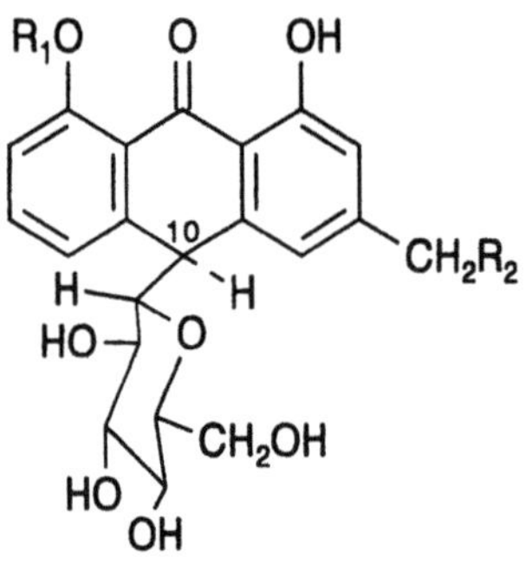

	R_1	R_2
Aloin A	H	OH
(−)-11-Desoxyaloin	H	H
Aloinoside A	H	O-α-L-rhamnose
Cascaroside A	β-D-glucose	OH
Cascaroside C	β-D-glucose	H

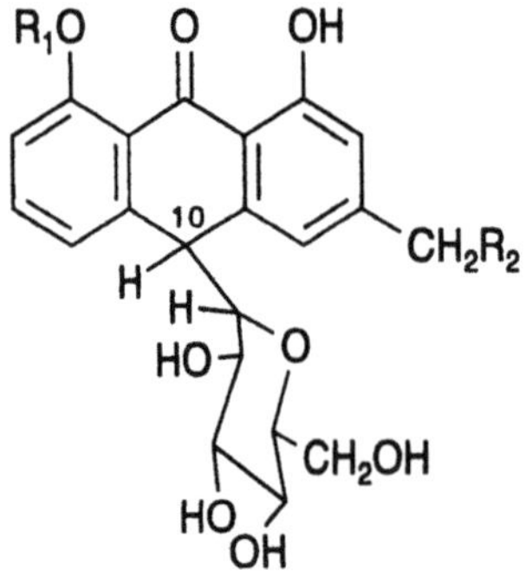

	R_1	R_2
Aloin B	H	OH
(−)-11-Desoxyaloin	H	H
Aloinoside B	H	O-α-L-rhamnose
Cascaroside B	β-D-glucose	OH
Cascaroside D	β-D-glucose	H

	R_1	R_2
Glucofrangulin A	α-L-rhamnose	β-D-glucose
Glucofrangulin B	β-D-apiose	β-D-glucose
Frangulin A	α-L-rhamnose	H
Frangulin B	β-D-apiose	H
Frangula emodin	H	H

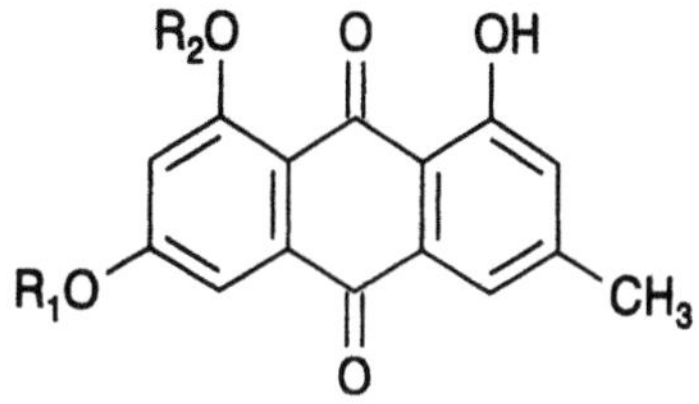

	R_1	R_2
Rheum emodin	CH_3	OH
Aloe emodin	CH_2OH	H
Rhein	COOH	H
Chrysophanol	CH_3	H
Physcion	CH_3	OCH_3

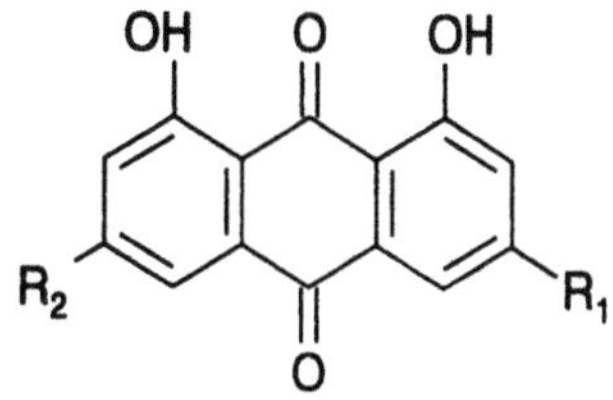

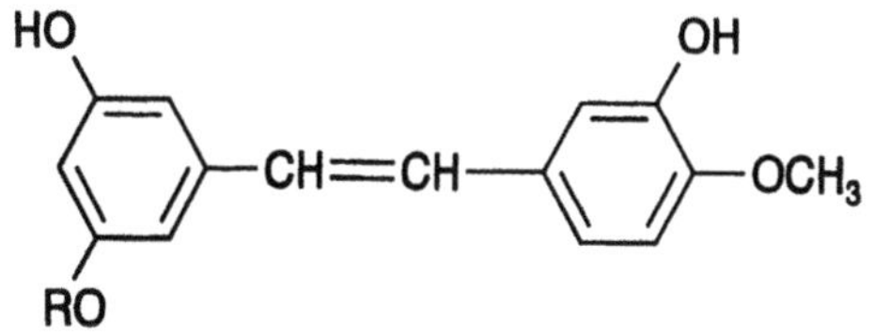

Rhaponticoside R = β-D-glucose
Rhapontigenin R = H

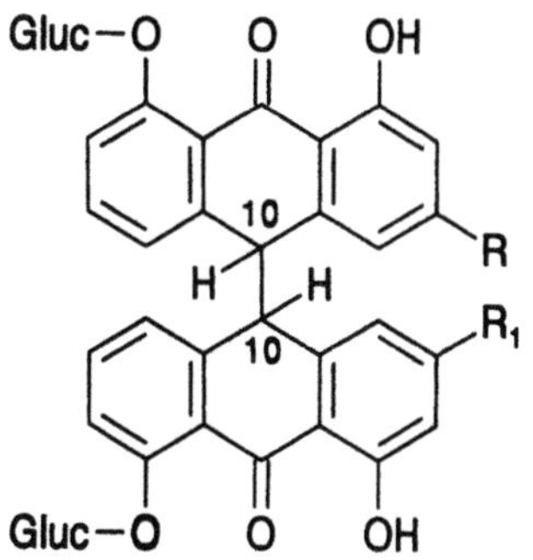

Sennoside A: R, R_1 = COOH (+)-form
Sennoside B: R, R_1 = COOH mesoform

Sennoside C: R = COOH R_1 = CH_2OH (+) form
Sennoside D: R = COOH R_1 = CH_2OH mesoform

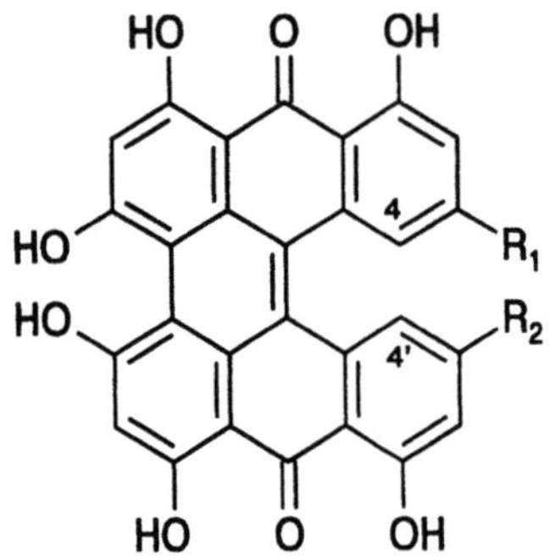

Protohypericin
$R_1 = R_2 = CH_3$
Protopseudohypericin
$R_1 = CH_3 \quad R_2 = CH_2OH$

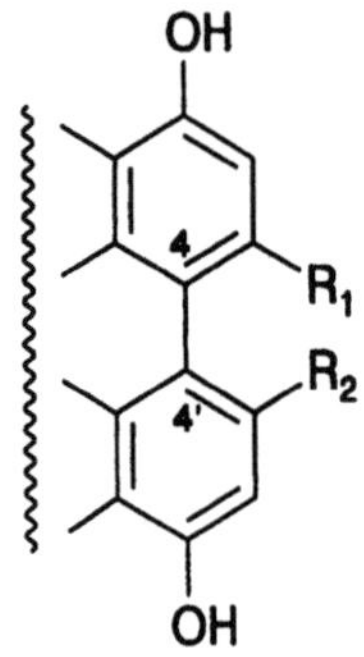

4-4′: Hypericin
$R_1 = R_2 = CH_3$
4-4′: Pseudohypericin
$R_1 = CH_3 \quad R_2 = CH_2OH$

2.7 Chromatograms

Aloes

Drug sample
1 Aloe capensis (type I)
2 Aloe capensis (type II)
3 Aloe barbadensis (Curacao aloe)
4 Aloe perryi (Socotrine aloe)
5 Aloe of Kenian origin
6 Aloe of Ugandan origin
(methanolic extracts, 5 µl)

Reference Compound
T1 aloin
T2 7-hydroxyaloin
T3 aloin ($R_f \sim 0.45$) ▶ aloe emodin ($R_f \sim 0.95$)

Solvent system ethyl acetate-methanol-water (100:13.5:10)

Detection
Fig. 1 Without chemical treatment → A UV-365 nm, B UV-254 nm
Fig. 2 10% ethanolic KOH reagent (No. 35) → C UV-365 nm, D vis

Aloe species are characterized by aloin A/B, aloe-emodin and the non-laxative aloeresins (aloesin A–C). In addition some aloes contain aloinosides and substituted aloins (5- or 7-hydroxyaloin A/B).

Fig. 1 **Aloe capensis** (1,2)

A Cape Aloe (1) is characterized by the yellow fluorescent zone of aloin ($R_f \sim 0.5$/T2) and aloe-emodin (solvent front). The zones of aloeresins such as aloesin A and B ($R_f \sim 0.55$ and $R_f \sim 0.25$, respectively) fluoresce light blue.
Trade samples of Cape aloe (2) can show besides the yellow fluorescent aloin and aloe-emodin, additional yellow zones of the aloinosides A/B (R_f 0.25–0.3) and additional glycosides (e.g. $R_f \sim 0.75$). The blue fluorescent zones are less prominent than in sample 1 (e.g. aloe resins).

B All major compounds, such as aloins or aloinosides and specifically the aloesins show quenching in UV-254 nm.
Note: 7-hydroxyaloin (T2) a characteristic compound in Curacao aloes (3) is absent in Cape aloes (1,2).

Fig. 2 **TLC synopsis of aloes** (1–6)

C Treatment with KOH reagent intensifies the yellow fluorescence of aloin and aloinosides as well as the blue fluorescence of the aloe resins. Aloe-emodin shows a typical red Bornträger reaction in UV-365 nm.

Aloe resins	Aloin	Aloinosides	Aloesins	Remarks
1 Cape aloe	+	++	+	Cape and Curacao aloes are differentiated
2 Cape aloe	+	−−	++	by the "isobarbaloin-test" of KLUNGE
3 Curacao aloes	++	−−	++	(see section 2.3) which gives yellow or
4 Socotrine aloes	+	+	++	wine red colour, respectively
5 Kenya aloes	++	+	(+)	Socotrine and Curacao aloes show a dark
6 Uganda aloes	+	(+)	(+)	zone directly below aloin, e.g. 7-hydroxyaloin in 3

D All Aloe (1–4) samples show aloin as prominent yellow zones (vis.). The samples 2 and 4 contain, in addition, aloinosides (yellow/R_f 0.25–0.3), and a dark violet-red zone (vis.) characterizes Curacao (3) and Scocotrine aloe (4). This zone directly below aloin can be identified in (3) as 7-hydroxyaloin.

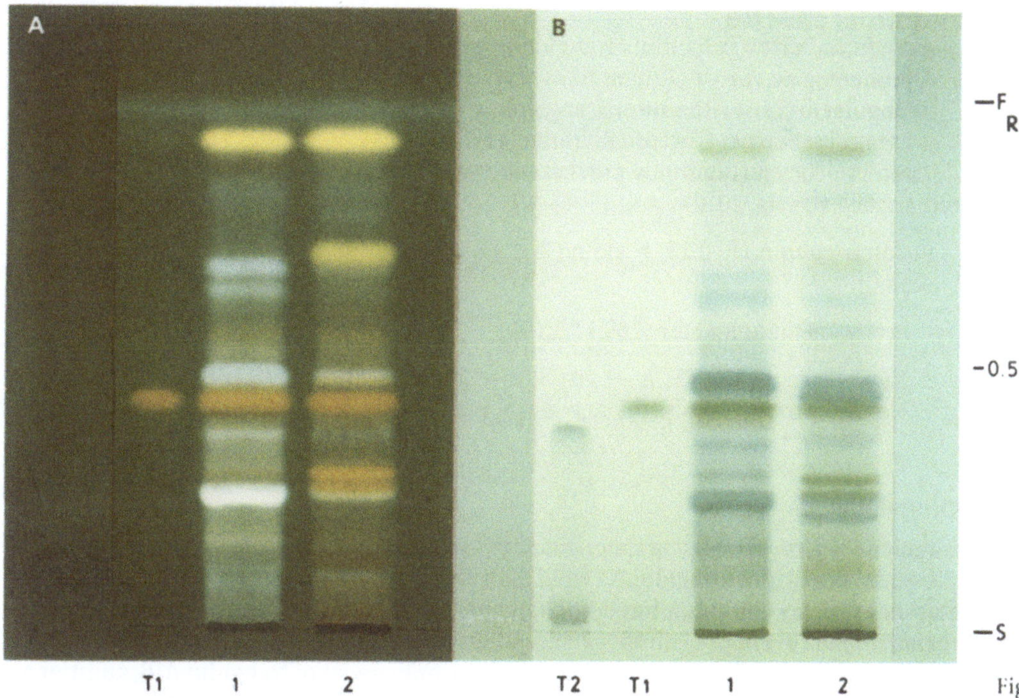

Fig. 1

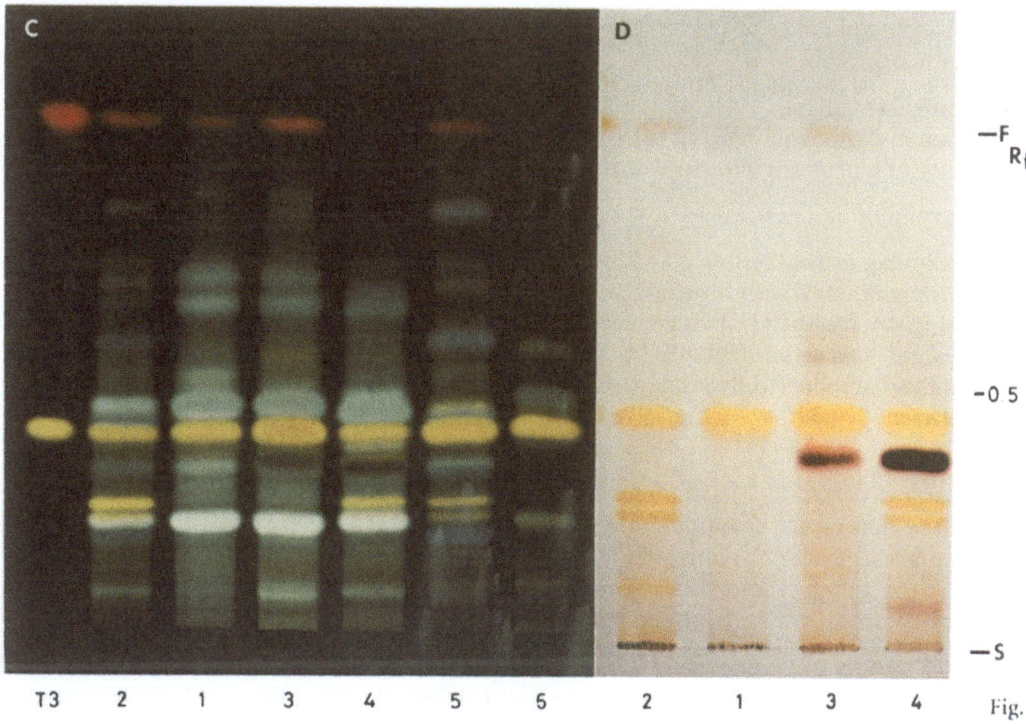

Fig. 2

Rhamnus species

Drug sample
1 Frangulae cortex (Rhamnus frangula)
2 Oreoherzogiae cortex (Rhamnus alpinus ssp. fallax)
3 Frangulae fructus (Rhamnus frangula)
4 Rhamni carthartici fructus (Rhamnus carthartícus)
5–7 Cascarae cortex (Rhamnus purshianus-trade samples)
(methanolic extracts, 20 µl)

Reference compound
T1 glucofrangulin A (R_f 0.25) ▶ aloin (R_f 0.45) ▶ frangulin A (R_f 0.75) ▶ emodin (front)
T2 aloin

Solvent system
ethyl acetate-methanol-water (100:13.5:10)

Detection
Fig. 3 KOH reagent (No. 35) A → vis; B, C → UV-365 nm
Fig. 4 Natural products-polyethylene glycol reag. (NP/PEG No. 28) D, E → UV-365 nm

---

Fig. 3 Anthraquinones

A **Frangulae cortex** (1) is characterized by two pairs of red-brown anthraquinone glycosides (vis.): glucofrangulin A (R_f 0.2), B (R_f 0.3) and frangulin A (R_f 0.75), B (R_f 0.8). Aglycones such as emodin, physcion and chrysophanol move with the solvent front.
Oreoherzogiae cortex (2) counts as an adulterant of Frangulae cortex: glucofrangulin A/B present in considerably lower concentration, only traces of frangulin A/B, additional anthraquinone glycosides such as physcion-rutinoside ($R_f \sim 0.3$) and emodin-glucoside ($R_f \sim 0.5$) dominate. A yellow zone at $R_f \sim 0.2$ in both samples (1,2) is due to flavonol glycosides see Fig. 4 D.

B All anthraquinones of Frangulae and Oreoherzogiae cortex (1,2) show a bright orange-red fluorescence in UV-365 nm.

C **Frangulae fructus** (3) shows only traces of frangula-emodin at the solvent front.
Rhamni carthartici fructus (4). Four to five orange-red zones are detectable in the R_f range of glucofrangulin ($R_f \sim 0.25$), frangulin ($R_f \sim 0.8$) and above.

Fig 4 Flavonoids and cascarosides

D **Frangulae cortex** (1): one green fluorescent flavonoid glycoside ($R_f \sim 0.2$) and the zones of frangulin A/B with brown fluorescence.
Frangulae fructus (3): two yellow orange fluorescent flavonol glycosides (R_f 0.15/0.45).
Rhamni cathartici fructus (4): a band of prominent orange-yellow fluorescent xanthorhamnins (triglycosides, see 2.5 Drug List) between the start and $R_f \sim 0.25$, and between $R_f \sim 0.75$ up to the solvent front. Xanthorhamnin ($R_f \sim 0.2$) is found in (3) and (4).

E **Cascarae cortex** (5–7) samples are characterized by anthrone glycosides: two pairs of yellow fluorescent cascarosides A/B (R_f 0.05–0.15) and cascarosides C/D (R_f 0.2–0.25). The cascarosides A/B dominate. The amount of yellow fluorescent aloin (T2), deoxyaloin (R_f 0.65) and the red-brown fluorescent aglycones emodin, aloe-emodin, chrysophanol (solvent front) varies. Four blue fluorescent naphthalide derivatives are detectable in the R_f range 0.3–0.45.

Note: Cascarosides A–C also fluoresce bright yellow when treated with the KOH reagent.

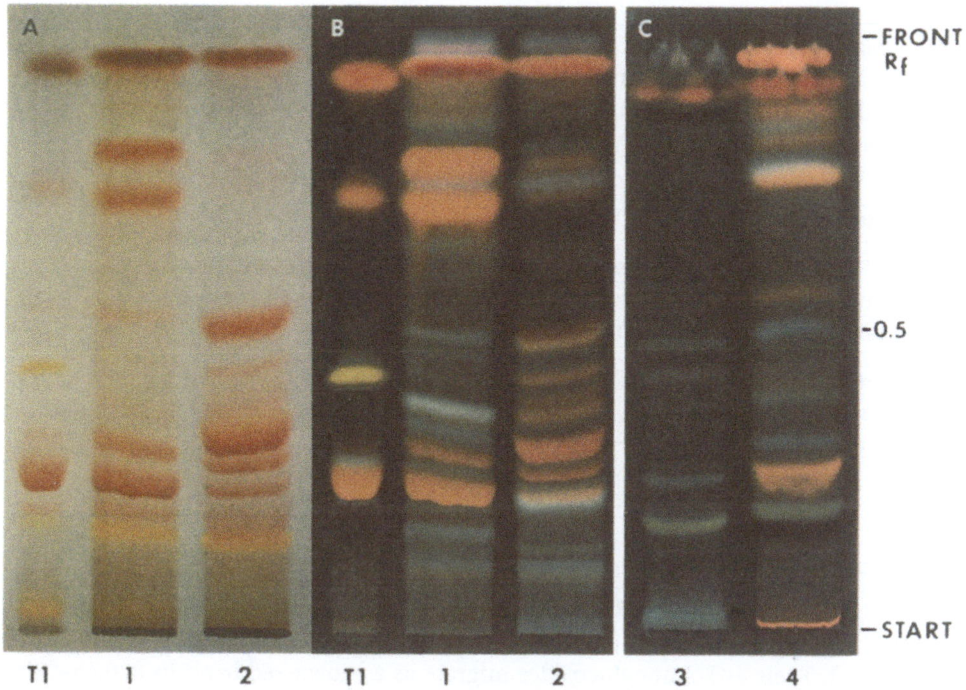
A
B
C
-FRONT
Rf
-0.5
-START
T1
1
2
T1
1
2
3
4

Fig. 3

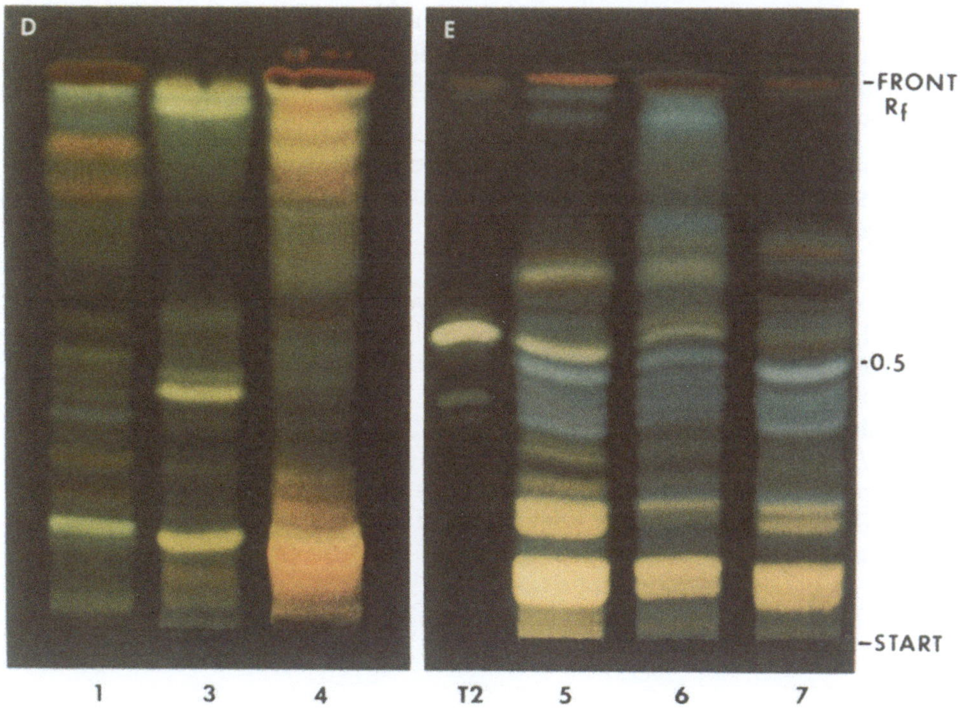
D
E
-FRONT
Rf
-0.5
-START
1
3
4
T2
5
6
7

Fig. 4

Rhei radix

Drug sample
1 Rhei palmati radix (methanolic extract, 20 µl)
2 Rhei rhapontici radix (methanolic extract, 20 µl)
3 Rhei palmati radix (hydrolysate, 10 µl)
4 Rhei rhapontici radix (hydrolysate, 10 µl)

Reference compound
T1 rhein
T2 rhaponticoside
T3 emodin ($R_f \sim 0.4$)

Solvent system
Fig. 5 ethyl acetate-methanol-water (100:13.5:10) → glycosides
Fig. 6 light petroleum-ethyl acetate-formic acid (75:25:1) → aglycones

Detection
A Without chemical treatment → UV-365 nm
B Phosphomolybdic acid/H_2SO_4 reagent (PMS No. 34) → vis
C Without chemical treatment → UV-254 nm
D Without chemical treatment → UV 365 nm

Fig. 5 **Glycosides**

A **Rhei radix** (1) is characterized in UV-365 nm by the prominent yellow fluorescent authraquinone aglycone zone (emodin, aloe-emodin, physcion, chrysophanol) at the solvent front. Their 8-O-monoglucosides migrate as a brown-red band to R_f 0.45–0.55. The corresponding diglycosides are present as minor compounds in the R_f range 0.1–0.3. The polar aglycone rhein (T1) at $R_f \sim 0.4$ is overlapped by blue fluorescent zones.
Rhei rhapontici radix (2) contains anthraquinone aglycones and monoglucosides in low concentration only. In addition the prominent violet-blue fluorescent stilbene derivatives rhaponticoside/deoxyrhaponticoside (R_f 0.45–0.55/T2) are present. They overlap the antraquinone monoglucoside zone.

B Treatment with the PMA reagent produces light yellow zones of anthraquinones (1) and a characteristic dark blue band of rhaponticoside/deoxyrhaponticoside (T2) and rhapontigenin (solvent front) in sample 2.

Fig. 6 **Aglycones**
C,D The aglycone mixtures (3,4) obtained by HCl hydrolysis of Rheum extracts (1,2) are separated in the lipophilic solvent system and evaluated in UV-254 nm and UV-365 nm. All aglycones show fluorescence quenching in UV-254 nm and uniformly yellow or orange-brown fluorescence in UV-365 nm.
Rhei palmati radix (3). Aloe-emodin and rhein (R_f 0.15–0.25/T1), emodin ($R_f \sim 0.3$/T3), chrysophanol and physcion (R_f 0.6–0.7) are characteristic aglycones.
Rhei rhapontici radix (4). The hydrolysate shows a qualitatively similar, but quantitatively different aglycone pattern with traces of rhein (T1) only. In addition blue fluorescent stilbene aglycones are found at R_f 0.05–0.1.

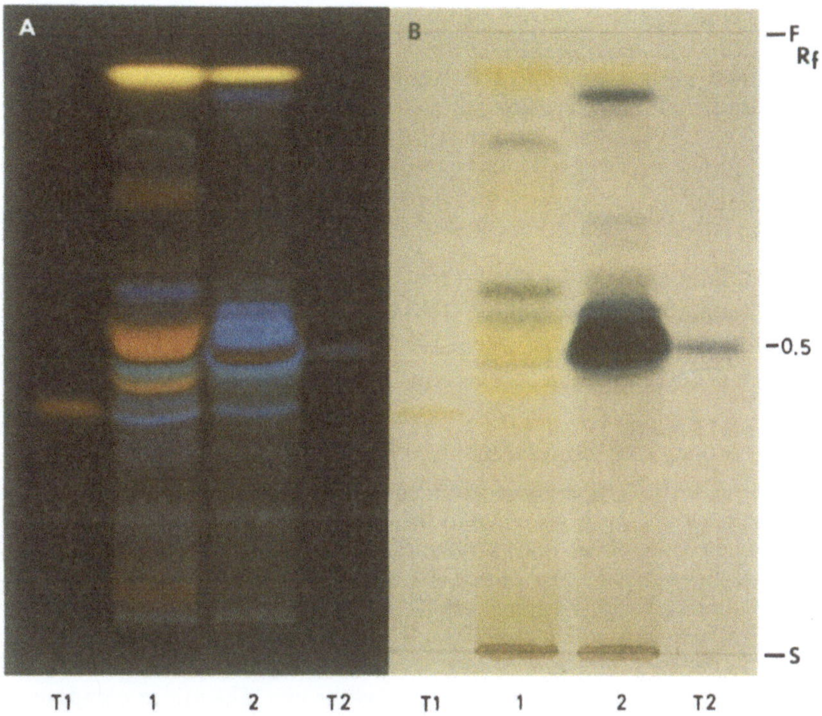

Fig. 5

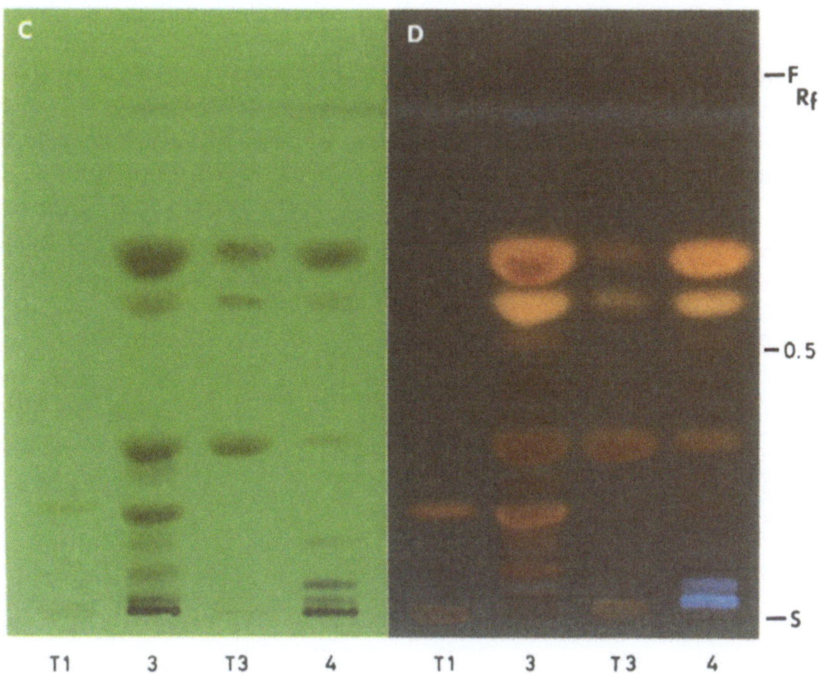

Fig. 6

Sennae folium, fructus

Drug sample 1 Sennae fructus (methanolic extract, 20 µl)
2 Sennae folium (methanolic extract, 20 µl)

Reference compound T1 sennoside A*)
T2 sennoside B*)

Solvent system n-propanol-ethyl acetate-water-glacial acetic acid (40:40:29:1)

Detection Fig. 7 HNO_3-potassium hydroxide reagent (HNO_3/KOH No.30) → vis
Fig. 8 A HNO_3-potassium hydroxide reagent (HNO_3/KOH No.30) → UV-365 nm
B Sodium metaperiodate reagent (see 2.3 Detection) → UV-365 nm

---

Fig. 7 **Sennae fructus** (1) and **folium** (2)
Treatment of the TLC plate with concentrated HNO_3, heating for approximately 30 min at 150°C and spraying with KOH reagent produces six to eight brownish and yellow zones (vis) in the R_f range 0.1 up to the solvent front.
The dark-brown zones are due to the sennosides B,A (R_f 0.25 and R_f 0.4) and the sennosides D,C (R_f 0.5 and R_f 0.7). The yellow zones indicate anthraquinone aglycones (e.g. rhein/$R_f \sim 0.8$; emodine/solvent front) and their glucosides ($R_f \sim 0.3$/$R_f \sim 0.6$).

Fig. 8A Evaluation under UV-365 nm light is more sensitive. The main brown zones (vis.) of Sennae extracts (1,2) now appear light brown to orange-brown. The minor compounds of the R_f range 0.5–0.9 are also more easily detectable.
The two dianthron glycosides, sennoside A (R_f 0.4/T1) and sennoside B (R_f 0.25/T2) are the major compounds in Sennae fructus (1) and S. folium (2).
In Sennae folium extract (2) a R_f value depression of sennoside A and specifically of sennoside B occur, caused by the mucilages also extracted from the plant material with 50% methanol. To avoid this effect the circular TLC method can be used (see Fig. 9).
Sennoside D ($R_f \sim 0.55$) is more highly concentrated in Sennae folium extracts (2) than in Sennae fructus extracts (1). Sennoside C can be localized at $R_f \sim 0.7$. Rhein is detectable as a yellow zone at $R_f \sim 0.8$ and its 8-O-glucoside is found between sennoside D and C.

B Direct treatment of the TLC plate with the sodium metaperiodate reagent and heating for 5 min under observation at 100°C reveals green-yellow or dark brownish zones when evaluated under UV-365 nm. It is a fast detection method, but less sensitive compared with the HNO_3/KOH reagent.

---

*) The commercial reference compound "sennoside A" contains small amounts of sennoside C and D. The reference compound "sennoside B" shows, in addition, sennoside A as minor component.

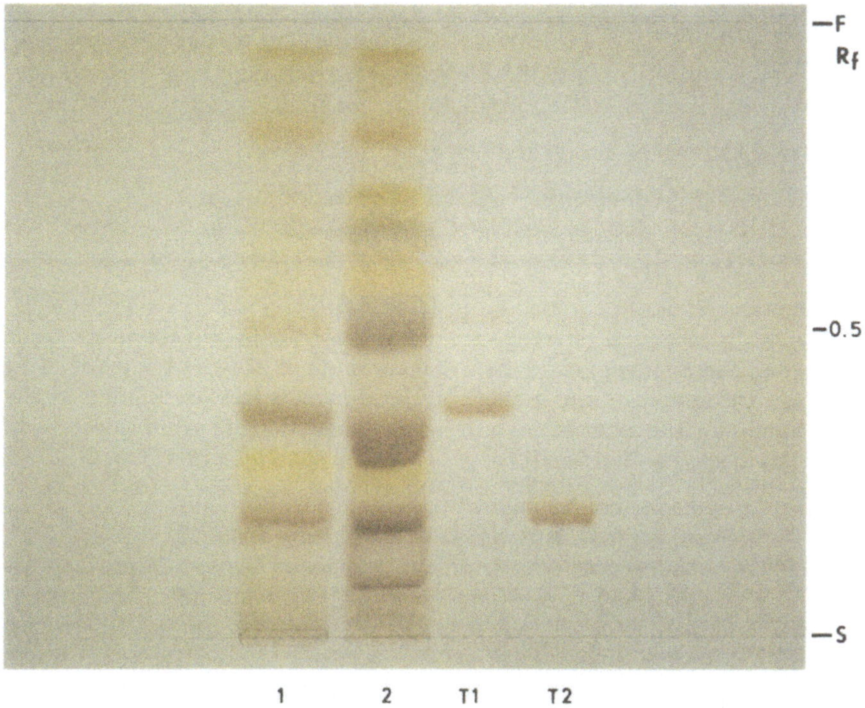

Fig. 7

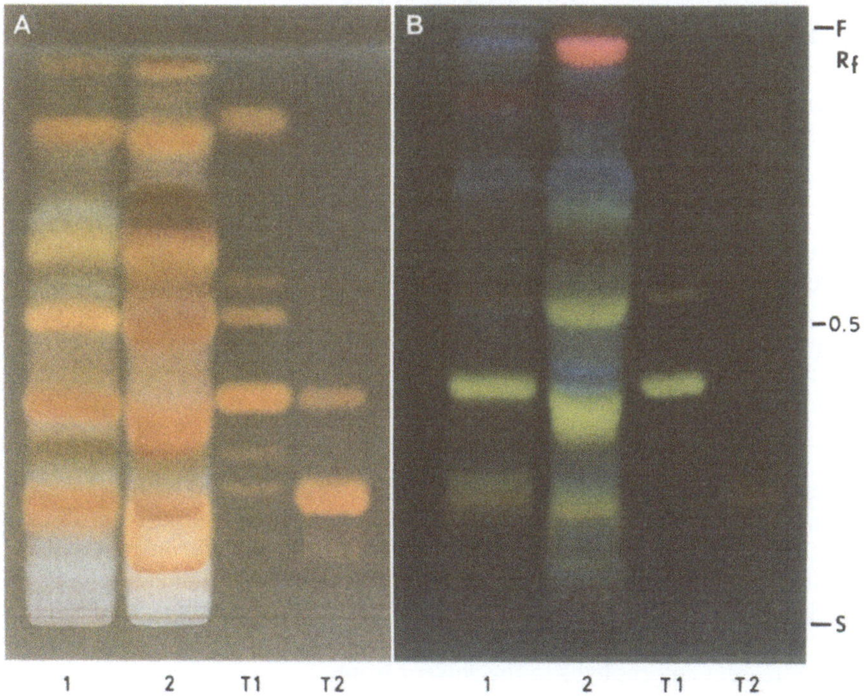

Fig. 8

Circular TLC (CTLC) in comparison to ascending TLC of Senna extracts

Drug sample, segment Sennae folium (upper segment) *A* sennoside A *D* sennoside D
Sennae fructus (lower segment) *B* sennoside B *Al* aloin *Rh* rhein

Solvent system n-propanol-ethyl acetate-water-glacial acetic acid (40:40:29:1)

Detection Fig. 9 CTLC Sodium metaperiodate reagent (see 2.3 Detection) → vis
Asc. TLC HNO_3-potassium hydroxide reagent (HNO_3/KOH No. 30) → vis

---

Description The CTLC in general is a convenient method to achieve good separations over the short distance of 5–6 cm. Extracts and reference compounds are applied in the inner circle (start) in an overlapping mode, to make sure that compounds are clearly identified by references. Ballast substances of the extracts such as mucilagines are diluted in the circular separation lines. The disturbance and R_f value depression of sennoside A,B are reduced (preparation see 2.4 Circular TLC).

Fig. 9 The sennosides are detected as bright yellow-brown bands with sodium metaperiodate (CTLC) and as darker brown zones with the HNO_3-KOH reagent (asc. TLC).
The CTLC of Sennae folium und Sennae fructus shows as two prominent circles sennoside A and B (→ test *A/B*) in the inner parts of both segments. The bands of sennoside D (→*D*) and C are found slightly below the aloin test (→ test *Al*). Rhein (test *Rh*) is clearly seen in Sennae fructus extracts. The influence of mucilagines on the R_f value of sennoside B results in a dwelling circle (CTLC) and causes an R_f value depression in the picture of the ascending TLC (compare with Figs. 7,8).

Hyperici herba

Drug sample 1 Hyperici herba (Hypericum perforatum) (methanolic extracts, 25 µl)
2 Hyperici herba (commercial trade sample)

Reference compound T1 hypericin
T2 rutin (R_f 0.35) ▶ chlorogenic acid (R_f 0.4) ▶ hyperoside (R_f 0.5) ▶ isochlorogenic acid

Solvent system Fig. 10 A,B ethyl acetate-formic acid-glacial acetic acid-water (100:11:11:26)
C toluene-ethyl formate-formic acid (50:40:10)

Detection A,B Natural products-polyethylene glycol reagent (NP/PEG No. 28);
A UV-365 nm, B vis.
C 10% pyridine in ethanol → vis

Fig. 10A **Hyperici herba** (1,2) is characterized in UV-365 nm after treatment with NP/PEG reagent by the prominent red-violet fluorescent zones of the non-laxative dehydrodianthrons, the hypericins (R_f 0.75–0.8), five bright yellow fluorescent flavonolglycosides (R_f 0.35–0.7) and blue fluorescent phenol carboxylic acids such as chlorogenic acid (R_f ~ 0.4/T2). The flavonolglycosides are identified as rutin (R_f ~ 0.35/T2), hyperoside (R_f ~ 0.5/T2), isoquercitrin (R_f ~ 0.6) and quercitrin (R_f ~ 0.7). The aglycones, e.g. quercetin, migrate with the red fluorescent chlorophylls to the solvent front.

B Hypericins are seen as green-brown and the flavonolglycosides as orange-yellow zones (vis).

C Variation of the solvent system and the detection with pyridine reagent reveals a broad band of red zones in the R_f range 0.5–0.6 (T1). Red zones at R_f 0.9–0.95 show chlorophyll compounds.

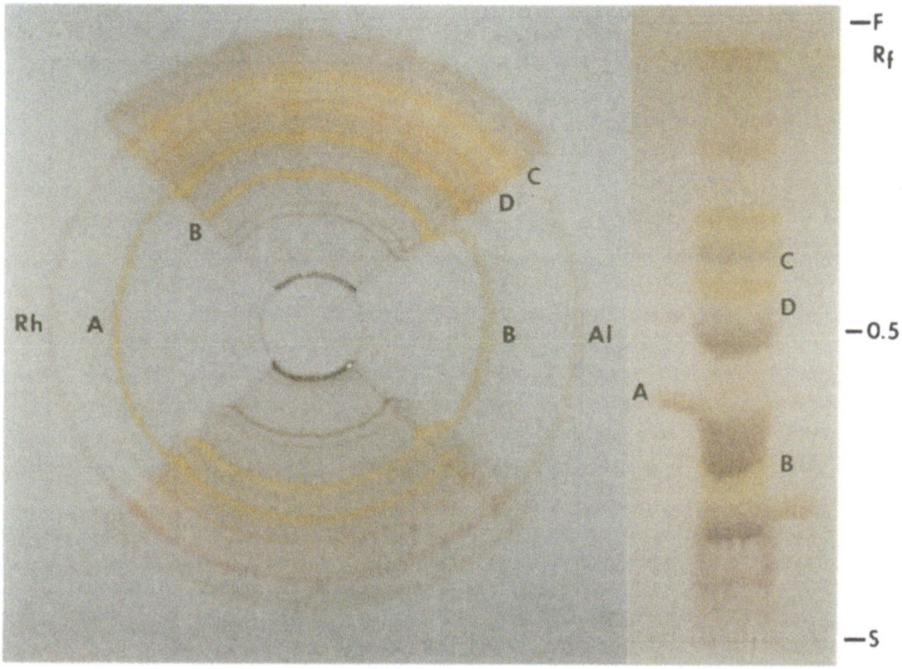

Fig. 9

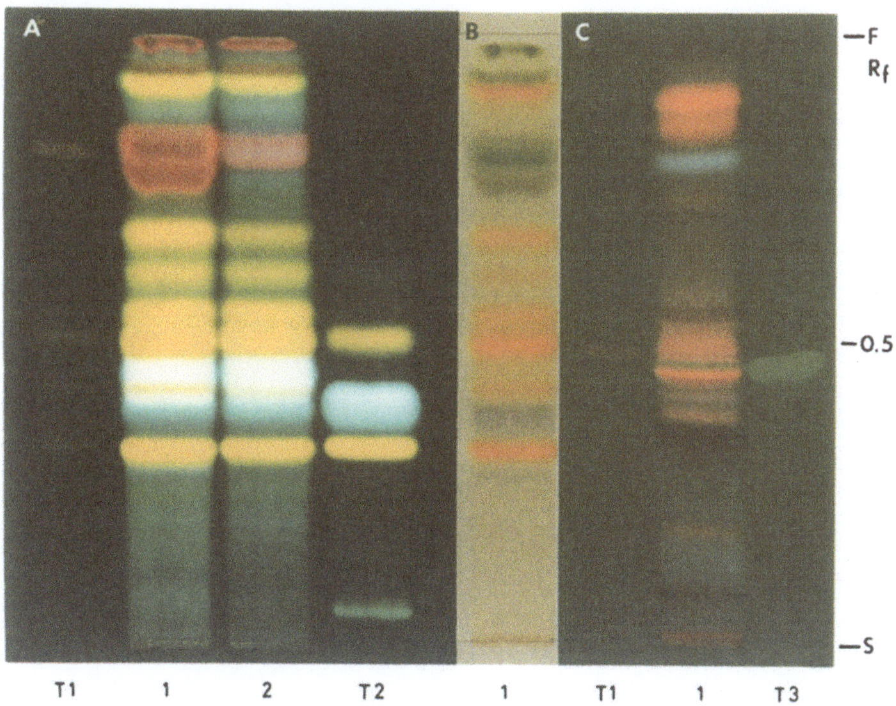

Fig. 10

3 Bitter Drugs

Most of the bitter principles in important official drugs possess a terpenoid structure, representing derivatives of monoterpenes (secoiridoids), sesquiterpenes, diterpenes and triterpenes.

3.1 Preparation of Extracts

General method, methanolic extract

Powdered drug (1 g) is extracted for 15 min with 10 ml methanol on the water bath. The mixture is filtered and the filtrate is evaporated to 1–1.5 ml; 20–30 µl is used for TLC investigations.

Enrichment

A total of 2 ml of the methanolic extract is evaporated to dryness and dissolved in 3 ml of water and 10 ml of n-butanol (saturated with water) is added. After shaking for 3–5 min, the butanol layer is separated and evaporated to a volume of 1 ml, and 30–40 µl is used for TLC investigations.

Exceptions

Humuli lupuli strobulus: Dried powdered drug (1 g) is extracted for 24 h with 15 ml cold ether. The filtrate is allowed to stand for 12 h in the refrigerator, precipitated waxy materials are removed by filtration and the fitrate evaporated to dryness at room temperature. The residue is dissolved in 1 ml methanol and 20–40 µl is used for TLC investigations.
Freshly harvested drug (1 g) is extracted for 2 h at room temperature with 10 ml 70% methanol. The filtrate is evaporated to about 3 ml, and 20–40 µl is used for TLC.

Drugs with cucurbitacins: Powdered drug (1 g) is extracted for 15 min with chloroform or ethanol on a water bath. The filtrate is evaporated to 1–1.5 ml (mainly cucurbitacin glycosides). Extraction with water results mainly in cucurbitacin aglycones; 20–30 µl is used for TLC investigations.

3.2 Thin-Layer Chromatography

Reference solutions

From all standard compound 0.1% methanolic solutions are prepared; 10–20 µl is used for TLC.

Adsorbent

Silica gel $60F_{254}$-precoated TLC plates (Merck, Germany).

Chromatography solvents

- ethyl acetate-methanol-water (77:15:8) — General screening system → e.g. Gentianae radix, Centaurii herba, Condurango cortex, Harpagophyti radix

• acetone-chloroform-water (70:30:2)	amarogentin	→ Gentianae radix
• dichlormethan-acetone (85:15)	absinthin	→ Absinthii herba
• chloroform-methanol (95:5)	quassin	→ Quassiae lignum
	marrubiin	→ Marrubii herba
• chloroform-methanol (95:10)	cucurbitacins	→ Bryoniae radix
• chloroform-methanol-water (60:40:4)	aucubin	→ Verbasci flos
• chloroform-acetone (40:30)	cnicin	→ Cardui benedicti herba
• chloroform-acetone (60:20)	cynaropicrin	→ Cynarae herba
• ethyl acetate-dioxane-water (30:10:0.3)	oleuropein	→ Oleae folium
• iso-octane:isopropanol:formic acid (83.5:16.5:0.5)	humulone	→ Humuli lupuli strobuli

3.3 Detection

- UV-254 nm — Compounds with conjugated double-bond systems show quenching effects (e.g. quassin, humulon, lupulon, neohesperidin).
- UV-365 nm — No characteristic fluorescence, with the exception of flavonoid glycosides in Aurantii pericarpium extracts.
- Spray reagents — (see Appendix A)

- Vanillin-sulphuric acid reagent (VS No. 42)

Evaluation after about 10 min at 100°C	(vis)
neohesperidin, naringin, harpagoside	red-violet
gentiopicroside, swertiamarin	brown-red
condurangin A–C	blue-green
foliamenthin, menthiafolin, quassin	blue
marrubiin, absinthin, cnicin	blue
aucubin, catalpol	grey, red-grey

- Anisaldehyde-sulphuric acid reagent (AS No. 3)
  Visualization after about 10 min at 100°C: Similar colours (vis) to those obtained with VS reagent and additional fluorescence in UV-365 nm.

- Liebermann-Burchard reagent (LB No. 25)
  The TLC plate is sprayed with freshly prepared solution, heated for 10 min at 100°C and inspected in UV-365 nm or vis
  Absinthin → sand-brown colour in UV-365 nm; dark brown in vis
  Cnicin → light grey in UV-365 nm; weak grey in vis

- Fast red salt B (FRS No. 16)
  Immediately after spraying, phenolic or reducing substances turn yellow, orange or red (vis)
  Amarogentin (orange); gentiopicroside (red); humulone (yellow); lupulone (red).

- 10% $FeCl_3$ solution
  The TLC plate is inspected immediately after spraying. The hop bitter principles and oleuropein turn yellow-brown to yellow-green (vis).

- Vanillin-phosphoric acid reagent (VPA No. 41)
  The TLC plate is sprayed with freshly prepared solution, heated for 10 min at 100°C and inspected in vis and UV-365 nm.
  The cucurbitacins are blue or red-violet (vis.) and fluoresce blue-pink, yellow and green in UV-365 nm.

- Natural products-polyethylene glycol reagent (NP/PEG No. 28)
  The TLC plate is sprayed with freshly prepared solution and inspected in UV-365 nm. The flavonoid glycosides and phenolcarboxylic acids (Aurantii pericarpium, Cynarae herba) show an orange, green or blue to blue-green fluorescence in UV-365 nm.

3.4 Drug List

Drug/plant source Family/pharmacopoeia	Bitter principles – Bitterness index (BI) Main compounds	
Terpenoid bitter principles **Monoterpenes (C-10)**		
Centaurii Herba Centaury Centaurium erythraea RAFN. (syn. C. minus MOENCH) Gentianaceae DAB 10, ÖAB, Helv VII, MD	Secoiridoid glycosides: swertiamarin (75%), gentiopicroside (gentiopicrin); swerosid, centapicroside (traces) BI plant, 2000–4700 BI flowers, 6000–12 000 ► Flavonoids, xanthones, triterpenes	**Fig. 3**
Gentianae radix Gentian root Gentiana lutea L. Gentianaceae DAB 10, ÖAB, Helv. VII, BP 88, MD, China	2%–4% secoiridoid glycosides: gentiopicroside (~2.5%; BI 12 000) amarogentin (0.025%–0.4%; BI 58 × 10^6) Oligosaccharides: gentianose (2.5%–5%; BI 120/fresh root), gentiobiose (1%–8%; BI 500/dry root) BI of the drug, 10 000–30 000 0.1% xanthons: gentisin, isogentisin	**Fig. 3**
Menyanthidis folium Trifolii fibrini folium Buckbean leaf Menyanthes trifoliata L. Menyanthaceae DAC 86, ÖAB, MD	>1% secoiridoid glycosides: foliamenthin, menthiafolin, 7′,8′-dihydrofoliamenthin, sweroside Verbenalin type: loganine BI folium; 4000–10 000	**Fig. 4**
Euphrasiae herba Euphrasy herb Euphrasia species E. stricta E. rostkoviana group Scrophulariaceae	Iridoid glycosides: aucubin, catalpol, euphroside, ixoroside ► Lignan: dehydrodiconiferyl alcohol-4-β-D-glucoside ► Flavonoids: quercetin and apigenin glucosides	**Fig. 5**

	Drug/plant source Family/pharmacopoeia	Bitter principles - Bitterness index (BI) Main compounds
Fig. 5	**Galeopsidis herba** Hemp nettle Galeopsis segetum NEK. Lamiaceae	Iridoid glycosides: harpagoside, 8-O-acetylharpagide, antirinoside
Fig.5	**Plantaginis folium** Ribwort leaf, Plantain Plantago lanceolata L. Plantaginaceae ÖAB, Helv. VII	Iridoid glycosides: aucubin (0.3%–2.5%) catalpol (0.3%–1.1%)
Fig. 5,6	**Verbasci flos** Mullein flowers Verbascum densiflorum BERTOL. Scrophulariaceae DAC 86, ÖAB, Helv. VII	Iridoid glycosides: aucubin, 6-β-xylosylaucubin, catalpol, catalpol-6-β-xyloside, methyl-, isocatalpol ► Saponins: verbascosaponin (~0.04%). ► 1.5%–4% flavonoids: (see 7.1.7, Fig. 1,2).
Fig. 5	**Veronicae herba** Male speedwell wort Veronica officinalis L. Scrophulariaceae	0.1%–1% iridoid glycosides: catalpol, veronicoside (2-benzoylcatalpol), verproside (6-protocatechuoylcatalpol) ► Flavonoids: luteolin glycosides (0.7%) chlorogenic, caffeic acid; saponins
Fig. 5,6	**Rehmanniae radix** Rehmannia glutinosa (GÄRTN) LIBOSCH. Scrophulariaceae Jap XI, China	Iridoid glycosides: aucubin, catalpol (0.3%–0.5%) rehmanniosides A–C, D (0.02%), ajugol (0.04%)
Fig. 5B	**Harpagophyti radix** Grapple plant root Harpagophytum procumbens (BURCH) DC. and H. zeyheri DECNE. Pedaliaceae	0.5%–3% iridoid glycosides: harpagoside (bitter), isoharpagoside, harpagid (sweet), procumbid BI of the drug, (600) 2000–5000
Fig. 5B	**Scrophulariae herba** **Scrophulariae radix** Figwort Scrophularia nodosa L. Scrophulariaceae	1%–2% iridoid glycosides Substitute for Harpagophyti radix, but lower amount (~50%) of harpagoside
Fig. 9	**Oleae folium** Olive leaf Olea europaea L. Oleaceae MD	iridoid glycosides: oleuropein (oleuropeoside 6%–9%) 6-oleuropeylsaccharoside ► Flavonoids: Luteolinglykosides

Drug/plant source Family/pharmacopoeia	Bitter principles - Bitterness index (BI) Main compounds	
Sesquiterpenes (C-15)		
Absinthii herba Wormwood Artemisia absinthium L. Asteraceae DAB 10, ÖAB 90, MD BHP 83	Sesquiterpene lactones: ~0.3% (leaves), ~0.15% (flowers) Absinthin (~0.2%) and anabsinthin Artabsin (0.1% in freshly harvested plants) BI of the drug, 10 000–25 000 BI of absinthin, about 12 700 000 ► ess. oil 1.5% e.g. thujon	**Fig. 7**
Cardui benedicti herba **Cnici herba** Blessed thistle Cnicus benedictus L. Asteraceae DAC 86, ÖAB 90, MD	Sesquiterpene lactons (~0.25%), (germacran type): cnicin, salonitenolid and artemisiifolin BI of the drug, 800–1800 ► Essential oil (0.03%–0.1%) citral, citronellal cinnamic acid, acetylene derivatives	**Fig. 8**
Cynarae herba Artichoke Cynara scolymus L. Asteraceae MD (leaves)	Sesquiterpene lactones (0.5%–6%) Cynaropicrin (40%–80%; BI 40×10^4) and/or grosheimin ► Caffeic acid derivatives: chlorogenic, and 1,3-dicaffeoyl quinic acid (cynarin). ► Flavonoids (0.1%–1%): scolymoside, cynaroside, luteolinglycosides	**Fig. 13**
Diterpenes (C-20)		
Marrubii herba White horehound Marrubium vulgare L. Lamiaceae ÖAB 90, BHP 83	Bitter principle: 0.3%–1% (labdan type) marrubiin (0.1%–1%); marrubiol, marrubenol, vulgarol premarrubiin (0.13%)	**Fig. 9**
Triterpenes (C-30)		
Quassiae lignum Quassia wood Quassia amara L. "SURINAM" Picrasma excelsa PLANCH. Simarubaceae MD	Secotriterpenes (simarubalides) ~0.25%: quassin, neoquassin and 18-hydroxy-quassin (0.1%–0.15%). BI of the drug, 40 000–50 000 BI of quassin/neoquassin, 17×10^6	**Fig. 10**
Cucurbitacins (C-30)		
Bryoniae radix Bryony root Bryonia alba L. and B. cretica ssp. dioica PLANCH. Cucurbitaceae, MD	Tetracyclic triterpenes cucurbitacin glucosides I,L,E and dihydro- cucurbitacins E,B and aglycones Bryonia alba and B. dioica: qualitatively similar contents of cucurbitacins	**Fig. 11,12**

	Drug/plant source Family/pharmacopoeia	Bitter principles – Bitterness index (BI) Main compounds
	Colocynthidis fructus Citrullus colocynthis (L.) SCHRAD, Cucurbitaceae	Tetracyclic triterpenes Cucurbitacin glucosides E,I,L
	Gratiolae herba Gratiola officinalis L. Scrophulariaceae	Tetracyclic triterpenes Cucurbitacin glucosides E,I,L and aglycones
	Iberidis semen Bitter Candy Iberis amara L. Brassicaceae	Tetracyclic triterpenes (0.2%–0.4%) Cucurbitacin glucosides E,I, and aglycones Cucurbitacin K,J (traces)
	Ecballii fructus Ecballium elaterium (L.) A.RICH. Cucurbitaceae	Tetracyclic triterpenes Cucurbitacin glucosides E,B,I,L and aglycones
	Tayuyae radix Cayaponia tayuya LOGN. Cucurbitaceae	Tetracyclic triterpenes Cucurbitacin glucoside B and aglycones
	Drugs containing non-terpenoid bitter principles	
Fig. 1	**Aurantii pericarpium** Seville orange peel Citrus aurantium L. ssp. aurantium Rutaceae DAB 10, MD, Japan, China	Flavanone glycosides: neohesperidin, naringin (see Fig. 23, Chap. 7 Flavonoid Drugs) Triterpene: limonin (mainly in seeds, BI 10^6), BI of the flavanone glycosides, about 500 000 BI of the drug, 600–1500 → (see Fig. 17/18, Chap. 6 Aetherolea)
Fig. 14	**Humuli lupuli strobulus** Hops Humulus lupulus L. Moraceae (Cannabaceae) DAB 10, BHP 83	Acyl phloroglucides: humulone (“α-acids, 3%–12%) Lupulone (“β-acids, 3%–5%) unstable compounds, hop bitter acids
	Pregnane type (steroids)	
Fig. 1	**Condurango cortex** Condurango bark Marsdenia cundurango REICHB.f. Asclepidiaceae DAC 86, ÖAB 90, Helv. VII, MD, Japan	1%–2% digitanol glycosides: complex mixture of C-21-steroidglycosides Condurangine A, A_1, B, C, C_1, D, E A: -20-carbonyl, linked to pentasaccharide B: -2-hydroxyl A, A_1, C, C_1 are diesters with acetic acid and cinnamic acid BI of the drug, about 15 000

3.5 Formulae

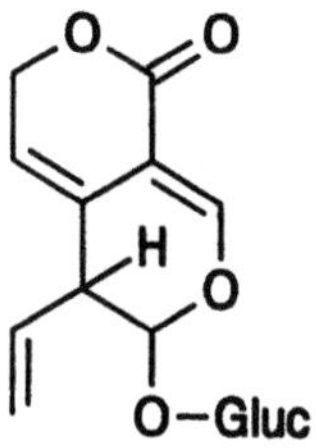

Gentiopicroside

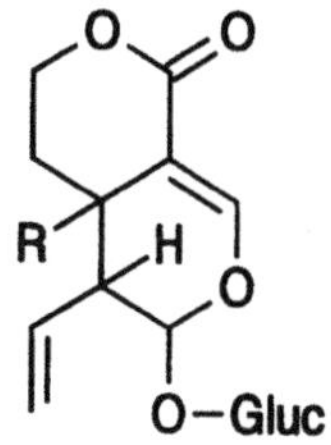

Swertiamarin R = OH
Sweroside R = H

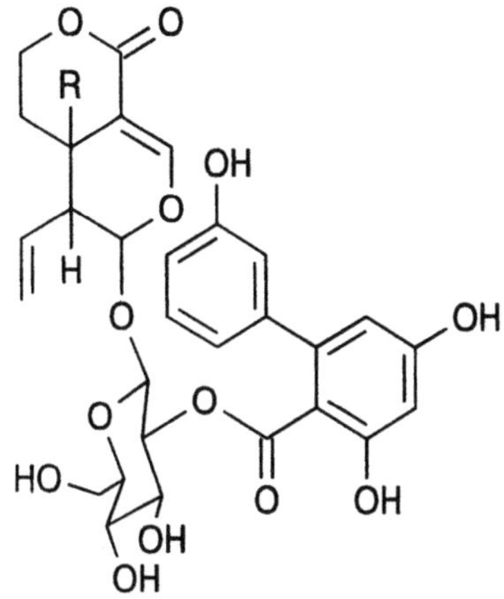

Amarogentin R = H
Amaroswerin R = OH

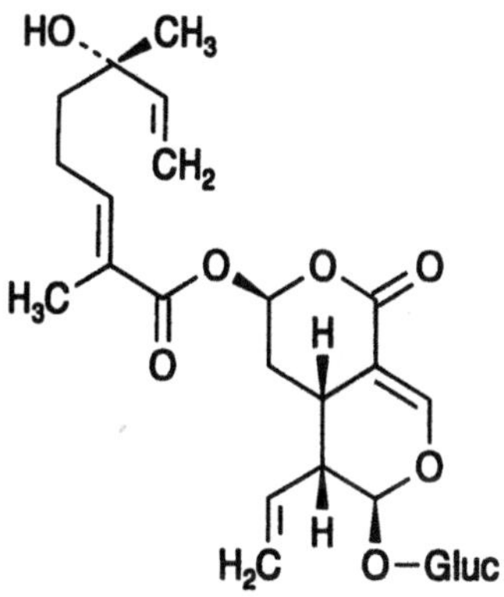

Menthiafolin

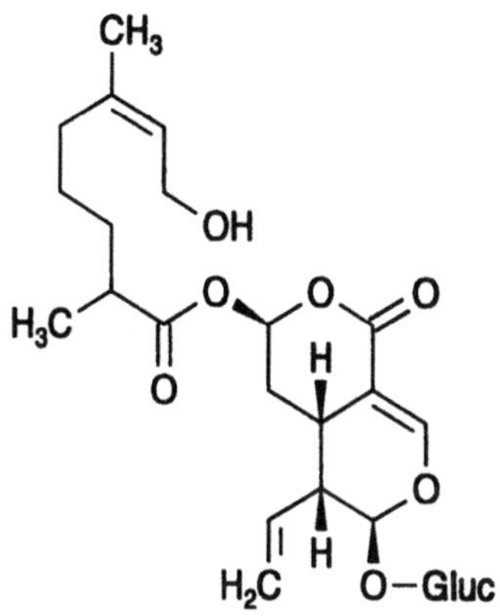

Dihydrofoliamenthin

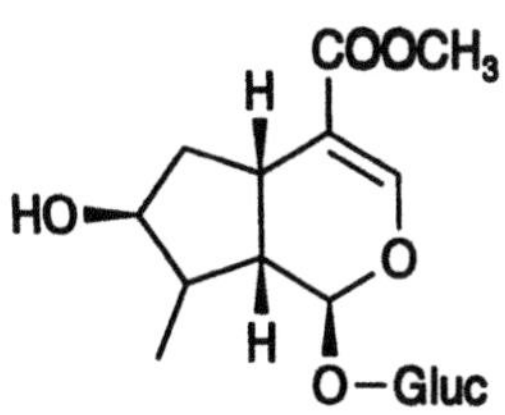

Loganine

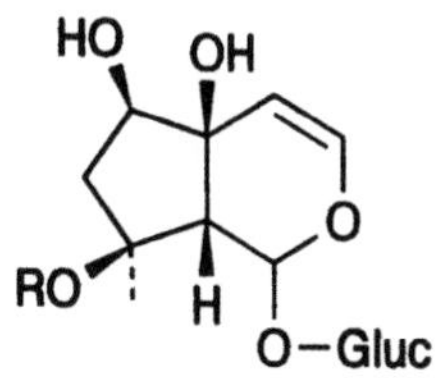

Harpagoside:
R = trans-cinnamoyl
Harpagid: R = H

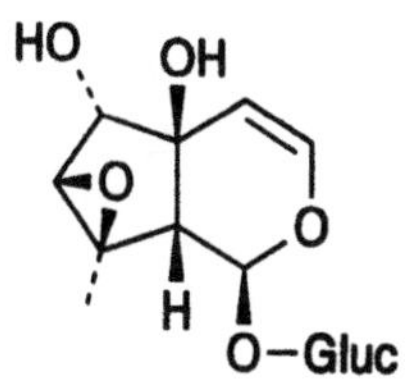

Procumbid

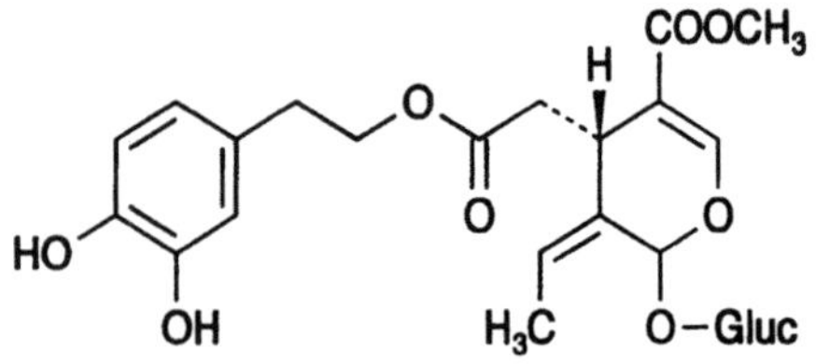

Oleuropein

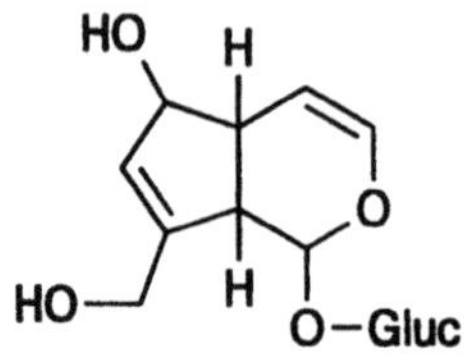

Aucubin

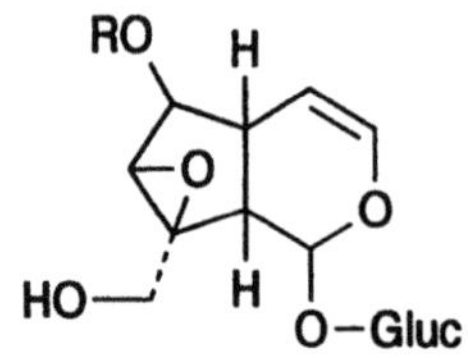

Catalpol R = H

Veronicoside R =

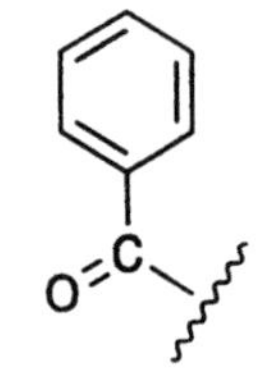

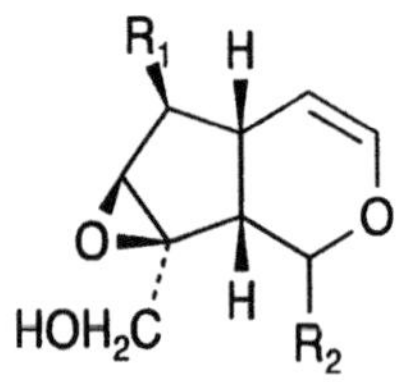

Rehmannioside

A: R_1 = OH; R_2 = O-Mel

B: R_1 = O-Gal; R_2 = O-Glu

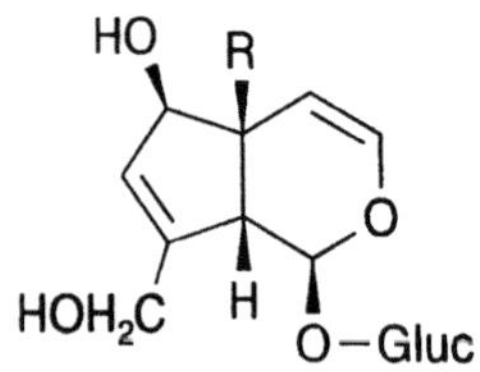

Rehmannioside D

R = O-Soph

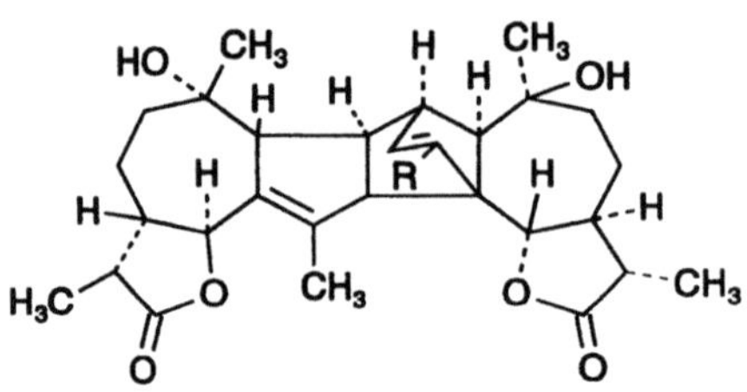

Absinthin R = CH_3

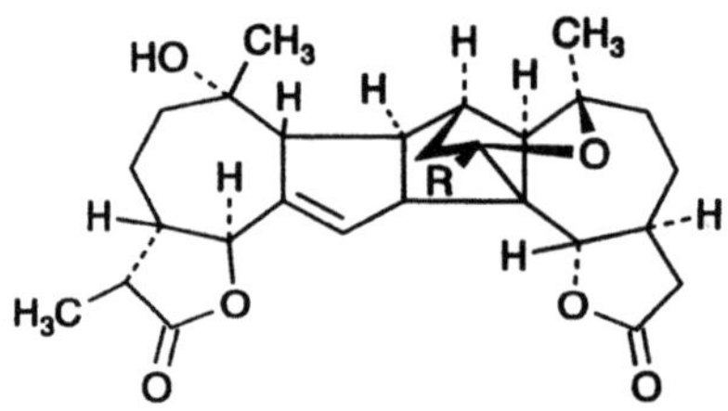

Anabsinthin R = CH_3

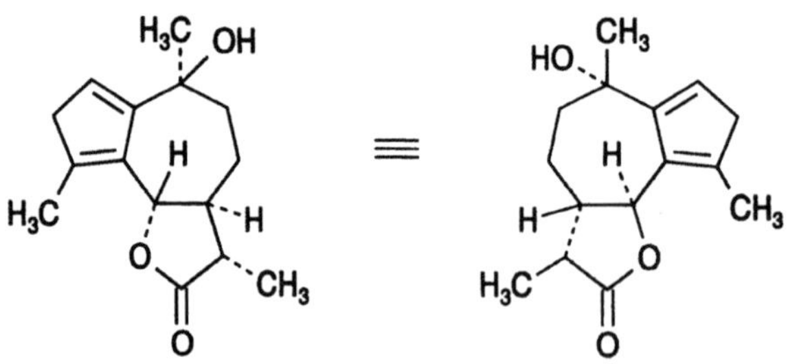

Artabsin

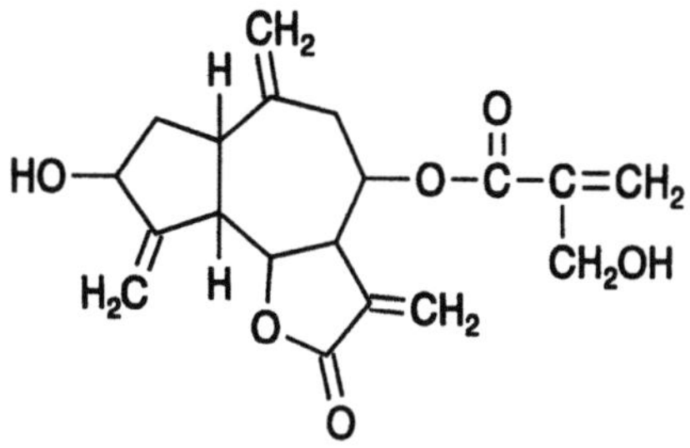

Cynaropicrin

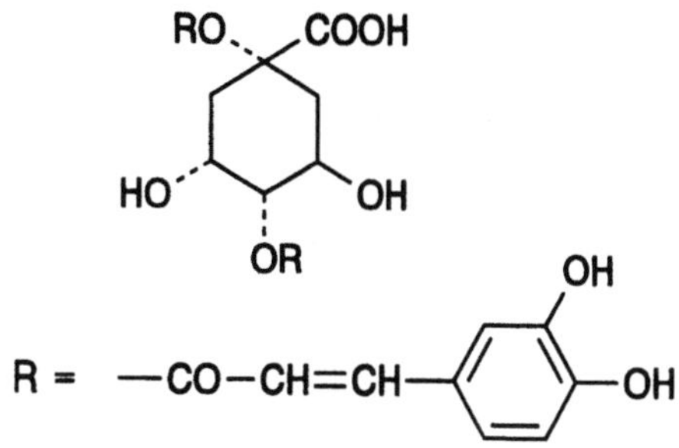

Cynarin

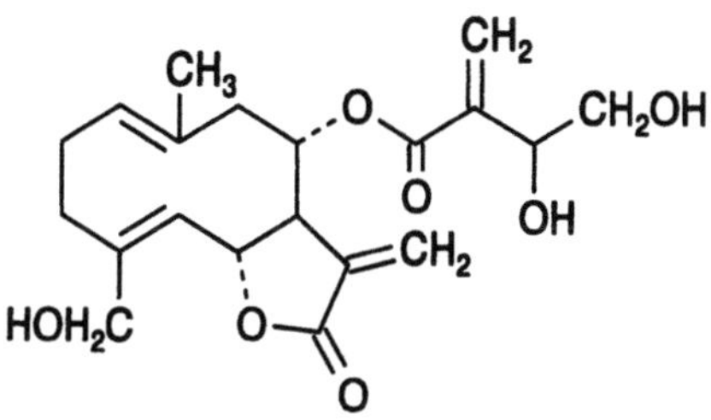

Cnicin

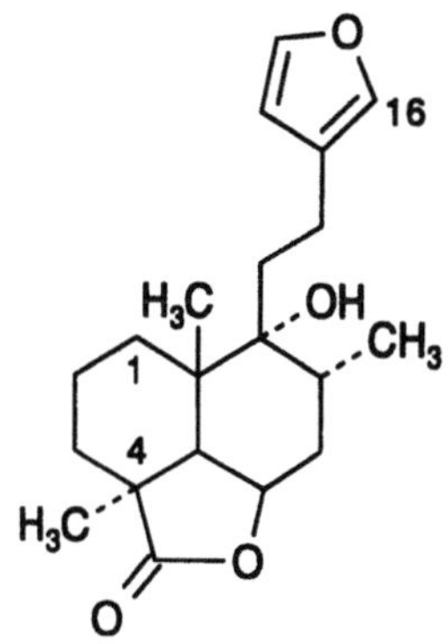

Marrubiin

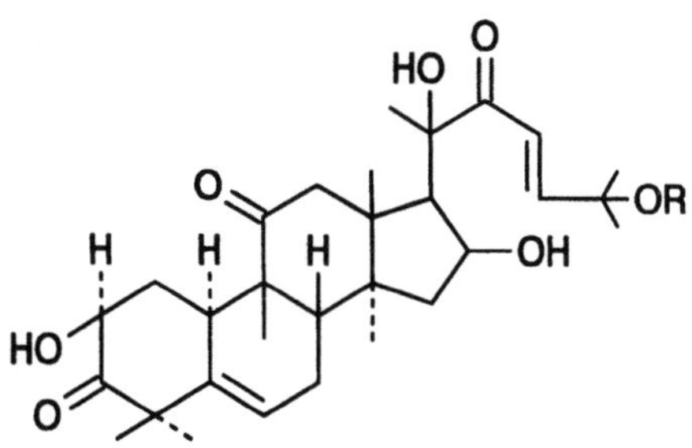

Cucurbitacin
B: R = Ac
D: R = H

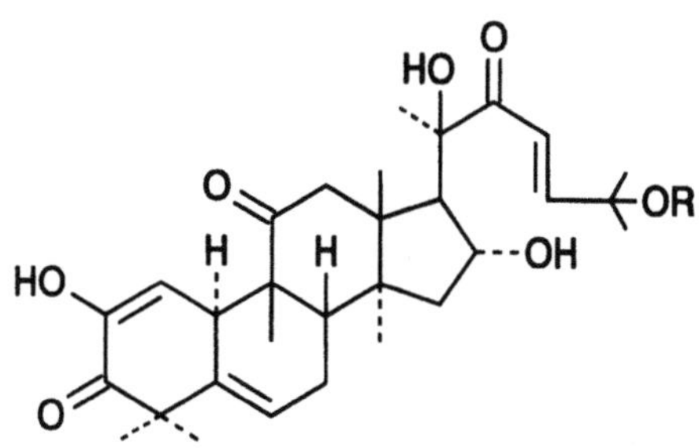

Cucurbitacin
I: R = H
E: R = Ac

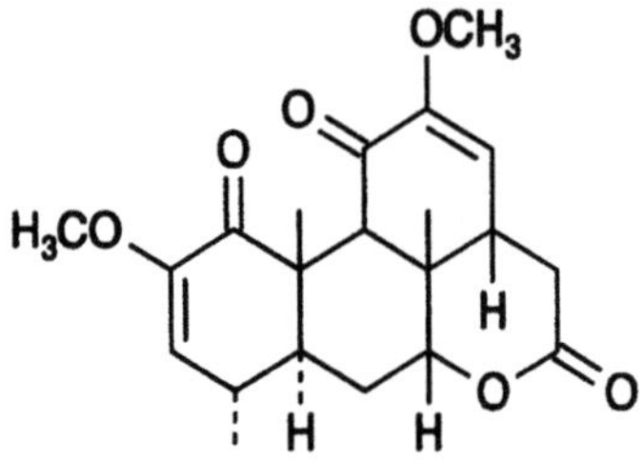

Quassin

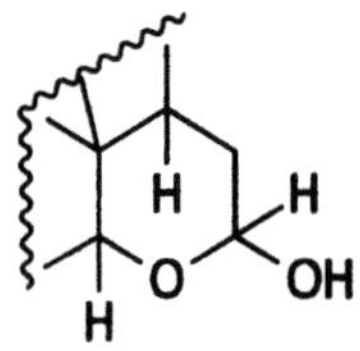

Neoquassin

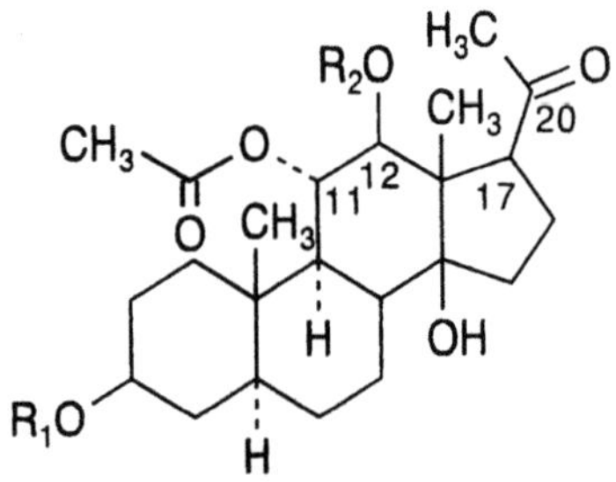

Condurangenine A R_1 = H

R_2 = 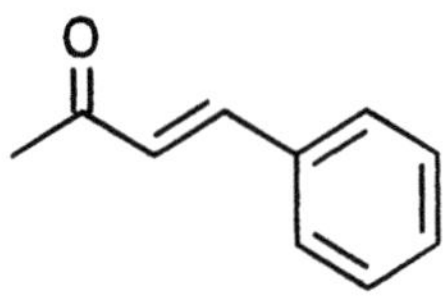

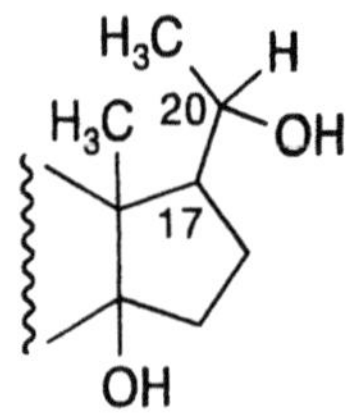

Condurangenine C

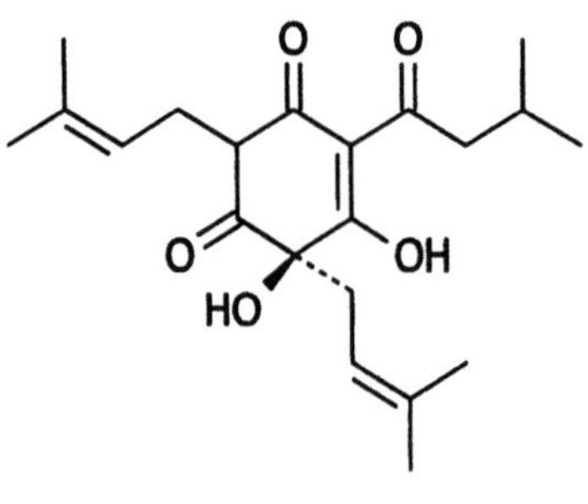

Humulone

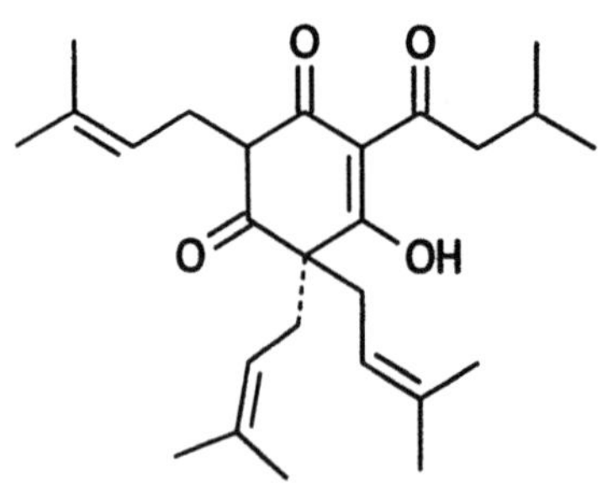

Lupulone

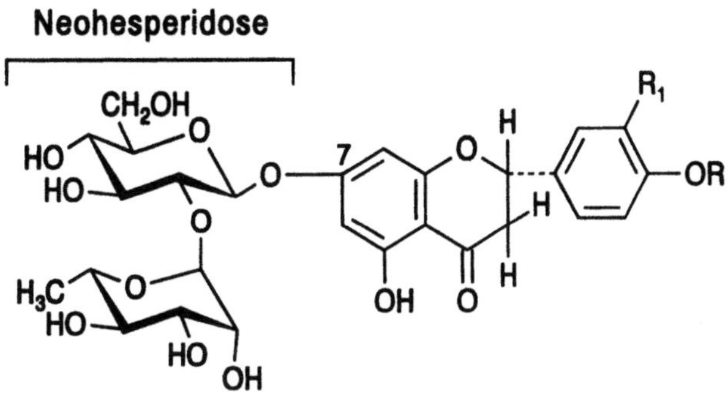

Naringin R = R_1 = H
Neohesperidin R = CH_3; R_1 = OH

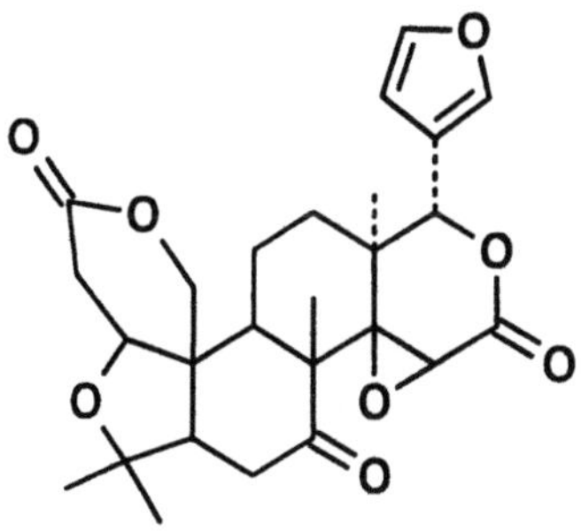

Limonin

3.6 TLC Synopsis of Bitter Drugs

Drug sample
1 Aurantii pericarpium
2 Harpagophyti radix
3 Gentianae radix
(methanolic extracts, 25 μl)
4 Centaurii herba
5 Condurango cortex
6 Menyanthidis folium

Reference T neohesperidin

Solvent system Fig. 1 ethyl acetate-methanol-water (77:15:8) → system I
Fig. 2 ethyl acetate-glacial acetic acid-formic acid-water (100:11:11:26) → system II

Detection Fig. 1 Vanillin sulphuric acid reagent (VS No. 42) → vis
Fig. 2 Anisaldehyde sulphuric acid reagent (AS No. 3) → vis

Fig. 1 **Aurantii pericarpium** (1): two characteristic red-orange zones of flavonoid glycosides → naringin/neohesperidin (bitter), rutin/eriocitrin (non-bitter) at R_f 0.4–0.5.
▶ see Fig. 23, 7.1.8 Flavonoid Drugs.

Harpagophyti radix (2): two prominent violet-red zones of iridoid glycosides → harpagoside (bitter/ R_f ~ 0.5) isoharpagoside, harpagid (sweet!) and procumbid R_f ~ 0.2).
▶ see Fig. 5, 3.7 Bitter Drugs, comparison with Scrophulariae herba and radix.

Gentianae radix (3): a major red-brown and a minor zone of secoiridoid glycosides → gentiopicroside (R_f ~ 0.45) and swertiamarine directly below.
▶ see Fig. 3, 3.7 Bitter Drugs (detection of amarogentin).

Centaurii herba (4): a yellow-brown prominent zone of swertiamarin at R_f ~ 0.4 as well as gentiopicroside directly above. Two yellow zones at R_f 0.25–0.3 are due to flavonoid glycosides.
▶ see Fig. 3, 3.7 Bitter Drugs.

Condurango cortex (5): a dark blue-black band of condurangins in the R_f range 0.4–0.55 (a complex mixture, see drug list) and eight dark blue-violet zones between R_f ~ 0.6 up to the solvent front.

Menyanthidis folium (6): three bright blue zones of the secoiridoid glycosides foliamenthin, menthafolin, dihydrofoliamenthin in the R_f range 0.6–0.8; additional yellow-brown flavonoid glycosides in the R_f range 0.2–0.5.
▶ see Fig. 4, 3.7 Bitter Drugs (loganine).

Note: Dark brown-black zones in the R_f range 0.05–0.2 are due to free sugars.

Fig. 2 Generally slightly lower R_f values and minor variations in colours of the main bitter principle compounds in comparison to those of Fig. 1 are recorded.

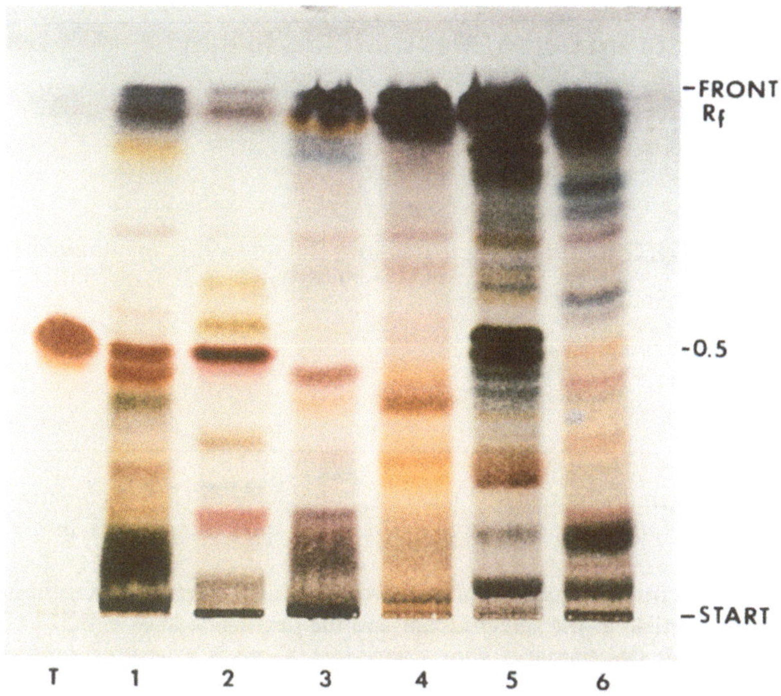

Fig. 1

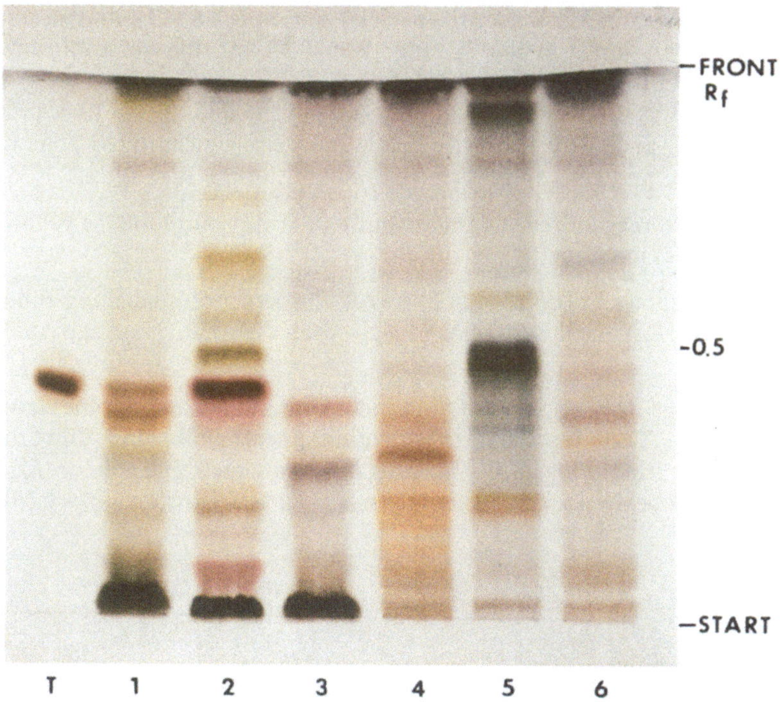

Fig. 2

3.7 Chromatograms

Gentianae radix, Centaurii herba, Menyanthidis folium

Drug sample	1 Gentianae radix 2 Centaurii herba 3 Menyanthidis folium	(methanolic extracts, 20 µl)
Reference compound	T1 amarogentin T2 gentiopricroside	T3 loganine T4,T5 bitter principle fractions (foliamenthin, menthafolin)
Solvent system	Fig. 3,4 ethyl acetate-methanol-water (77:15:8)	
Detection	A UV-245 nm (without chemical treatment) B vanillin sulphuric acid (VS No. 42) → vis C fast red salt reagent (FRS No. 17) → vis	

Fig. 3 A The secoiridoid glycosides amarogentin ($R_f \sim 0.8$/T1), gentiopicroside ($R_f \sim 0.45$/extract 1) and swertiamarin ($R_f \sim 0.4$/extract 2) give fluorescence-quenching zones in UV-254 nm.
Besides gentiopicroside **Gentianae radix** (1) shows two prominent quenching xanthone zones with gentisin/isogentisin at the solvent front and the gentioside at $R_f \sim 0.3$.
In **Centaurii herba** (2) the swertiamarin zone dominates; there is a weaker zone of flavonol glycosides at $R_f \sim 0.2$.

B After VS reagent **Gentianae radix** (1) generates the gentiopicroside as a brown-violet zone at R_f 0.45 (T2), amarogentin as a weak brown-violet zone at R_f 0.8 (T1), nonspecific blue, violet or brown-green zones in the R_f range 0.25–0.95 and the gentiobioside/gentianoside as major green-brown zone at R_f 0.1–0.2.
Centaurii herba (2) contains swertiamarin as main bitter principle, found as a pronounced brown-blue zone at $R_f \sim 0.4$ directly below the weak concentrated zone of gentiopicroside (T2). Flavonoid glycosides form yellow bands in the R_f range 0.2–0.35.

C Specific treatment with FRS reagent reveals amarogentin (T1) and xanthones as yellow-orange coloured zones (vis) in extract 1.

Fig. 4 A **Menyanthidis folium** (3) shows in UV-254 nm five weak fluorescence-quenching zones of secoiridoide glycosides (R_f 0.4/0.55/0.70), flavonol glycosides ($R_f \sim 0.1$) and aglycones (front).

B Treatment with the VS reagents generates two prominent blue and two minor blue zones in the R_f range 0.55–0.8. They represent foliamenthin, menthafolin and dihydrofoliamenthin (T4,T5). The iridoid loganine (T3) migrates as a violet-blue zone to $R_f \sim$ 0.45. Brown zones directly below are due to compounds such as sweroside ($R_f \sim 0.35$).

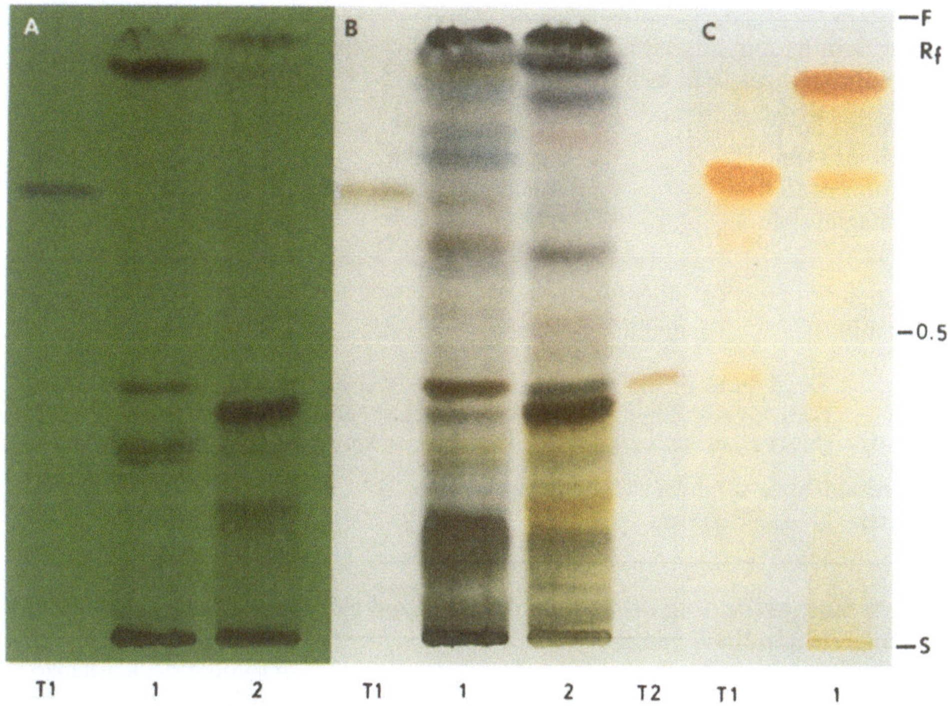

Fig. 3

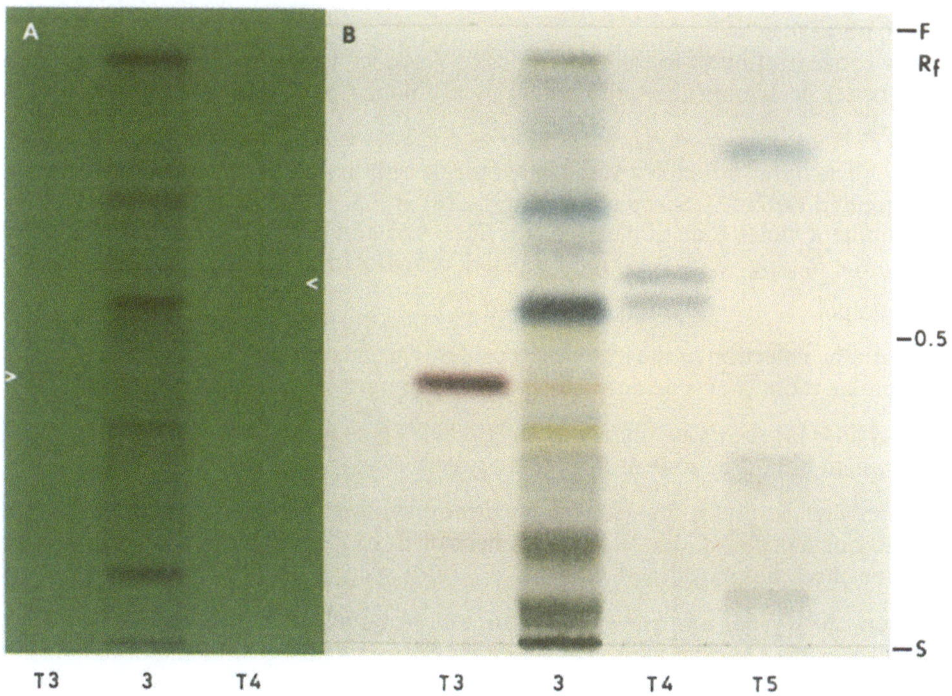

Fig. 4

TLC Synopsis, Drugs with Iridoid Glycosides

Drugs with bitter and non bitter iridoid glycosides
(methanolic extracts, 40 µl; n-BuOH extract, 30 µl)

Drug sample

1 Verbasci flos
1a Verb. flos (BuOH)
2 Veronicae herba
3 Plantaginis lanc. herba
4 Euphrasiae herba
5 Galeopsidis herba
6 Rehmanniae radix
6a Rehmanniae radix (BuOH)
7 Harpagophti radix
8 Scrophulariae herba
9 Scrophulariae radix

Reference compound

T1 catalpol
T2 aucubin
T3 glucose
T4 melittoside

Solvent system

Fig. 5A chloroform-methanol-water (60:40:4) → system 1
Fig. 5B ethyl acetate-methanol-water (77:15:8) → system 2
Fig. 6A–C chloroform-methanol-water (60:40:4) → system 1

Detection

A–C Anisaldehyde sulphuric acid reagent (AS No. 3)
A,B → vis C → UV 365 nm

Fig. 5 A System 1: Most of the drug extracts are characterized by iridoid glycoside compounds which migrate into the R_f range 0.45–0.75.
The extracts 1–6 contain catalpol (T1), their derivatives, e.g. veronicoside, a 2-benzoyl-catalpol, aucubin (T2) and/or derivatives (e.g. aucubin-xyloside) in varying concentrations. They all react with AS reagent as grey, blue or violet zones (vis).
Galeopsidis herba (5) shows harpagoside at $R_f \sim 0.6$ and harpagoside derivatives in the lower R_f range 0.3–0.45.
The low concentration of iridoid glycosides of **Verbasci flos** (1) and **Rehmanniae radix** (5) are better detectable after enrichment by n-butanol extraction, as demonstrated in Fig. 6A–C.

B System 2: The bitter principles of the harpagoside type are better separated in system 2. **Harpagophyti radix** (7), **Scrophulariae herba** (8) and **S. radix** (9) are characterized by the prominent violet zone of harpagoside ($R_f \sim 0.5$) and two to three additional violet zones in the R_f range 0.25–0.45 (e.g. harpagid, procumbid).

Note: catalpol (T1) would migrate to $R_f \sim 0.25$.

Fig. 6 System 1: The detection of the iridoid glycosides (e.g. aucubin) in Verbasci flos (1a) and **Rehmanniae radix** (6a) is achieved by n-butanol extraction (enrichment see 3.1)

A **Verbasci flos** (1a) shows aucubin and catalpol (T2/T1) as grey zones at R_f 0.4–0.5 as well as prominent blue-grey zone at $R_f \sim 0.4$ (e.g. verbascosaponine).

B **Rehmanniae radix** (6a) is characterized by three grey, almost equally concentrated zones in the R_f range 0.25–0.4, due to glucose, melittoside and aucubin (T2–T4) and a weak grey zone of rehmanniosides at $R_f \sim 0.6$.

C Detection in UV-365 nm shows aucubin and catalpol with brown and greenish fluorescence (T1–T2). The rehmanniosides appear as a light-brown band at $R_f \sim 0.6$.

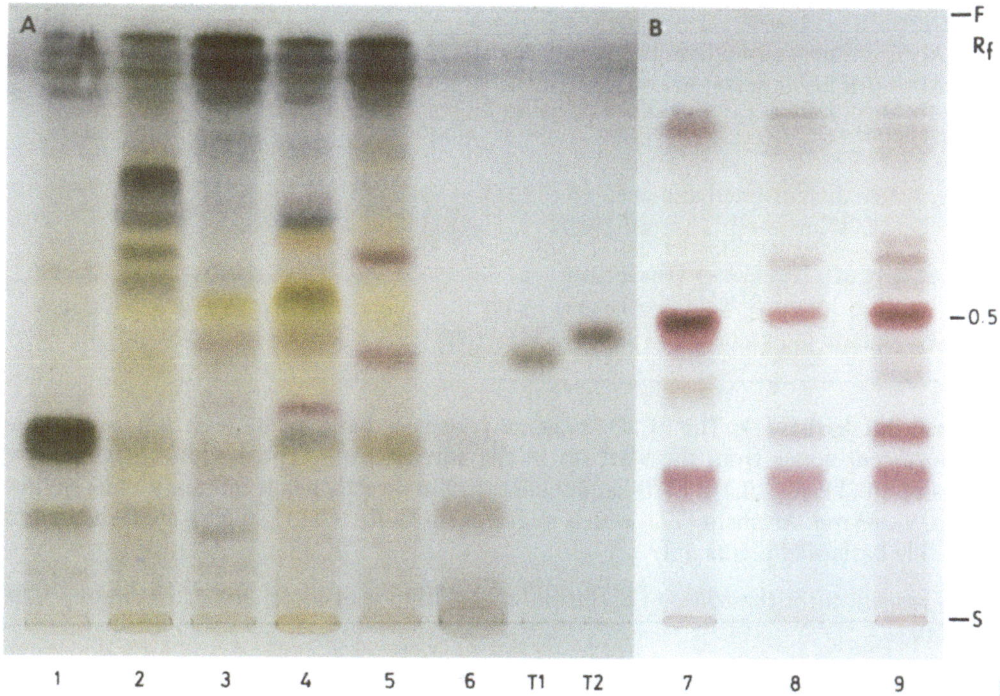
A
B
F
Rf
0.5
S
1
2
3
4
5
6
T1
T2
7
8
9

Fig. 5

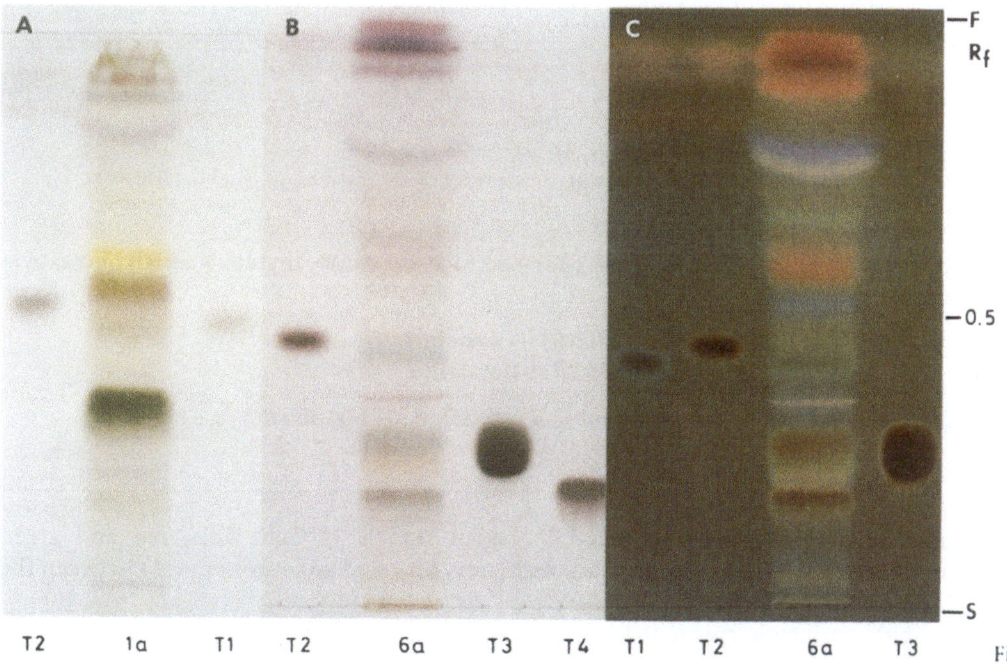
A
B
C
F
Rf
0.5
S
T2
1a
T1
T2
6a
T3
T4
T1
T2
6a
T3

Fig. 6

Absinthii herba

Drug sample 1 Absinthii herba (methanolic extract, 30 µl)
2 Absinthii herba (essential oil, 1:9, 5 µl)

Reference compound T1 absinthin T2 artabsin T3 thujone

Solvent system Fig. 7 A,B dichloromethane-acetone (85:15) → system 1
C toluene-ethyl acetate (93:7) → system 2

Detection A 50% H_2SO_4 (No.37) → UV-365 nm
B Vanillin sulphuric acid (VS No.42) → vis
C Phosphate molybdic acid (PMS No.34) → vis

Fig. 7A **Absinthii herba** (1). The H_2SO_4 reagent reveals a band of at least ten white-blue fluorescent zones from the start up to the solvent front. The sesquiterpene lactone absinthin (T1, $R_f \sim 0.3$) and its isomer anabsinthin directly below fluoresce white-yellow in UV-365 nm. Artabsin (T2), which migrates up to $R_f \sim 0.6$, is highly concentrated in freshly harvested plants only.

B VS reagent turns the zones of absinthin/anabsinthin grey-violet and artabsin grey-blue (vis).

C **Absinthii aetheroleum** (2). After treatment with PMS reagent the essential oil shows in system 2 seven to eight blue terpene zones in the R_f range 0.15 up to the solvent front. A major zone of thujyl alcohols (thujol) is followed by the violet-blue thujone zone at $R_f \sim 0.45$ (T3) and thujyl esters and terpenehydrocarbons at the solvent front.

Cnici herba

Drug sample 1 Cnici herba (methanolic extract, 30 µl)
2 Cnici herba (essential oil, 1:9, 5 µl)

Reference compound T1 cnicin T2 absinthin
T3 essential oil mixture: linalool ($R_f \sim 0.25$) ► carvon ► thymol ► linalyl acetate ► anethole ($R_f \sim 0.85$)

Solvent system Fig. 8 A,B acetone-chloroform (30:40) → system 1
C toluene-ethyl acetate (93:7) → system 2

Detection A,B Liebermann Burchard reagent (LB No. 25); A → UV-365 nm B → vis
C Vanillin sulphuric acid (VS No. 42) → vis

Fig. 8A,B **Cnici herba** (1). Detection with LB reagent reveals 14 light blue, red and green fluorescent zones (UV-365 nm) and weak grey, blue and violet zones (vis.) between the start and solvent front.
The bitter principle cnicin at $R_f \sim 0.4$ (T1) is seen as a light yellow-green zone in UV-365 nm and as light grey-blue zone in vis. The volatile oil components give a pominent blue zone at the solvent front (vis.). They are separated in system 2 (→C).

C **Cnici aetheroleum** (2). The terpenes show with VS reagent seven to nine blue to red-violet zones: four in the R_f range of terpene alcohols (R_f 0.15–0.25, linalool/T3), citral, cinnamic acid (R_f 0.4–0.5), citronellal ($R_f \sim 0.6$) and terpene hydrocarbons (front).

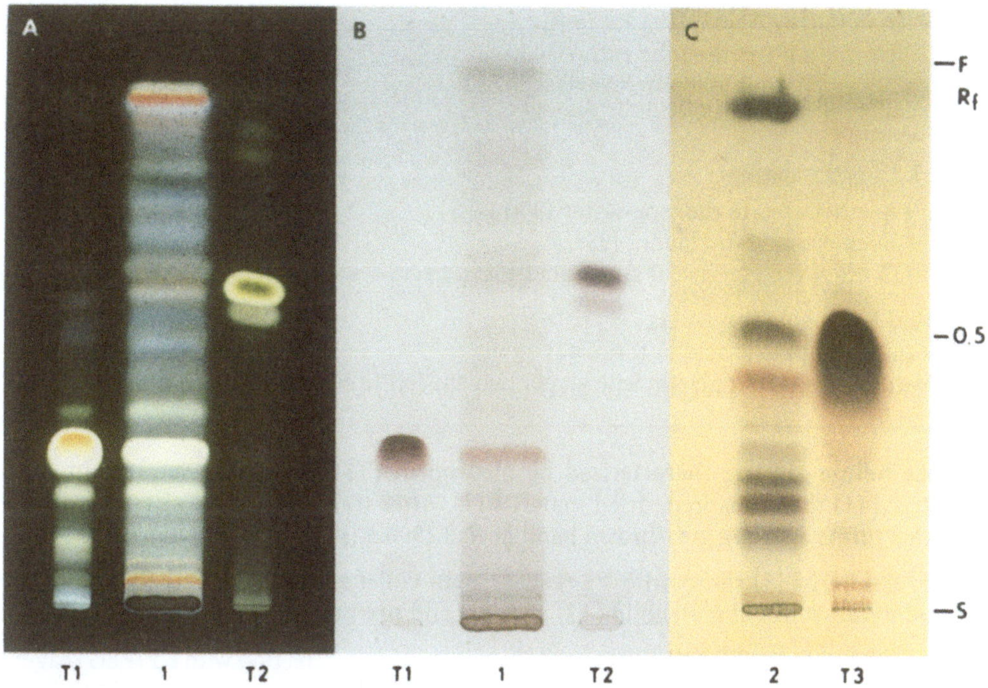

Fig. 7

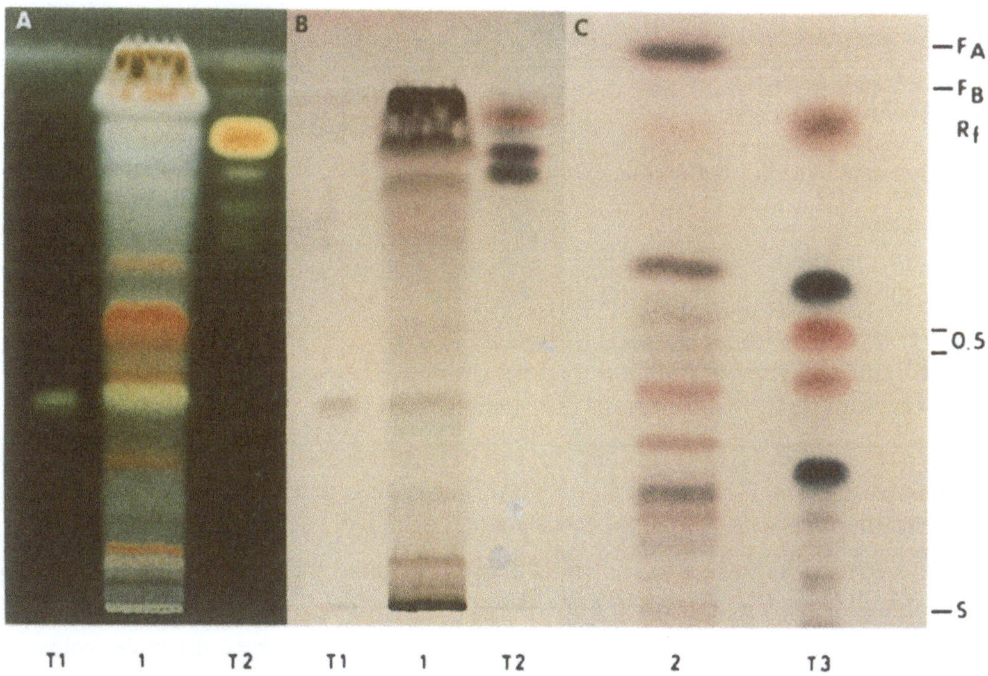

Fig. 8

Oleae folium, Marrubii herba

Drug sample 1,2 Oleae folium (methanolic extracts, 30 µl)
3 Marrubii herba (methanolic extracts, 30 µl)

Reference compound T1 oleuropein
T2 marrubiin

Solvent system Fig. 9 A ethyl acetate-dioxane-water (30:10:0.3)
B chloroform-methanol (95:5)
C ethyl acetate-glacial acetic acid-formic acid-wate (100:11:11:26)

Detection A 10% $FeCl_3$ solution → vis
B Vanillin sulphuric acid (VS No. 42) → vis
C Natural products-polyethylene glycol reagent (NP/PEG No. 28) → UV-365 nm

---

Fig. 9A **Oleae folium** (1,2) is characterized by oleuropein (T1), more concentrated in fresh material (1) than in stored, dried material (2). After treatment with $FeCl_3$ reagent the extract forms a strong grey-brown band at R_f 0.25–0.3 (vis).

B **Marrubii herba** (3) shows with VS reagent eight violet zones (e.g. diterpenes) with the pronounced zones of marrubiin (T2) at R_f 0.9 and premarrubiin at $R_f \sim 0.5$.

C Separation of Marrubii herba extract (3) in solvent C and detection with NP/PEG reagent reveals six blue fluorescent zones (e.g. caffeic acid derivatives) between R_f 0.15 and $R_f \sim 0.8$ and two weak green-yellow flavonoid glycosides at R_f 0.5–0.65.

Quassiae lignum

Drug sample 1 Quassiae lignum (methanolic extract, 40 µl)

Reference compound T1 quassin

Solvent system Fig. 10 chloroform-methanol (95:5)

Detection A UV-254 nm (without chemical treatment)
B UV-365 nm (without chemical treatment)
C Vanillin sulphuric acid reagent (VS No. 42) → vis

Fig. 10A **Quassiae lignum** (1) extract shows the bitter-tasting quassin (T1) as a prominent quenching zone at $R_f \sim 0.65$ in UV-254 nm.

B In UV-365 nm ten to 12 blue and violet fluorescent zones from the start up to $R_f \sim 0.85$ are detectable in UV-365 nm. Quassin does not fluoresce.

C Treatment with VS reagent needs at least 15 min at 110°C to form the violet-coloured zone of quassin at $R_f \sim 0.65$ (vis.), which is accompanied by a blue zone directly above.

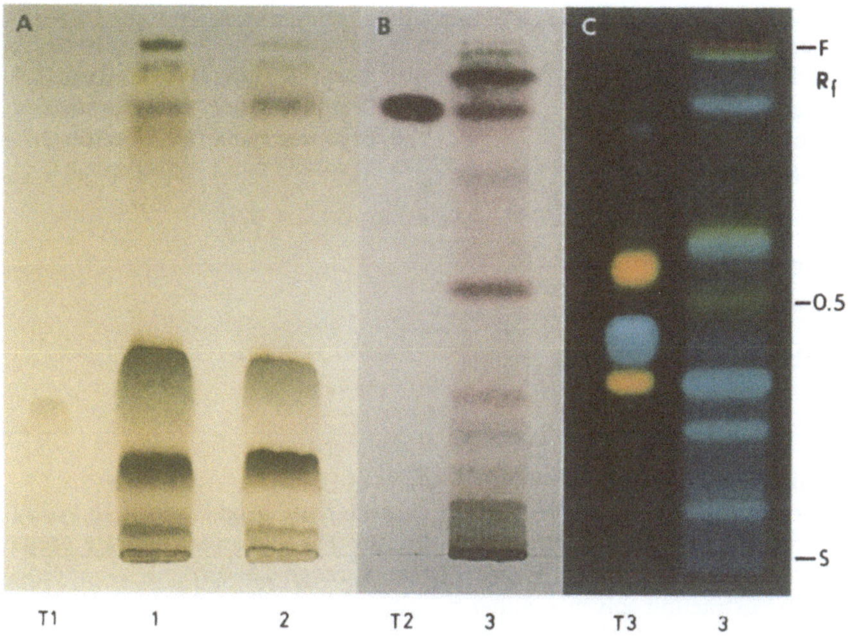

Fig. 9

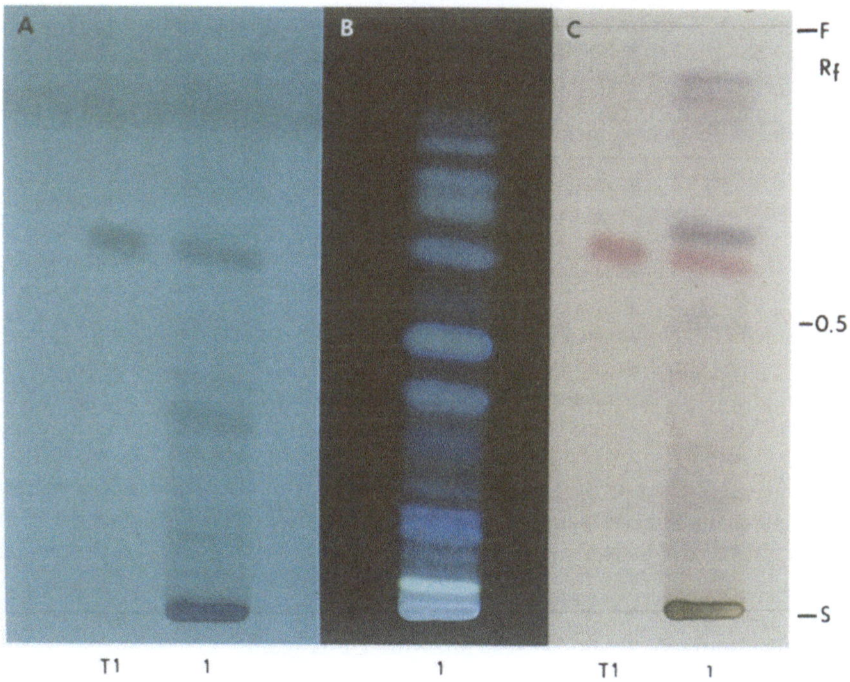

Fig. 10

TLC Synopsis, Drugs with Cucurbitacins

Drug sample
1 Colocynthidis fructus ($CHCl_3$ extract)
2 Gratiolae herba ($CHCl_3$ extract)
3 Iberidis semen ($CHCl_3$ extract)
4 Ecballii fructus ($CHCl_3$ extract)
5 Tayuyae radix ($CHCl_3$ extract)
6 Bryoniae radix ($CHCl_3$ extract)
7 Bryoniae radix (EtOH extract)
8 Bryoniae radix (water extract)
(extracts, 20–30 µl)

Reference T cucurbitacin B-glucoside (R_f ~ 0.22) ▶ cucurbitacin B (R_f ~ 0.9)

Solvent system Figs. 11, 12 chloroform-methanol (95:10)

Adsorbent Silica gel HPTLC plates (Merck, Germany) → 10 cm

Detection Vanillin phosphoric acid reagent (VP No. 41)
A UV-365 nm B vis

Fig. 11 **Cucurbitacin drugs**
The $CHCl_3$ extracts 1–6 show with VS reagent characteristic bright yellow to yellow-green and red fluorescent cucurbitacins in UV-365 nm. The glycosides migrate preferably into the R_f range 0.1–0.4, the aglycones into the R_f range 0.5–0.9.
Depending on the extraction solvents, either the glucosides or the aglycones are dominant in the extracts, as shown with **Bryoniae radix** (6–8).
Glucosides derived from 23,24-dihydrocucurbitacin show yellow to yellow-green fluorescence; those derived from 23-cucurbitacins give red-orange zones. They very often appear as pairs with dominant yellow fluorescence. The most common glucosides and aglycones are the cucurbitacins E,I,L and B. They are present in varying concentrations in the extacts 1–8:

Cucurbitacin glucosides	Cucurbitacin aglycones
L R_f ~ 0.14	L R_f ~ 0.67
I R_f ~ 0.16	I R_f ~ 0.7
E R_f ~ 0.29	E R_f ~ 0.72
B R_f ~ 0.29	B R_f ~ 0.9

The total contents of cucurbitacins are generally lower in the extracts 1–4 than in Tayuae (5) and Bryoniae radix extracts (6–8).
Tayuae radix (5) shows predominantly yellow and red-orange fluorescent cucurbitacin zones above R_f 0.45 with additional blue fluorescent zones of flavonoids in the R_f range 0.05–0.25.
Bryoniae radix (6–8) The chloroform extract 6 contains the cucurbitacin aglycones and glucosides I,E,L in almost equal concentration. While glucosides dominate in the ethanolic extract 7, the water extract 8 contains more aglycones due to preceeded enzymatic degradation.

Note: Sterines also fluoresce red in UV-365 nm with **V P** reagent.

Fig. 12 All cucurbitacins are seen with VP reagent as weak yellow-brown and blue violet zones (vis.).

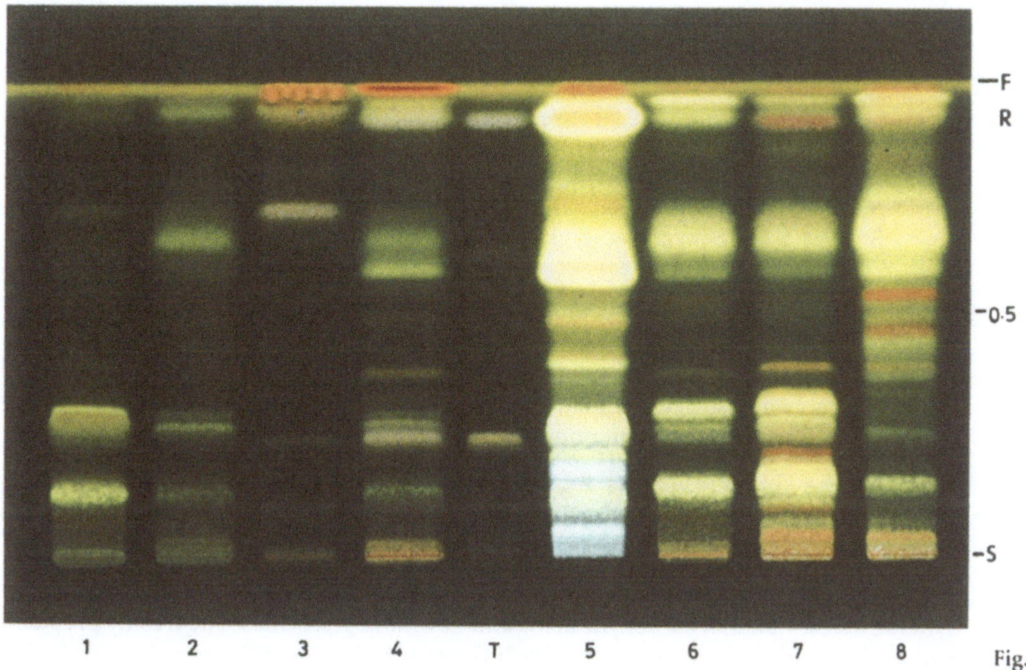

Fig. 11

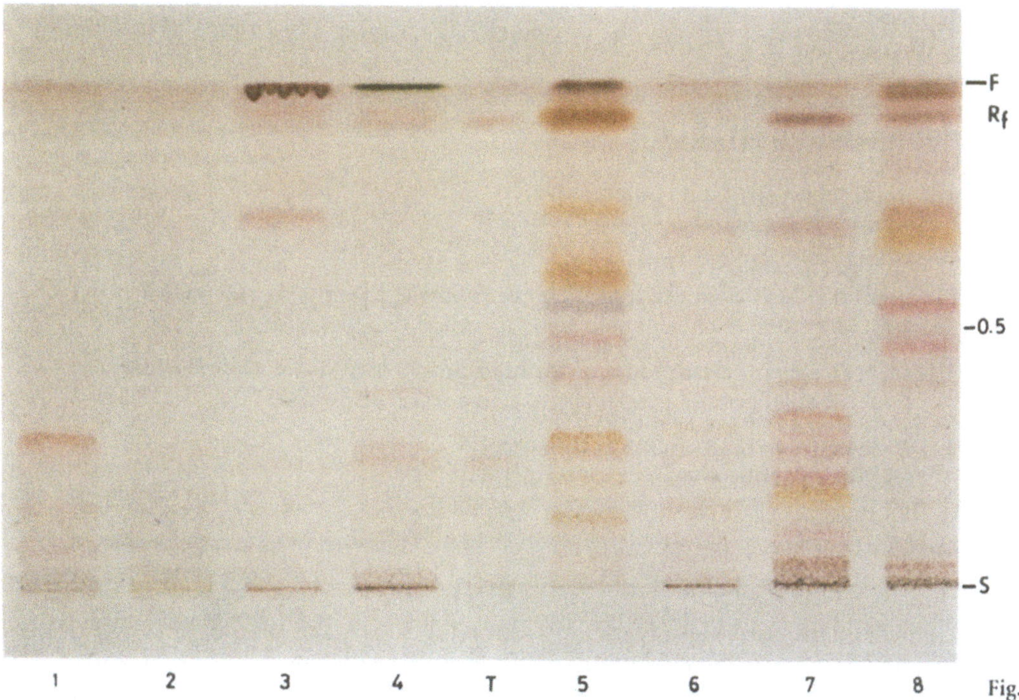

Fig. 12

Cynarae herba

Drug sample 1–4 Cynarae herba (freshly prepared or stored methanolic extracts, 20 μl)

Reference compound
T1 cynaropicrin — T2 T1 and degradation products
T3 luteolin-7-O-glucoside — T4 cynarin
T5 chlorogenic acid (R_f ~ 0.45) ▶ isochlorogenic acid (R_f ~ 0.8) ▶ caffeic acid (R_f ~ 0.9)

Solvent system Fig. 13 A chloroform-acetone (60:20) → for bitter principle
B ethyl acetate-formic acid-glacial acetic acid-water (100:11:11:26) acids

Detection A Anisaldehyde sulphuric acid reagent (AS No. 3) → vis
B Natural products-polyethylene glycol reagent (NP/PEG No. 28) → UV-365 nm.

Fig. 13 Freshly prepared and stored alcoholic extracts of **Cynarae herba** (1–4) can show varying TLC pattern of bitter tasting compounds and caffeoyl quinic acids.

A **Cynarae herba** – bitter principles. A freshly prepared methanolic extract (1) is characterized by the major violet zone of cynaropicrin (R_f ~ 0.3/T1). Degradation products are formed in alcoholic solutions or during storage process as seen in extract (2) and reference compound T2.

B **Cynarae herba** – phenol carboxylic acids and flavonoid glycosides. The freshly prepared methanolic extract 3 shows with NP/PEG reagent in UV-365 nm a band of blue fluorescent caffeoyl quinic acids such as chlorogenic acid (R_f ~ 0.45), cynarin (R_f ~ 0.65/T4), isochlorogenic and caffeic acid (R_f 0.8–0.9/T5) overlapped by the yellow fluorescent flavonoid luteolin-7-O-glucoside at R_f ~ 0.6 (T3).
Extract 4 shows less cynarin, due to isomerisation during extraction and in solution.

Humuli lupuli strobulus

Drug sample 1 Humuli lupuli strobulus (ether extract)
2 Humuli lupuli strobulus (MeOH extract)

Reference compound
T1 lupulon (R_f ~ 0.25) T2 humulon (R_f ~ 0.5)
T3 rutin (R_f ~ 0.4) ▶ chlorogenic acid (R_f ~ 0.5) ▶ hyperoside (R_f ~ 0.6)

Solvent system Fig. 14 A–C n-heptan-isopropanol-formic acid (90:15:0.5)
D ethyl acetate-glacial acetic acid-formic acid-water (100:11:11:26)

Detection A UV-365 nm (without chemical treatment)
B UV-254 nm (without chemical treatment)
C Fast blue salt (FBS No. 15) → vis
D Natural products-polyethylene glycol reagent (NP/PEG No. 28) → UV-365 nm

---

Fig. 14A–C **Humuli lupuli strobulus.** Fresh hop extract 1 shows the phloroglucine derivatives lupulon (T1, R_f ~ 0.25) and humulon (T2, R_f ~ 0.5) with light-blue fluorescence in UV-365 nm (A), as strong quenching zones in UV-254 nm (B) and as red or orange zones after FBS reagent (C). Both compounds are unstable and transformed to "bitter acids", then found at lower R_f values also as blue fluorescent, quenching and red-orange zones (→T1/T2).

D The methanolic extract (2) mainly contains the orange fluorescent rutin, hyperoside and the blue chlorogenic acid according to the test mixture T3 and an additional yellow-green flavonol monoglycoside at R_f 0.7 (NP/PEG reagent UV-365 nm).

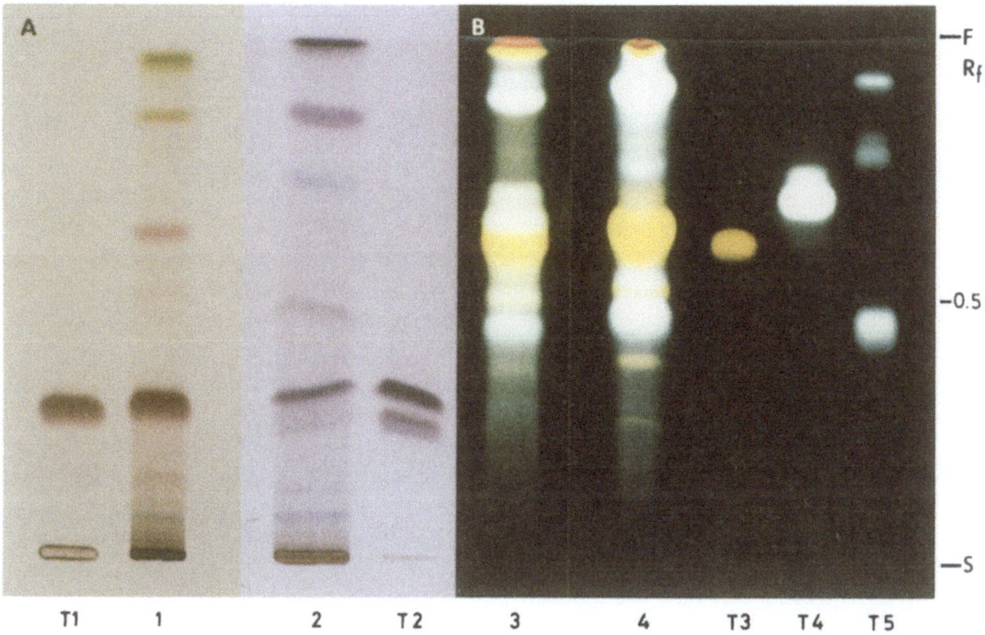
A
B
F
Rf
0.5
S
T1
1
2
T2
3
4
T3
T4
T5

Fig. 13

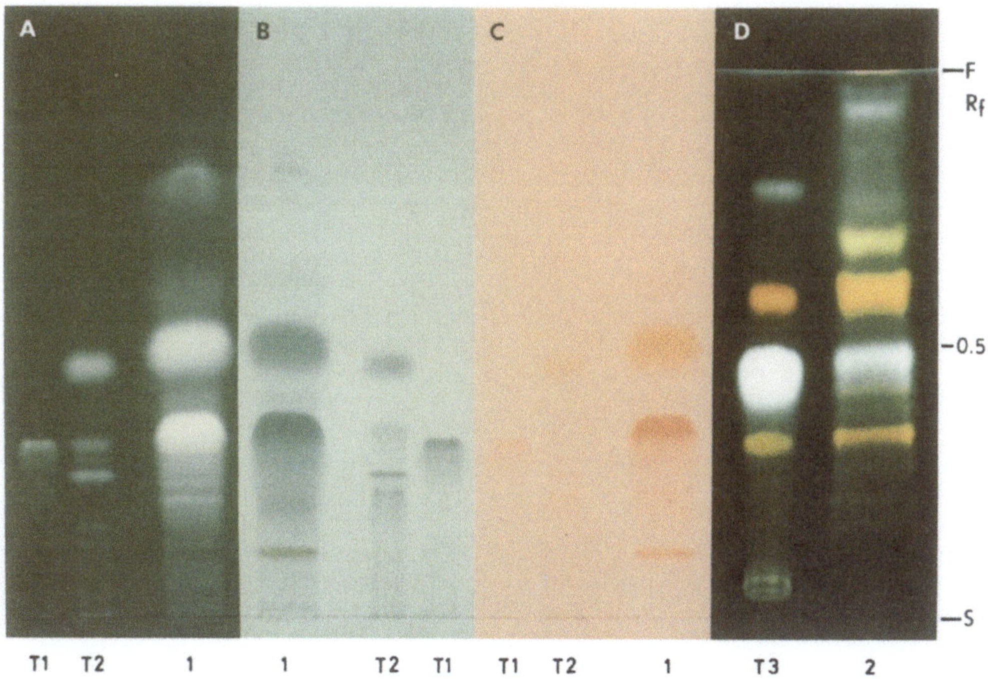
A
B
C
D
F
Rf
0.5
S
T1
T2
1
1
T2
T1
T1
T2
1
T3
2

Fig. 14

4 Cardiac Glycoside Drugs

These drugs contain steroid glycosides which specifically affect the dynamics and rhythm of the insufficient heart muscle.
The steroids are structurally derived from the tetracyclic 10,13-dimethylcyclopentanoperhydrophenanthrene ring system. They possess a γ-lactone ring (cardenolides) or a δ-lactone ring (bufadienolides) attached in the β-position at C-17. The typical sugar residues are derived from deoxy and/or C-3-O-methylated sugars, and they are linked glycosidically via the C-3-OH group of the steroid skeleton.

4.1 Preparation of Extracts

General method, cardenolide extract

A total of 2 g (>1% total cardenolides) or 10 g (<0.1% total cardenolides) of powdered drug are extracted by heating for 15 min under reflux with 30 ml 50% ethanol, with the addition of 10 ml 10% lead-(II)-acetate solution. After cooling and filtration, the solution is extracted by shaking with three 15-ml quantities of dichloromethane/isopropanol (3:2); shaking must be gentle to avoid emulsion formation.
The combined lower phases are filtered over anhydrous sodium sulphate and evaporated to dryness. The residue is dissolved in 1 ml dichloromethane/isopropanol (3:2) and used for chromatography.
► All cardiac glycoside drugs can be extracted by this method.

Exception

A simplified extraction procedure can be used for Hellebori radix, Xysmalobii radix and Strophanthi semen.
Strophanthi semen: Finely ground seeds (2 g) are defatted by heating for 1 h under reflux with light petroleum. The defatted and dried seed powder (1 g) is extracted for 5 min with 10 ml ethanol at about 60°C. The filtrate is used directly for chromatography.
Hellebori radix, Xysmalobii radix: Powdered drug (1 g) is extracted by heating under reflux with 10 ml ethanol for 30 min on a water bath. The filtrate is used directly for chromatography.

4.2 Thin-Layer Chromatography

Reference solutions

Commercial reference compounds:
A total of 5 mg is dissolved in 2 ml methanol at 60°C: digoxin, digitoxin, lanatosides A,B,C; k-strophanthin, g-strophanthin, uzarin, hellebrin, proscillaridin.
Convallatoxin: 3 mg is dissolved in 1 ml 80% ethanol on a water bath.

Gitoxin: 10 mg is dissolved in 3 ml methanol with the addition of 0.01 ml pyridine at 60°C.

Standard compounds from proprietary pharmaceuticals:

- **Digitalis glycosides:** Ten tablets or dragées (average 0.1–0.25 mg per tablet or dragée) are powdered in a mortar and then extracted by heating in a flask at 60°C for 5 min with 5–15 ml (depending on the weight of powder) dichloromethane/ethanol (1:1). The clear filtrate is evaporated to about 2 ml and 20 μl of this solution is used for chromatography.
- **Strophanthus glycosides:** Ten tablets are powdered and extracted with 10 ml methanol for 5 min on the water bath; 20 μl of each filtrate is used for chromatography.
- **Scilla glycosides:** Twenty dragees are powdered and extracted with 10 ml methanol for 5 min at about 60°C; 20 μl of each clear filtrate is used for chromatography.
- **Uzara glycosides:** Five dragées of Uzara (total glycosides of Xysmalobii radix) are finely powdered and extracted with 10 ml methanol for 5 min at 60°C; 20 μl of the clear filtrate is used for chromatography. Uzara tincture can be used directly for TLC comparison.

Adsorbent Silica gel 60 F_{254}-precoated plates (Merck, Germany)

Sample concentration 30–50 μl drug extracts, depending on the total cardiac glycoside concentration.
5 μl reference compound solutions.
20 μl reference solutions prepared from pharmaceuticals.

Chromatography solvents Ethyl acetate-methanol-water (100:13.5:10) ≙ (81:11:8)
→ a generally applicable solvent system for cardiac glycosides
Ethyl acetate-methanol-ethanol-water (81:11:4:8).
→ the addition of ethanol increases the R_f values of strongly polar compounds, e.g. k-strophantoside
Chloroform-methanol-water (35:25:10) – lower phase.
→ for Hellebori radix

4.3 Detection

- Without chemical treatment
  UV-254 nm very weak fluorescence quenching of all cardiac glycosides
  UV-365 nm no fluorescence at all

- Spray reagents (see Appendix A)

- Specific detection of the γ-lactone ring of cardenolides:
  - Kedde reagent (Kedde No. 23)
    Immediately on spraying, cardenolides generate a pink or blue-violet (vis) colour. The colour fades after a few minutes, but can be regained by repeated spraying. Bufadienolides do not react.
    Remarks: Reagents such as Legal (alkaline sodium nitroprusside solution), Baljet (alkaline picric acid solution) or Raymond reagent (alkaline m-dinitrobenzene solution) also give red, red-orange or violet (vis) cardenolide-specific colours.

- General detection methods for cardenolides and bufadienolides

  - Antimony-(III)-chloride reagent ($SbCl_3$ No. 4)
    A TLC plate (20 × 20 cm) has to be sprayed with a minimum of 10 ml $SbCl_3$ reagent and heated at 100°C for about 8–10 min; evaluation is done in vis and UV-365 nm (see Table 1). Changes are observed in the fluorescence response if the sprayed plate is allowed to stand for a longer time. In visible light, the zones appear mainly grey, violet or brown.

  - Chloramine-trichloroacetic acid reagent (CTA No. 9)
    Blue, blue-green, or yellow-green fluorescent zones are observed in UV-365 nm, similar to those obtained with $SbCl_3$ reagent. Only weak, nonspecific colours are seen in visible light.

  - Sulphuric acid reagent (concentrated H_2SO_4 No. 37)
    The TLC plate is sprayed with about 5 ml reagent and then heated for 1–3 min at 80°C under observation. Blue, brown, green and yellowish fluorescent zones are seen in UV-365 nm; the same zones appear brown or blue in daylight.

  - Anisaldehyde sulphuric acid reagent (AS No. 3)
    Bufadienolides in extracts of Hellebori radix, e.g. hellebrin, show a prominent blue colour (vis).

Table 1. Fluorescence of Cardiac Glycosides

Cardiac glycoside	Fluorescence in UV-365 nm $SbCl_3$ reagent 8 min/100°C
K- and g-strophanthidine derivatives	
K-strophantoside, k-strophanthidin-β, cymarin, helveticoside, erysimoside, g-strophanthin, convallatoxin	orange, pale brown or yellow-green
Digitalis glycosides	
Digitoxin, acetyl digitoxin	
purpurea glycoside A, lanatoside A	dark blue or dark brown
gitoxin, digoxin	
purpurea glycoside B, lanatoside B/C	light blue
Oleander glycosides	
oleandrin, adynerin	light blue
Bufadienolides	
Proscillaridin, scillaren A, glucoscillaren	yellow-brown
scilliroside, glucoscilliroside	pale green
hellebrin, helleborogenone	yellow

4.4 Drug List

	Drug/plant source Family/pharmacopoeia	Main constituents *)for minor constituents see 4.5 Formulae and Tables
Fig. 3,4	**Digitalis lanatae folium** White foxglove leaves Digitalis lanata EHRH. Scrophulariaceae DAB 10, ÖAB 90, MD	0.5%–1.5% total cardenolides, ~60 glycosides*) Lanatosides A and C (~50%) lanatosides B, D, E as well as digoxin and digitoxin DAB 10: Digitalis lanata powder standardized at 0.5% digoxin activity
Fig. 3,4	**Digitalis purpureae folium** Red foxglove leaves Digitalis purpurea L. Scrophulariaceae DAB 10, ÖAB, Helv VII, BP 88, USP XX, Japan, MD	0.15%–0.4% total cardenolides, ~30 glycosides*) Purpurea glycosides A and B (~60%), digitoxin (~12%), gitoxin (~10%) and gitaloxin (~10%) DAB 10: Digitalis purpurea powder standardized at 1% digitoxin activity
Fig. 5	**Oleandri folium** Oleander leaves Nerium oleander L. Apocynaceae DAB 10	1%–2% total cardenolides, ~15 glycosides*) Oleandrigenin (16-acetylgitoxigenin): O-L-oleandroside (oleandrin), O-glucoside, O-D-diginoside (nerigoside), O-gentiobioside (gentiobiosyl oleandrin). Adynerigenin-D-diginoside (adynerin) Digitoxigenin-D-digitaloside (odoroside H), -D-diginoside (odoroside A). Oleagenin-D-diginoside (oleaside A), oleasides B-F ► Flavonoids: e.g. rutin (0.5%)
Fig. 6	**Xysmalobii radix** Uzara root Xysmalobium undulatum (L.) R. BROWN Asclepidiaceae	1%–2% total cardenolides Glycosides of uzarigenin and xysmalogenin (5,6-dehydrodigitoxin); as main compounds the diglucosides uzarin and xysmalobin Uzarigenin differs from digitoxin by trans linkage of rings A and B
Fig. 7	**Strophanthi grati semen** Strophanthus seeds Strophanthus gratus (WALL et HOOK) BAILL. Apocynaceae DAC 86, MD	4%–8% total cardenolides 90% g-strophanthin (g-strophanthidin-rhamnoside), strogoside, small quantities of sarmentosides A, D, E

Drug/plant source Family/pharmacopoeia	Main constituents *)for minor constituents see 4.5 Formulae and Tables	
Strophanthi kombé semen Strophanthus seeds Strophanthus kombe OLIVER Apocynaceae MD	5%–10% total cardenolides k-Strophanthidin-glycosides: cristalline glycoside mixture, "k-strophanthin": 80% k-strophanthoside, k-strophanthin-β (10%–15%), erysimoside (15%–25%) Minor glycosides: cymarin, cymarol, helveticosol, periplocymarin, helveticoside	**Fig. 7**
Cheiranthi cheirii herba Wallflower, Violier Cheiranthus cheiri L. Brassicaceae	0.01%–0.015% total cardenolides Cheirotoxin (k-strophanthidin-gulomethylosido-glucoside); desglucocheirotoxin, cheiroside A (uzarigenin-fucosido-glucoside)	**Fig. 8**
Erysimi herba (►) Grey wall-flower Erysimum species, e.g. E. crepidifolium ROHB. E. diffusum EHRH. Brassicaceae (Cruciferae)	0.2%–1.8% total cardenolides depending on species five to ten glycosides: erysimoside (glucohelveticoside) and/or helveticoside are always reported ► A drug derived from various species with an enormous variation in cardenolide compounds	**Fig. 8**
Adonidis herba Adonis Adonis vernalis L. Ranunculaceae DAB 10 DAB 10: Adonis powder standardized at 0.2% cymarin activity	0.25%–0.8% total cardenolides, ~20 glycosides*) k-Strophanthidin-glycosides: cymarin (0.02%), desglucocheirotoxin, k-strophanthin-β, k-strophanthoside Adonitoxigenin glycosides: adonitoxin (0.07%) A-acetyl rhamnoside, A-glucoside, A-xyloside ► Flavone-C-glycosides: adonivernith, vitexin	**Fig. 9,10**
Convallariae herba Lily of the valley Convallaria majalis L. Convallariaceae DAB 10, ÖAB 90, MD DAB 10: standardized at 0.2% convallatoxin activity	0.2%–0.5% total cardenolides, ~40 glycosides*) k-Strophanthidin-glycosides: convallatoxin, convallalloside (4–40%), derglucocheirotoxin k-Strophanthidol-glyosides: convallatoxol, convallotoxoloside. Periplogenin and sarmentogenin-glucosides Convallatoxin is the main glycoside in drugs of western and northern European origin (40%–45%). In middle European drugs, lokundjoside (bipindogenin-rhamnoside, 1–25%) predominates	**Fig. 9,10**

Drug/plant source Family/pharmacopoeia	Main constituents *)for minor constituents see 4.5 Formulae and Tables

Bufadienolides

Fig. 11,12	**Hellebori radix** Hellebore root Helleborus niger L. Helleborus viridis L. and other Helleborus ssp. Ranunculaceae MD	The bufadienolide pattern and their amount varies, depending on species and drug origin Hellebrin as main glycoside, e.g. in H. viridis and H. odorus (<0.5%); not always present (e.g. H. niger)
Fig. 13,14	**Scillae bulbus** Squill Classified white or red Urginea maritima (L.) BAKER = Aggregate of six species (different polyploidy). ► BP 88 (new name) Drimia maritima (L.) STEARN Drimia indica (ROXB) Hyacinthaceae (Liliaceae)	0.1%–2.4% total bufadienolides, ~15 glycosides White variety: average 0.2%–0.4% Proscillaridin, scillaren A, glucoscillaren (aglycone: scillarenin) Scilliphaeoside, scilliglaucoside Red variety: <0.1% Scilliroside and glucoscilliroside (algycone: scillirosidin); proscillaridin and scillaren A as in the white variety DAB 10: squill powder standardized at 0.2% proscillaridin activity

4.5 Formulae and Tables

Digitalis lanatae and Digitalis purpureae folium		R_1	R_2	R_3
Cardenolide aglycones 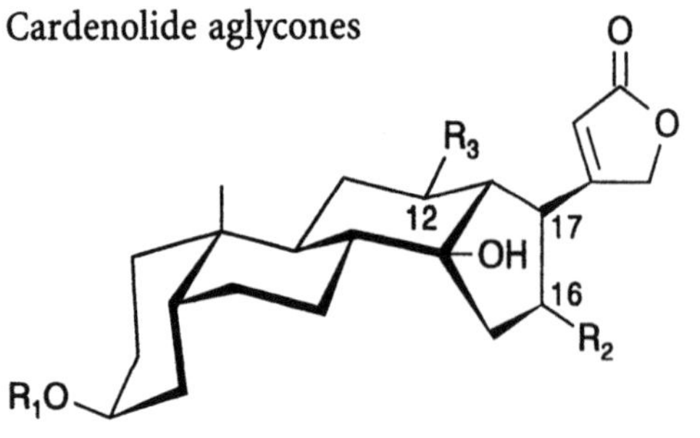	Digitoxigenin	H	H	H
	Gitoxigenin	H	OH	H
	Digoxigenin	H	H	OH
	Diginatigenin	H	OH	OH
	Gitaloxigenin	H	O-CHO	H

Cardenolide	R_1	R_2	R_3	D. lanata	D. purpurea
Digitalinum verum	Gl-Dtl	OH	H	x	x
Glucogitoroside	Gl-Dx	OH	H	x	–
Glucodigifucoside	Gl-Fuc-	H	H	x	–
Glucoverodoxine	Gl-Dtl-	O-CHO	H	x	x
Glucolanadoxine	Gl-Dx	O-CHO	H	x	–
Glucoevatromonoside	Gl-Dx	H	H	x	–
Digitoxin	Dx-Dx-Dx-	H	H	–	x
Gitoxin	Dx-Dx-Dx-	OH	H	–	x
Digoxin	Dx-Dx-Dx-	H	OH	x	(x)
Gitaloxin	Dx-Dx-Dx-	O-CHO	H	–	x
Lanatoside A	Gl-Acdx-Dx-Dx-	H	H	x	–
Lanatoside B	Gl-Acdx-Dx-Dx-	OH	H	x	–
Lanatoside C	Gl-Acdx-Dx-Dx-	H	OH	x	–
Purpureaglycoside A	Gl-Dx-Dx-Dx-	H	H	–	x
Purpureaglycoside B	Gl-Dx-Dx-Dx-	OH	H	–	x
Glucogitaloxin	Gl-Dx-Dx-Dx-	O-CHO	H	–	x

Gl = Glucose
Dtl = Digitalose
Fuc = Fucose
Dx = Digitoxose
Acdx = Acetyldigitoxose

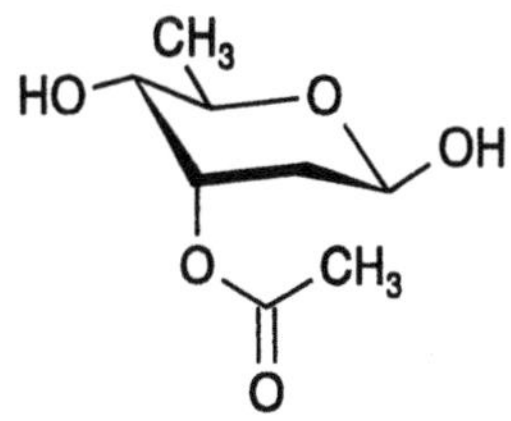

3-Acetyldigitoxose

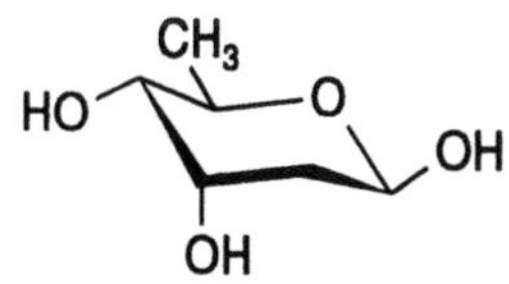

Digitoxose

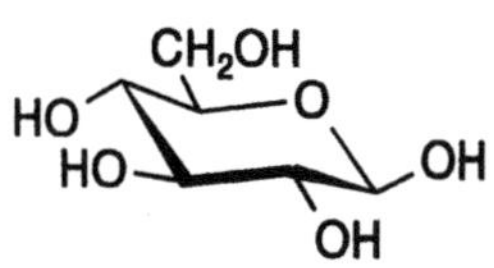

β-D-Glucose

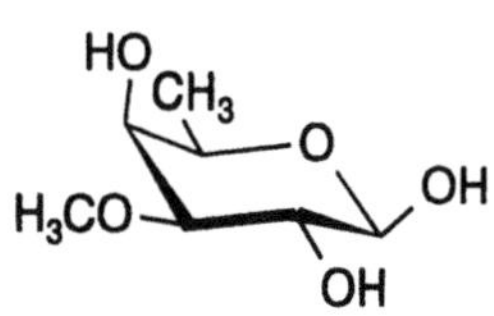

Digitalose

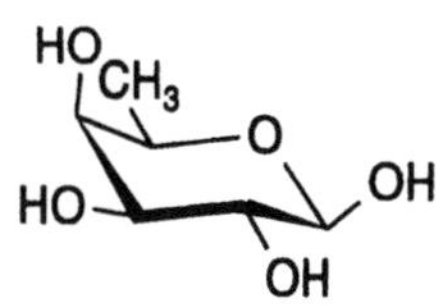

Fucose

Nerium oleander

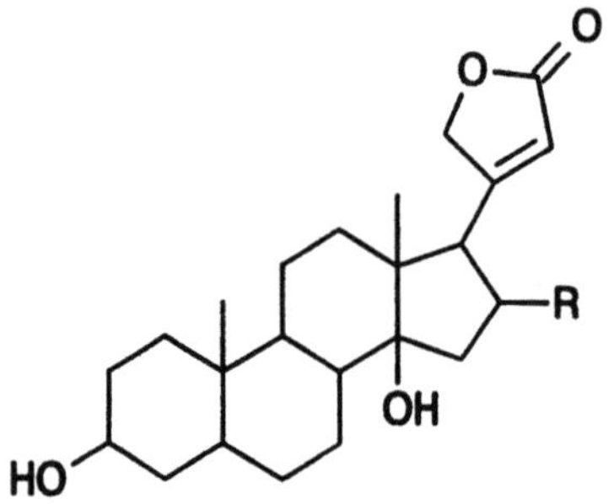

Oleandrigenin R = $OCOCH_3$
Digitoxigenin R = H
Gitoxigenin R = OH

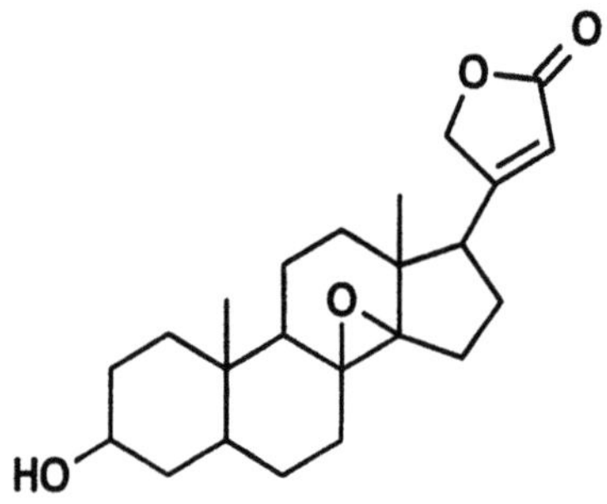

Adynerigenin

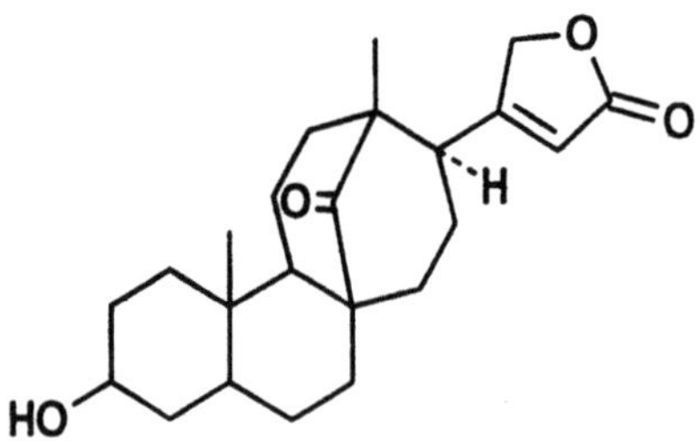

Oleagenin

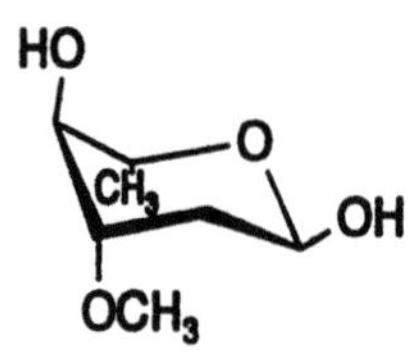

L-Oleandrose

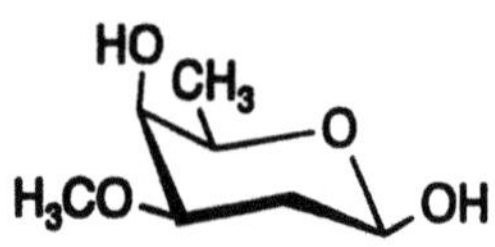

D-Diginose

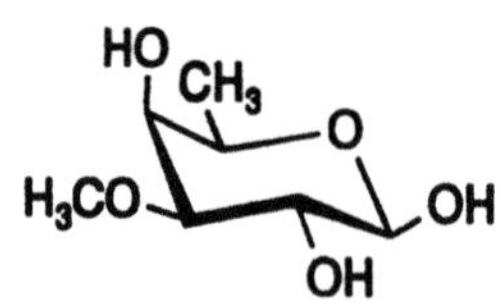

D-Digitalose

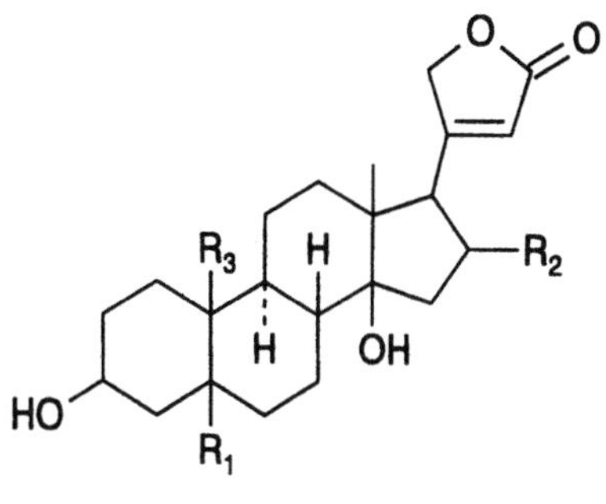

Table 1
Cardiac glycosides
in Adonidis herba, Cheiranthii herba,
Strophanthi kombé semen, Erysimi herba

R_1	R_2	R_3	
OH	H	CHO	**Adonidis herba**
k-Strophanthidin (*S*)			Cymarin (*S*-cymaroside) desglucocheirotoxin (*S*-gulomethyloside) k-Strophanthidin-β, k-strophanthoside
H	OH	CHO	Adonitoxin (*A*-rhamnoside), *A*-2-O-acetyl-rhamnoside,
Adonitoxigenin (*A*)			*A*-3-O-acetylrhamnoside, and glucosides and xylosides.
H	OH	CH_2OH	Adonitoxigenol (-rhamnoside).
OH	OH	CHO	Strophadogenin (-diginoside).
OH	H	CHO	**Cheiranthi cheiri herba**
k-Strophanthidin (*S*)			Cheirotoxin (*S*-gulomethylosyl-D-glucoside) desglucocheirotoxin
			Strophanthi kombé semen
OH	H	CHO	Cymarin (*S*-cymaroside), helveticoside (*S*-β-D-digitoxide)
k-Strophanthidin (*S*)			erysimoside (*S*-digitoxoside-glucoside), k-strophanthin-β, k-strophanthoside
OH	H	CHO	**Erysimum species**
k-Strophanthidin			Helveticoside, erysimoside (see Stroph. Kombé semen)

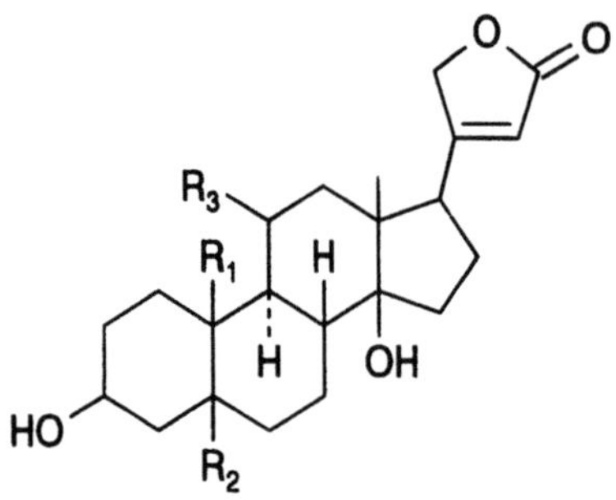

Table 2
Cardenolides in Convallariae herba

Aglycone	R_1	R_2	R_3
(1) Strophanthidin	CHO	OH	H
(2) Strophanthidol	CH_2OH	OH	H
(3) Periplogenin	CH_3	OH	H
(4) Bipindogenin	CH_3	OH	OH
(5) Sarmentogenin	CH_3	H	OH

Aglycones	Glycosides + Rhamnose	Gluc-Rham	Gulomethylose	Allomethylose
(1)	Convallotoxin 4%–40%	Convalloside 4%–24%	Desglucocheirotoxin 3%–15%	Strophalloside 1.2%
(2)	Convallatoxol 10%–20%		Desglucocheirotoxol 2%–5%	Strophanolloside 2%
(3)	Periplorhamnoside 0.5%–3%			
(4)	Lokundioside 1%–25%			
(5)	Rhodexin A 2%–3%			

Cardenolides in Xysmalobii radix

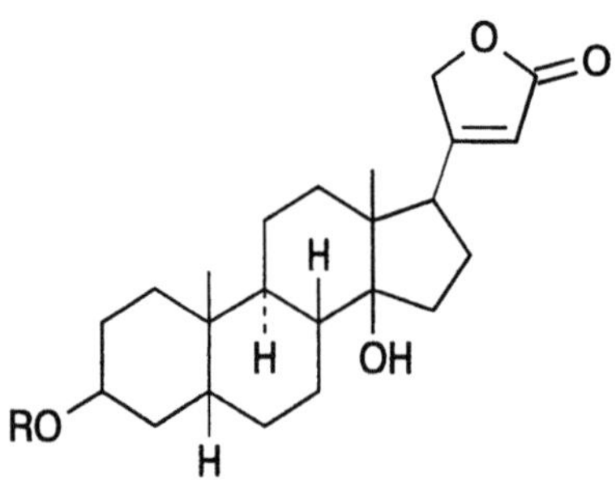

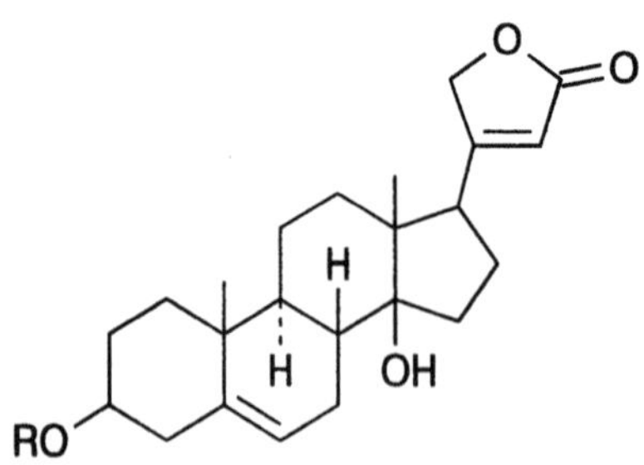

Uzarigenin R = H
Uzarin R = Gluc-Gluc

Xysmalogenin R = H
Xysmalorin R = Gluc-Gluc

Bufadienolides
Hellebori radix

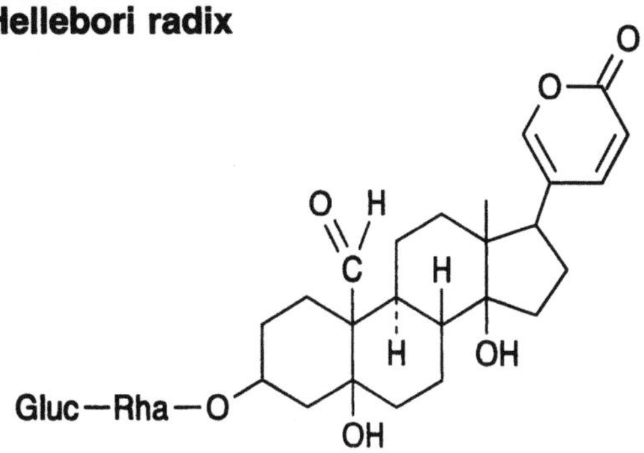

Hellebrin

Scillae bulbus

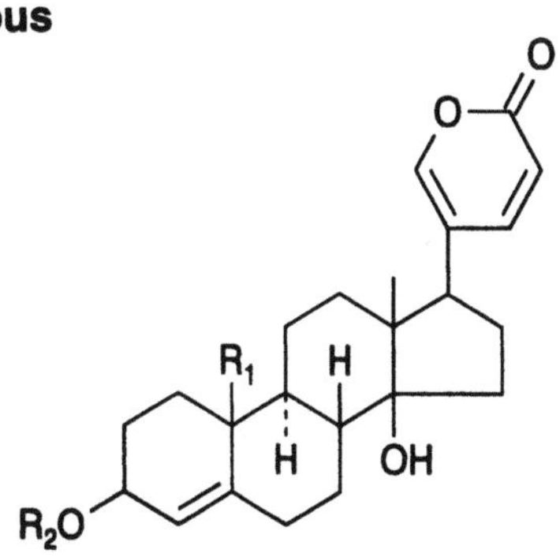

	R_1	R_2
Scillarenin	CH_3	H (Aglycon)
Proscillaridin A	CH_3	Rham
Scilliphaeoside	H	Rham
Scillaren A	CH_3	Gluc-Rham
Glucoscillaren A	CH_3	Gluc-Gluc-Rham

4.6 TLC Synopsis of Cardiac Glycosides

Reference compound

1	g-strophanthin	8	digoxin
2	"k-strophanthin"	9	gitoxin
3	convallatoxin	10	digitoxin
4	cymarin	11	cymarol
5	lanatoside A	12	peruvoside
6	lanatoside B	13	oleandrin
7	lanatoside C		(1–13, 10 µl)

Solvent system Fig. 1,2 ethyl acetate-methanol-water (81:11:8)

Detection Fig. 1 Kedde reagent (No. 23) → vis
Fig. 2 Chloramine-trichloracetic acid reagent (CTA No. 9) → UV-365 nm

---

Fig. 1 **Kedde reagent** (vis.)
Immediately after spraying, the cardiac glycosides generate blue to red-violet, fairly stable colours (vis.) with the exception of peruvoside.

Digitalis glycosides
Their colours are indicative of the structural type:

digoxin and lanatoside C → red-violet
gitoxin and lanatoside B → blue-violet
digitoxin and lanatoside A → blue

Fig. 2 **CTA reagent (UV-365 nm)**
All cardiac glycosides show light blue, blue-green or yellow-green fluorescent zones.

Strophanthus, Convallaria and Thevetia glycosides → blue-green fluorescence
cymarin, cymarol, convallatoxin, peruvoside, g- and k-strophanthin.
"k-strophanthin" is a glycoside mixture; for TLC analysis see Fig. 9, 4.7, Cardiac Glycoside Drugs.

Digitalis and Oleander glycosides → intense light-blue fluorescence
with the exception of digitoxin, which shows a yellow-green fluorescence.

After CTA treatment, chromatograms of some standard substances show additional zones in UV-365 nm, due to degradation products and impurities.

Note: Spraying with concentrated H_2SO_4 results in UV-365 nm detection with similar fluorescent zones: 1 (yellow) 2, 3, 4, 11 (greenish blue) 5, 6, 7, 8, 9, 10, 12, 13 (blue)

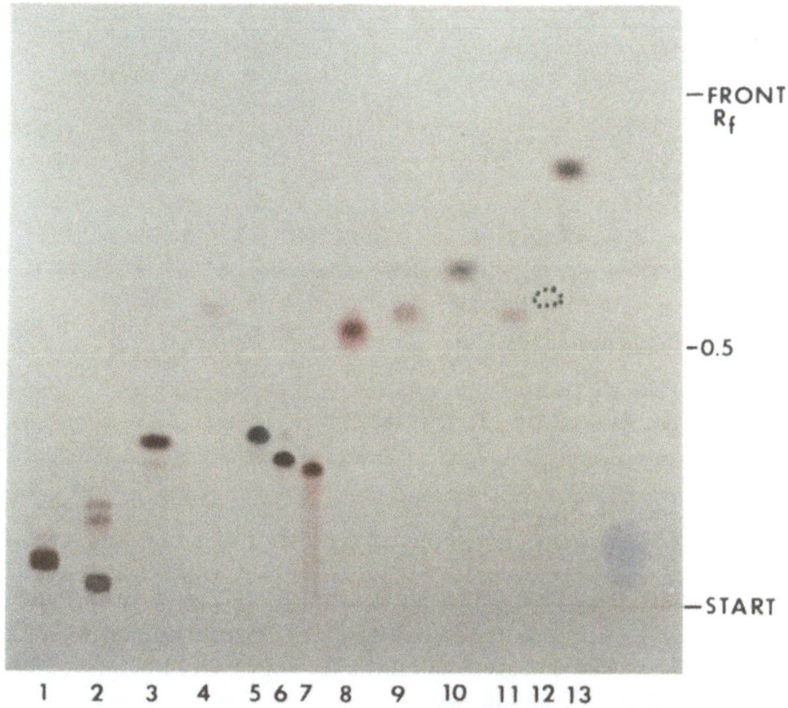
FRONT
Rf
0.5
START
1 2 3 4 5 6 7 8 9 10 11 12 13

Fig. 1

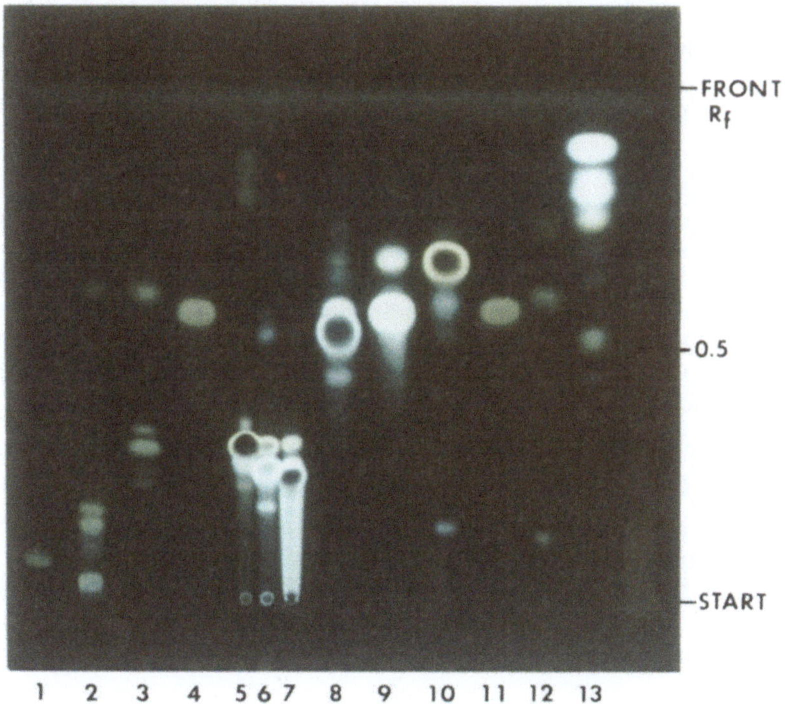
FRONT
Rf
0.5
START
1 2 3 4 5 6 7 8 9 10 11 12 13

Fig. 2

4.7 Chromatograms

Digitalis folium

Drug sample 1,1a Digitalis lanatae folium (trade samples)
2,2a Digitalis purpureae folium (trade samples)
(cardenolide extracts, 20–40 µl)

Reference compound

T1 lanatoside C (Tc)	T4 gitoxin
T2 digitoxin	Ta lanatoside A
T3 digoxin	Tb lanatoside B

Solvent system Figs. 3,4 ethyl acetate-methanol-water (81:11:8)

Detection Fig. 3 A,B Kedde reagent (No. 23) → vis
Fig. 4 C,D $SbCl_3$ reagent (No. 4) → C vis D UV-365 nm

Fig. 3A **Digitalis lanatae folium** (1) and **D. purpureae folium** (2) both show their major zones in the lower R_f range 0.2–0.4 with seven violet-blue cardenolide zones in 1 and five in sample 2.
Digitalis lanatae folium (1) is characterized by the lanatosides A–C at R_f 0.3–0.4 with lanatoside A as the principal cardenolide, followed by smaller quantities of lanatoside B and C (T1) directly below. The cardenolide zone in the R_f range 0.2 can be prominent (sample 1) or of low concentration (sample 1a).
Digitalis purpureae folium (2) is characterized by the major zone of purpurea glycoside A with a slightly lower R_f value than lanatoside C (T1). Purpurea glycoside B is found as a minor zone directly below purpurea glycoside A, followed by a cardenolide zone at R_f ~ 0.2.
Samples 1 and 2 contain digitoxin (T2) and either traces of gitoxin (T4) or digoxin (T3) in the R_f range 0.6–0.75.

B In Digitalis extracts, generally lanatoside A or purpurea glycoside A are found as major cardenolides, and the lanatoside B/C and purpurea glycoside B in considerably lower concentration. Additional cardenolide zones which are detectable in the R_f range 0.2–0.25 (e.g., glucogitaloxin) can be present in low concentration, as demonstrated with samples 1a and 2a. In this case, zones of more lipophilic cardenolides are seen in the upper R_f range 0.5–0.8 (1a,2a). This can be due to a fermentation process in the plant material during storage. The plant enzymes (digilanidase and purpidase) preferentially remove the terminal glucose residues.

Fig. 4C In visible light a similar TLC fingerprint (compared with Kedde detection) of corresponding grey to violet-grey cardenolide zones is given.

D In UV-365 nm, however, a spectrum of about 20 blue fluorescent zones from R_f 0.05–0.95 is seen, with a specific dark-blue fluorescence of lanatoside A and purpurea glycoside A. Green (vis) or red (UV-365 nm) zones at R_f ~ 0.85 are due to chlorophyll, and yellow zones at the solvent front are due to flavonoids or anthraquinones (e.g. digilutein), which to some extent overlay the cardenolide genins.

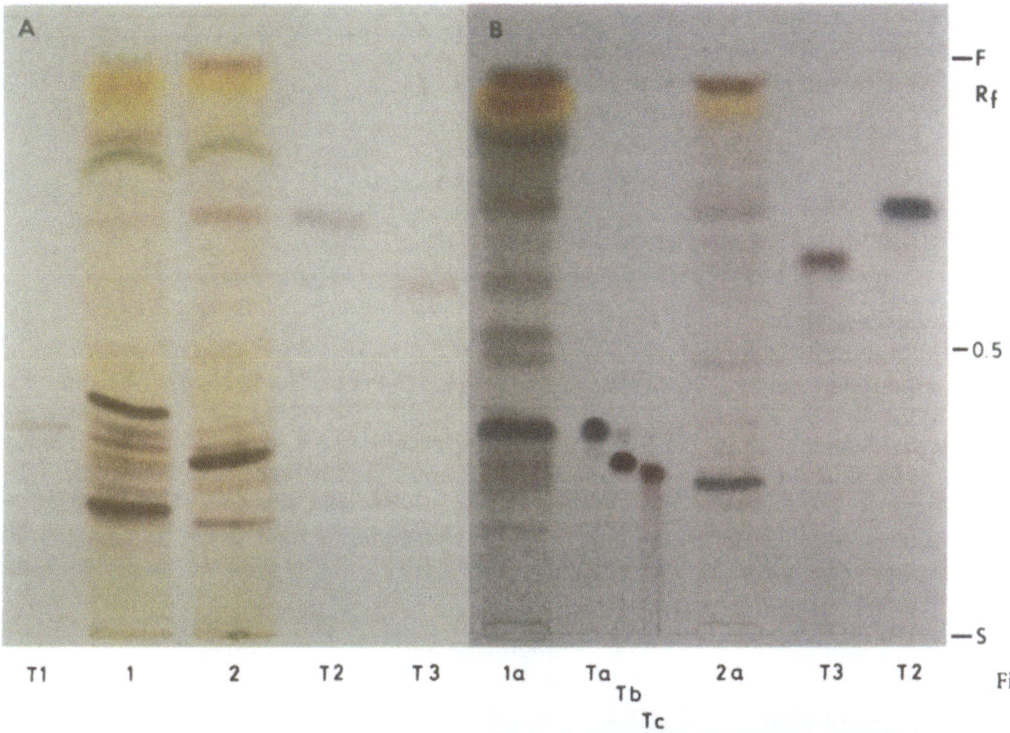
A
B
F
Rf
0.5
S
T1
1
2
T2
T3
1a
Ta
Tb
Tc
2a
T3
T2

Fig. 3

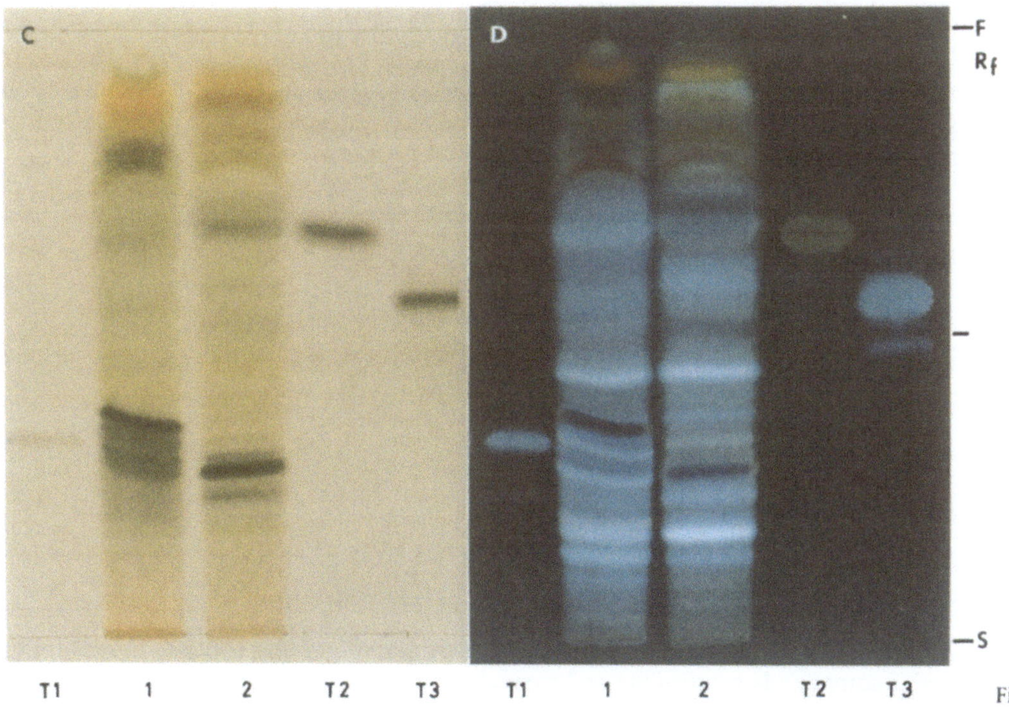
C
D
F
Rf
S
T1
1
2
T2
T3
T1
1
2
T2
T3

Fig. 4

Nerii (Oleandri) folium

Drug sample 1 Nerii folium (cardenolide extract, 20 µl)
2 Nerii folium (MeOH extract 1 g/5 ml, 10 µl)

Reference compound T1 oleandrin
T2 "oleander glycosides"
T3 adynerin
T4 rutin ($R_f \sim 0.4$) ▶ chlorogenic acid ($R_f \sim 0.5$) ▶ hyperoside ($R_f \sim 0.6$) → flavonoid test mixture

Solvent system Fig. 5A,B ethyl acetate-methanol-water (81:11:8)
C ethyl acetate-formic acid-glacial acetic acid-water (100:11:11:26)

Detection A Kedde reagent (No. 23) → vis.
B Chloramine-trichloroacetic acid reagent, (CTA No. 9) → UV-365 nm
C Natural products-polyethylene glycol reagent (NP/PEG No. 28) → UV-365 nm

Fig. 5A,B **Nerii (oleandri) folium** (1). The cardenolides reveal in the R_f range 0.1–0.9 a minimum of 13 Kedde positive red-violet zones (vis.) or up to 16 blue fluorescent zones after CTA reagent in UV-365 nm, with oleandrin (T1/$R_f \sim 0.85$) and adynerin (T3/$R_f \sim 0.75$) as major cardenolides in the upper R_f range. The cardenolide zones in the R_f range 0.1–0.8 are due to glycosides of oleandrigenine, digoxigenine and oleagenine, as shown in the following table:

Oleandrigenin:		Digitoxigenin:		Oleagenin:	
nerigoside	$R_f \sim 0.7$	odoroside A	$R_f \sim 0.8$	oleaside A	$R_f \sim 0.75$
glucosyloleandrin	$R_f \sim 0.4$	odoroside H	$R_f \sim 0.55$	oleaside E	$R_f \sim 0.1$
glucosylnerigoside	$R_f \sim 0.35$				
gentiobiosyloleandrin	$R_f \sim 0.15$				

C A characteristically high amount of rutin, chlorogenic acid and traces of other flavonoid glycosides ($R_f \sim 0.1/0.65$) represent the flavonoid phenolcarboxylic acid pattern of methanolic extract (2).

Uzarae (Xysmalobii) radix

Drug sample 1 Xysmalobii radix
2 Uzara extract (commercially available) (ethanolic extracts, 30 µl)

Reference compound T1 uzarin
T2 uzarigenin glucoside
T3 uzarigenin
T3 uzarigenin
T4 lanatoside B (A,C)
T5 xysmalorin

Solvent system Fig. 6 ethyl acetate-methanol-water (81:11:8)

Detection A,B Chloramine trichloroacetic acid reagent (CTA No. 9) → UV-365 nm
C $SbCl_3$ reagent (No. 4) → vis.

Fig. 6A,B **Xysmalobii radix** (1) and the pharmaceutical preparation 2 show with CTA reagent the major compounds uzarin (T1) and xysmalorin (T5) in one prominent blue to yellow-blue fluorescent zone at R_f 0.1–0.15, followed by seven blue, lower-concentrated zones with uzarigenin monoglucoside (T2) at $R_f \sim 0.35$ and uzarigenin (T3) at $R_f \sim 0.8$.

C Treatment with $SbCl_3$ reagent reveals mainly the blue-violet (vis) zone of uzarin and xysmalorin, with traces of the corresponding monoglucoside in the R_f range of the lanatoside B test (T4).

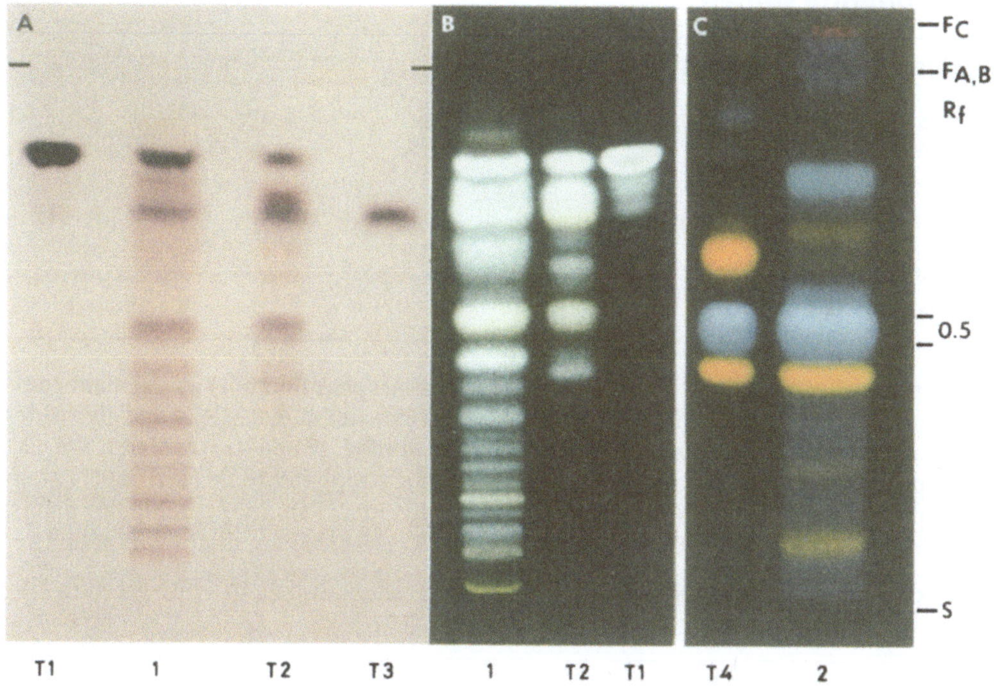
A
B
C
FC
FA,B
Rf
0.5
S
T1
1
T2
T3
1
T2
T1
T4
2

Fig. 5

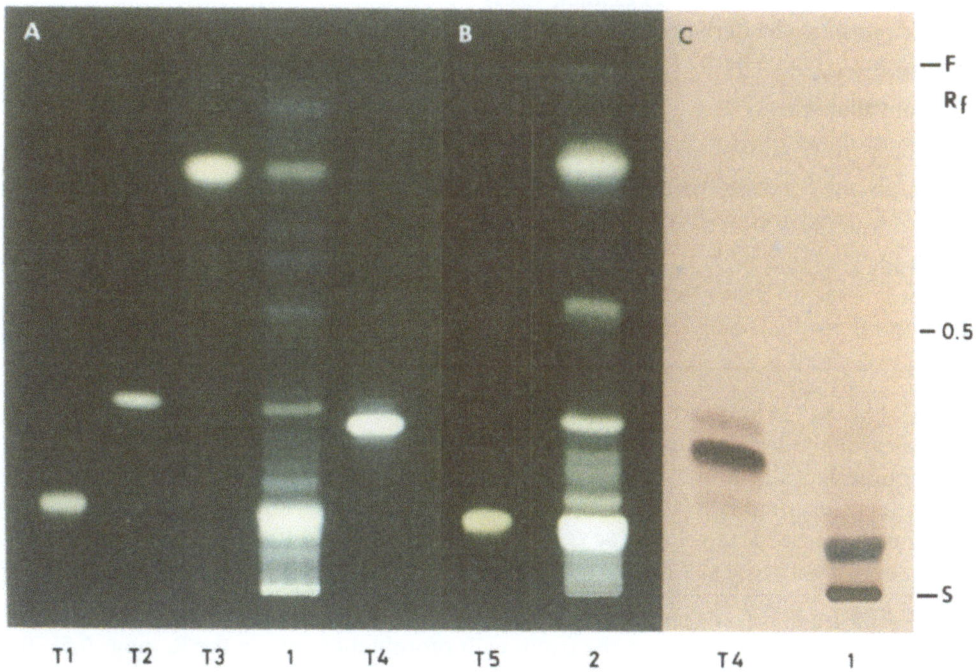
A
B
C
F
Rf
0.5
S
T1
T2
T3
1
T4
T5
2
T4
1

Fig. 6

Strophanthi semen

Drug sample 1 Strophanthi grati semen 2 Strophanthi kombé semen (ethanolic extracts, 20 µl)

Reference compound
T1 g-strophanthin
T2 k-strophanthin
T3 k-strophanthin-β
T4 erysimoside
T5 helveticoside
T6 cymarin
T7 k-strophantoside

Solvent system Fig. 7 ethyl acetate-methanol-water (81:11:8)

Detection
A Kedde reagent (No. 23) → vis
B $SbCl_3$ reagent (No. 4) → UV-365 nm

Fig. 7A **Strophanthi grati semen** (1) is characterized by g-strophanthin (T1) as the main compound at $R_f \sim 0.1$, with smaller amounts of sarmentosides at R_f 0.25–0.4 and glycosides above and below g-strophanthin. **Strophanthi kombé semen** (2) contains the "k-strophantin-glycoside mixture", which consists of k-strophanthoside ($R_f \sim 0.05$/T7), k-strophanthin β ($R_f \sim 0.25$/T3) and erysimoside ($R_f \sim 0.2$/T4). They form major bands, while helveticoside ($R_f \sim 0.55$/T5) and cymarin ($R_f \sim 0.6$/T6) are minor compounds.

B All k-strophanthidin glycosides (T3–T7) fluoresce yellow-brown to white-green in UV.

Erysimi herba, Cheiranthi herba

Drug sample
1 Erysimi herba (cardenolide extract, 40 µl)
2 Cheiranthi cheiri herba (cardenolide extract, 40 µl)
3 Cheiranthi cheiri herba (methanol extract 1 g/10 ml, 10 µl)

Reference compound
T1 convallatoxin
T2 "k-strophanthin"
T3 rutin ($R_f \sim 0.4$) ▸ chlorogenic acid ($R_f \sim 0.5$) ▸ hyperoside ($R_f \sim 0.55$)

Solvent system
Fig. 8A,B ethyl acetate-methanol-water (81:11:8)
C ethyl acetate-formic acid-glacial acetic acid-water (100:11:11:26)

Detection
A Kedde reagent (No. 23) → vis
B $SbCl_3$ reagent (No.4) → UV-365 nm
C Natural products-polyethylene glycol reagent (NP/PEG No. 28) → UV-365 nm

---

Fig. 8A **Erysimi herba** (1) and **Cheiranthi herba** (2) both show k-strophanthidin glycosides as major cardenolides, seen as violet-red zones (vis) in the R_f range 0.15–0.5.
Erysimi herba (1) has two cardenolide glycoside zones above and two zones below test T1 with e.g. erysimoside ($R_f \sim 0.2$) and helveticoside ($R_f \sim 0.5$).
Cheiranthi herba (2) develops one zone above and three to four zones below the R_f range of the convallatoxin test T1, e.g. cheirotoxin, desglucocheirotoxin and cheiroside A.

B In UV-365 nm ($SbCl_3$ reagent) a band of white-blue and yellow-brown fluorescent zones appears in (2) from the start to the front, with prominent zones in the R_f range 0.55–0.6 (e.g. cymarin) and $R_f \sim 0.7$ up to the solvent front. The light-yellow zones of the lower R_f range are due to quercetin and kaempferol glycosides.

C The separation in solvent system C and detection with NP/PEG reagent reveals three yellow-orange fluorescent **flavonoids** at R_f 0.2–0.45, with quercetin-3-O-rhamnosyl-arabinoside and kaempferol-3-O-robinosyl-7-rhamnoside as major zones.

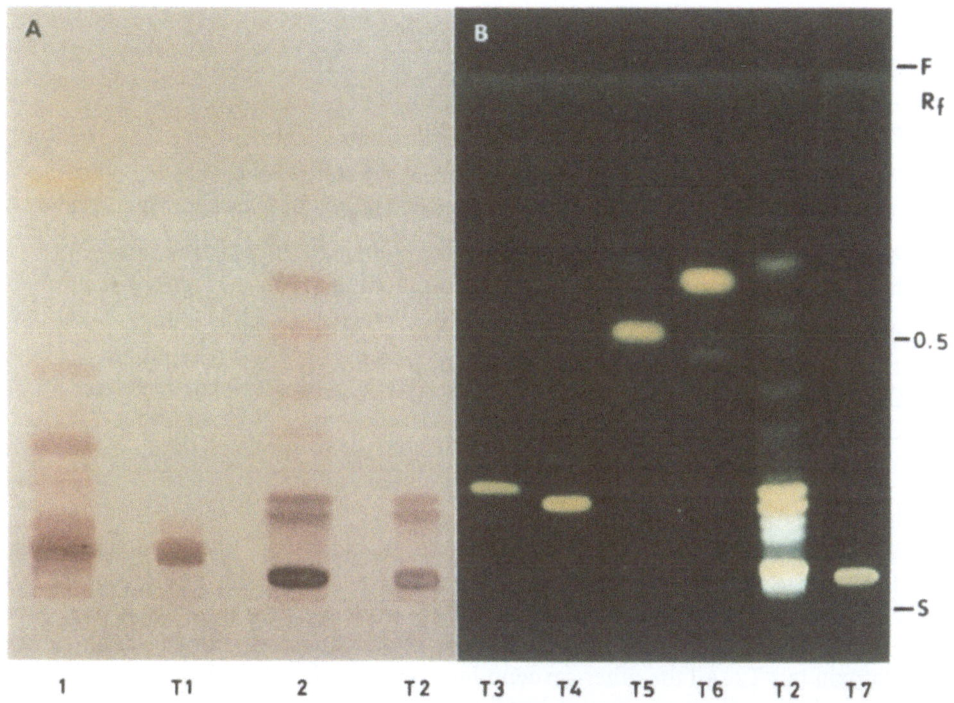

Fig. 7

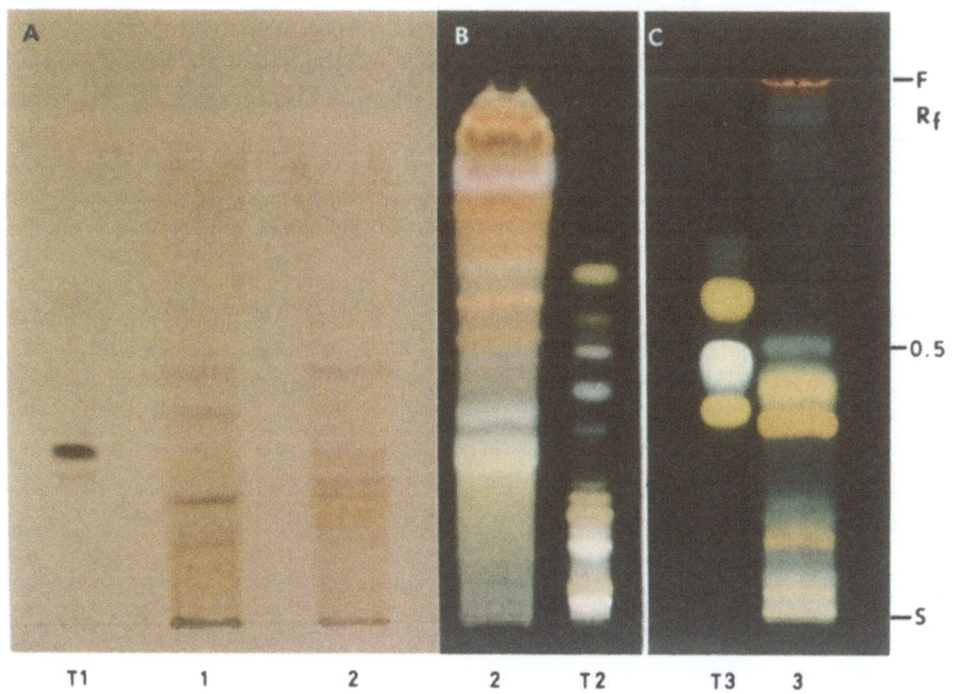

Fig. 8

Adonidis herba, Convallariae herba

Drug sample
1 Adonidis herba (cardenolide extract, 40 µl)
2 Convallariae herba (cardenolide extract, 50 µl)
1a Adonidis herba (MeOH extract 1 g/10 ml, 10 µl)
2a Convallariae herba (MeOH extract 1 g/10 ml, 10 µl)

Reference compound
T1 adonitoxin
T2 convallatoxin
T3 rutin ($R_f \sim 0.4$)► chlorogenic acid ($R_f \sim 0.5$) ► hyperoside ($R_f \sim 0.6$) ► isochlorogenic acid

Solvent system
Fig. 9A, Fig. 10C,D ethyl acetate-methanol-water (100:13.5:10) → cardenolides
Fig. 9B ethyl acetate-glacial acetic acid-formic acid-water (100:11:11:26) → flavonoids

Detection
A Kedde reagent (No. 23) → vis.
B Natural products-polyethylene glycol reagent (NP/PEG No.28) → UV-365 nm
C $SbCl_3$ reagent (No. 4) → vis
D $SbCl_3$ reagent (No. 4) → UV-365 nm

Fig. 9A (Kedde, vis) → Cardenolides

Adonidis herba (1). The major cardenolides such as desglucocheirotoxin, adonitoxigenin-rhamnoside (adonitoxin/T1), as well as its xyloside and glucoside, are found as violet-red zones in the R_f range 0.4–0.55, while cymarin migrates to $R_f \sim 0.6$. **Convallariae herba** (2) shows only two weak violet zones in the R_f range of the convallatoxin test T2. All the other cardenolide glycosides (~20) which are reported for Adonidis and Convallariae herba are hardly detectable because of their low concentrations (see detection C and D).

B (NP/PEG reagent UV-365 nm) → Flavonoids

The methanolic extract of **Adonidis herba** (1a) has a characteristically high amount of the C-glycosylflavone adonivernith ($R_f \sim 0.4$), which is accompanied by two minor flavonoid glycosides at $R_f \sim 0.35$ and 0.45.

The flavonoid content of the methanolic extract of **Convallariae herba** (2a) is low. Undefined yellow-green flavonoid glycoside zones in the R_f range 0.25–0.45, blue fluorescent phenol carboxylic acids and red chlorophyll zones (front) are detectable.

Fig. 10 ($SbCl_3$ reagent vis/UV-365 nm) → Cardenolides

C,D Besides cardenolides zones (R_f 0.4–0.6), **Adonidis herba** (1) generates various other compounds reacting to $SbCl_3$ with intense dark-blue colours (vis.) and as dark, almost black zones in UV-365 nm. The cardenolides are seen as grey-blue or brown-green zones (vis) and as intense light-blue fluorescent zones (UV-365 nm), mainly in the R_f range 0.45–0.75.

Convallariae herba (2) shows weak grey zones in the R_f range 0.25–0.95 (vis), whereas in UV-365 nm a band of green to brown zones appears in the R_f range 0.25–0.95.

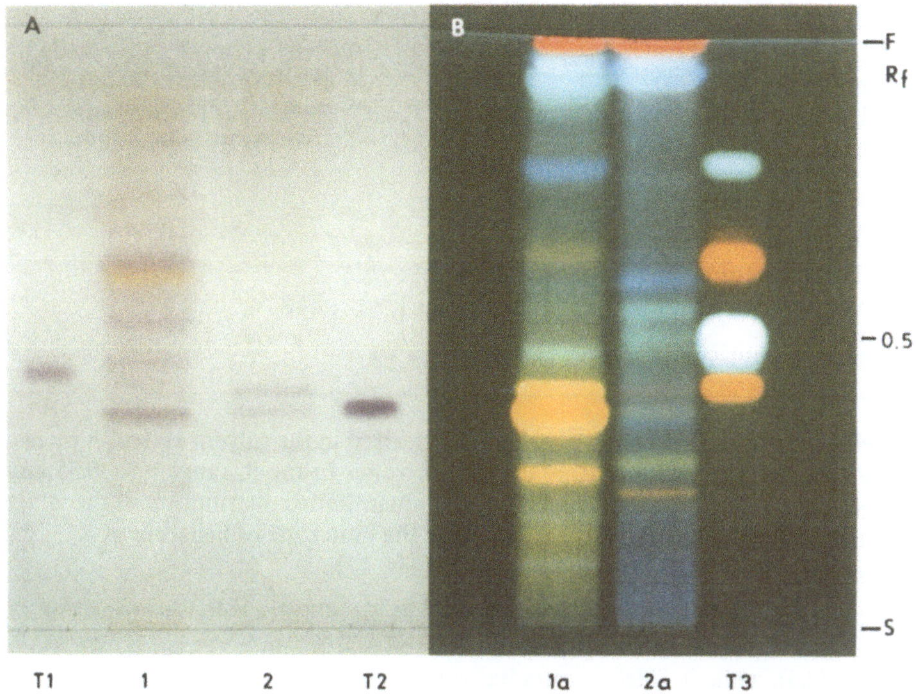

Fig. 9

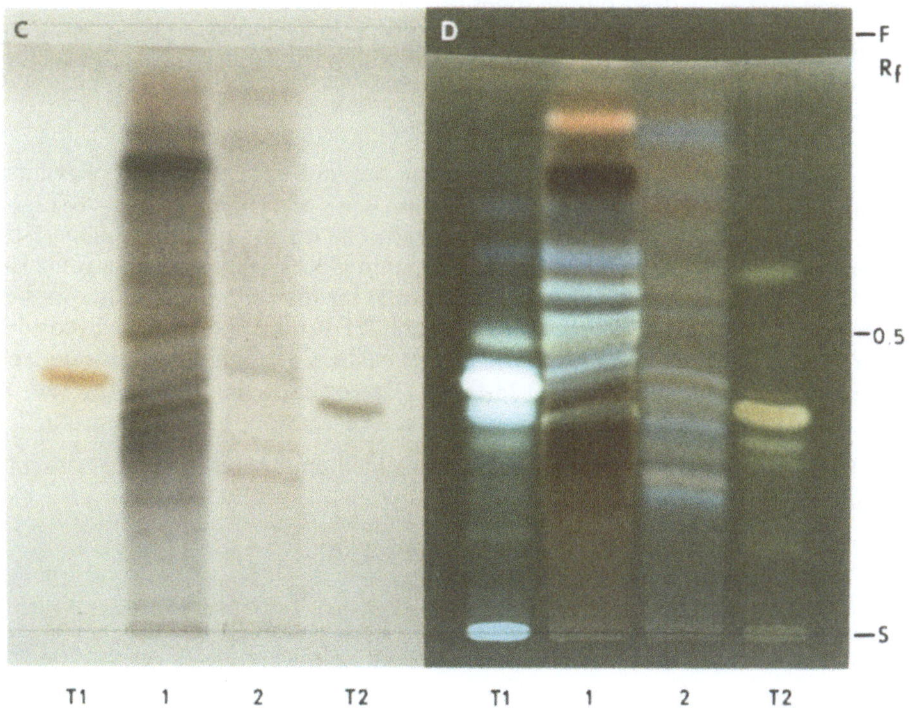

Fig. 10

Helleborus species

Radix sample 1 Helleborus purpurascens
2 Helleborus dumetorum
3 Helleborus atrorubens
4,5 Helleborus odorus (different origin)
6 Hellebori nigri radix-trade sample (≙ H.istriacus)
7 Helleborus macranthus
(1–7 ELOH-extracts, 30 µl)

Reference compound T1 hellebrin
T2 helleborogenone

Solvent system Fig. 11 ethyl acetate-methanol-water (81:11:8) system A
Fig. 12 chloroform-methanol-water (35:25:10) → lower phase, system B

Detection Anisaldehyde sulphuric acid reagent (AS No. 3) → vis

---

Fig. 11 The **Helleborus** species (1–7) are generally characterized in the solvent system A by two dark-blue zones at R_f ~ 0.2 and R_f ~ 0.4, yellow zones in the R_f range 0.05–0.35 and violet-blue zones at R_f ~ 0.9. The qualitative and quantitative distribution of the zones varies. The detection with the AS reagent reveals the blue zone of hellebrin at R_f ~ 0.2 (T1) and an additional blue zone at R_f ~ 0.4 (see Fig. 12).

High concentration of hellebrin could be observed in the samples of **H. purpurascens, H. dumentorum, H. atrorubens** and **H. macranthus (1,2,3,7)** and in lower concentration in **H. odoratus (4,5)**, while in "Hellebori radix" trade sample **6** hellebrin is absent.

The bufadienolide aglycone helleborogenone is found at R_f ~ 0.8 (e.g. sample 7) as a grey zone, followed by mainly deep-purple zones (e.g. hellebrigenin).

The prominent yellow zones at R_f ~ 0.3 (e.g. samples 2–5) and weaker yellow bands in the R_f range 0.05–0.25 are due to saponin glycosides derived e.g. from the sapogenine spirostan-5,25 (27) dien-1β,3β,11α-triol.

Fig. 12 The TLC run of Helleborus extracts 1–7 in solvent system B results in higher R_f values for hellebrin (R_f ~ 0.4/T1) and separates the blue zone of system A (R_f ~ 0.4) into two blue zones with R_f 0.5–0.55 due to desglucohellebrin and β-ecdysone/5-α-hydroxyecdysone. Additional yellow zones of saponin glycosides in the R_f range 0.05–0.25 are detectable. A higher R_f value (R_f ~ 0.45–0.5) for the yellow saponine zones of system A (R_f ~ 0.25–0.3) is achieved. Spirostan-5,25(27) dien-1β,3β,11α-triol glycosides are not present in H. purpurascens (1), instead dracoside A, a saponin glycoside derived from spirostan-5,25(27) dien-1β,3β,23,24-tetrol (Dracogenin), is seen in the low R_f range.

Remark: Hellebrin, but none of the saponins, has a medium fluorescence quenching in UV-254 nm.

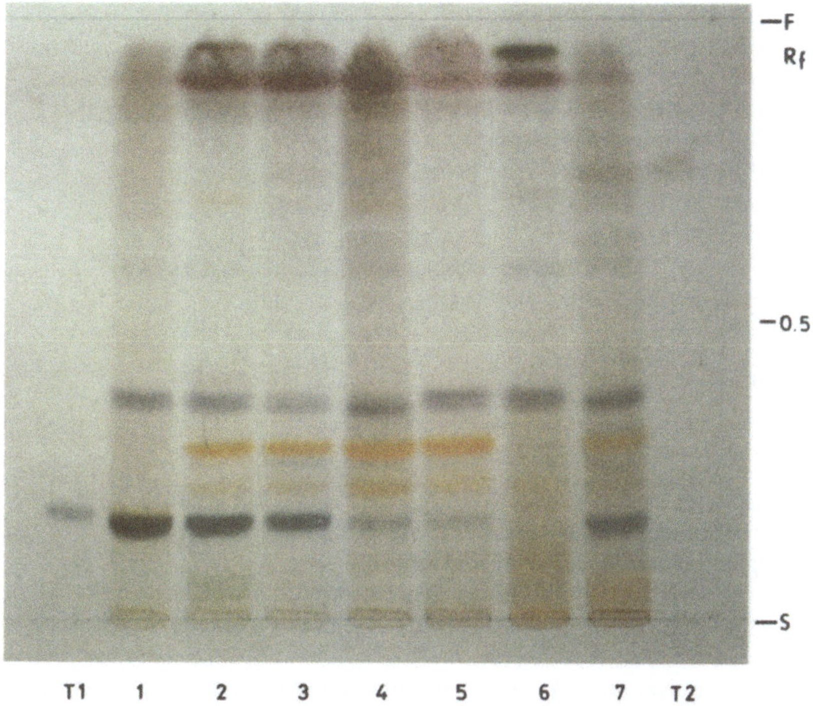

Fig. 11

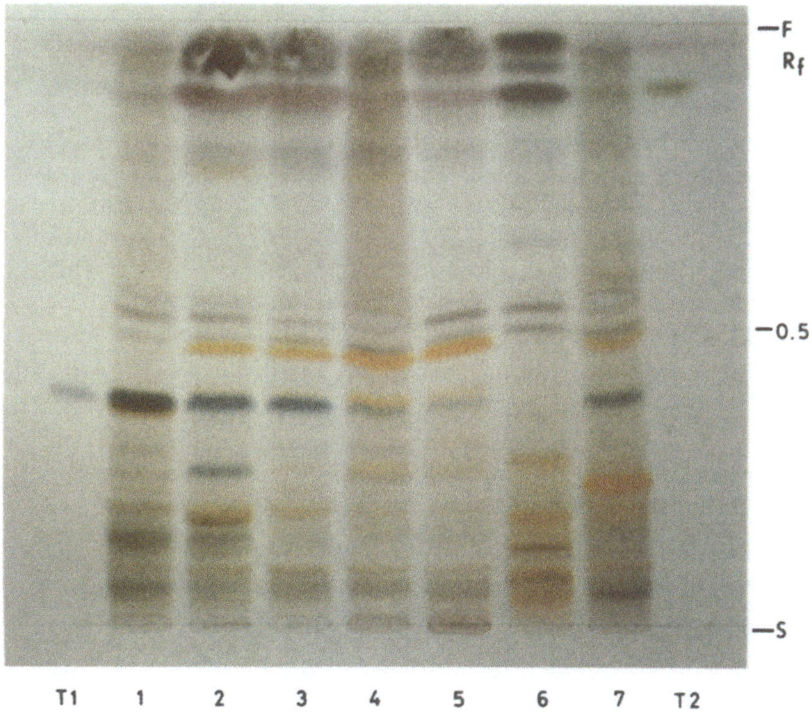

Fig. 12

Scillae bulbus

Drug samples
1 Scillae bulbus (red variety; trade sample)
2 Scillae bulbus (red variety; commercial extract)
3 Scillae bulbus (white variety; trade sample)
4 Scillae bulbus (white variety; trade sample)
5 Scillae bulbus (white variety; commercial extract)
(Bufadienolid extracts, 30 µl)

Reference T proscillaridin

Solvent system Fig. 13,14 ethyl acetate-methanol-water (81:11:8)

Detection
Fig. 13 $SbCl_3$ reagent (No. 3) → vis
Fig. 14 $SbCl_3$ reagent (No. 3) → UV 365 nm

Fig. 13 vis

Scilla extracts are characterized by predominantly blue bufadienolide zones. The contribution and amount of bufadienolides vary according to the classification of white or red squill variety of Urginea maritima (Drimia maritima).
The extracts of the red variety only show weak blue bands with proscillaridin (R_f ~ 0.6/T) and scillaren A (R_f ~ 0.4). The zone of scilliroside is seen as a weak green-yellow zone directly below the blue scillaren A zone (1,2).
In the white squill extract (sample 3,4) the highly concentrated zone of proscillaridin is found at R_f ~ 0.6, whereas in extract 5 the scillaren A zone at R_f ~ 0.4 predominates.

Fig. 14 UV-365 nm

All Scilla extracts show a variety of intense, light-yellow, yellow-brown, green or light-blue to almost white fluorescent zones.
The extracts 1,3 and 4 have major compounds in the upper R_f range, while the extracts 3 and 5 show those in the R_f range 0.2–0.6.

Scillae bulbus var. rubra (1,2). In both extracts ten to 12 intense yellow-green or blue fluorescent zones are found in varying concentrations in the R_f range 0.35–0.95.
Besides proscillaridin (R_f ~ 0.6/a) and scillaren A (R_f ~ 0.45/b), the bufadienolid aglycone scillirosidin (R_f ~ 0.8), its monoglycoside scilliroside (R_f ~ 0.4/c) and the diglycoside glucoscilliroside (R_f ~ 0.2) are characteristic compounds for red squill. Scillirosidin is more highly concentrated in 1; glucoscilliroside is present in 2 only.

Scillae bulbus var. alba (3–5). The white squill samples (3,4) contain predominantly proscillaridin, seen as a major light-brown fluorescent zone at R_f ~ 0.6 (T/a). The glycoside scillaren A (R_f ~ 0.4/b) dominates the standardized commercial extract 5. Three additional cardenolide zones (e.g. glucoscillaren R_f ~ 0.2) are detectable as yellow-brown fluorescent zones. Scillirosidin glycosides are absent in white squill.

Remark: The bufadienolid glycosides found in the lower R_f range are easily cleaved during storage into bufadienolides with fewer sugar moities (e.g. proscillaridin).

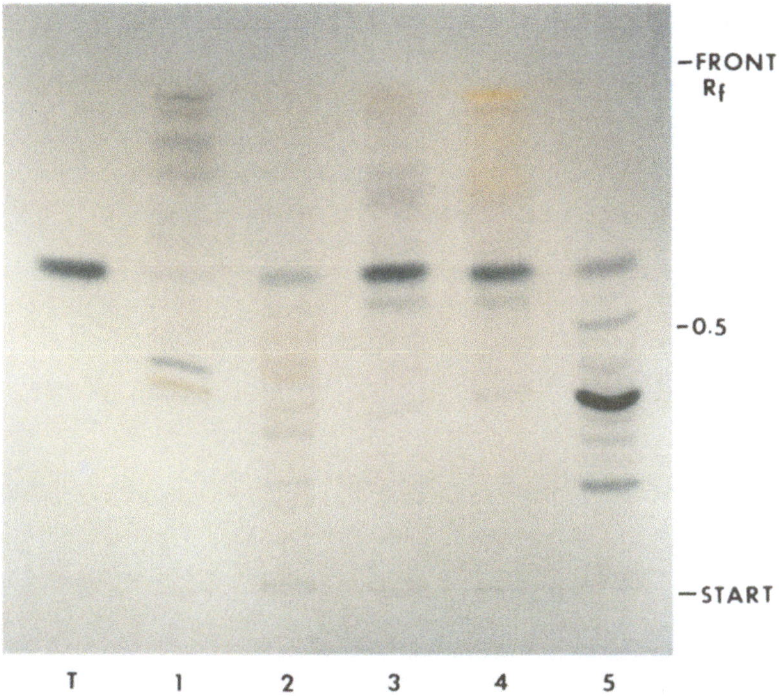

Fig. 13

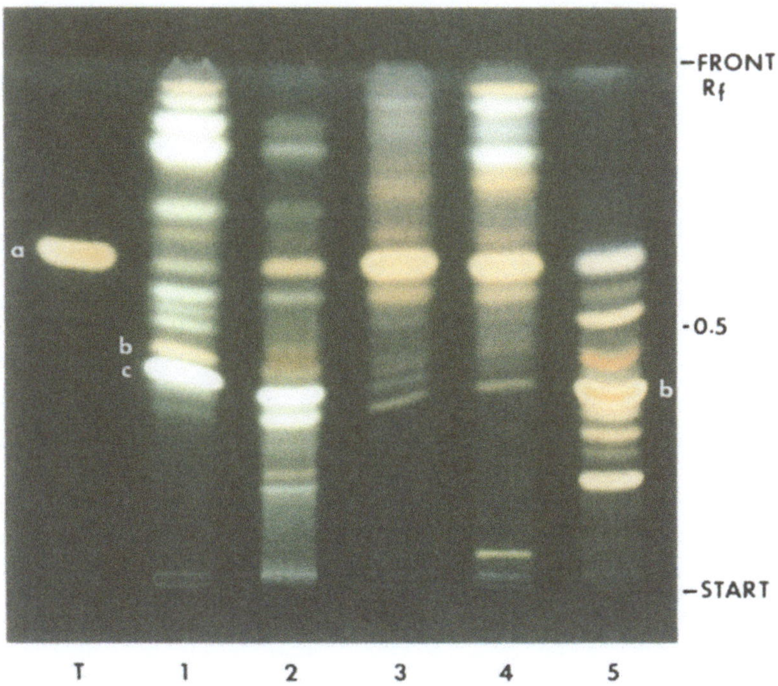

Fig. 14

5 Coumarin Drugs

The active principles of there drugs are benzo-α-pyrones, which are further classified as:

- Non-condensed coumarins
  substituted with OH or OCH_3 at positions C-6 and C-7, less common at C-5 and C-8.
  e.g. umbelliferon (7-hydroxy-coumarin in Angelicae radix, Heraclei radix), scopoletin (6-methoxy-7-hydroxy-coumarin in Scopoliae radix), fraxin, isofraxidin and fraxetin (Fraxini cortex) and herniarin (Herniariae herba)

- C-prenylated coumarins
  e.g. rutamarin (Rutae herba), umbelliprenin (Angelicae radix), ostruthin (Imperatoriae radix)

- Furanocoumarins
  an additional furan ring is fused at C-6 and C-7 (psoralen-type) or C-7 and C-8 (angelicin-type).
  e.g. imperatorin, bergapten, angelicin (Angelicae, Imperatoriae, Pimpinellae radix) xanthotoxin (Ammi majoris fructus), psoralen (Rutae herba)

- Pyranocoumarins
  an additional pyran ring is fused at C-7 and C-8 (seselin-type)
  e.g. visnadin, samidin (Ammi majoris fructus)

- Dimeric coumarins
  e.g. daphnoretin (Daphne mezerei cortex)

5.1 Preparation of Extracts

General method methanolic extract

Powdered drug (1 g) is extracted with 10 ml methanol for 30 min under reflux on the water bath. The filtrate is evaporated to about 1 ml, and 20 μl is used for TLC investigation.

5.2 Thin-Layer Chromatography

Reference solutions Coumarins are prepared as 0.1% methanolic solutions; 5–10 µl is used for TLC.

Adsorbent Silica gel 60 F_{254}-precoated TLC plates (Merck, Germany)

Chromatography solvent

- Toluene-ether (1:1, saturated with 10% acetic acid) ▶ coumarin aglycones

Toluene (50 ml) and ether (50 ml) are shaken for 5 min with 50 ml 10% acetic acid in a separating funnel. The lower phase is discarded, and the saturated toluene-ether mixture is used for TLC.

- Ethyl acetate-formic acid-glacial acetic acid-water (100:11:11:26) ▶ for glycosides

5.3 Detection

- UV-254 nm distinct fluorescence quenching of all coumarins.
- UV-365 nm intense blue or blue-green fluorescence (simple coumarins) yellow, brown, blue or blue-green fluorescence (furano- and pyranocoumarins).

The non-substituted coumarin fluoresces yellow-green in UV-365 nm only after treatment with KOH- reagent or ammonia vapour.
Chromones show less intense fluorescence, e.g. visnagin (pale blue), khellin (yellow-brown).

- Spray reagents (see Appendix A)
  - Potassium hydroxide (KOH No. 35)
    The fluorescence of the coumarins are intensified by spraying with 5%–10% ethanolic KOH. Concentrated ammonia vapour has the same effect.
  - Natural poducts-polyethylene glycol reagent (NP/PEG No. 28)
    This reagent intensifies and stabilizes the existing fluorescence of the coumarins. Phenol carboxylic acids fluoresce blue or blue-green (e.g. chlorogenic or caffeic acid)

5.4 Drug List

	Drug/plant source Family/pharmacopoeia	Main constitutents (see 5.5 Formulae)
	Drugs with simple coumarins	
Fig. 1,2	**Asperulae herba** Woodruff Galium odoratum (L.) SCOP. Rubiaceae	Unsubstituted coumarin (0.1%–0.3%) umbelliferone, scopoletin Flavonoid glycosides e.g. rutin; chlorogenic and caffeic acid.

Drug/plant source Family/pharmacopoeia	Main constitutents (see 5.5 Formulae)	
Meliloti herba Tall melilot Melilotus officinalis (L.) M. altissima THUIL. Fabaceae DAC 86	Unsubstituted coumarin (0.2%–0.45%), melilotoside, umbelliferone, scopoletin ► Flavonoids: quercetin, kaempferol biosides and triosides ► Caffeic acid and derivatives	
Toncae semen Tonca beans Dipteryx odorata WILLD. Fabaceae	Unsubstituted coumarin (2%–3%) umbelliferone	Fig. 1,2
Abrotani herba Southernwood Artemisia abrotanum L. Asteraceae BHP 83	Coumarins: umbelliferone, isofraxidin, scopoletin and -7-O-glucoside ► Flavonoids: quercetin glycosides, e.g. rutin ► Chlorogenic and isochlorogenic acids	Fig. 4
Fabianae herba Pichi-Pichi Fabiana imbricata RUIZ. et PAV. Solanaceae	Coumarins: scopoletin and its -7-O-primveroside (= fabiatrin), isofraxidin and its 7-O-glucoside ► Flavonoids: rutin, quercetin-3-O-glucoside ► Chlorogenic and isochlorogenic acids	Fig. 4
Fraxini cortex Ash bark Fraxinus excelsior L. Fraxinus ornus L. Oleaceae MD, China	Coumarins: fraxidin (~0.06%), isofraxidin (~0.01%), fraxetin, fraxin (fraxetin-glucoside), fraxinol (~0.05%)	Fig. 5,6
Mezerei cortex Mezereon bark Daphne mezereum L. Thymelaeaceae	Coumarins: daphnetin, daphnin (7,8-dihydroxy-coumarin-7-O-glucoside), umbelliferone and derivatives (triumbellin), scopoletin (traces)	Fig. 5,6

Drugs from the family Solanaceae
All drugs contain the same coumarins and alkaloids, but differ in concentrations

Scopoliae radix Scopolia rhizome Scopolia carniolica L.	Coumarins: scopoletin, and -7-O-glucoside Alkaloids: hyoscyamine (see Fig. 27,28; Chap. 1 Alkaloid Drugs)	Fig. 7
Belladonnae radix Belladonna root Atropa belladonna L. DAB 10, MD	Coumarins: scopoletin, and -7-O-glucoside Alkaloids: hyoscyamine, scopolamine (see Fig. 27,28, Chap. 1 Alkaloid Drugs)	

Drug/plant source Family/pharmacopoeia		Main constitutents (see 5.5 Formulae)
	Mandragorae radix Mandrake Mandragora officinarum L. M. autumnalis BERTOL.	Coumarins: scopoletin, and 7-O-glucoside Alkaloids: hyoscyamine, scopolamine (see Fig. 27,28, Chap. 1 Alkaloid Drugs)

Drugs from Apiaceae

Fig. 8	**Ammi majoris fructus** Ammi fruit Ammi majus L. Apiaceae, MD	Furanocoumarins (fc): bergapten (5-methoxy 2′,3′:7,6-fc), imperatorin xanthotoxin (8-methoxy-2′,3′:7,6-fc)
	Ammi visnagae fructus Ammeos visnagae fructus Ammi visnaga fruits Ammi visnaga (L.) LAM. Apiaceae DAB 10, MD	Coumarins of the visnagan group samidin, dihydrosamidin and visnadin Furanochromones (2%–4%): khellin (0.3%–1%), visnagin, khellinol, khellol, khellol glucoside, ammiol
Fig. 9–12	**Angelicae radix** Angelica root Angelica archangelica L. ssp. Angelica var. sativa RIKLI DAC 86, ÖAB 90, MD China/Japan: different Angelica species Apiaceae	Furanocoumarins (fc): angelicin (2′,3′:7,8-fc), bergapten (5-methoxy 2′,3′:7,6-fc), imperatorin, oxypeucedanin hydrate, xanthotoxin (8-methoxy-2′,3′:7,6-fc) xanthotoxol (8-oxy-2′,3′:7,6-fc) Coumarins: umbelliferone, umbelliprenin, osthenol (7-oxy-8-(8,8-di-methylallyl)- coumarin)
	Angelicae silvestris radix Wild angelica Angelica silvestris L. Apiaceae	Counts as adulterant of Angelica archangelica: isoimperatorin (5-oxy-(γ-γ-di-methyl-allyl- 2′,3′:7,6-fc)), oxypeucedanin-hydrate, umbelliferone
Fig. 9–12	**Imperatoriae radix** Masterwort Peucedanum ostruthium L. Apiaceae	Furanocoumarins: oxypeucedanin and its hydrate, imperatorin, isoimperatorin; ostruthol (angelic acid ester of oxypeucedanin hydrate); ostruthin (6-(3-methyl-6-dimethyl- 2,5-hexene)-7-oxycoumarin)
Fig. 9–12	**Levistici radix** Lovage Levisticum officinale KOCH Apiaceae DAC 86, ÖAB, Helv. VII	Coumarins: bergapten, umbelliferone generally lower coumarin content than Angelicae and Imperatoriae radix Phtalide: 3-butylidenephtalide (ligusticum- lacton)

Drug/plant source Family/pharmacopoeia	Main constituents (see 5.5 Formulae)	
Pimpinellae radix Burnet root Pimpinella major (L.) HUDS Pimpinella saxifraga L. Apiaceae	Furanocoumarins (fc) 0.07%: sphondin, bergapten, isobergapten, pimpinellin, isopimpinellin Coumarins: umbelliferone, scopoletin Umbelliprenin, xanthotoxin (P. saxifraga) Angelicin, isooxypeucedanin (P. major)	**Fig. 11–12**
Pastinacae radix Adulterant Pastinaca sativa L. Apiaceae	Low furanocoumarin content: bergapten, imperatorin, isopimpinellin, xanthotoxin	
Heraclei radix Hogweed root Heracleum sphondylium L. Apiaceae	Furanocoumarins (1.0%): sphondin, isopimpinellin, pimpinellin, bergapten, isobergapten Umbelliferone, umbelliprenin, scopoletin	**Fig. 11,12**

Drugs with coumarins and other constituents as major compounds

Rutae herba Rue Ruta graveolens L. Rutaceae DAC'86, MD	Coumarins: scopoletin, umbelliferone, bergapten, isoimperatorin, psoralen, xanthotoxin, rutacultin, rutamarin, daphnoretin, daphnoretin methyl ether ► Flavonol glycoside rutin ► Alkaloids: γ-fagarine, kokusagenine (furanoquinoline-type)	**Fig. 13,14**
Herniariae herba Rupturewort Herniaria glabra L. Herniaria hirsuta L. Caryophyllaceae DAC 86, ÖAB, MD	Coumarins: herniarin, umbelliferone ► Flavonol glycosides: rutin, narcissin ► Saponins: Herniaria saponins I/II (aglycone medicagenic, 16-hydroxy-medicagenic acid)	**Fig. 15,16**

5.5 Formulae

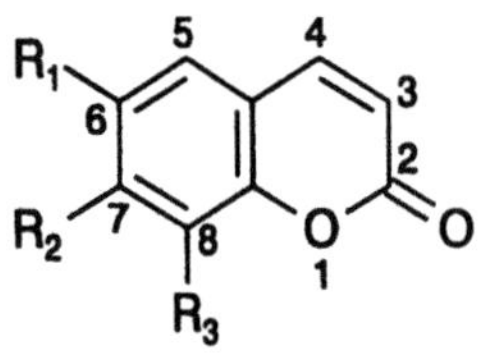

R1	R2	R3	
H	H	H	Coumarin
H	OH	H	Umbelliferone
H	OCH_3	H	Herniarin
H	OH	OH	Daphnetin
OH	OH	H	Aesculetin
OCH_3	OH	H	Scopoletin
OCH_3	OH	OH	Fraxetin
OCH_3	OH	OCH_3	Isofraxidin
OCH_3	OH	O-gluc	Fraxin
H	OH	$-CH_2-CH=C(CH_3)_2$	Osthol
H	OCH_3	$-CH_2-CH=C(CH_3)_2$	Osthenol

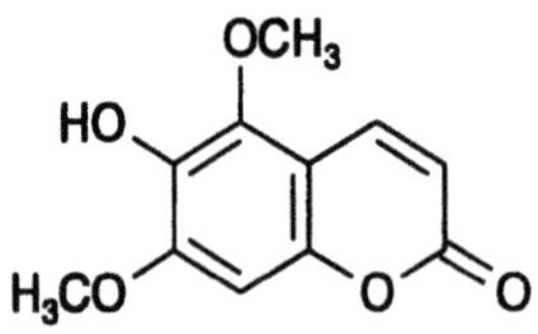

Fraxinol

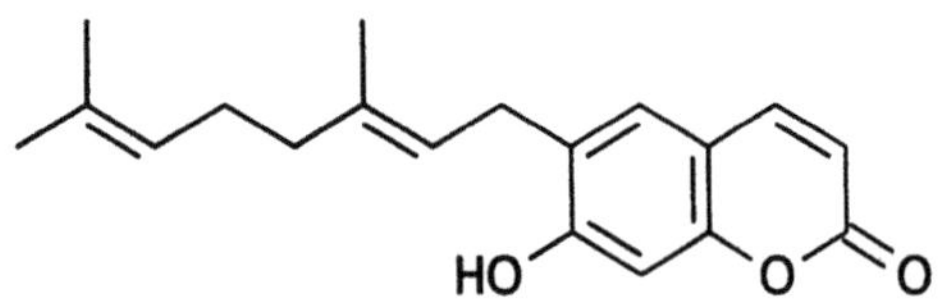

Ostruthin

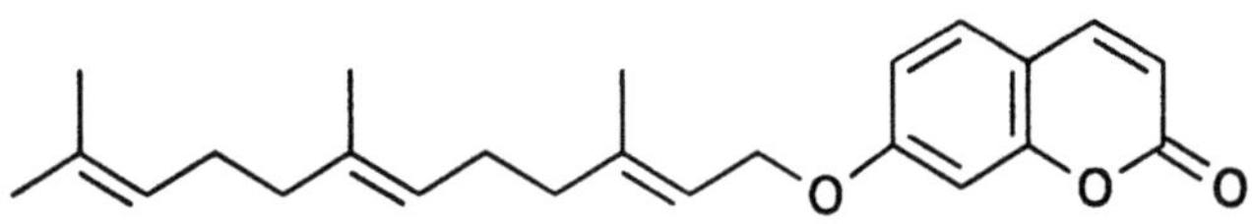

Umbelliprenin

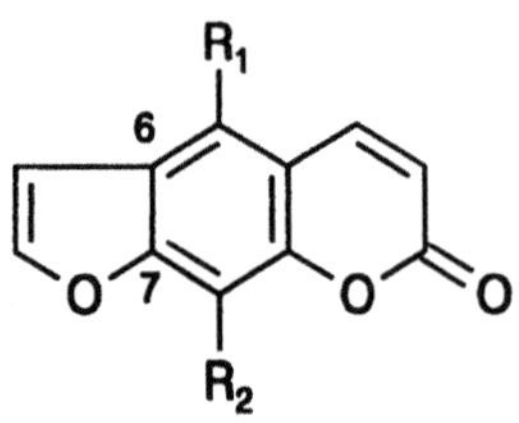

7,6 Furanocoumarins R1	R2	
H	H	Psoralen
H	OCH_3	Xanthotoxin
H	OH	Xanthotoxol
OCH_3	H	Bergapten
OH	H	Bergaptol
OCH_3	OCH_3	Isopimpinellin
H	$-OCH_2-CH=C(CH_3)_2$	Imperatorin
$-OCH_2-CH=C(CH_3)_2$	H	Isoimperatorin
$-OCH_2-CH\overset{O}{-}C(CH_3)_2$ (epoxide)	H	Oxypeucedanin
$-OCH_2-CH(OH)-C(OH)(CH_3)_2$	H	Oxypeucedanin hydrate

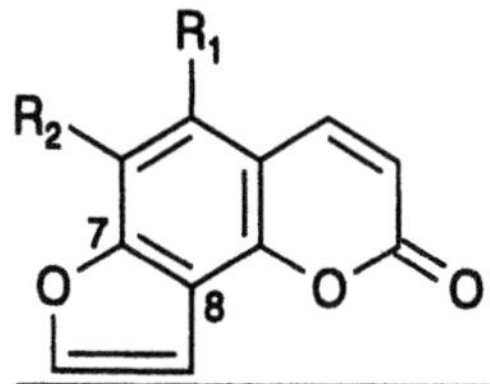

7,8-Furanocoumarins	R_1	R_2	
	H	H	Angelicin
	OCH_3	H	Isobergapten
	H	OCH_3	Sphondin
	OCH_3	OCH_3	Pimpinellin

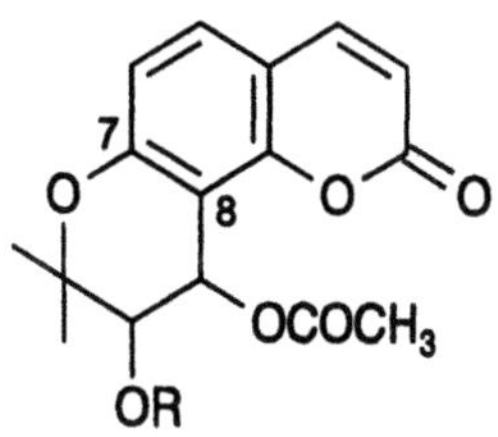

Pyranocoumarins	R	
	$-CO-CH=C(CH_3)_2$	Samidin
	$-CO-CH_2-CH(CH_3)_2$	Dihydrosamidin
	$-CO-CH(CH_3)-C_2H_5$	Visnadin

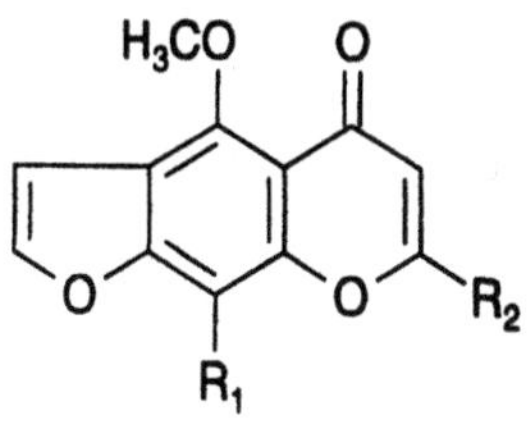

Furanochromones	R_1	R_2	
	H	CH_3	Visnagin
	OCH_3	CH_3	Khellin
	H	CH_2OH	Khellol

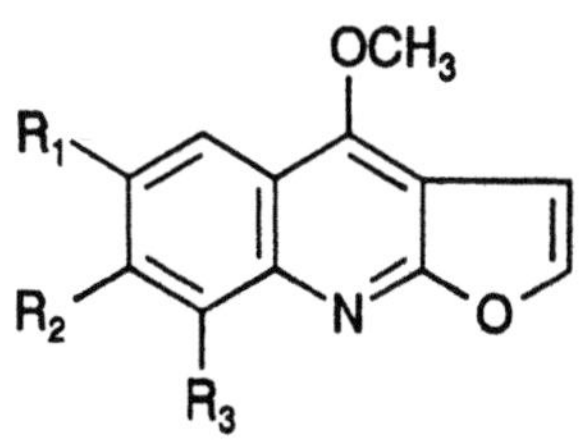

Furanoquinolines	R_1	R_2	R_3	
	H	H	CH_3	γ-Fagarine
	OCH_3	OCH_3	H	Kokusaginin

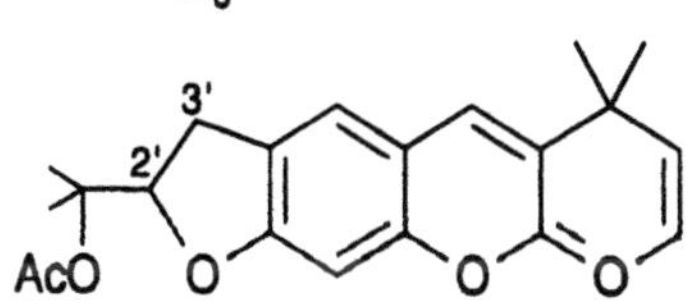

Rutamarin

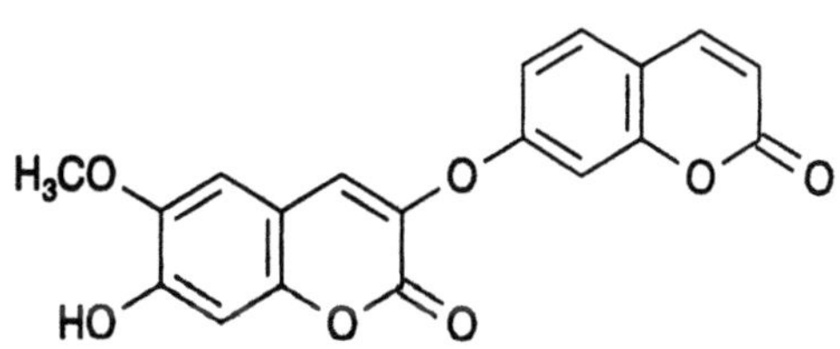

Daphnoretin

5.6 Chromatograms

Asperulae, Meliloti herba; Toncae semen

Drug sample
1 Meliloti herba (ethyl acetate extract/chlorophyll free, 20 µl)
1a Meliloti herba (methanolic extract 1 g/10 ml, 20 µl)
2 Asperulae herba (ethyl acetate extract/chlorophyll free, 20 µl)
2a Asperulae herba (methanolic extract 1 g/10 ml, 20 µl)
3 Toncae semen (methanolic extract 0.5 g/10 ml20 µl)

Reference compound
T1 scopoletin
T2 coumarin
T3 umbelliferone

Solvent system
Fig. 1 toluene-ether (1:1/saturated with 10% acetic acid) → aglycones
Fig. 2 ethyl acetate-glacial acetic acid-formic acid-water (100:11:11:26) → glycosides

Detection
Fig. 1 A UV-254 nm (without chemical treatment)
B 5% ethanolic KOH reagent (No. 35) → UV-365 nm
Fig. 2 A UV-254 nm (without chemical treatment)
B Natural products-polyethylene glycol reagent (NP/PEG No. 28) → UV-365 nm

---

Fig. 1A Methanolic extracts of **Meliloti herba** (1), **Asperulae herba** (2) and **Toncae semen** (3) contain the unsubstituted coumarin (T2), which is seen as a prominent quenching zone at $R_f \sim 0.6$ (UV-254 nm).

B In UV-365 nm, the unsubstituted coumarin, in contrast to coumarins with -OH, $-OCH_3$ substituents or furano- or pyrano-coumarins, shows a typical green-blue fluorescence only after treatment with KOH reagent. Scopoletin (T1) and umbelliferone (T3) are present in low concentrations only, seen as blue fluorescent zones at R_f 0.25 and 0.4, respectively (1–3).

Fig. 2A **Meliloti herba** (1,1a) and **Asperulae herba** (2,2a) show prominent quenching zones of flavonoid glycosides and caffeic acid derivatives in different patterns and amounts. The coumarins (T1–T3), flavonoid aglycones and caffeic acid migrate almost up to the solvent front. Flavonoid glycosides and chlorogenic acid are found from $R_f \sim 0.05$ to 0.5. Extracts of **Meliloti herba** (1,1a) show mainly one prominent zone at $R_f \sim 0.5$, while **Asperulae herba** (2,2a) has four almost equally concentrated zones in the R_f range 0.25–0.5.

B Treatment with the NP/PEG reagent reveals bright orange-red, yellow-green and blue fluorescent zones in UV-365 nm.
Meliloti herba (1,1a) has a characteristic TLC pattern of quercetin and kaempferol biosides and triosides in the lower R_f range, seen as three pairs of red-orange and yellow-green fluorescent zones at R_f 0.05–0.4, as well as a weak blue fluorescent zone at $R_f \sim 0.45$ and 0.8. In extracts of **Asperulae herba** (2,2a) blue fluorescent caffeic acid derivatives, e.g. chlorogenic acid ($R_f \sim 0.45$), isochlorogenic acids (R_f 0.7–0.8), caffeic acids (R_f 0.9) dominate. A prominent orange-green flavonoid trioside at $R_f \sim 0.05$ and rutin at $R_f \sim 0.35$ are detectable. Red-orange fluorescent zones (1a, 2a) at the solvent front are due to chlorophyll compounds.

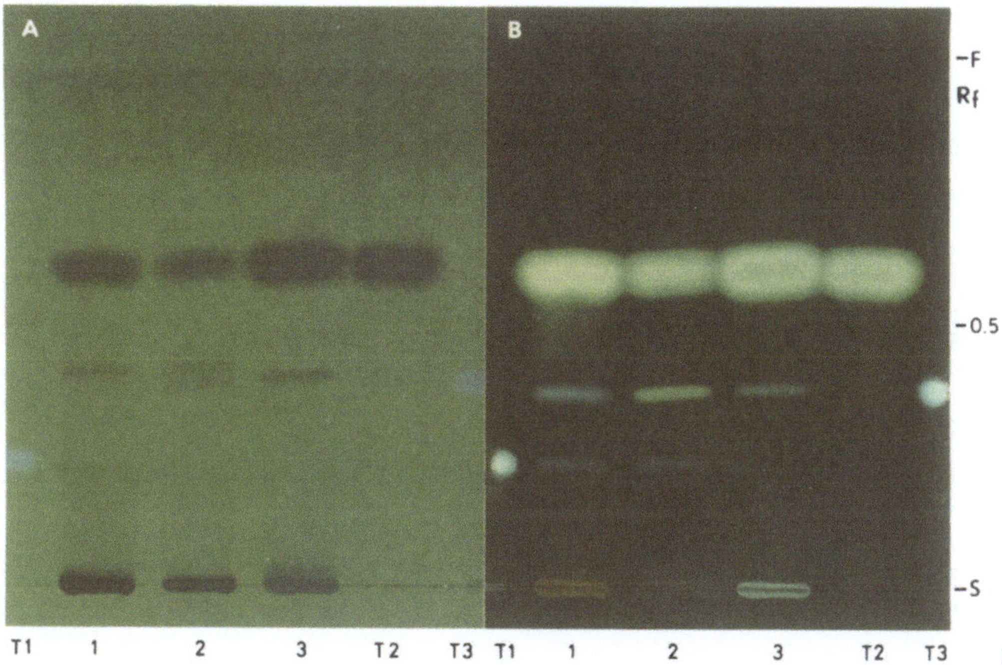
A
B
-F
Rf
-0.5
-S
T1 1 2 3 T2 T3 T1 1 2 3 T2 T3

Fig. 1

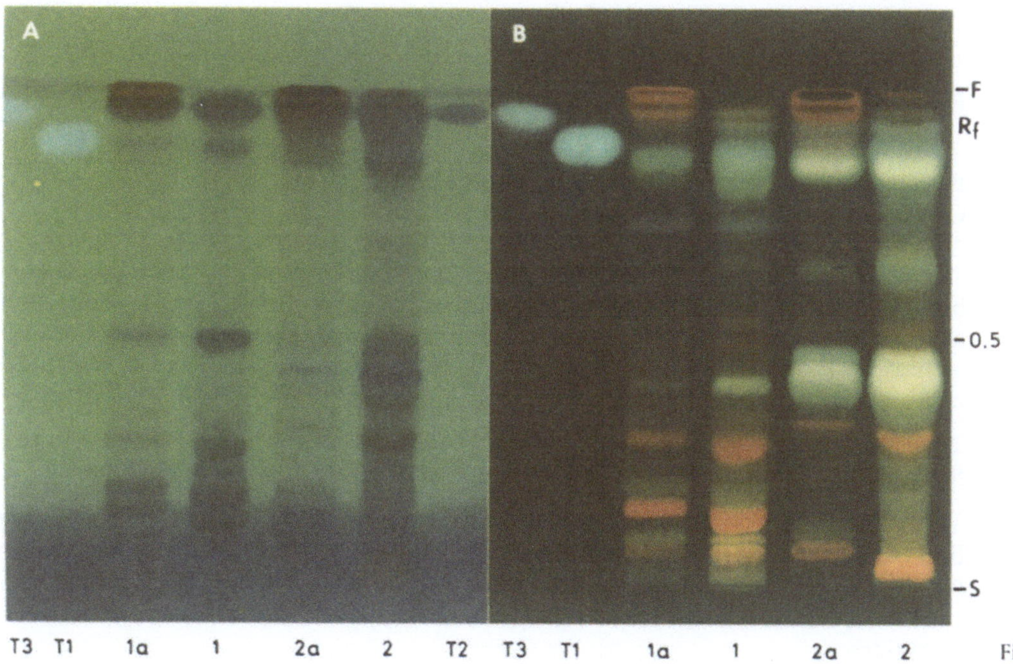
A
B
-F
Rf
-0.5
-S
T3 T1 1a 1 2a 2 T2 T3 T1 1a 1 2a 2

Fig. 2

Coumarins – Chromatographic Standards

Standard

1 daphnoretin	4 umbelliferone	7 imperatorin	10 isopimpinellin
2 scopoletin	5 herniarin	8 ferulic acid *	11 isobergapten
3 isofraxidin	6 xanthotoxin	9 caffeic acid*	12 oxypeucedanin

(methanolic solutions, 5–10 µl)

Solvent system Fig. 3 toluene-ether (1:1/saturated with 10% acetic acid)

Detection UV-365 nm (without chemical treatment)

Fig. 3 Characteristic fluorescence of coumarins in UV-365 nm:

bright blue: daphnoretin, scopoletin, isofraxidin, umbelliferone
blue-green: xanthotoxin, isobergapten, oxypeucedanin
yellow-green: isopimpinellin
violet-blue: herniarin

**Remark*: Coumarin drugs often contain phenol carboxylic acids, e.g. ferulic acid and caffeic acid, which also show blue fluorescence

Abrotani herba, Fabiani herba

Drug sample 1 Abrotani herba 2 Fabiani herba (= Pichi-Pichi)
(methanolic extracts, 20 µl)

Reference compound
T1 chlorogenic acid (R_f 0.45)
T2 rutin (R_f 0.4) ► chlorogenic acid ► hyperoside (R_f 0.55) ► isochlorogenic acid

Solvent system Fig. 4 ethyl acetate-glacial acetic acid-formic acid-water (100:11:11:26)

Detection
A UV-365 nm (without chemical treatment)
B Potassium hydroxide reagent (KOH No. 35) → UV-365 nm
C Natural products-polyethylene glycol reagent (NP/PEG No. 28) → UV-365 nm

Fig. 4A The methanolic extracts of **Abrotani herba** (1) and **Fabiani herba** (2) are characterized by the prominent blue fluorescent coumarin aglycone zone at $R_f \sim 0.95$ (scopoletin, isofraxidin and umbelliferone). They are differentiated by the violet-blue fluorescent isofraxidin-7-O and scopoletin-7-O-glucosides in the range of 0.4–0.45 and an additional zone at R_f 0.7 in sample 1 and the scopoletin-7-O-primveroside at R_f 0.15 in extract 2.

B The coumarin-7-O-glucosides of **Abrotani herba** (1) are seen with KOH reagent as two fluorescent zones at R_f 0.4–0.45. The aglycones at $R_f \sim 0.95$ become bright blue.

C Treatment with the NP-PEG reagent shows in **Abrotani herba** (1) a broad band of intense bluish-white fluorescent zones of coumarins and phenolcarboxylic acids in the R_f range 0.35 up to the solvent front, which overlay the orange fluorescent flavonoid glycosides (R_f 0.4 (rutin), R_f 0.6–0.65 e.g. hyperosid, isoquercitrin) and the coumarin-7-O-glucosides.
The caffeic acid derivatives with chlorogenic acid at $R_f \sim 0.45$ (T1) and isochlorogenic acids at R_f 0.7–0.8 are more concentrated in Abrotani herba (1) than in **Fabiani herba** (2). The later shows rutin at $R_f \sim 0.4$ as a prominent orange zone and the violet-blue zone of scopoletin-7-O-primveroside at $R_f \sim 0.15$. Blue coumarin and orange fluorescent flavonoid aglycones move with the solvent front.

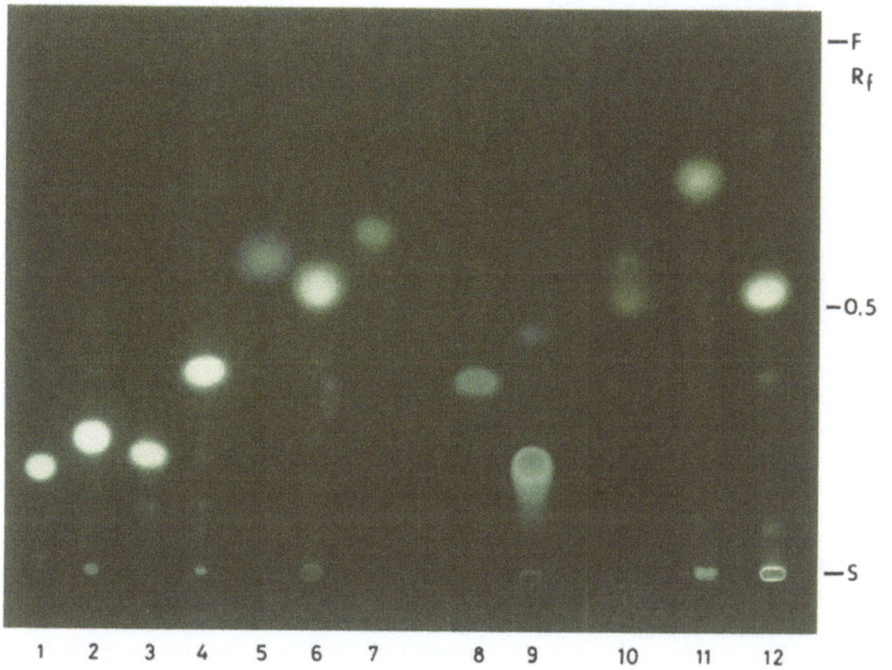

Fig. 3

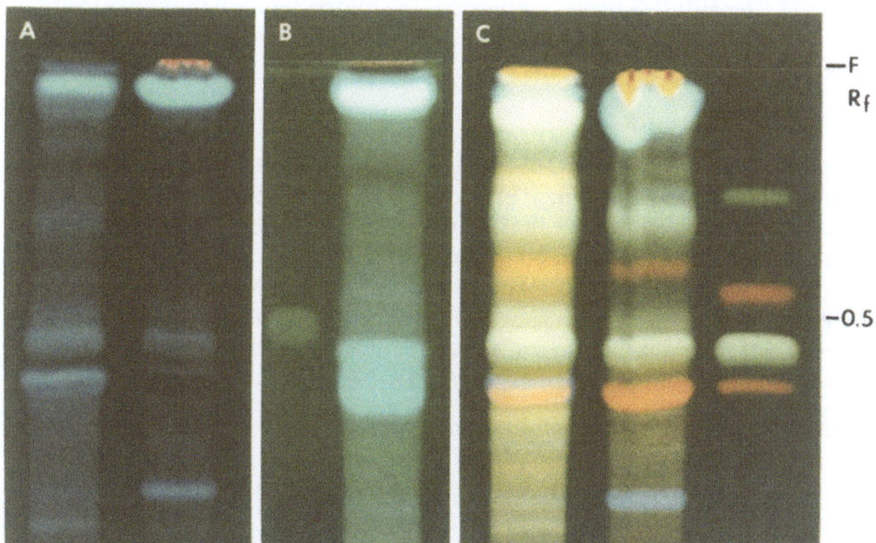

Fig. 4

Fraxini cortex, Mezerei cortex

Drug sample
1 Mezerei cortex
2 Fraxini cortex
(methanolic extracts, 20 µl)

Reference compound
T1 daphnetin — T3 fraxetin
T2 fraxin — T4 scopoletin

Solvent system
Fig. 5 toluene-ether (1:1/saturated with 10% acetic acid) → system 1 (aglycones)
Fig. 6 ethyl acetate-formic acid-glacial acetic acid-water (100:11:11:26) → system 2 (glycosides, polar compounds)

Detection
A UV-365 nm (without chemical treatment)
B Natural products-polyethylene glycol reagent (NP/PEG No. 28) → UV-365 nm
C 10% ethanolic KOH (No. 35) → UV-365 nm

Fig. 5 Solvent system 1 for aglycones

A **Mezerei cortex** (1). In UV-365 nm, two prominent and two minor blue fluorescent coumarin zones are found in the R_f ranges 0.1–0.25 and 0.45–0.65, respectively.
Fraxini cortex (2) has two prominent blue fluorescent coumarin aglycones such as fraxidin, fraxinol and isofraxidin in the R_f range 0.1–0.25, directly below the scopoletin test T4.

B **Mezerei cortex** (1). In addition to four bright-blue fluorescent coumarin zones, treatment with NP/PEG reagent reveals the yellow-brown zone of daphnetin at $R_f \sim 0.3$ (T1).
Fraxini cortex (2). The NP/PEG reagent intensifies the fluorescence of the zones from the start up to $R_f \sim 0.25$ and shows the additional yellow-brown fraxetin (T3) at $R_f \sim 0.05$. The glucoside fraxin (T2) remains at the start.

Fig. 6 Solvent system 2 for glycosides

B (NP-PEG reagent, UV-365 nm). The characteristically polar compound in **Mezerei cortex** (1) is triumbellin seen as a prominent blue zone at $R_f \sim 0.55$ as well as five to six weak blue-violet fluorescent umbelliferone derivatives in the R_f range 0.2–0.4. The yellow-brown zone of daphnetin (T1) moves up to the solvent front.
The coumarin glycosides of **Fraxini cortex** (2) are detected with NP/PEG reagent as four to five intense, bright-blue fluorescent zones (UV-365 nm) in the R_f range 0.35–0.75, such as fraxin ($R_f \sim 0.25$/T2), and the coumarin aglycones at R_f 0.8–0.95 with fraxetin (yellow/$R_f \sim 0.8$, T3), isofraxidin and scopoletin (blue/R_f 0.8–0.95, T4).

C With KOH reagent all coumarins of Fraxini sample 2 show a blue to violet-blue fluorescence in UV-365 nm.

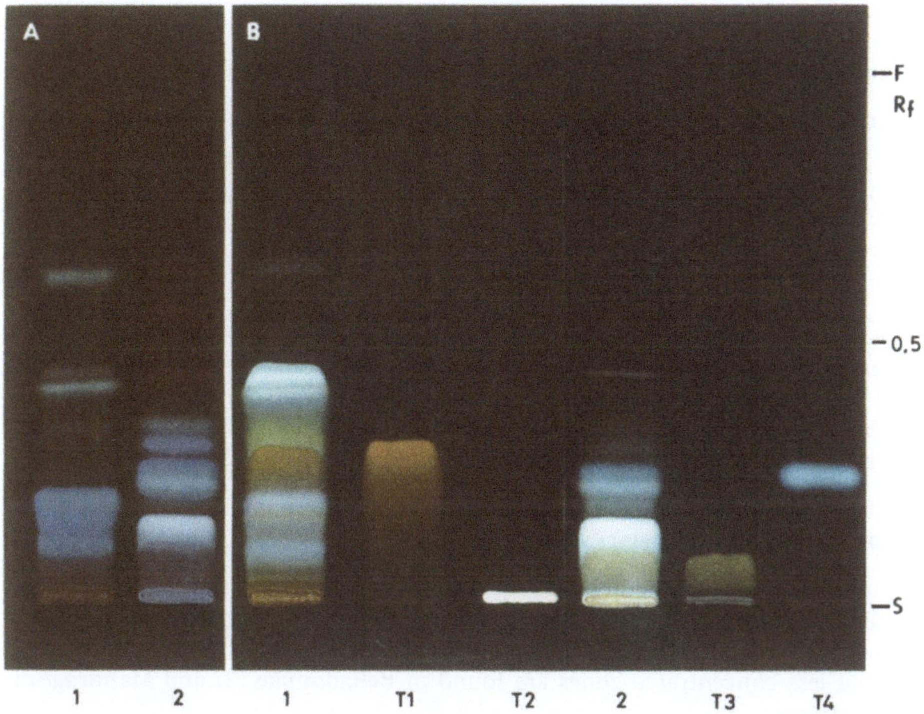

Fig. 5

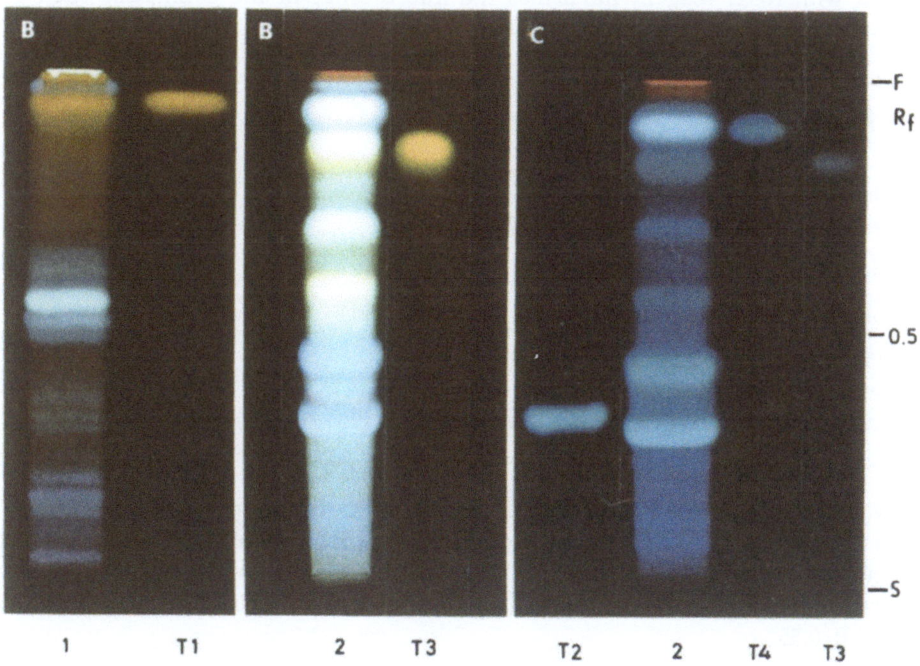

Fig. 6

Scopoliae, Belladonnae, Mandragorae radix

Drug sample 1 Scopoliae radix; 2 Belladonnae radix; 3 Mandragorae radix (methanolic extracts, 20 µl)

Reference T1 scopoletin; T2 chlorogenic acid

Solvent system
A toluene-ether (1:1/saturated with 10% acetic acid) → system A
B,C ethyl acetate-glacial acetic acid-formic acid-water (100:11:11:26) → system B

Detection
A,C 10% ethanolic KOH reagent (No. 35) → UV-365 nm
B UV-365 nm (without chemical treatment)

Description The **Solanaceae root** extracts (1–3) are characterized not only by the alkaloids hyoscyamine and scopolamine (see Chapter 1, Figs. 27, 28), but also by the coumarins scopoletin (T1) and the scopoletin-7-O-glucoside ("scopolin"). The alkaloid content dominates in Belladonnae radix, while Scopoliae radix has a high coumarin content.

Fig. 7A In the non-polar solvent system A scopoletin (T1) migrates to $R_f \sim 0.3$, while its glucoside scopolin remains at the start.

B In the polar system B scopoletin moves almost with the solvent front ($R_f \sim 0.95$). The scopoletin glucoside is found at $R_f \sim 0.4$ directly below chlorogenic acid (T2/$R_f \sim 0.45$). **Scopoliae radix** (1) shows five to six blue fluorescent zones from the start up to R_f 0.5. Similar, but less concentrated zones are found in **Belladonnae** (2) and **Mandragorae radix** (3) extracts.

C Treatment with KOH reagent intensifies the fluorescence of the coumarins such as scopoletin at the solvent front and scopoletin-7-O-glucoside at $R_f \sim 0.45$.

Ammi fructus

Drug sample 1 Ammi visnagae fructus; 2 Ammi majoris fructus (methanolic extracts, 20 µl)

Reference T1 khellin; T2 visnagin

Solvent system Fig. 8 toluene-ether (1:1/saturated with 10% acetic acid)

Detection
A UV-254 nm (without chemical treatment)
B 10% ethanolic KOH reagent (No. 35) → UV-365 nm
C Natural products-polyethylene glycol reagent (NP/PEG No. 28) → UV-365 nm

Fig. 8 A, B **Ammi visnagae fructus** (1) is identified by the furanochromones khellin (T1) and visnagin (T2), which are found at R_f 0.2–0.25 as prominent quenching zones (UV-254 nm → A) and as brown (T1) and blue (T2) fluorescent zones in UV-365 nm (→B). Four additional, blue-white fluorescent zones (e.g. furanocoumarins) are detectable between R_f 0.4 and R_f 0.55.

C **Ammi majoris fructus** (2) is characterized by furanocoumarins, seen as ten to 12 blue fluorescent zones between the start and $R_f \sim 0.65$, with prominent zones in the lower R_f range. Bergapten, xanthotoxin (e.g. scopoletin) and imperatorin move into the R_f range 0.45–0.55, and visnadin to R_f 0.6. Ammi visnagae fructus (1) shows from the start up to $R_f \sim 0.6$ white-blue zones (e.g. visnagin, T2).

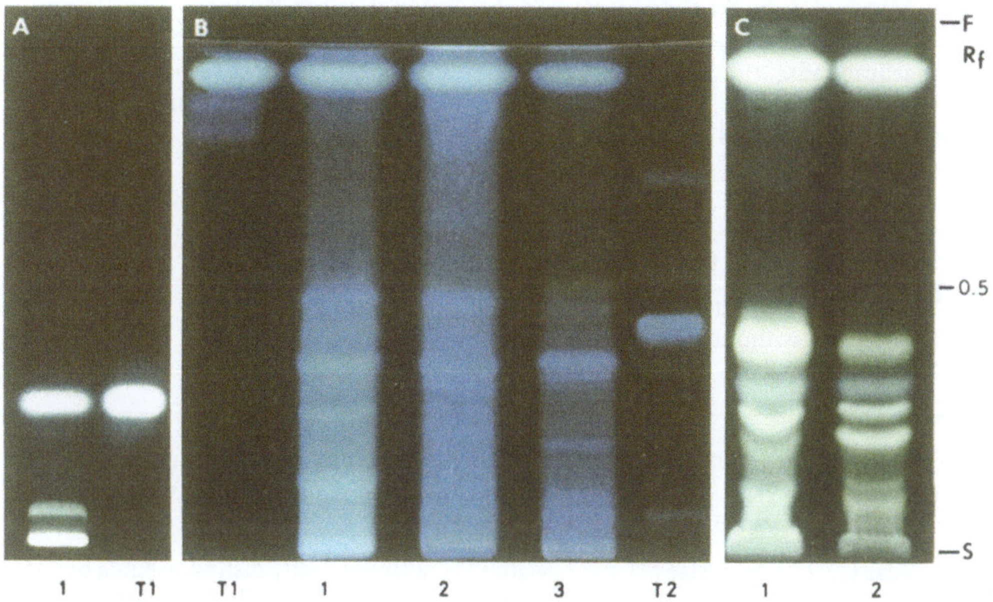

Fig. 7

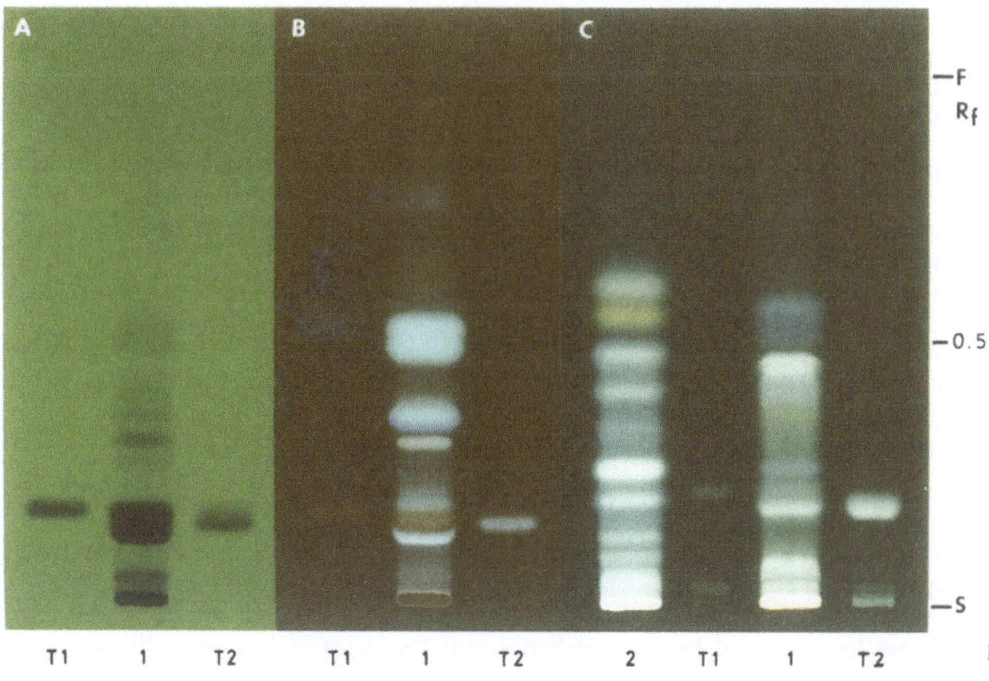

Fig. 8

TLC Synopsis of Apiaceae Roots, Furanocoumarins (FC)

Drug sample 1 Angelicae radix; 2 Imperatoriae radix; 3 Levistici radix; 4 Heraclei radix; 5 Pimpinellae radix (P. saxifrage); 6 Pimpinellae radix (P. major)
(methanolic extract, 20 µl)

Reference compound T1 xanthotoxin; T2 imperatorin; T3 umbelliferone ($R_f \sim 0.45$) ▶ xanthotoxin ($R_f \sim 0.5$)

Solvent system Fig. 9,10 toluene-ether (1:1 saturated with 10% acetic acid)

Detection
A UV-254 nm (without chemical treatment)
B UV-365 nm (without chemical treatment)
C 10% ethanolic KOH (No. 35) → UV-365 nm

Description Apiaceae roots are generally characterized by a high number of structurally similar furanocoumarins (fcs). All fcs show quenching in UV-254 nm and blue, violet or brown fluorescence in UV-365 nm.

Fig. 9A **Angelicae** (1), **Imperatoriae** (2) and **Levistici** (3) radix are easily distinguishable by their different patterns and numbers of fcs zones in UV-254 nm. Sample (2) has the highest, and sample (3) the lowest fcs content.
Angelicae radix (1): a major fcs zone at $R_f \sim 0.65$ as well as six to seven less concentrated zones (R_f 0.05–0.6).
Imperatoriae radix (2): two almost equally strong fcs bands (R_f 0.6/0.75), a major fcs band (R_f 0.45–0.55) and six less concentrated zones between R_f 0.05 and 0.4.
Levistici radix (3): low content of fcs zones (R_f 0.05–0.6) and a phtalide ($R_f \sim 0.8$).

B **Angelicae** (1) and **Imperatoriae** (2) **radix** both show characteristic bands of ten to 15 violet, blue or brown fluorescent zones from the start up to $R_f \sim 0.6$ and $R_f \sim 0.7$, respectively. Levistici radix (3) has weak, pale-blue zones at R_f 0.25–0.3 and $R_f \sim 0.8$.
Angelicae radix (1): a band of 12 mostly violet-blue fluorescent zones (R_f 0–0.6), with two major violet fcs zones directly above and below the xanthotoxin test (T1) → for details see Figs. 11+12
Imperatoriae radix (2): the bright blue fluorescent ostruthin at $R_f \sim 0.45$ dominates, followed by two characteristically brown fluorescent fcs zones (e.g. imperatorin, T2) → see also Figs. 11+12
Levistici radix (3): the pale blue ligusticum lactone dominates

Fig. 10B **Imperatoriae radix** (2) and **Heraclei radix** (4) are mainly distinguishable by a different pattern of characteristic fcs zones in the R_f range 0.45–0.75.
Imperatoriae radix (2): the prominent bright blue ostruthin ($R_f \sim 0.5$) is followed by imperatorin (T2).
Heraclei radix (4): the blue spondin, directly below the xanthotoxin test T1 is followed by four weaker blue and brown (<) fluorescent fcs zones of isopimpinellin (R_f 0.54), pimpinellin (R_f 0.65), isobergapten, bergapten.

C With KOH reagent **Pimpinellae radix** (5,6) develops six blue or violet fluorescent zones between the start and $R_f \sim 0.5$. In this R_f range Pimpinellae (5,6) and Heraclei radix (4) show similar TLC features (e.g. scopoletin, $R_f \sim 0.3$; spondin, $R_f \sim 0.5$; isopimpinellin, $R_f \sim 0.55$), but (4) has highly concentrated fcs zones in the R_f range 0.55–0.75 (e.g. pimpinellin, bergapten, isobergapten), while the fcs isobergapten or isopimpinellin are detectable in enriched extracts of Pimpinellae radix only.

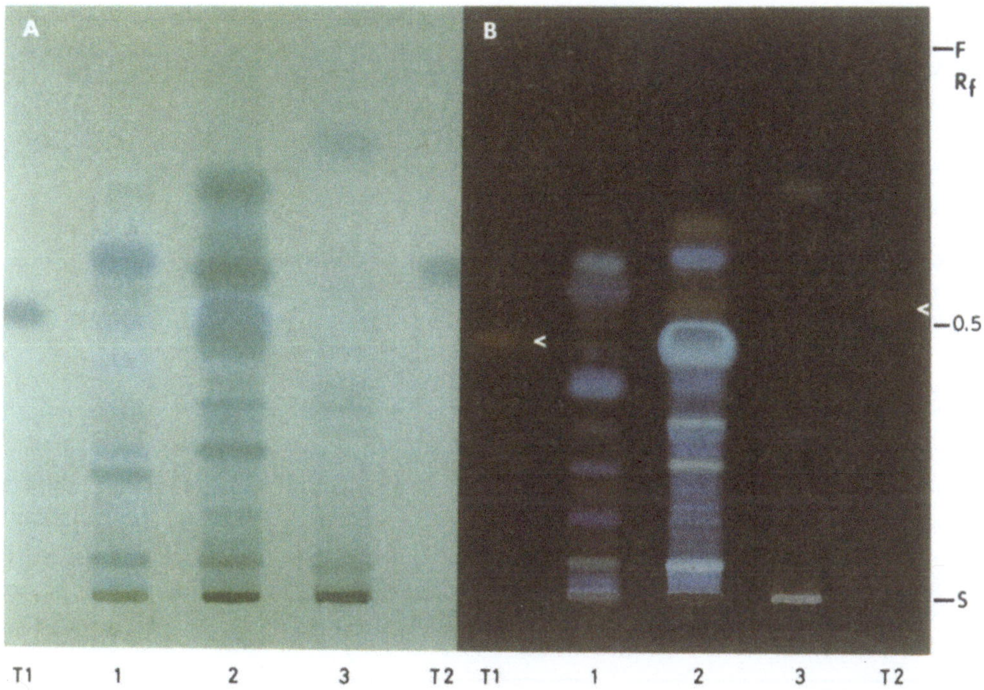

Fig. 9

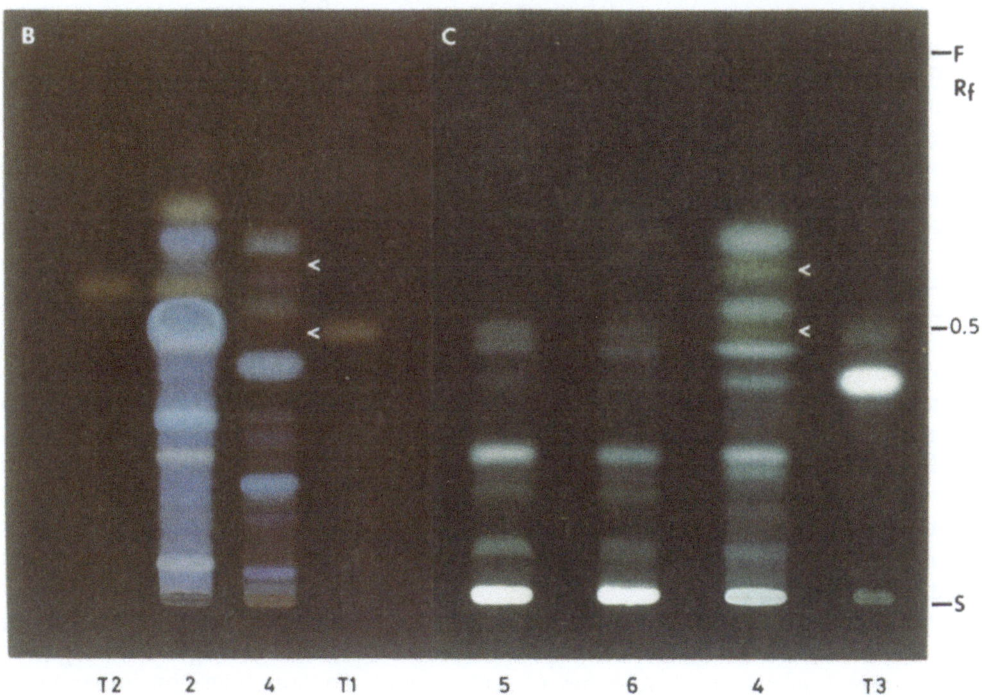

Fig. 10

Imperatoriae, Angelicae and Levistici radix

Drug sample 1 Imperatoriae radix
2 Angelicae radix
3 Levistici radix (methanolic extracts, 30 µl)

Reference compound T1 imperatorin
T2 xanthotoxin
T3 umbelliferone

Solvent system Fig. 11, 12 toluene-ether (1:1/saturated with 10% acetic acid)

Detection A UV-254 nm (without chemical treatment)
B UV-365 nm (without chemical treatment)
C 10% ethanolic KOH (No. 35) → UV-365 nm

Description More than 15 structurally similar furanocoumarins (fcs) and coumarins are identified compounds of Imperatoriae radix (1) and Angelicae radix (2). Levistici radix (3) differs from 1 and 2 by a generally lower coumarin content. Because of their structural similarity, fcs are found as overlapping zones specifically in the R_f range 0.5–0.65. In the table below, the known fcs in the drug samples 1–3 are listed:

Fig. 11, 12

Coumarins	1	2	3	~ R_f value
Umbelliprenin		x		0.8
Bergapten	x	x	x	0.6
Imperatorin (T1)	x	x		0.6
Ostruthin	x			0.6
Xanthotoxin (T2)		x		0.55
Angelicin (brown)			x	0.5
Umbelliferone (T3)	x	x	x	0.45
Scopoletin	x	x	x	0.25
Oxypeucedanin hydrate		x	x	0.15
Plant acids	x	x	x	0.1–0.4

To distinguish the three Apiaceae roots, the different amount of quenching zones of the samples 1–3 can be used (→A). To characterize the single drug extract, the contribution of the blue, violet and brown fluorescent coumarin zones before (→B) and after spraying with the KOH reagent (→C) has to be considered. The originally pale-brown zones such as imperatorin T1 and xanthotoxin T2 become light brown, and the blue fluorescent zones become bright blue after spraying. They then form a band of strongly fluorescent zones from the start up to R_f ~ 0.8, often overlapping the brown adjacent zones, as seen with the sample 2.

In the R_f range 0.55–0.65, the main blue fluorescent ostruthin and the brown imperatorin zone dominate the TLC picture of Imperatoriae radix (1), while in Angelicae radix 2 two prominent bright fluorescent zones of that R_f range contain angelicin, imperatorin, xanthotoxin, bergapten and osthenol.

Levistici radix (3). The KOH detection (→C) reveals five to seven clearly visible zones in the R_f range 0.25–0.9, with three zones at R_f 0.25–0.4 (e.g. umbelliferone/T3), two zones in the R_f range of T1 and T2 with bergapten close to the imperatorin test T1 and the pale blue fluorescent 3-butylidenphthalide (ligusticum lactone) at R_f ~ 0.85.

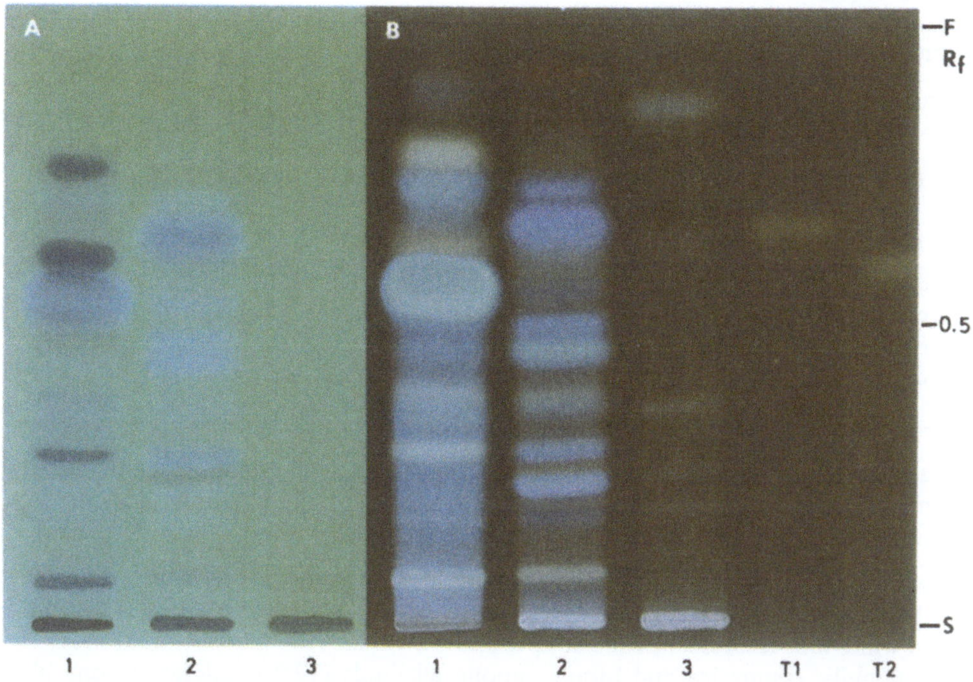

Fig. 11

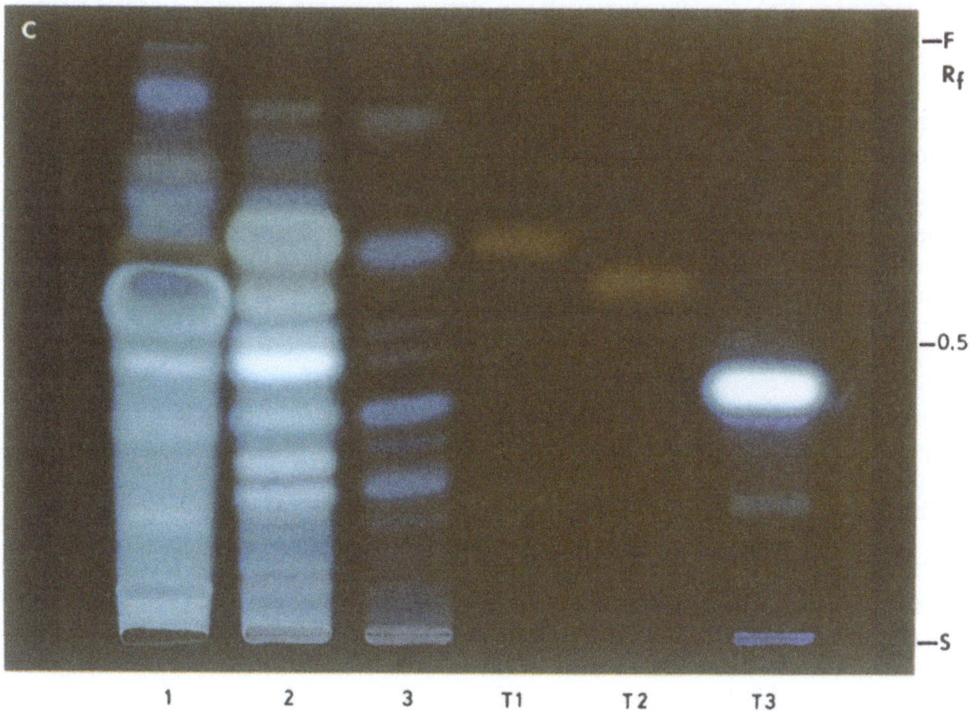

Fig. 12

Rutae herba

Drug sample 1 Rutae herba (methanolic extract, 20 µl)

Reference compound

T1 rutarin	T3 kokusaginine	T5 xanthotoxin
T2 γ-fagarine	T4 scopoletin	T6 umbelliferone

Solvent system Fig. 13 A ethyl acetate-glacial acetic acid-formic acid-water (100:11:11:26) → system I
B toluene-ether (1:1/saturated with 10% acetic acid) → system II
Fig. 14 C,D toluene-ether (1:1/saturated with 10% acetic acid) → system II

Detection
A 5% ethanolic KOH reagent (No. 35) → UV-365 nm
B UV-365 nm (without chemical treatment)
C UV-254 nm (without chemical treatment)
D Dragendorff reagent (DRG No. 13) → vis

Description **Rutae herba** can be characterized by its coumarins (e.g. rutamarin), the furanocoumarins (e.g. bergapten, psoralen) as well as the furanoquinoline alkaloids (e.g. kokusagine, γ-fagarine) and the flavonol glycoside rutin.

Fig. 13A **Rutae herba** (1) generates with KOH reagent in solvent system I a band of ten to 12 violet and blue fluorescent zones from R_f ~ 0.2 up to the solvent front. Rutarin (T1) forms a major white-blue fluorescent zone at R_f ~ 0.35.
The lipophilic coumarins and furanoquinolin alkaloids (T2/T3) migrate in one blue fluorescent major zone up to the solvent front.
The dark zone (>) directly above rutarin derives from the flavonol glycoside rutin, which develops a bright orange fluorescence in UV-365 nm when treated with the Natural product/PEG reagent (Appendix A, No. 28; Rutin see 7.1.7, Fig. 4).

B Separation of **Rutae herba** (1) extract in solvent system II yields more than ten blue fluorescent zones from the start up to R_f ~ 0.9. Red zones are chlorophyll compounds.

Compounds	*R_f value*
dimethyl-allyl-herniarin	~0.9
isoimperatorin rutamarin psoralen, bergapten	} as partly overlapping zones R_f range 0.55–0.85
xanthotoxin test	~0.55 (T5)
umbelliferone	~0.4 (T6)
kokusaginin	~0.35 (T3)
scopoletin	~0.3 (T4)
γ-fagarine	~0.25 (T2)
daphnetin	~0.2

Fig. 14C All coumarin and alkaloid zones show prominent quenching in UV-254 nm.

D With Dragendorff's reagent, the alkaloids T2 and T3 form brown zones (vis). Coumarins also can give a weak, nonspecific reaction ("false positive Dragendorff reaction") due to the α,β-insaturated lactone structure.

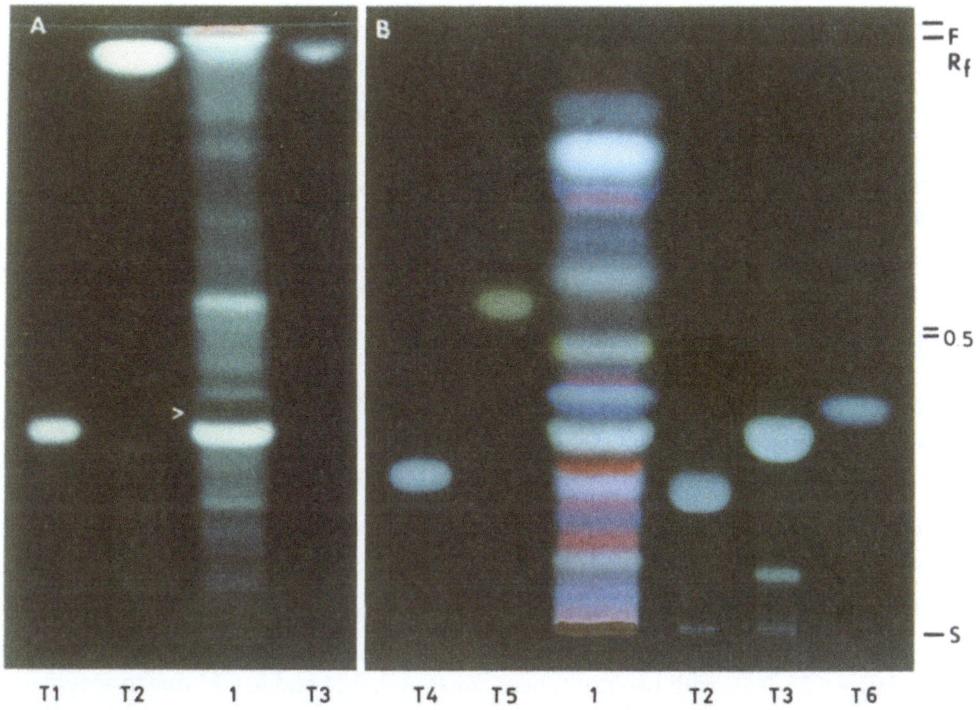

Fig. 13

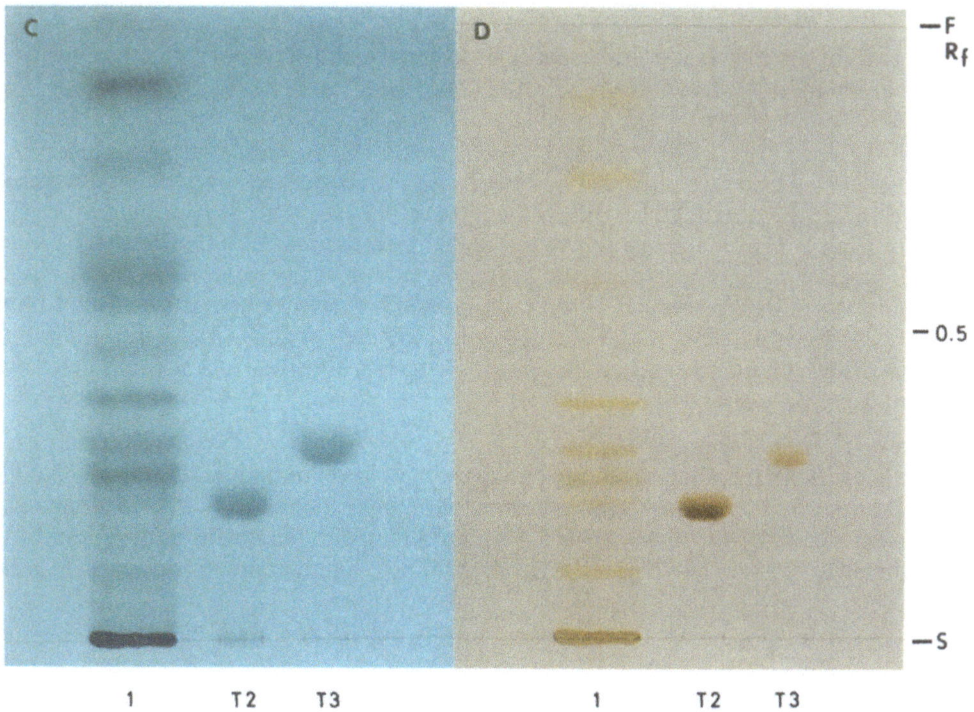

Fig. 14

Herniariae herba

Drug sample 1 Herniariae herba (H. glabra)
2 Herniariae herba (H. hirsuta)
3 Herniariae herba (trade sample) (methanolic extracts, 20 µl)

Reference compound T1 herniarin
T2 rutin ($R_f \sim 0.4$) ▶ chlorogenic acid ($R_f \sim 0.45$) ▶ hyperoside ($R_f \sim 0.6$)
T3 aescin

Solvent system Fig. 15 A,B toluene-ether (1:1/saturated with 10% acetic acid) ▶ system I
Fig. 16 C,D ethyl acetate-glacial acetic acid-formic acid-water (100:11:11:26) ▶ system II

Detection
A UV-365 nm (without chemical treatment)
B 5 % ethanolic KOH reagent (No. 35) ▶ UV-365 nm
C Natural products-polyethylene glycol reagent (NP/PEG No. 28) ▶ UV-365 nm
D Anisaldehyde-sulphuric acid reagent (AS No. 3) ▶ vis

Description **Herniariae herba** can be identified not only by its coumarins but also by its flavonoid and saponin pattern.

Fig. 15 **Coumarins** (solvent system I)

A Methanolic extracts of Herniariae herba (1) and (2) show five violet fluorescent zones in the R_f range 0.25–0.55, with two prominent zones at R_f 0.25 and R_f 0.55 due to scopoletin ($R_f \sim 0.25$) and herniarin (R_f 0.55/T1), respectively. In the commercial sample 3, both coumarins are hard to detect.

B KOH reagent intensifies the fluorescence of herniarin and scopoletin to a bright blue. Both compounds are detectable in samples (1)–(3).

Fig. 16 **Flavonoids** (solvent system II)

C The extracts 1–3 generate with NP/PEG reagent three to five orange fluorescent quercetin and yellow isorhamnetin glycosides in the R_f range 0.2–0.4, with the prominent orange zone of rutin (T2) at R_f 0.4 and the yellow zone of narcissin at R_f 0.45 directly above. The number of additional orange and yellow flavonoid glycoside zones in the R_f range 0.25–0.35 varies in the samples 1–3. Trade sample 3 shows a marked number of blue fluorescent zones in the R_f range of chlorogenic acid ($R_f \sim 0.45$/T2) and above.
The lipophilic coumarins, such as herniarin, migrate as blue fluorescent zones up to the solvent front.

D **Saponins** (solvent system II)
Detection with AS reagent reveals one or two prominent yellow flavonoid zones at R_f 0.4–0.45 and up to five small, dark yellow-brown zones in the R_f range 0.1–0.3 due to saponin glycosides derived from medicagenic, 16-hydroxy-medicagenic acid and gypsogenin, with the main zones in and below the R_f range of the reference compound aescin (T3/ $R_f \sim 0.2$).

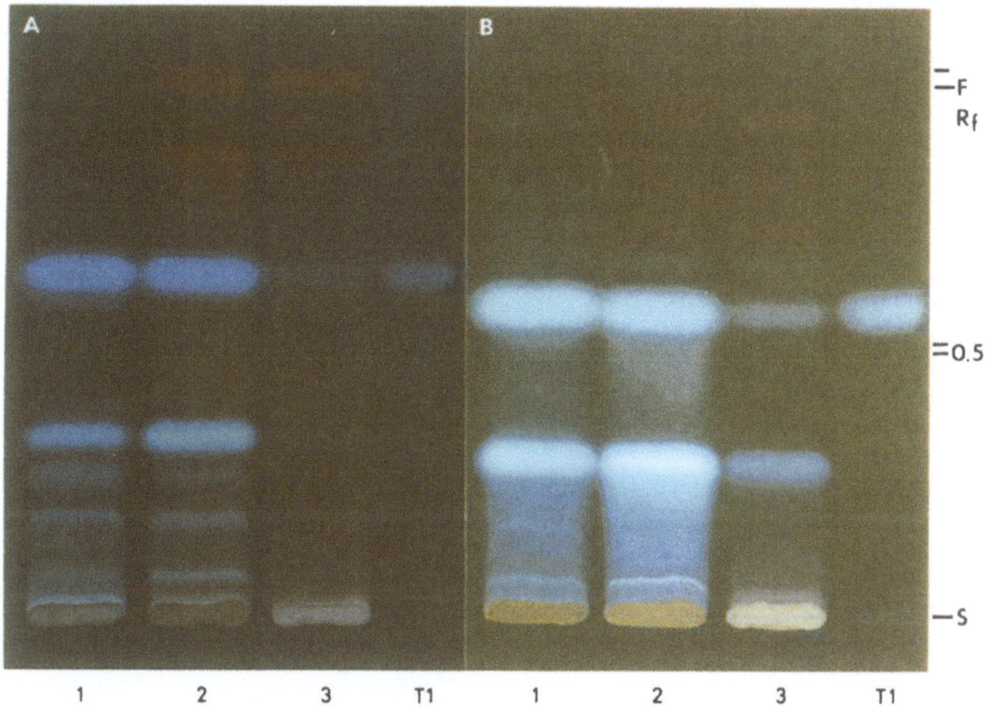

Fig. 15

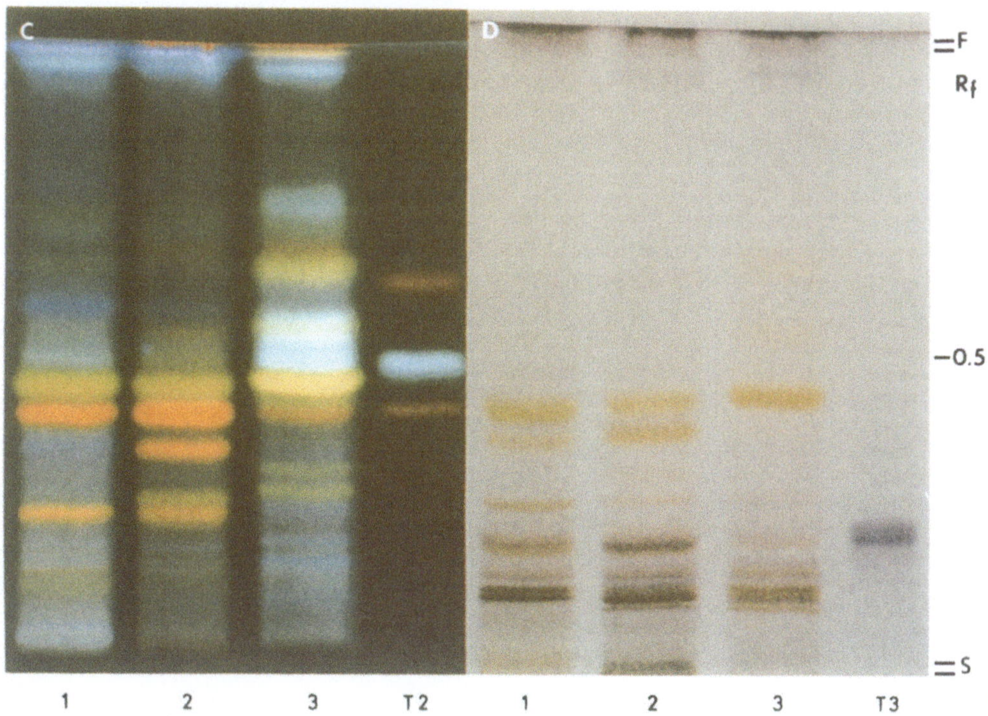

Fig. 16

6 Drugs Containing Essential Oils (Aetherolea), Balsams and Oleo-Gum-Resins

Essential oils are volatile, odorous principles consisting of terpene alcohols, aldehydes, ketones and esters (>90%) and/or phenylpropane derivatives. Aetherolea are soluble in ethanol, but only to a very limited extent in water. They are mostly obtained by steam distillation of plant material.
Balsams, e.g. tolu balsam, are exudates obtained by incision into stems or trunks of plants or trees, respectively. They are water-insoluble resinous solids or viscous liquids with an aromatic odour, their constituents being 40%–60% of balsamic esters.
The oleo-gum-resin myrrh contains resins, gums and 7%–17% of volatile oil and is about 50% water soluble.

6.1 Determination of Essential Oils

Steam Distillation

A steam distillation is carried out with a modified distillation apparatus, as described in many pharmacopoeias (e.g. Cocking and Middleton Ph. Eur.).
The quantity of drug used must be sufficient to yield 0.1–0.3 ml essential oil. Therefore, 10–50 g sample weight and 200–500 ml water are needed, depending on the nature of the drug to be examined. Normally 1 ml xylene is added in the distillation flask before starting the distillation. The rate of distillation has to be adjusted to 2–3 ml/min. The distillate is collected in a graduated tube using xylene to take up the essential oil. For quantitative analysis of the essential oil, a blank xylene value has to be determined in a parallel distillation in the absence of the vegetable drug.
Table 1 shows the conditions for the quantitative determination of essential oils according to the German pharmacopoeia DAB 10.
For the qualitative investigation of an essential oil by TLC, the distillation period can be reduced to 1 h and can be performed in most cases without xylene. The resulting oil is diluted in the graduated tube with xylene (1:9) and used directly for TLC investigation. Essential oils with a density greater than 1.0, such as eugenol-containing oils, still need xylene for distillation.

Micromethods

Microsteam distillation

A 50-ml Erlenmeyer flask is connected with a glass tube (U-shaped, 10- to 15-cm long, 5 mm in diameter). 1 g powdered drug and 10 ml water are then heated to boiling in the flask and a distillation via the U-tube is performed slowly until about 1 ml of the water-essential oil mixture has been collected in the test tube. With 1 ml pentane, the lipophilic compounds are dissolved by shaking in the upper phase; the lipophilic solution is removed with a pipette and 20–100 µl is used for TLC investigation.

Table 1

Drug	Content of essential oil (ml/100 g)	Sample weight (g)	Water (ml)	Time (hr)	Rate (ml/min)
Absinthii herba	0.3	50	300	3	2–3
Anisi fructus	2.0	25	200	2	2–3
Anthemidis flos	0.7	30	300	3	3–5.5
Aurantii pericarpium	1.0	20	250	1.5	2–3
Carvi fructus	4.0	10	200	1.5	2–3
Curcumae rhizoma	3.5	10	200	3	3–4
Foeniculi fructus	4.0	10	200	2	2–3
Juniperi fructus	1.0	20	200	1.5	3–4
Matricariae flos	0.4	50	500[a]	4	3–4
Melissae folium	0.05	40	400	2	2–3
Menthae folium	1.2	50	500	2	3–3.5
Salviae offic. folium	1.5	50	500	1.5	2–3
Salviae trilobae folium	1.8	50	500	1.5	2–3
Thymi herba	1.2	20	300	2	2–3

[a] distilled from 1% NaCl solution.

▶ The microsteam distillation method gives a preliminary indication of the composition of the essential oil, but TLC of different sample concentrations is needed.

Thermomicro-distillation (TAS Method)

The use of a so-called TAS oven (Desaga, Germany) allows the direct application of vegetable drug compounds that volatilize at a fairly high temperature (220°–260°C) without decomposition.
A glass cartridge is filled at the tapered end with a small amount of quartz wool, followed by about 50 mg powdered drug and about 50 mg starch. The cartridge is sealed with a clamp and placed in the oven block of the TAS apparatus, which is heated to about 220°C. The open end of the cartridge points directly to the TLC plate. Volatile compounds at the given temperature then distil onto the starting zone of the TLC plate in about 1.5 min. Immediately afterwards, the plate can be placed in a solvent system sufficient for TLC separation.

▶ All components of essential oils and some other volatile compounds, e.g. coumarins, are obtained by this method.

Extraction with dichloromethane (DCM extract)

1 g powdered drug is extracted by shaking for 15 min with 10 ml dichloromethane. The suspension is filtered and the clear filtrate evaporated to dryness. The residue is dissolved in 1 ml toluene, and 30–100 µl is used for TLC.

▶ This method also extracts other, often interfering lipophilic substances.

Cinnamoyl pigments

Curcumae rhizoma: 1 g powdered drug is extracted by shaking for 5 min with 5 ml MeOH at about 60°C; 10 µl of the filtrate is used for TLC.

Myrrha

0.5 g powdered Oleo-gum-resin is extracted by shaking for 5 min with 5 ml 96% ethanol, 20 µl of the supernatant or filtrate is used for TLC.

Balsam

0.5 g peru balm is dissolved in 10 ml ethyl acetate, and 10 µl of this solution is used for TLC. For tolu balm, 10 µl of a 1:10 dilution in toluene is used for TLC.

6.2 Thin Layer Chromatography

Essential oil

1 ml essential oil is diluted with 9 ml toluene; 5 µl is used for TLC.

Reference solutions

Solutions of commercially available compounds are prepared in toluene (1:30). 3 µl (≙ 100 µg) of each reference solution is used. These quantities applied to the TLC plate are sufficient for detection of essential oil compounds. Thymol and anethole are detectable in quantities of 10 µg and less.

Alcohols:	borneol, geraniol, linalool, menthol
Phenols:	thymol, carvacrol
Aldehydes:	anisaldehyde, citral, citronellal
Ketones:	carvone, fenchone, menthone, piperitone, thujone
Oxides:	1,8-cineole
Esters:	bornyl acetate, geranyl acetate, linaly acetate, menthyl acetate
Phenylpropanoids:	anethole, apiole, allyltetramethoxybenzene, eugenol, myristicin, safrole

Adsorbent

Silica gel 60F $_{254}$-precoated TLC plates (Merck, Germany)

Chromatography solvents

Toluene-ethyl acetate (93:7)
This system is suitable for the analysis and comparison of all important essential oils.

The pharmacopoeias describe various other solvent systems for individual drugs or their essential oils:

Chloroform:	Curcumae xanth. rhizoma, Melissae folium
Dichloromethane:	Anisi –, Carvi –, Caryophylli –, Foeniculiaeth. Lavandulae and Rosmarini aeth.; Salviae fol. and Juniperi fructus;
Toluene-ethyl acetate (90:10):	Eucalypti aeth.
Toluene-ethyl acetate (95:5):	Menthae piperitae aeth.
Chloroform-toluene (75:25):	Absinthii herba, Matricariae flos, Thymi herba

6.3 Detection

- Without chemical treatment

UV-254 nm Compounds containing at least two conjugated double bonds quench fluorescence and appear as dark zones against the light-green fluorescent background of the TLC plate.
► all phenylpropane derivatives (e.g. anethole, safrole, apiole, myristicin, eugenol)

▶ or compounds such as thymol and piperitone.

UV-365 nm No characteristic fluorescence of terpenoids and propylphenols is noticed.

- Spray reagents (see list Appendix A)
- Anisaldehyde-sulphuric acid (AS No.3)
  10 min/110°C; evaluation in vis.: essential oil compounds show strong blue, green, red and brown colouration. Most of the compounds develop fluorescence under UV-365 nm.
- Vanillin-sulphuric acid (VS No.42)
  10 min/110°C; evaluation in vis.: colourations very similar to those obtained with the AS reagent, but no fluorescence at all under UV-365 nm.
  Exceptions: Anisaldehyde and thujone only give very weak daylight colour with AS or VS reagent and should be treated with PMA or concentrated H_2SO_4. Fenchone needs special treatment (see below).
- Phosphomolybdic acid (PMA No.34)
  Immediately after spraying, evaluation in vis.: the constituents of essential oils show uniform blue zones on a yellow background, with the exeption of thujone, anisaldehyde and fenchone.
  Thujone: The TLC plate has to be heated for 5 min at 100°C. Thujone then shows an intense blue-violet colour in the visible.
  Anisaldehyde appears blue with PMA reagent only when present in concentrations higher than 100 µg; at lower concentrations, its colour response varies from whitish to pale green (vis.). When sprayed with concentrated H_2SO_4 and heated at about 100°C for 3–5 min, anisaldehyde appears red (vis.).
  Fenchone: The TLC plate has to be sprayed first with PMA reagent, then with a solution of 0.5 g potassium permanganate in 5 ml concentrated sulphuric acid. After heating for 5 min at 100°C, fenchone appears dark blue (vis.).
  Spraying of the TLC plate with concentrated H_2SO_4 and heating for 3–5 min at 110°C yields a lemon-yellow (vis.) zone of fenchon, but only when applied in quantities greater than 100 µg;

6.4 List of Essential Oil Drugs, Gums and Resins

Reference compounds: Fig. 1,2, Sect. 6.6
Chromatograms of essential oils with phenylpropanoids: Fig. 3–11, Sect. 6.7
Chromatograms of essential oils with terpenoids: Fig. 12–28, Sect. 6.7

	Drug/plant source/family/ pharmacopoeia	Content of essential oil/main constituents THC = Terpene hydrocarbon(s)
Fig. 3,4	**Anisi fructus** Anise Pimpinella anisum L. Apiaceae DAB 10 (oil), PhEur III, ÖAB 90 (oil), Helv VII (oil), BP'88 (oil), MD	2%–6% essential oil Trans-anethole (80%–90%), methyl chavicol (=estragol; 1–2%), anisaldehyde (1%), ester of 4-methoxy-2-(1-propenyl)-phenol (→5%), γ-himachalen Adulteration: Conii maculati fructus (alkaloid coniin); Aethusae cynap. fructus

Drug/plant source/family/ pharmacopoeia	Content of essential oil/main constituents THC = Terpene hydrocarbon(s)	
Anisi stellati fructus Star anise Illicium verum HOOK. fil Illiciaceae ÖAB 90, MD,	5%-8% essential oil Anethole (85%-90%), terpineol, phellandren and up to 5% THC (limone, α-pinene) Adulteration: Illicium anisatum L. (fructus mostly safrole, cineole and linalool)	**Fig. 4**
Foeniculi fructus Fennel seed Foeniculum vulgare MILL. ssp. vulgare Apiaceae DAB 10 (oil), ÖAB 90 (oil), Helv VII (oil), MD, Japan	ssp. vulgare-french bitter fennel: 4%-6% essential oil Trans-anethole (60%-80%), (methyl chavicol), anisaldehyde, (+)-fenchone (12-22%), THC ssp. vulgare *var. dulce:* french sweet or roman fennel: 2%-6% essential oil Trans-anethole (50%-60%), (methyl chavicol), anisaldehyde, (+)-fenchone (0.4%-0.8%), THC (e.g. limonene, β-myrcene)	**Fig. 3,4**
Basilici herba Basil Ocimum basilicum L. Lamiaceae	0.1%-0.45% essential oil Methyl chavicol (up to 55%) and linalool *or* linalool (up to 70%) and methyl chavicol (chemotype or geographic type?)	**Fig. 3**
Sassafras lignum Sassafras wood, root Sassafras albidum (NUTT) NEES var. molle (RAF) FERN Lauraceae MD	1%-2% essential oil Safrole (about 80%), eugenol (about 0.5%)	**Fig. 3**
Cinnamomi cortex Cinnamon bark Cinnamomum verum J.S. PRESL (syn. C. zeylanicum BLUME) Cinnamomum aromaticum NEES (syn. C. cassia BLUME) Chinese or cassia cinnamom Lauraceae DAB 10, ÖAB 90 (oil), Helv VII (oil), BP 88 (oil), MD, DAC 86, Japan, China (C. cassia)	0.5%-2.5% essential oil ► Ceylon cinnamon Cinnamic aldehyde (65%-80%), hydroxymethoxy- and methoxy cinnamic aldehyde, trans-cinnamic acid, eugenol (4%-10%), α-terpineol, THC (e.g. β-caryophyllene, α-pinene, limonene) 1%-2% essential oil ► Chinese cinnamon: Cinnamic aldehyde (75%-90%); only traces of eugenol (0-10%, chemotype) (Unsubstituted coumarin)	**Fig. 5,6**

	Drug/plant source/family/ pharmacopoeia	Content of essential oil/main constituents THC = Terpene hydrocarbon(s)
Fig. 5,6	**Caryophylli flos** Cloves Syzygium aromaticum (L.) MERR. et PERRY Myrtaceae DAB 10 (oil), ÖAB 90, Helv VII, BP 88 (oil), MD, Japan	14%–20% essential oil Eugenol (= 4-allyl-2-methoxyphenol; 72%–90%), aceteugenol (10%–15%), β-caryophyllene (>12%) Clove stalks contain 5%–6% essential oil, "mother cloves", anthophylli 2%–9% essential oil
Fig. 7,8	**Calami rhizoma** Sweet flag (Acorus) root Acorus calamus L. Araceae ÖAB 90, Helv VII, MD Acorus calamus L. var. americanus WULFF (diploid) var. calamus L. (triploid) var. angustata ENGLER (tetraploid)	1.7%–9.3% essential oil Triploid race (East Europe): 3% essential oil variable content of α-, β-asarones (3.3%–14%, average 8%), acoron (sesquiterpene ketone) Diploid race: 2.7%–5% essential oil, β-asarone absent or in low concentration; 30 compounds (e.g. isoeugenol, methyl-isoeugenol, 2.4.5. trimethoxy-benzaldehyde and artefacts formed during distillation). Helv. VII (maximum 0.5% asarone).
Fig. 8	**Asari radix** Hazelwort, wild nard Asarum europaeum L. Aristolochiaceae MD	0.7%–4% essential oil, α-asarone (= trans-isoasarone), methyl eugenol (0.5%–40%) or transisoelemicin (0.5%–70%) (chemovariant)
Fig. 9,10	**Petroselini fructus** Parsley fruits Petroselinum crispum (MILL.) ssp. crispum leaf parsley var. tuberosum BERNH. ex RCHB. Root parsley Apiaceae	3%–6% essential oil apiol, myristicin and 1-allyl-2,3,4,5-tetramethoxybenzene (ATMB); chemotype: 60%–80% apiol or 50%–75% myristicin or 50%–60% ATMB Remarks: Petroselini radix also contains essential oil (0.2%–0.3%) with apiol and myristicin
Fig. 9	**Myristicae semen** Nutmeg Myristica fragrans HOUTT. Myristicaceae	6%–10% essential oil (nutmeg) Phenylpropane derivatives: myristicin (50%–75%), safrole, eugenole, elemicin; THC: α-pinene, limonene, p-cymene terpene alcohols in low concentration, geraniol, borneol, linalool and α-terpineol

Drug/plant source/family/ pharmacopoeia	Content of essential oil/main constituents THC = Terpene hydrocarbon(s)	
Myristicae arillus Macis, Mace Myristica fragrans HOUTT Myristicaceae MD (oil), ÖAB 90, Helv VII, BP 88 (oil), USP XXI	4%–12% essential oil (mace); qualitatively similar to nutmeg oil; safrole is absent	
Ajowani fructus Ajowan fruits Trachyspermum ammi (L.) SPRAGUE Apiaceae	2.6%–4.5% essential oil Thymol (45%–60%) with small amounts of carvacrol	**Fig. 11**
Thymi herba Thyme Thymus vulgaris L. Lamiaceae DAB 10, Helv VII, MD ÖAB 90 (oil) Thymus zygis L. Spanish thyme Lamiaceae DAB 10, MD	1%–2.5% essential oil Thymol (30%–70%), carvacrol (3%–15%), thymol-monomethylether (1.5%–2.5%) 1,8-cineole (2%–14%); geraniol, linalool, bornyl and linalyl acetate (1%–2.5%) α-pinene Content and composition of essential oil similar to that of Thymus vulgaris, but higher amount of carvacrol and less thymolmonomethylether (0.3%).	**Fig. 11**
Serpylli herba Wild thyme Thymus pulegioides L. Lamiaceae	0.1%–0.6% essential oil Thymol (1%–4%), carvacrol (5%–33%) geraniol (3%–10%), linalool, linalyl acetate (20%–40%), cineole (→7%), borneol (1%–15%) and bornyl acetate (→5%)	**Fig. 11**
Carvi fructus Caraway fruits Carum carvi L. Apiaceae DAB 10 (oil), ÖAB 90 (oil), Helv VII, BP 88 (oil), MD	3%–7% essential oil (DAB 10: 4%) (s)(+)-Carvone (50%–85%), dihydrocarvone, carveol, dihydrocarveol, up to 50% limonene	**Fig. 12**
Menthae crispae folium Spearmint leaves Mentha spicata L. BENTH. var. crispa Lamiaceae MD, → DAC 86, BP 88 oil only	1%–2% essential oil L-Carvone (42%–80%) acetates of dihydrocarveol and dihydrocuminyl alcohol, THC (pinene, limonene, phellandrene)	**Fig. 12**

	Drug/plant source/family/ pharmacopoeia	Content of essential oil/main constituents THC = Terpene hydrocarbon(s)
Fig. 12	**Coriandri fructus** Coriander fruits Coriandrum sativum L. var. vulgare ALEF. large Indian coriander var. microcarpum DC. small Russian coriander Apiaceae ÖAB 90, BP 88 (oil), MD (oil) USP XX (oil)	0.2%–0.4% essential oil (Indian coriander) 0.8%–1% essential oil (Russian coriander) (+)-Linalool (50%–80%), small amounts of geraniol and geranyl acetate, about 20% THC (α-pinene, γ-terpinene ~10%, myrcene, limonene ~10% and camphene <5%) borneol (Russian coriander only) thymol (Indian coriander only)
Fig. 12	**Cardamomi fructus** Cardamoms Elletaria cardamomum (L.) MATON Zingiberaceae DAC 86, BP 88 (oil), MD (oil), Japan, USP XXI	Essential oil: fruits (3%–8%), seeds (4%–9%), pericarp (0.5%–1%) α-Terpinyl acetate (30%), 1,8-cineole (20%–40%); small amounts of borneol, linalool, linalyl acetate and limonene (2%–14%)
Fig. 13	**Menthae piperitae folium** Peppermint leaves Mentha piperita (L.) HUDS. Lamiaceae DAB 10 (oil), ÖAB 90 (oil), Helv VII (oil), Ph. Eur. III, BP 88 (oil), MD (oil), USP XXII	0.5%–4% essential oil Menthol (35%–45%), (−)menthone (10%–30%) with small amounts of isomenthone, menthyl acetate (3%–5%), menthofuran (2.5%–5%) pulegone, piperitone (1%), cineole (8%), pinene, limonene, jasmone (0.1%), sabinenhydrate
	M. arvensis L. var. piperascens (L.) HOLMES ex CHRISTY Lamiaceae MD, Japan, DAB 10 (oil)	1%–2% essential oil (corn mint oil), similar composition as the oil from Mentha piperita, but menthofuran and cineole are absent
	Mentha pulegium L. Lamiaceae MD	1%–2% essential oil (Pulegii folium aeth.) Pulegone (80%–95%) with small amounts of piperitone, menthol and THC adulterant of M. piperita, M. arvensis
Fig. 14	**Rosmarini folium** Rosemary leaves Rosmarinus officinalis L. Lamiaceae DAC 86, MD (oil) → DAB 10, ÖAB 90, Helv VII oil only	1%–2% essential oil (oil composition varies due to drug origin). 1,8-Cineole (15%–30%), borneol (10%–20%), bornyl acetate, α-pinene (up to 25%), camphene (15%–25%) ► rosmarinic acid, picrosalvin, carnosolic acid

Drug/plant source/family/ pharmacopoeia	Content of essential oil/main constituents THC = Terpene hydrocarbon(s)	
Melissae folium Balm leaves, honeyplant or Lemon balm Melissa officinalis L. Lamiaceae DAB 10, ÖAB 90, Helv VII, MD Cymbobogon nardus RENDLE Cymbobogon winterianus JOWITT Cymbobogon flexuosus TUND. et STEUD. Poaceae	0.01%–0.20% essential oil Citronellal (30%–40%), citral (20%–30%), citronellol, nerol, geraniol and THC (10%, e.g. β-caryophyllene) ► 4% rosmarinic acid *Melissa oil substitutes*: Java Citronella oil (Ceylon type) Citronellae aetheroleum Java Citronella oil (Java type) 0.5%–1.2% essential oil with citronellal (24%–25%) and gerianol (16%–45%) Lemon grass oil 53%–83% citral (West Indian type) 70%–85% citral (East Indian type) 80%–84% citral (Angola type; odourless)	**Fig. 15,16**
Lavandulae flos Lavender flowers Lavandula angustifolia MILL. Lamiaceae DAC 86, MD (oil) → DAB 10, ÖAB 90, Helv VII Lavandula latifolia MED. Lavandula hybrida REV.	1%–3% essential oil Linalyl acetate (30%–55%), linalool (20%–35%), with small quantities of nerol, borneol, β-ocimen, geraniol, cineole, caryophyllene-epoxide, camphene Spike lavender oil (0.5%–1%), linalool (30%–50%), cineole (>20%), ester absent "Lavandin oils" 20%–24% or 30%–32% linalyl acetate, linalool, THC Commercial "Lavandin" oils; mixture with oil of spike lavender possible	**Fig. 16**
Aurantii pericarpium Bitter orange peel Citrus aurantium L. ssp. aurantium Rutaceae DAB 10, MD, Japan, China ÖAB 90, Helv VII, BP 88	0.6%–2.2% essential oil (+)-Limonene (90%), linalool, linalyl acetate, neryl and citronellyl acetate, citral ► anthranile methylate, coumarins ► flavonoids: rutin, eriocitrin, naringin, neohesperidin, nobiletin, sinensetin (see Sect. 7.1.8 Fig. 23/24 p. 232)	**Fig. 17,18**
Aurantii flos Orange flowers Citrus sinensis (L.) PERSOON ENGL. Rutaceae Helv VII (oil), ÖAB 90 (oil)	0.2%–0.5% essential oil (oil of neroli) Linalyl acetate (8%–25%), linalool (about 30%), geraniol, farnesol, limonene ► anthranile methylate	**Fig. 17**

	Drug/plant source/family/ pharmacopoeia	Content of essential oil/main constituents THC = Terpene hydrocarbon(s)
Fig. 17,18	**Citri pericarpium** Lemon peel, limon Citrus limon (L.) BURM Rutaceae BP 88 (+ oil); oil only: DAB 10, ÖAB 90, MD	0.1%–6% essential oil (+S)-Limonene (90%), citral (3.5%–5%), small amounts of terpineol, linalyl and geranyl acetate Coumarins: geranylmethoxycoumarin, citroptene, bergamottin ► flavonoids: rutin, eriocitrin, neohesperidin (see Sect. 7.1.8).
	Citrus aurantium (L.) ssp. bergamia (RISSO et POIT.) ENGL. MD	"Bergamot oil" (fruit peel oil): linalyl acetate (35%–40%), linalool (20%–30%); dihydrocumin alcohol
	Citrus aurantium var. amara Rutaceae	"Oil of Petit Grain" (leaf oil) >60% linalyl acetate
Fig. 19,20	**Salviae folium** Sage leaves Salvia officinalis L. ssp. minor ssp. major Dalmatian sage	1.5%–2.5% essential oil Composition varies, depending on origin: thujone (22%–37%; ssp. minor or major), cineole (8%–24%), camphor (30%), borneol (5%–8%), bornyl acetate and THC (e.g. α-pinene) ► flavonoids: 1%–3% ► rosmarinic acid (2%–3%) Diterpene bitter pinciples: picrosalvin
	Salvia lavandulifolia VAHL Spanish sage Lamiaceae DAB 10, ÖAB 90, Helv VII (+oil), MD (+oil)	1%–1.5% essential oil cineole (20%), thujone (<1%), campher (26%)
Fig. 19,20	**Salviae trilobae folium** Greek sage Salvia triloba L. fil. Lamiaceae DAB 10, Helv VII, MD	2%–3% essential oil 1,8-Cineole (40%–70%), thujone (about 5%), borneol, bornyl acetate, THC ► Diterpene carnosol
Fig. 20	**Eucalypti folium** Eucalyptus, bluegum leaves Eucalyptus globulus LABILL. E. fruticetorum MUELLER E. smithii R.T. BAKER Myrtaceae DAB 10, USP XXI oil: ÖAB 90, Helv VII, BP'88, MD	1.5%–3.5% essential oil 1,8-Cineole (eucalyptol; 70%–90%); piperiton, α-pinene, phellandrene ► Non-official oils can contain cineole (40%–50%), piperitone (10%–20%) and/or phellandrene (40%–50%), e.g. Eucalyptus dives SCHAUER ► Non-rectified oils contain e.g. butyraldehyde and caprylaldehyde, which cause bronchial irritation

Drug/plant source/family/ pharmacopoeia	Content of essential oil/main constituents THC = Terpene hydrocarbon(s)	
Matricariae flos Chamomillae flos Camomile flowers Matricaria recutita (L.) Chamomilla recutita (L.) RAUSCHERT Asteraceae DAB 10, Ph. Eur. III, ÖAB 90, Helv VII, BP 88, MD	0.5%–1.5% essential oil Chamazulene (0%–15%) (−)-α-Bisabolol (10%–25%) Bisabolol oxide A, B, C (10%–25%); acetylenes (cis- and trans-ene-ine-dicycloether, 1%–40%); farnesene (15%) Bisabolon oxide A, spathulenol Flavonoids: see Sect. 7.1.7, Fig. 3	**Fig. 21**
Anthemidis flos Roman camomile flowers Chamaemelum nobile (L.) ALL. Asteraceae DAB 10, Ph. Eur. III, ÖAB 90, Helv VII, BP 88, MD	0.6%–2.4% essential oil esters of angelic, methacrylic, tiglic and isobutyric acids: n-butylangelat (34%); polyacetylenes Flavonoids: see Sect. 7.1.7, Fig. 3, 4	**Fig. 22**
Cinae flos Wormseed Artemisia cina O.C. BERG et C.F. SCHMIDT Asteraceae MD, ÖAB 9	2%–3% essential oil 1,8-Cineole (about 80%) with small amounts of α-terpineole, carvacrole, THC; sesquiterpene lactone: L-α-santonin (6%), α-hydroxy-santonin (artemisin) - bitter principle	**Fig. 22**
Curcumae rhizoma Turmeric Curcuma zanthorriza ROXB. Round turmeric Zingiberaceae DAB 10 Curcuma domestica VAHL Finger or long turmeric Zingiberaceae DAC 86, MD	3%–12% essential oil Zingiberene (30%), xanthorrhizol (phenolic sesquiterpene, 20%), cineol, borneol, camphor (1%–5%) 1%–2% pigments (curcumin, monodemethoxy-curcumin 0.3%–5% essential oil Sesquiterpenes ketone (65%; e.g. turmerone), zingiberene (about 25%), phellandrene, sabinene, borneol and cineole 3%–4% pigments curcumin, monodemethoxy curcumin, bisdemethoxycurcumin, di-p-coumaroylmethane	**Fig. 23,24**
Juniperi fructus Juniper berries Juniperus communis L. ssp. communis Cupressaceae DAB 10, ÖAB 90, Helv VII, MD	0.3%–1.5% essential oil with varying composition of terpinene-4-ol (~5%), terpineol, terpinyl acetate, borneol, bornyl acetate, caryophyllene, epoxydihydrocaryophyllene, camphor, α- and β-pinene (50%), myrcene	**Fig. 25**

Drug/plant source/family/ pharmacopoeia	Content of essential oil/main constituents THC = Terpene hydrocarbon(s)

Pine Oils

These are essential oils from the needles and branch tips of Abies, Picea and Pinus species (4%–10%, Pinaceae family).

Fig. 27,28	**Pini pumilionis aeth.** Mountain pine oil Pinus mugo TURRA ssp. mugo ssp. pumilio (HAENKE) FRANCO ÖAB, Helv VII, MD	3%–10% esters, calc. als bornyl acetate and bornyl formiate; α- and β-phellandrene (60%), α- and β-pinene (10%–20%), anisaldehyde
	Pinus silvestris aeth. Scots pine needle oil Pinus silvestris L. DAB 10	1.5%–5% esters calc. as bornyl acetate 10%–50% α-, β-pinene, limonene
	Piceae aeth. Pine needle oil Picea mariana B.S.P. Picea abies (L.) KARSTEN	37–45% bornyl acetate
	Pini silvestris aeth. Siberian spruce oil Abies sibirica LEDEB.	10% bornyl acetate, borneol 32%–44% bornyl acetate α-, β-pinene
Fig. 27,28	**Terebinthinae aetheroleum** T. rectificatum aeth. Turpentine oil Pinus palustris MILLER Pinus pinaster AITON et al. ÖAB 90, Helv VII, BP 88, MD (resin), Japan	Distillate of turpentine (oleoresin) from various Pinus ssp. 80%–90% THC (α-, β-pinene, limonene, phellandrene); autoxidation produces α-pinene peroxides and subsequently verbenol and pinol hydrate (=sorbenol)

Oleo-Gum-Resins

Fig. 25	**Myrrha** Gum myrrh Commiphora molmol ENGL and Commiphora ssp. Burseraceae DAB 10, ÖAB 90, Helv VII, MD, BHP 90	2%–10% essential oil, complex mixture Sesquiterpenes: germacran-type, furanoeleman, furanoeudesman type 2-methoxyfuranodien, curzerenone Cinnamic and cuminaldehyde, eugenol, m-cresol and alcohols; 25%–40% ethanol-soluble resin fraction with α-, β- and γ-commiphoric acids and esters

Drug/plant source/family/ pharmacopoeia	Content of essential oil/main constituents THC = Terpene hydrocarbon(s)	
Benzresins and balsams		
Benzoe tonkinensis Siam-benzoin, gum benjamin Styrax tonkinensis (PIERRE) CRAIB ex HARTWICH Styracaceae Ph. Eur. III, ÖAB 90, Helv VII, USP XXII, MD	25% free or combined acids, determined as benzoic acid (Ph. Eur. III). Coniferyl benzoate (60%–80%), coumaryl benzoate (10–15%), benzoic acid (10%–20%), vanillin (about 0.3%), α-siaresinolic acid (=19-hydroxyoleanolic acid)	Fig. 26
Benzoe sumatra Sumatra-benzoin Styrax benzoin DRYAND. Styracaceae BP 93, USP XXII, MD	Cinnamoyl benzoate and coniferyl benzoate (70%–80%), cinnamic acid esters, styracin, cinnamic acid (about 10%), cinnamic acid phenylpropyl ester (about 1%), vanillin (about 1%), sumaresinolic acid (= 6-hydroxyoleanolic acid)	
Tolutanum balsamum Tolu balsam Myroxylon balsamum (L.) HARMS var. balsamum Fabaceae Leguminosae HELV VII, USP XXII, MD	About 7.5% "cinnamein", a mixture of benzoyl benzoate (4%–13%) and cinnamoyl benzoate; (1–3%); about 80% resin (mostly cinnamic esters of toluresitannol), cinnamic acid, benzoic acid, vanillin, eugenol	Fig. 26
Peruvianum balsamum Peru balsam Myroxylon balsamum (L.) HARMS var. pereirae Fabaceae DAB 10, ÖAB 90, Helv VII, MD	50%–70% esters: benzoyl benzoate (25%–40%) and cinnamoyl benzoate (10%–25%) 20%–28% resin (mostly cinnamicesters of peresitannol), cinnamic acid (about 10%), benzoic and dihydrobenzoic acid, α-nerolidol (3%–5%)	

6.5 Formulae

α-Pinene

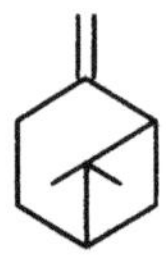

β-Pinene

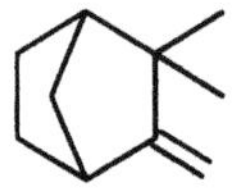

Camphene

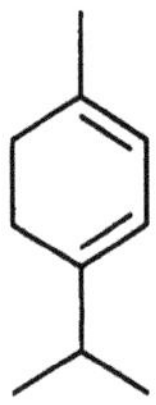

α-Terpinene

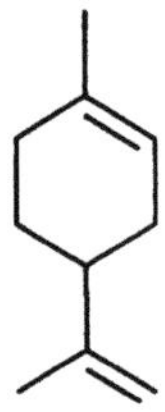

Limonene

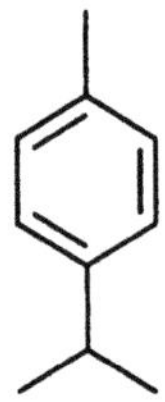

p-Cymene

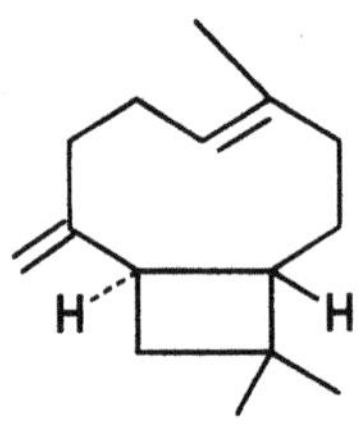

Caryophyllene

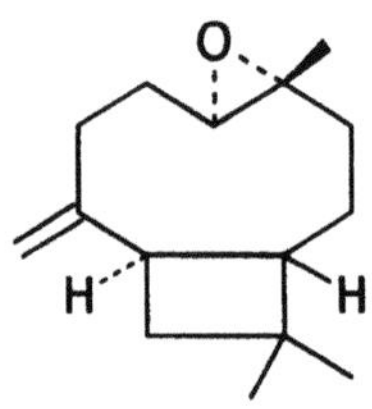

Caryophyllene epoxide

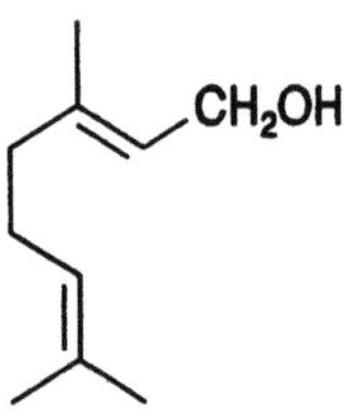

Gerianiol

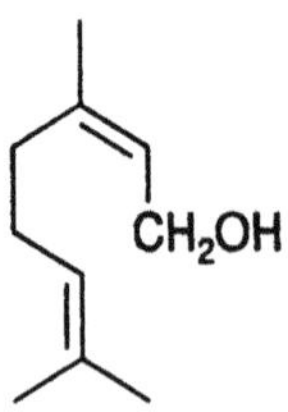

Nerol

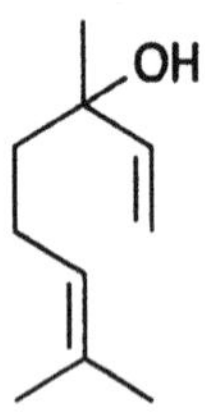

Linalool

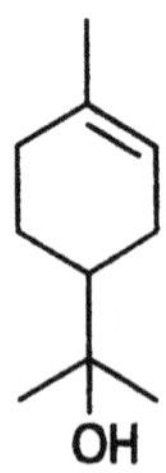

Terpineol

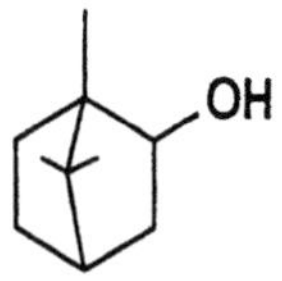

Borneol

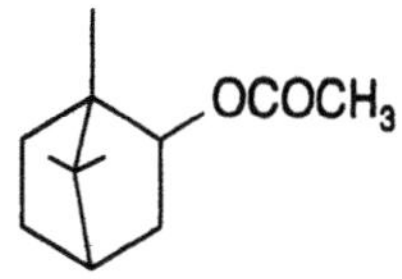

Bornyl acetate

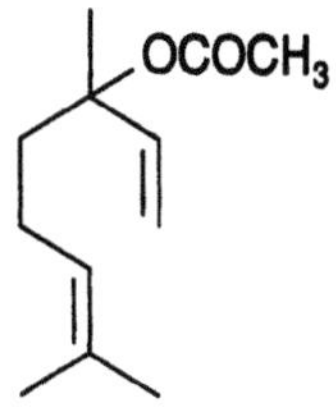

Linalyl acetate

Cineole

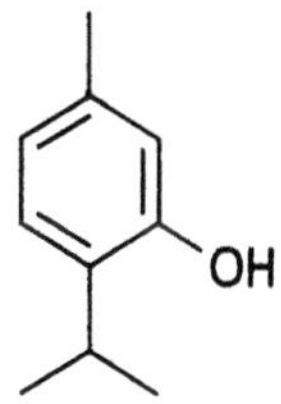

Thymol

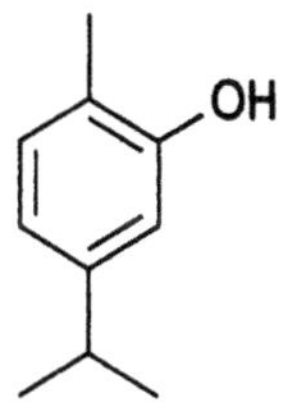

Carvacrol

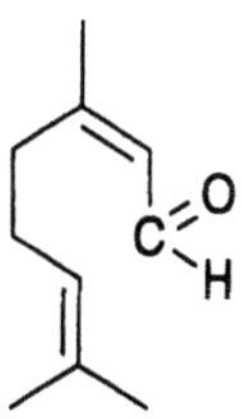

Citral

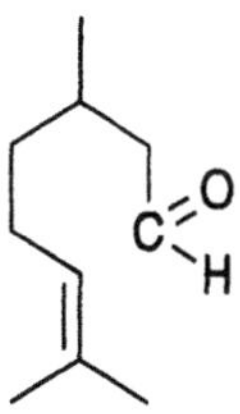

Citronellal

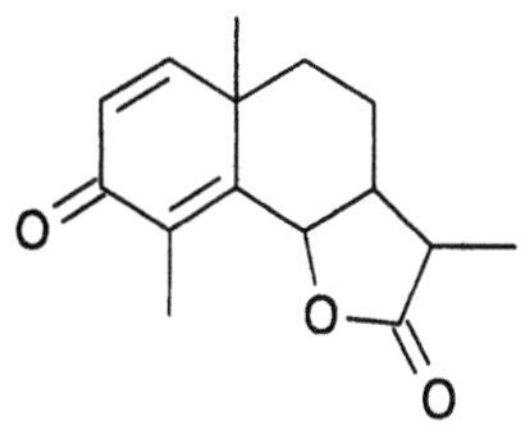

α-Santonin

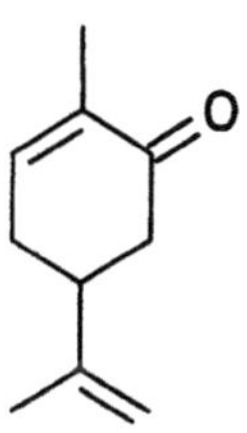

Carvone

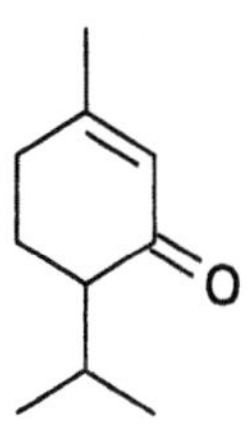

Piperitone

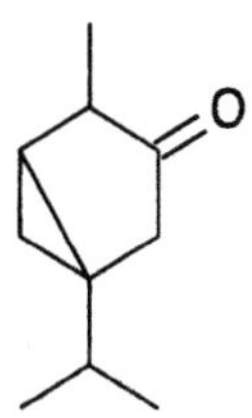

Thujone

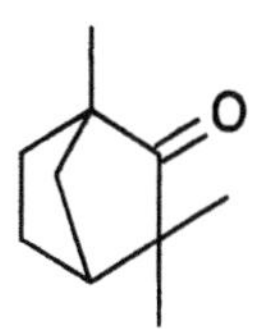

Fenchone

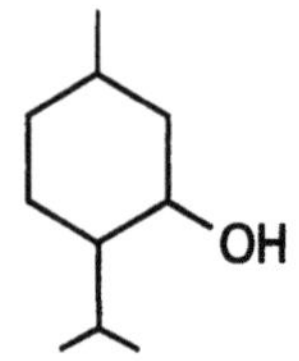

D(-)-Menthol

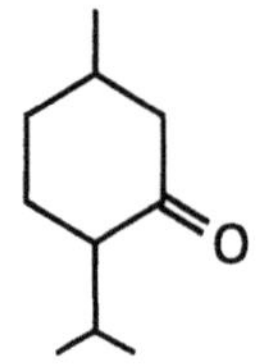

(-)-Menthone

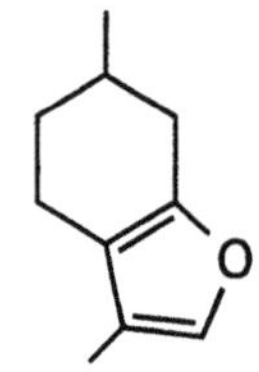

Menthofuran

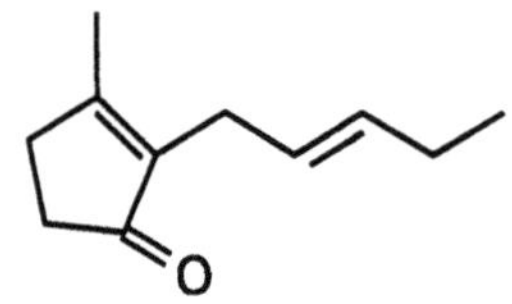

Jasmone

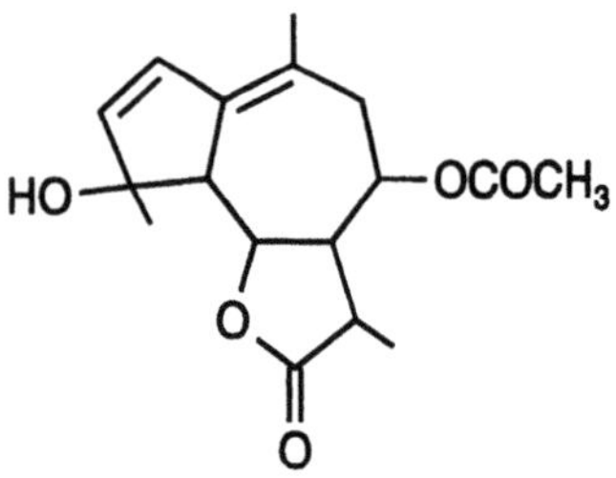

Matricin (Proazulene)

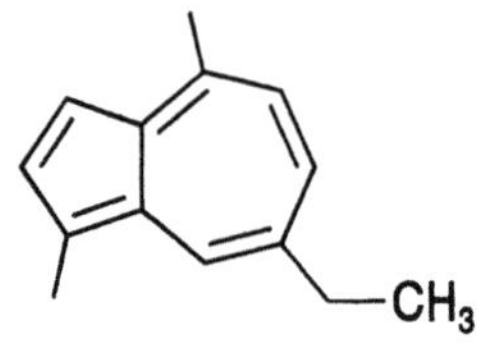

Chamazulene

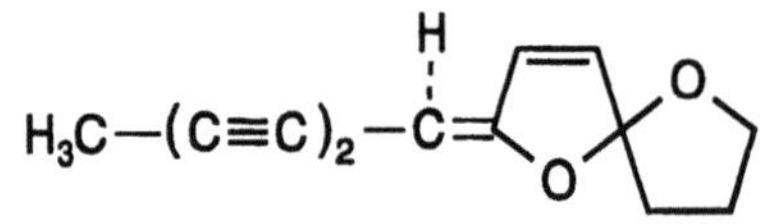

(cis-,trans-) Ene-Ine-Dicycloether

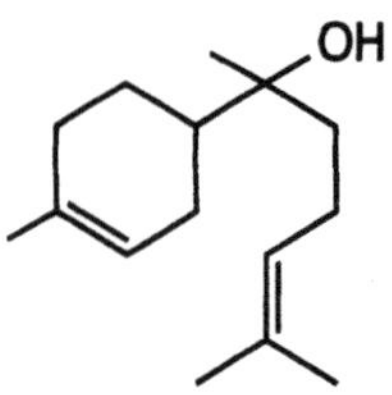

(-)-α-Bisabolol

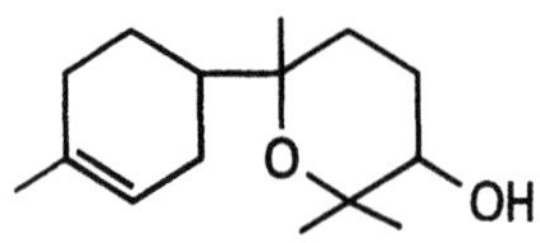

(-)-α-Bisabolol oxide A

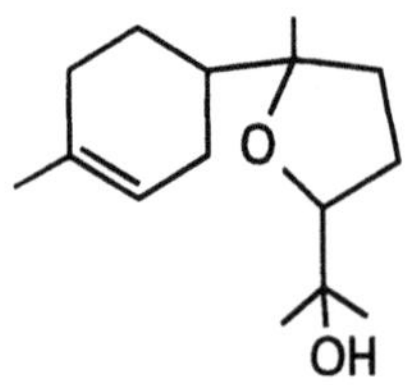

(-)-α-Bisabolol oxide B

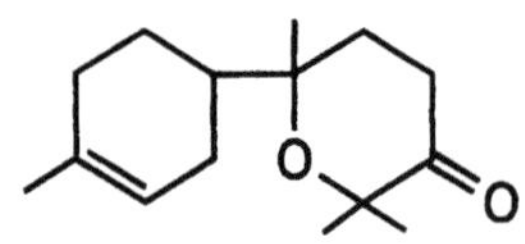

(-)-α-Bisabolon oxide A

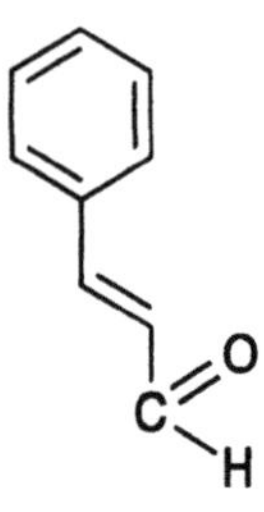

Cinnamic aldehyde

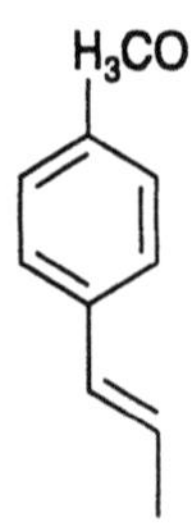

trans-Anethole

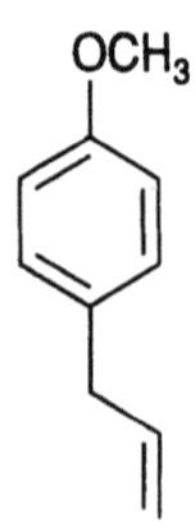

Methyl chaviol

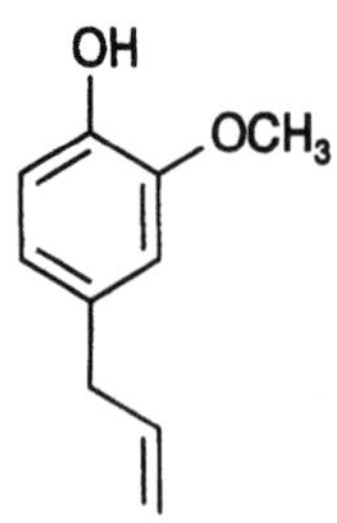

Eugenol

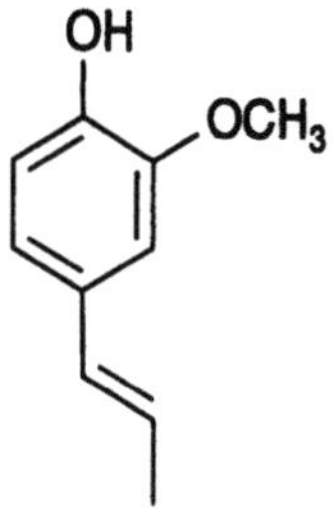

Isoeugenol

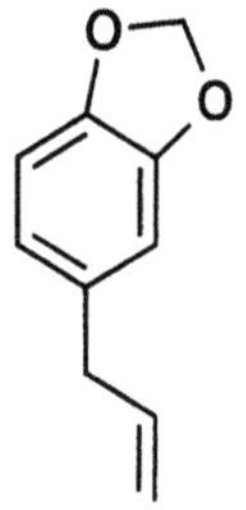

Safrole

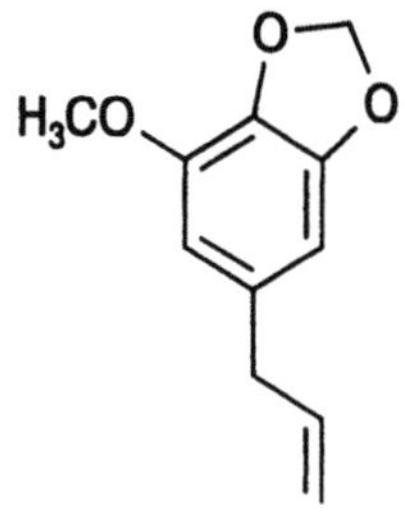

Myristicin

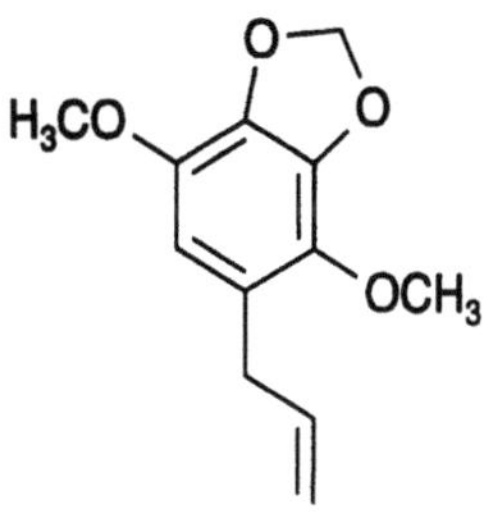

Apiol

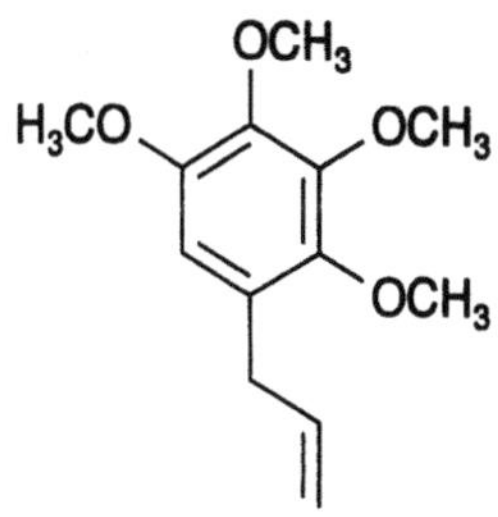

Allyltetramethoxybenzene

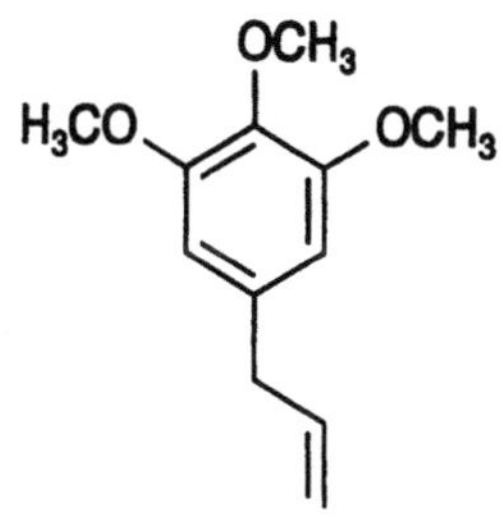

Elemicin

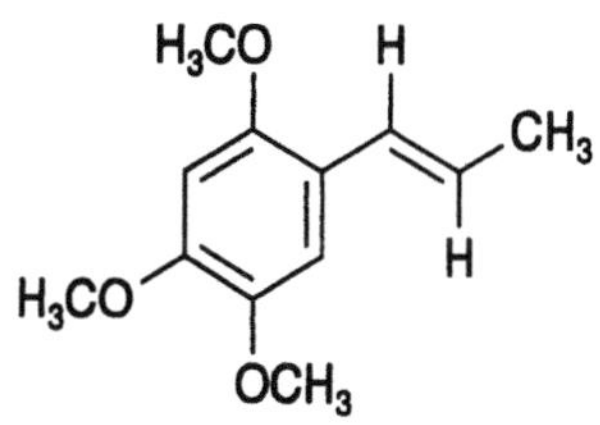

α-Asaron
(trans-Asaron)

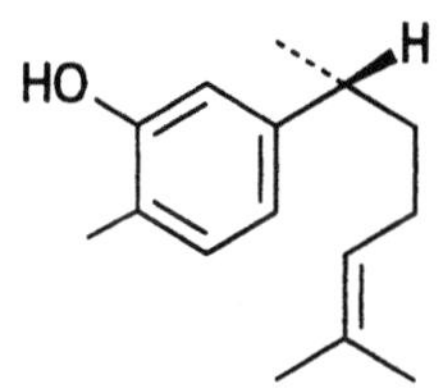

Xanthorrhizol

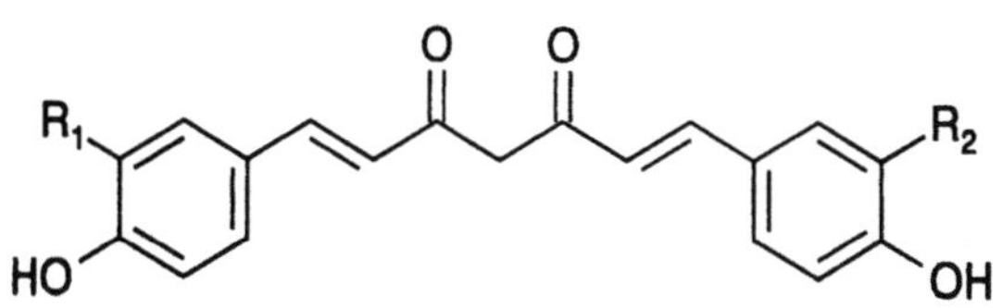

Curcumin	$R_1 = R_2 = OCH_3$
Bisdesmethoxycurcumin	$R_1 = R_2 = H$

6.6 Terpene and Phenylpropane Reference Compounds

	Reference compound[1]	R_f value	Colour
Fig. 1	Compounds applied in order of increasing R_f value and decreasing polarity		
	1 borneol	0.24	violet-blue
	2 linalool	0.30	blue
	3 piperitone	0.35	orange-red
	4 cineole	0.40	blue
	5 citral	0.42	blue-violet
	6 carvone	0.46	red-violet
	7 eugenol	0.47	yellow-brown
	8 thymol	0.52	red-violet
	9 citronellal	0.65	blue
	10 apiol	0.65	red-brown
	11 myristicin	0.75	red-brown
	12 anethole	0.85	red-brown
	13 safrole	0.87	red-brown
Fig. 2	**Monoterpene alcohols and their esters**		
	14 geraniol	0.22	blue
	15 geranyl acetate	0.64	blue
	16 nerol	0.24	blue
	17 neryl acetate	0.66	blue
	18 borneol	0.24	blue-violet
	19 bornyl acetate	0.65	blue-violet
	20 menthol	0.28	blue
	35 menthyl acetate	0.72	blue
	22 linalool	0.33	blue
	23 linalyl acetate	0.68	blue

Solvent system toluene-ethyl acetate (93:7)
Detection Vanillin-sulphuric acid reagent (VS No.42) →vis

After treatment with the VS reagent the monoterpene alcohols and their esters, cineole, the aldehyde citral and citronellal show blue or blue-violet colour in vis. The phenylpropane derivatives safrole, anethole, myristicin, apiol and eugenol are brown-red/violet, while thymol and carvon are red to red-violet; piperitone shows a typical orange colour.
Commercially available reference compounds often show additional zones at the start or in the low R_f range. This can be due to resinification, decomposition products or incompletely removed impurities.

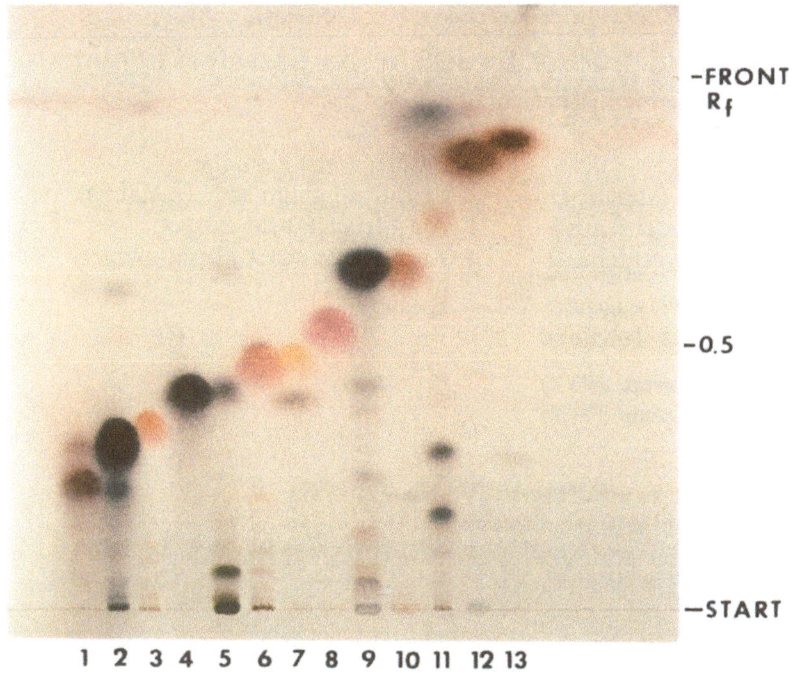
-FRONT
Rf
-0.5
-START
1 2 3 4 5 6 7 8 9 10 11 12 13

Fig. 1

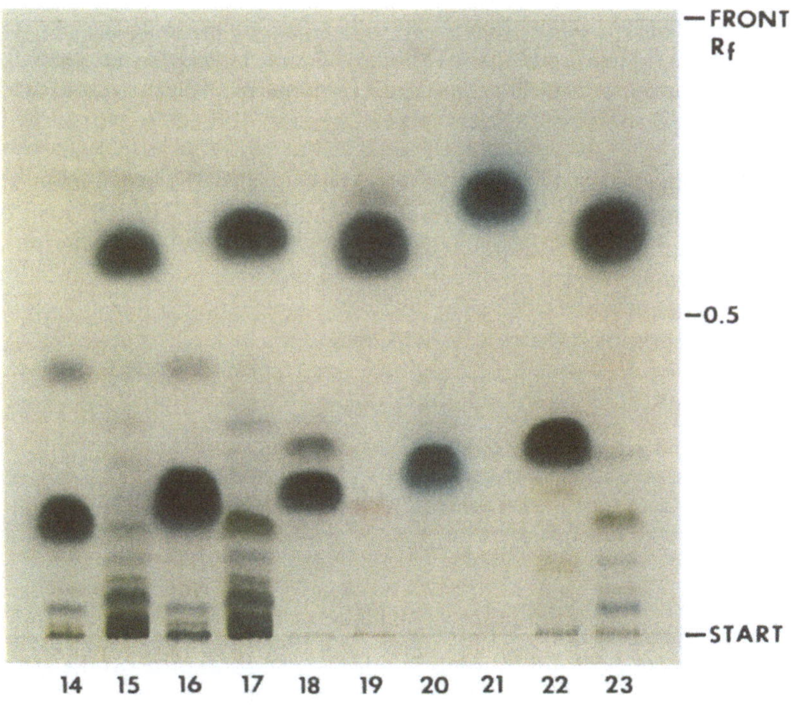
-FRONT
Rf
-0.5
-START
14 15 16 17 18 19 20 21 22 23

Fig. 2

6.7 Chromatograms

Anisi fructus, Foeniculi fructus, Basicili herba, Sassafras lignum

Essential oils with anethole/methylchavicol or safrole

Drug sample (essential oil)
1 Anisi fruct. aeth. (anise)
2 Anisi stellati fruct. aeth. (staranise)
3 Foeniculi fruct. aeth. (bitter fennel)
4 Foeniculi fruct. aeth. (sweet fennel)
5 Basilici herba aeth. (basil)
6 Sassafras lignum aeth. (sassafras)
7 Anisi fruct. (DCM-extract)
8 Anisi stellati fruct. (DCM-extract)

Reference compound
T1 anethole
T2 safrole
T3 eugenol
T4 fenchone

Solvent system
Fig. 3A+B toluene-ethyl acetate (93:7)
Fig. 4A+B toluene-ethyl acetate (93:7)
Fig. 4C toluene

Detection
Fig. 3A+B Vanillin-sulphuric acid reagent (VS No. 42) →vis
Fig. 4A Concentrated sulphuric acid →vis.
B Phosphormolybic acid/K permanganate (PMS/PM No. 34 + 36) →vis
C Vanillin-sulphuric acid (No.42) →vis

Fig. 3A, B The major constituent of the essential oils 1–6 is detectable VS reagent as a red-violet to brown-violet zone at R_f 0.9–0.95. In the essential oil of anise (1), staranise (2), bitter fennel (3) or sweet fennel (4) it is anethole (T1) with small amounts of the isomer methylchavicol, while basil (5) has predominantly methylchavicol which has the same R_f value as anethole. The prominent zone of sassafras oil (6) is safrole (T2). Anethole (T1) and safrole (T2) can be separated in the solvent toluene (see Fig. 4C), where safrole then shows a higher R_f value.
The blue zones in the R_f range 0.1–0.4 of the oils 1–6 are terpene alcohols (e.g. linalool at R_f 0.4) at a very low concentration in the samples 1–2, slightly higher in bitter fennel (3) and sweet fennel (4), while basil (5) shows three intensive blue terpene alcohols with linalool as a major compound. In basil oils, linalool can be the predominant compound with very little methylchavicol (chemo- or geotype). A red-violet zone at $R_f \sim 0.5$, as in samples 2–5, can occur (e.g. epoxidihydrocaryophyllene).

Fig. 4A Anethole at $R_f \sim 0.9$ and anisaldehyde at $R_f \sim 0.45$ with concentrated sulphuric acid immediately give a red to red-violet colour. Fenchone is detected as a yellow ochre zone at $R_f \sim 0.5$ after being heated at 110°C for about 5 min and at a concentration greater than >100 µg.

B Fenchone, if present in a lower concentration, can be detected by the PMA/PM reagent only. The dark blue-coloured zone of fenchone (T4) is seen in the sample of bitter fennel (3) (12%–22% fenchone), whereas a weak whitish zone is detected in sweet fennel (4) (0.4%–0.8% fenchone). Fenchone is absent in anise (1) or star anise.

C Detection with VS reagent (110°C/5 min) reveals in anise (7) and staranise (8) the grey-violet zones of anethole (T1) at R_f 0.8 and of triglycerides (in DCM extracts only) at R_f 0.2–0.3. In the R_f range above anethole, no prominent zone should be present. A high amount of safrole (T2) instead of anethole might indicate an adulteration with the poisonous Illicium anisatum (syn.I. religiosum).

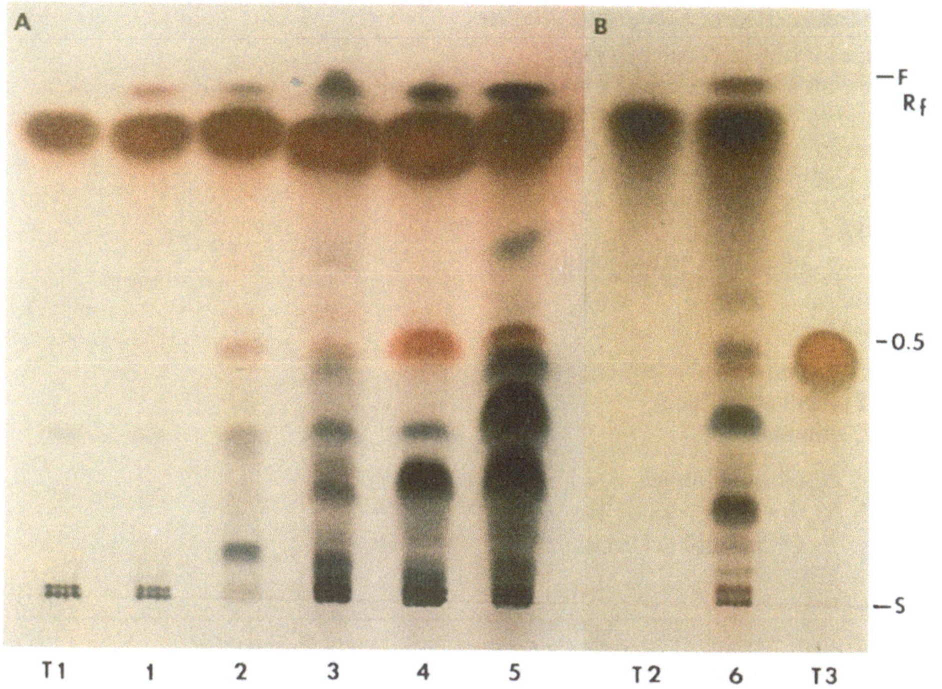

Fig. 3

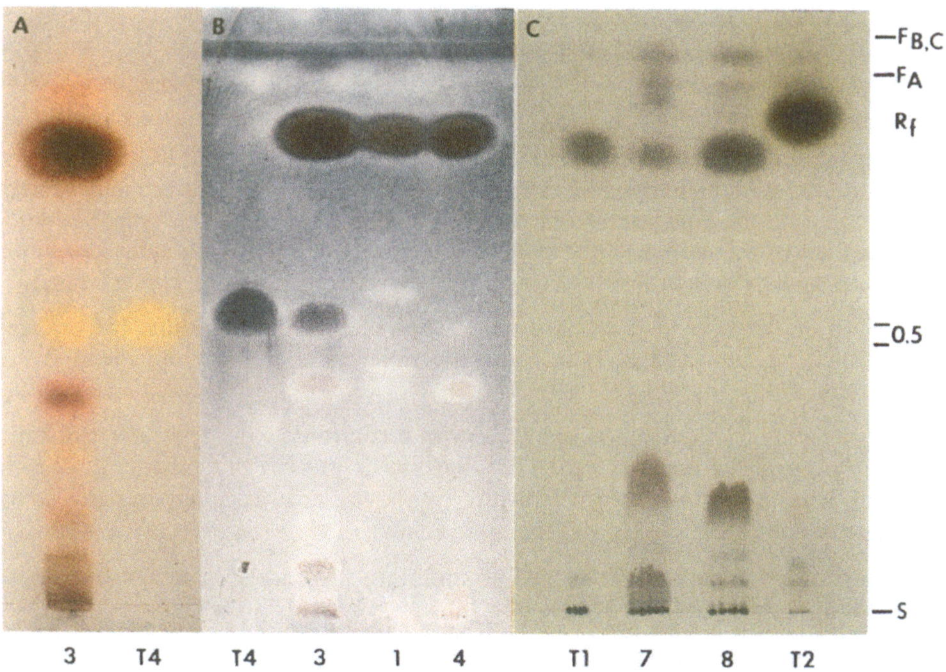

Fig. 4

Cinnamomi cortex, Caryophylli flos Essential oils with eugenol

Drug sample (essential oil)
1 Cinnamomi ceylanici cortex aeth.
2 Cinnamomi aromaticae cortex aeth.
3 Cinnamomi ceyl. cortex (DCM extract)
4 Cinnamomi aromat. cortex (DCM extract)
5 Caryophylli flos aeth.

Reference compound
T1 linalool
T2 cinnamic aldehyde (=cinnamaldehyde)
T3 eugenol
T4 coumarin

Solvent system
Fig. 5A–C toluene-ethyl acetate (93:7)
Fig. 6A+B dichloromethane
C toluene

Detection
Fig. 5A+C Vanillin-sulphuric acid (VS No. 42) → vis
B KOH reagent (KOH No.35) → UV-365 nm
Fig. 6A+C Vanillin-sulphuric acid (VS No. 42) → vis
B UV-254 nm

Fig. 5A **Cinnamon oils** (1,2) are characterized by cinnamic aldehyde (T2), seen as major grey-blue zone at R_f ~ 0.5 (VS reagent, vis).
Ceylon cinnamon oil (1) shows an additional violet-blue zone at R_f ~ 0.2, a blue zone at R_f ~ 0.4 (linalool/T1), and the terpene ester at R_f ~ 0.65.
Cassia cinnamon oil (2) has a prominent blue zone of terpene hydrocarbons (e.g. caryophyllene, α-pinene) at the solvent front as well as two minor blue zones at R_f 0.25–0.3.

B Development of DCM extracts in dichloromethane and detection with KOH reagent in UV-365 nm shows in the R_f range of cinnamic aldehyde at R_f ~ 0.5 a green and directly below at R_f ~ 0.45 the blue fluorescent zone of o-methoxycinnamic aldehyde. Besides a higher amount of the aldehyde, the cassia cinnamon bark (4) also contains coumarin (T4), which is found as a blue fluorescent zone below the aldehyde (see note below).

C Essential oil of **Caryophylli flos (clove oil,** 5) shows as major compound the orange-brown zone of eugenol (T3, R_f ~ 0.5) and the violet zone of β-caryophyllene at the solvent front.

Fig. 6A A TLC development of cinnamon oils (1,2) with dichloromethane separates cinnamic aldehyde (T2) and the eugenol zone (T3). Eugenol is present in Cinnamomi ceylani cortex only and is found as a brown zone directly above cinnamic aldehyde, followed by the blue ester zone (VS reagent, vis) (see note below).

B The phenyl propane derivatives as well as coumarin (T4) are seen in UV-254 nm as prominent quenching zones at R_f 0.45–0.55.

C In the solvent toluene eugenol is found in a lower R_f-range.

Note: Eugenol (<5%) is reported in Cinnamomi ceylanici cortex only, while coumarin is found in C. aromaticae cortex only. Very often the powdered trade samples of cinnamon bark are mixtures of both species and therefore both compounds are present.

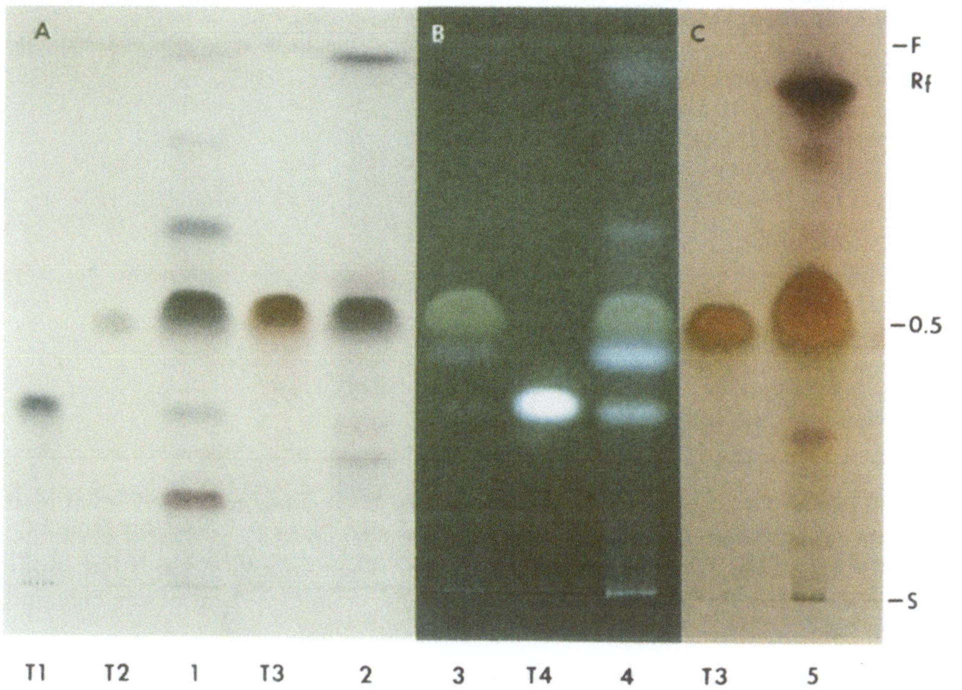
A
B
C
–F
Rf
–0.5
–S
T1
T2
1
T3
2
3
T4
4
T3
5

Fig. 5

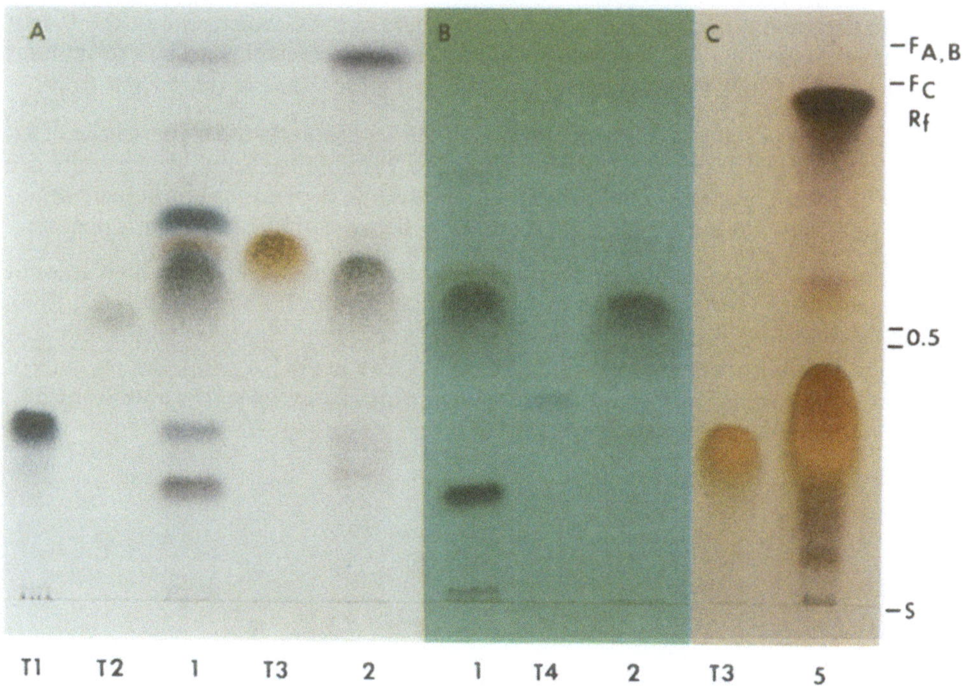
A
B
C
–F A,B
–F C
Rf
0.5
–S
T1
T2
1
T3
2
1
T4
2
T3
5

Fig. 6

Calami rhizoma, Asari radix Essential oils with asarone

Drug sample
1,2 Calami rhizoma (without bark)
3 Calami rhizoma (USA/with bark)
4 Calami extract (5:1/40% EtOH)
5 Asari europaeae radix
6 Asari canadensis radix

Reference compound
T1 trans-isoasarone
T2 eugenol
T3 bornyl acetate

Solvent system toluene-ethyl acetate (93:7)

Detection Fig. 7 A UV-365 nm B UV-254 nm
Fig. 8 Vanillin-H_2SO_4 reagent (VS No. 42) vis

Fig. 7A DCM extracts or TAS distillates of **Calami rhizoma** (1–3) show in UV-365 nm at least seven blue or violet-blue fluorescent zones from the start up to $R_f \sim 0.55$ and additional zones in the R_f range 0.75 and at the solvent front. Their concentration is low in the commercial extract (4). The zone at $R_f \sim 0.4$ in the samples 1–4 fluoresces blue and violet-blue, due to the α-β-asarone mixture (T1, violet-blue).

B The samples 1–3 show prominent quenching zones (UV-254 nm) from the start up to R_f 0.65, with two major zones at $R_f \sim 0.4$ (α-β-asarone, T1) and $R_f \sim 0.65$.

Fig. 8 Treatment with the VS reagent characterizes the chromatogram of **Calami rhizoma** samples 1–3 by a series of violet, blue and brown-violet zones (vis.), extending from $R_f \sim 0.05$ up to the solvent front. The asarone (T1) appears as a red-violet zone at $R_f \sim 0.4$. In the R_f range of eugenol (T2) all oils show one to two weak zones followed by a prominent blue zone at R_f 0.75 (R_f range of bornyl acetate) and a blue zone at the solvent front.

The TLC pattern of Calami rhizoma samples varies according to the origin of the drug, the vegetation period and the extraction method (sample 4). Some compounds are unstable and form artefacts. The amount of α-β-asarone depends on the genetic origin (di-, tri- or tetraploid) but should not exceed 0.5%, because of its carcinogenic potential.

DCM extracts or TAS distillates of **Asari europ. radix** (5) show a relatively high amount of asarones (T1), accompanied by four weaker blue zones in the R_f range 0.1–0.3, while in **Asari canadensis radix** (6) only traces of asarones are found. Sample 6 is characterized by a major dark-blue zone in the R_f range of bornyl acetate (T3), a yellow-brown zone of eugenol (T2) at $R_f \sim 0.5$ and five to six dark-blue zones from the start up to $R_f \sim 0.35$.

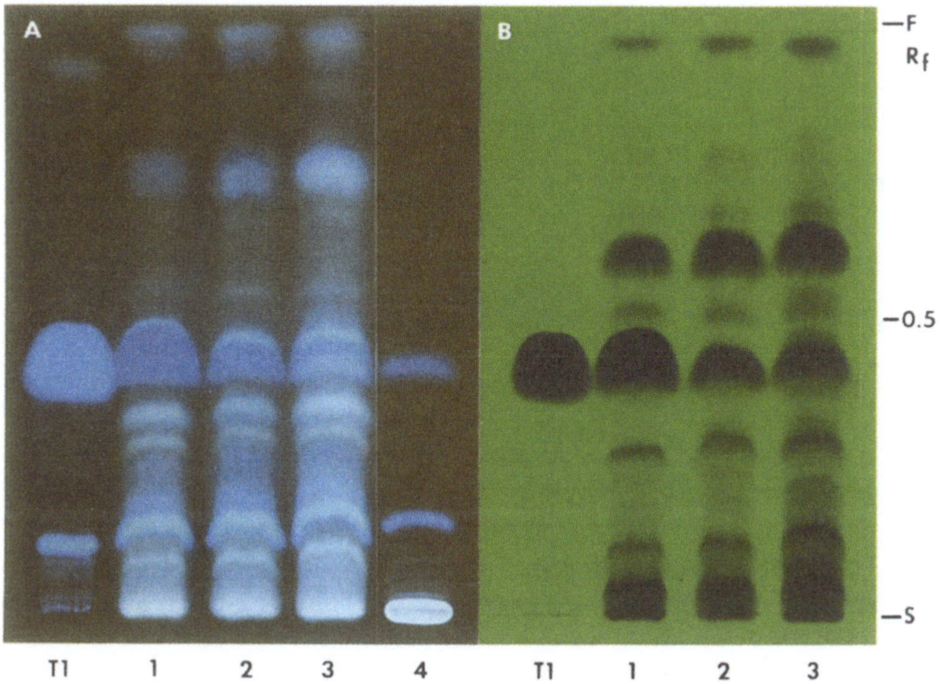

Fig. 7

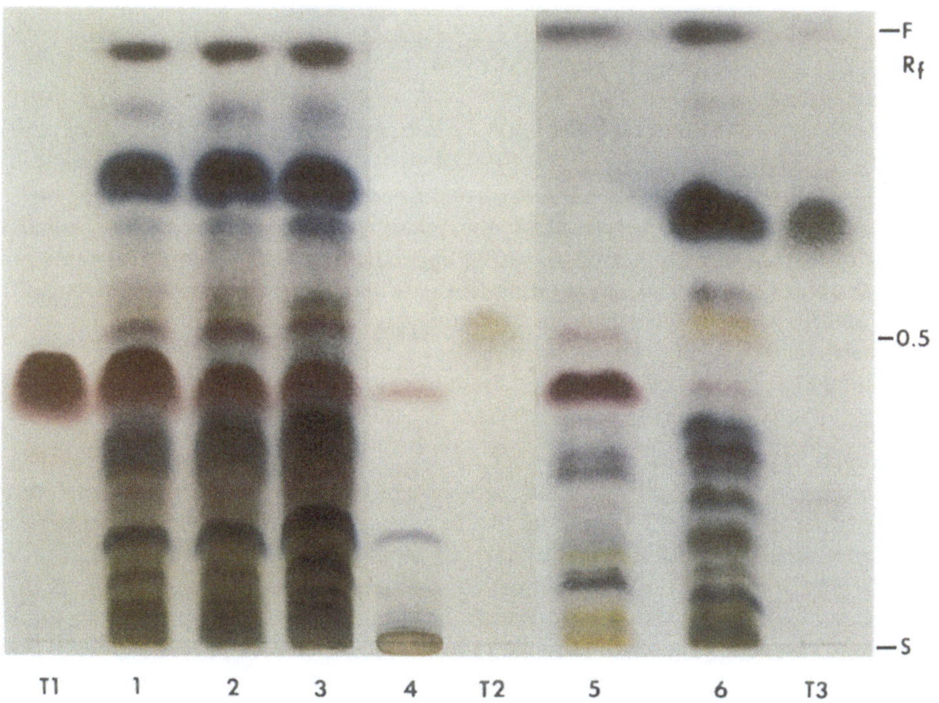

Fig. 8

Myristicae semen, Petroselini fructus Essential oils with apiole and myristicin

Drug sample (essential oil)	1 Myristicae aeth. (commercial oil) 2 Petroselini fructus (myristicin race) 3 Petroselini fructus (steam distillate) 4 Petroseli fructus (apiol race) 5–7 Petroselini aeth. (commercial oils)
Reference compound	T1 myristicin T2 eugenol T3 apiol T4 allyltetramethoxybenzene (ATMB) T5 elemicin
Solvent system	Figs. 9, 10 toluene-ethyl acetate (93:7)
Detection	Fig. 9A UV-254 nm B Vanillin-sulphuric acid reagent (VS No. 42) →vis Fig. 10A+B Vanillin-sulphuric acid reagent (VS No. 42) →vis

Fig. 9 All phenylpropane derivatives of **Myristicae** (1) and **Petroselini aeth.** (2) are seen as quenching zones in UV-254 nm (→A) and as brown to red-brown-coloured zones with VS reagent in vis. (→B).

Compound		R_f	**Essential oil Fig. 9**	**Fig. 10**
safrole	(T13, p. 166)	0.95	1 Myristicae aeth.	6 Petroselini aeth.
myristicin	T1	0.80	1 Myristicae aeth.	2–7 Petroselini aeth.
apiol	T3	0.75	2 Petroselini aeth.	3–7 Petroselini aeth.
eugenol	T2	0.55	1 Myristicae aeth.	6–7 Petroselini aeth.
allyltetramethoxy-benzene	T4	0.45	2 Petroselini aeth.	3–7 Petroselini aeth.
elemicin	T5	0.40	2 Petroselini aeth.	3–7 Petroselini aeth.

Myristicae aeth. (1) is characterized by the major zone of myristicin (T1), smaller amounts of safrole directly above, traces of eugenol (T2) and two to three zones of terpene alcohols (R_f 0.15–0.25). Depending on the origin of the oil (semen or macis), the amount of THC at the solvent front can be more highly concentrated and at the same time safrole can be absent.
Petroselini aeth. (2). This oil shows myristicin (T1) as its major compound (myristicin race).

Fig. 10 **TLC synopsis of parsley oils**
Petroselinum can occur as chemical race (chemotype), in which the predominant compound is either myristicin (2,3) or apiol (4). In rare cases allyltetramethoxy benzene is the major compound. Commercial parsley oils from cultivated plants (5,6,7) contain myristicin and apiol in various, sometimes in approximately equal concentrations (6). The parsley oils 2–7 also show slight variations of minor phenylpropanoids in the R_f range 0.4–0.5 (eugenol/T2, allytetramethoxy benzene/T4, elemicin/T5)

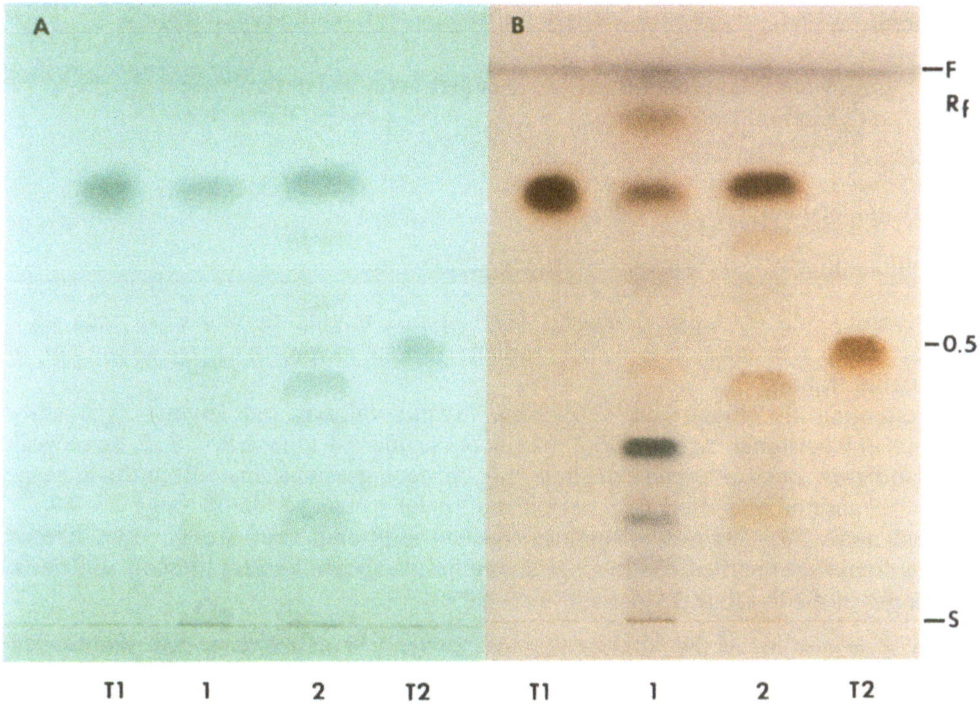

Fig. 9

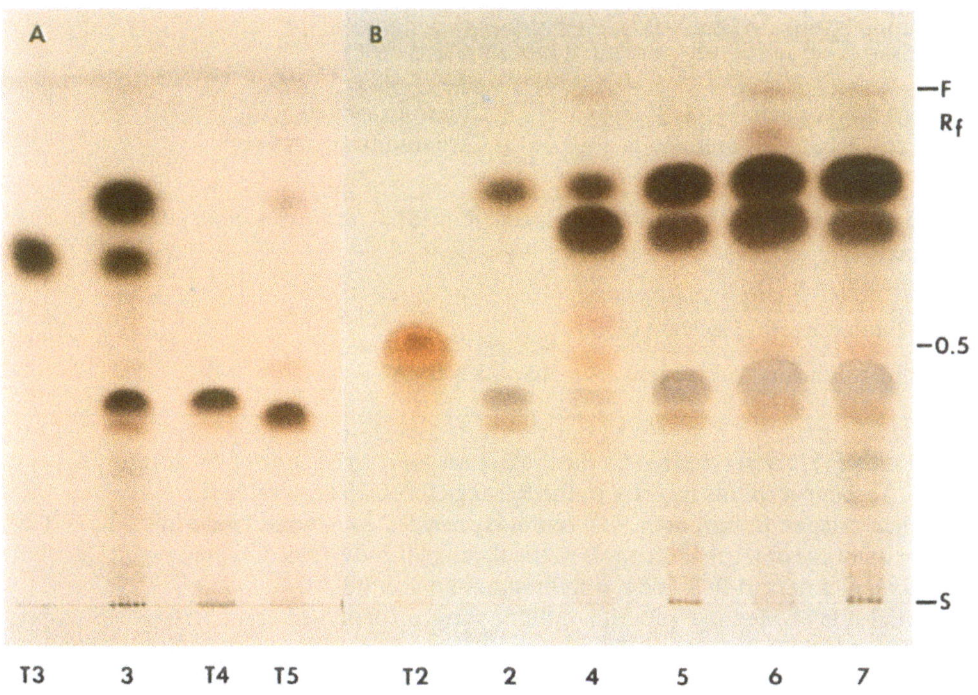

Fig. 10

Ajowani fructus, Thymi and Serpylli herba Essential oils with thymol/carvacrol

Drug sample (essential oil) 1 Ajowani fructus aeth. 3,6 Serpylli herba aeth.
2,4,5,7,8 Thymi herba aeth.

Reference T1 thymol T2 carvacrol

Solvent system Fig. 11 toluene-ethyl acetate (93:7)

Detection Vanillin-sulphuric acid reagent (VS No. 42) →vis

Fig. 11 The essential oil of **Ajowani fructus** (1) contains mainly thymol (T1), seen as a characteristic red zone at $R_f \sim 0.5$. Indian Ajowan is known as an adulterant of Petroselini fructus.
The essential oils **Thymi aeth.** (2,4) from Thymus vulgaris and Thymus zygis show thymol and its isomer carvacrol (5,7) (see note) as one red zone at $R_f \sim 0.55$, three weak blue and grey zones of terpene alcohols (e.g. borneol, geraniol, linalool) in the R_f range 0.15–0.35 and terpene esters (e.g. bornyl and linalyl acetate) in the R_f range 0.7–0.8.
Serpylli aeth. (3,6) (Thymus pulegioides) has two additional terpene ester zones directly above thymol. A rectified commercial thyme oil (8) shows, besides thymol, additional red zones in the R_f range 0.3–0.4 and 0.65–0.95.

Note: A separation of the isomers thymol/carvacrol is achieved by two-dimensional TLC with toluene-ethyl acetate (93:7) in the first and toluene-carbon tetrachloride-o-nitrotoluene (33:33:33) in the second dimension.

Carvi, Coriandri, Cardamomi fructus
Menthae crispae folium Essential oils with terpenes

Drug sample (essential oil) 1 Carvi fructus aeth. 4 Coriandri semen aeth.
2 Menthae crispae folium aeth. 5 Cardamomi fructus aeth.
3 Coriandri fructus aeth.

Reference compound T1 carvone T4 α-terpineol (R_f 0.25) ► terpinyl acetate (R_f 0.75)
T2 linalool
T3 cineole

Solvent system Fig. 12 toluene-ethyl acetate (93:7)

Detection Vanillin-sulphuric acid reagent (VS No. 42)→vis A–C

Fig. 12A **Carvi aeth.** (1) is characterized by the intense raspberry-red zone of D-carvone (T1) at $R_f \sim 0.5$. Terpene alcohols migrate in the R_f range 0.2–0.25 (e.g. carveol).
Menthae crispae folium aeth. (2) contains, besides L-carvone (red-violet, $R_f \sim 0.5$), higher amounts of terpene alcohols in the R_f range 0.2–0.3 (e.g. dihydrocuminyl alcohol) and terpene esters at R_f 0.7 (e.g. dihydrocuminyl acetate).
Essential oils of Menthae piperitae folium show a totally different terpeneoid pattern (see fig. 13 p. 178).

B **Coriandri fructus** (3) and **C. semen** (4). Linalool (T2) is the major compound in both essential oils. Commercial seed oil can have a higher amount of linalool and in addition geraniol (R_f 0.2) and geranyl acetate (R_f 0.7), detected as grey zones.

C **Cardamon oil** (5) shows the prominent blue zone of α-terpinyl acetate ($R_f \sim 0.75$/T4), cineole (R_f 0.5/T3) and three minor terpene alcohols such as borneol, terpineol (R_f 0.2–0.25), linalool ($R_f \sim 0.35$/T2) and limonene at the solvent front.

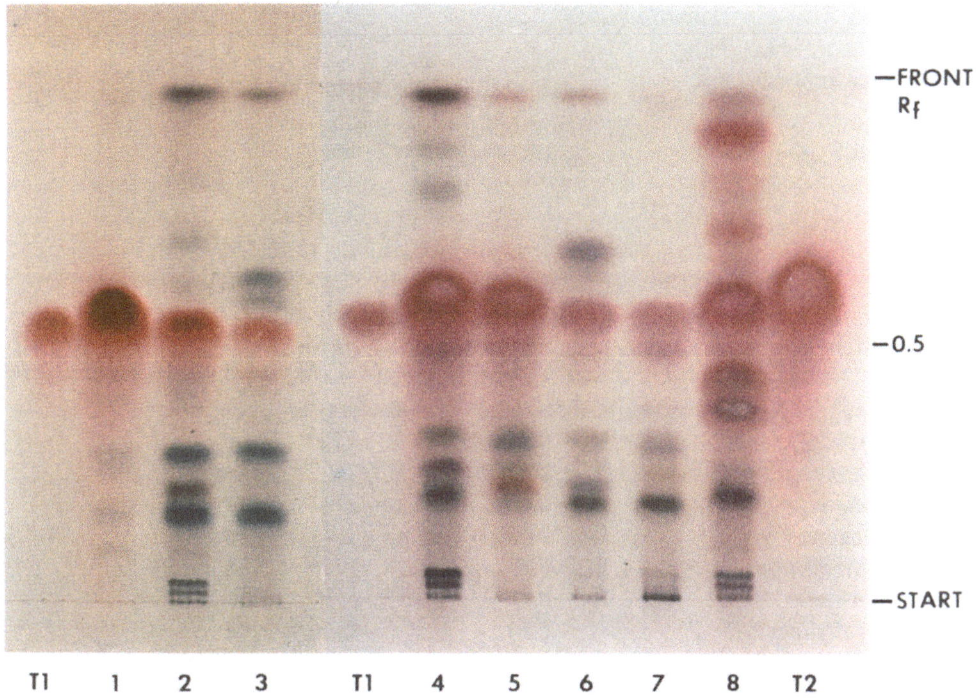

—FRONT
Rf
—0.5
—START
T1
1
2
3
T1
4
5
6
7
8
T2

Fig. 11

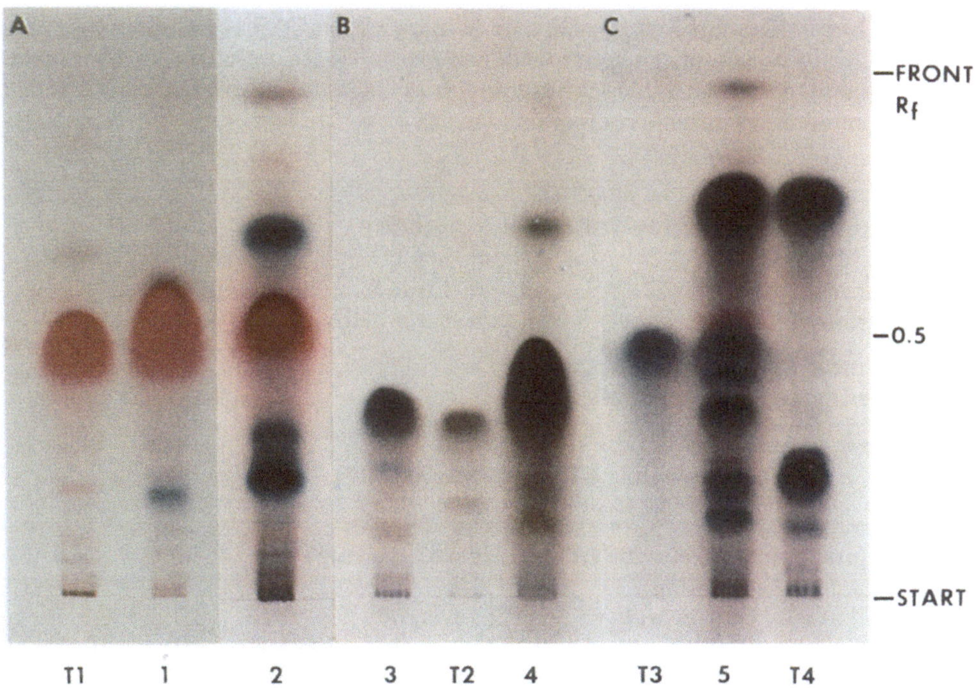

A
B
C
—FRONT
Rf
—0.5
—START
T1
1
2
3
T2
4
T3
5
T4

Fig. 12

Menthae folium (Lamiaceae)

Drug sample (essential oil)	1 Menthae piperitae aeth. 2 Menthae arvensis aeth.	
Reference compound	T1 menthol T3 menthyl acetate	T2 menthone/isomenthone T4 menthofuran
Solvent system	Fig.13A,B toluene-ethyl acetate (93:7) C dichlormethane (100)	
Detection	A Vanillin-sulphuric acid reagent (VS No. 42) →vis B Phosphomolybdic acid reagent (PMA No. 34) →vis C Anisaldehyde-sulphuric acid reagent (AS No. 3) →vis	

Fig. 13A Official peppermint oil 1 is characterized by the following terpenes:

I	menthol	$R_f \sim 0.30$	(T1)	blue
II	piperitone	$R_f \sim 0.35$		orange
III	cineole	$R_f \sim 0.40$		blue
IV	pulegone (?)	$R_f \sim 0.48$		blue
V	isomenthone	$R_f \sim 0.55$	(T2)	blue-green
VI	menthone	$R_f \sim 0.70$	(T2)	blue-green
VII	menthyl acetate	$R_f \sim 0.75$	(T3)	blue
▶	THC	solvent front		violet-blue
▶	menthofuran	below THC	(T4)	red-violet

B Even with low concentrations of terpenes, such as menthyl acetate (T3) or THC in sample 1, the PMA reagent produces intense, uniform blue-black-coloured zones.

C For **Menthae arvensis aeth.** (2) the German pharmacopoeia DAB 10 describes the separation in dichloromethane. The prominent terpenes I–VII are detected with AS reagent. Cineole and menthofuran are absent. Menthofuran (T4) is detectable in freshly distilled peppermint oil only (instable compound).

Rosmarini and Melissae folium (Lamiaceae)

Drug sample (essential oil)	1 Rosmarini aeth. 2 Rosmarini fol. (MeOH extract)	3 Melissae fol. (MeOH extract) 4,5 Melissae aeth.
Reference compound	T1 1,8-cineole T3 rosmarinic acid	T2 borneol T4 citral
Solvent system	Fig. 14A+C toluene-ethyl acetate (93:7) B toluene-ethyl fomiate-formic acid (50:40:10)	
Detection	A+C Vanillin-sulphuric acid reagent (VS No. 42) →vis B Natural products reagent (NP/PEG No. 28)→ UV-365 nm	

Fig. 14A **Rosmarini aeth.** (1) shows with VS reagent six mainly blue zones (vis.) in the R_f range 0.25–0.45 with cineole as the major zone (T1). Due to plant origin the amount of terpene alcohols in the R_f range below cineole differs (e.g. borneol, T2 > 20%).

B A methanolic extract of **Rosmarini folium** (2) and **Melissae folium** (3) contains up to 5% rosmarinic acid (T3).

C Oil of Melissa balm (4,5) shows as main blue zone citronellal at R_f 0.75, citral at R_f 0.45 and terpene alcohols at R_f 0.15–0.3. The quality of the oils varies, as explained in Fig. 15.

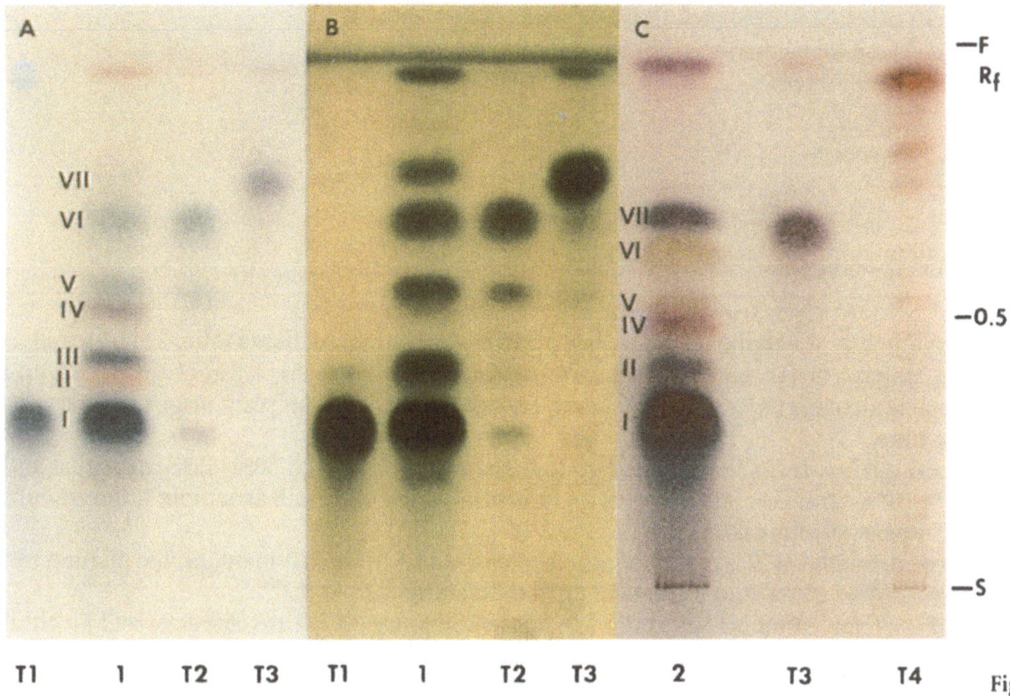
A
B
C
—F
Rf
VII
VI
V
IV
III
II
I
VII
VI
V
IV
II
I
—0.5
—S
T1
1
T2
T3
T1
1
T2
T3
2
T3
T4
Fig. 13

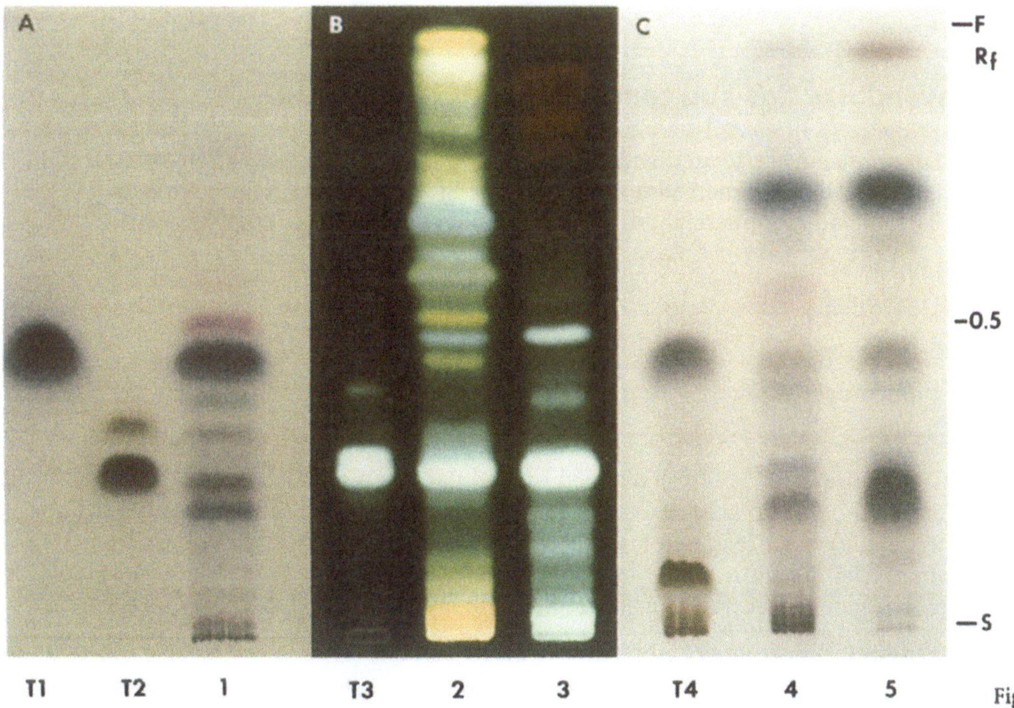
A
B
C
—F
Rf
—0.5
—S
T1
T2
1
T3
2
3
T4
4
5
Fig. 14

Melissae folium and substitutes (Lamiaceae)

Commercial oils	1 Melissae fol. aeth. 2 Citronellae aeth.	3 Citri aetheroleum 4 Lemon grass oil
Reference compound	T1 citronellal T2 citral	
Solvent system	Fig. 15 toluene-ethyl acetate (93:7)	
Detection	Vanillin-sulphuric acid (VS No. 42) →vis	

Fig. 15 The amount of volatile oil gained by steam distillation of **Melissae folium** (1), as well as the amount of the oil constituents citronellal (T1/R_f ~ 0.75), citral (T2/R_f ~ 0.5 and terpene alcohols (R_f 0.2–0.4, e.g. nerol, citronellol) depends on plant origin and harvesting time.

Good-quality drugs should yield up to 0.2% essential oil with 30%–40% citronellal and 20%–30% citral (see Fig. 14, track 2). In lower-quality oils, such as sample 1, the amount of terpene alcohols dominate.

Java citronella oil (2) resembles official melissa oil 1 in its chromatographic picture, but has a higher content of citronellal (T1) and geraniol (R_f ~ 0.2).

Commercial **lemon oil** (see note) (3) and **lemon grass oil** (4) are characterized by citral (R_f ~ 0.5/T2). The oils 2 and 4 are used as substitutes of Melissae aetheroleum.

Note: A TLC comparison between the different qualities of distilled and squeezed lemon oils is given in Figs. 17 and 18.

Lavandulae flos and commercial oils (Lamiaceae)

Essential oil	1 Lavandulae flos (steam distillate) 2 Lavandin oil (commercial oil) 3 Barrême oil (commercial oil)	4 French Mt. Blanc oil (commercial oil) 5 Spike Lavender (commercial oil) 6 Lavender oil (L. angustifolium)
Reference compound	T1 linalyl acetate T2 linalool	T3 linalool ▶ linalyl acetate
Solvent system	Fig. 16 toluene-ethyl acetate (93:7)	
Detection	Vanillin-sulphuric acid reagent (VS No. 42)→vis	

Fig. 16A **Lavandulae aeth.** (1) of fresh distilled Lavandulae flos is characterized by the prominent blue zones of linalyl acetate (R_f ~ 0.75/T1), linalool (R_f ~ 0.3/T2) and a further terpene alcohol at R_f ~ 0.2 (e.g. nerol, geraniol).

Commercial lavandin oil (2) contains cineole, a blue zone directly above linalool, in almost equal concentration as linalool and linalyl acetate.

B **Lavandin** (2), **Barrême** (3), **French Mt. Blanc** (4) and **lavender oil** (6) are qualitatively alike in the main zones, with quantitative differences in the amount of linalyl acetate, linalool, cineole and epoxidihydrocaryophyllene at R_f 0.5–0.55 characteristic red-violet zone in the commercial oil samples 2–6. **Spike lavender oil** (5) has an almost equal linalool and cineol content. Linalyl acetate is absent.

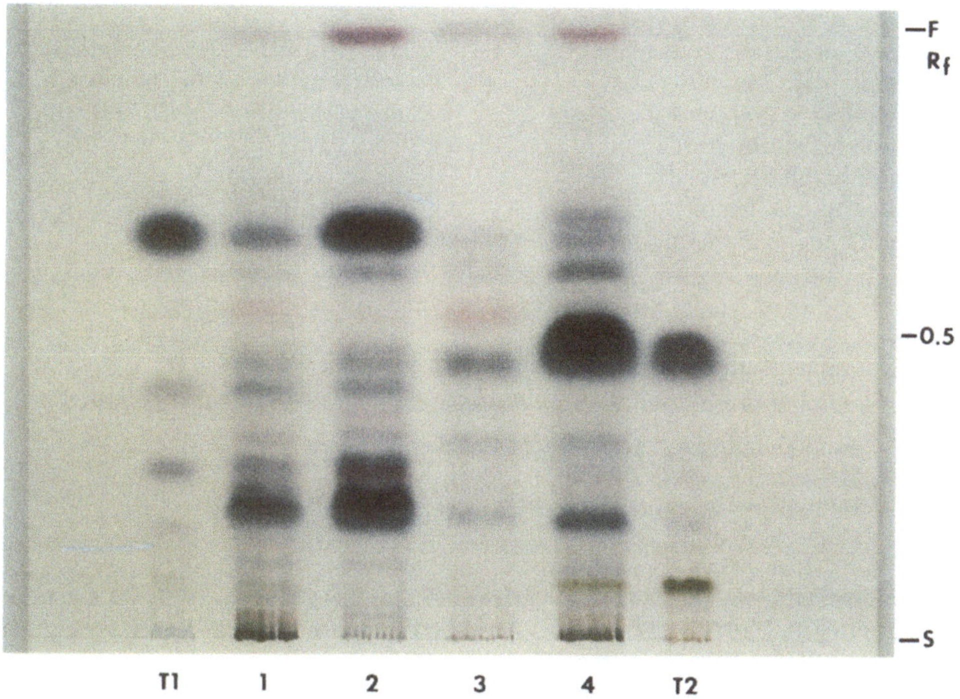

Fig. 15

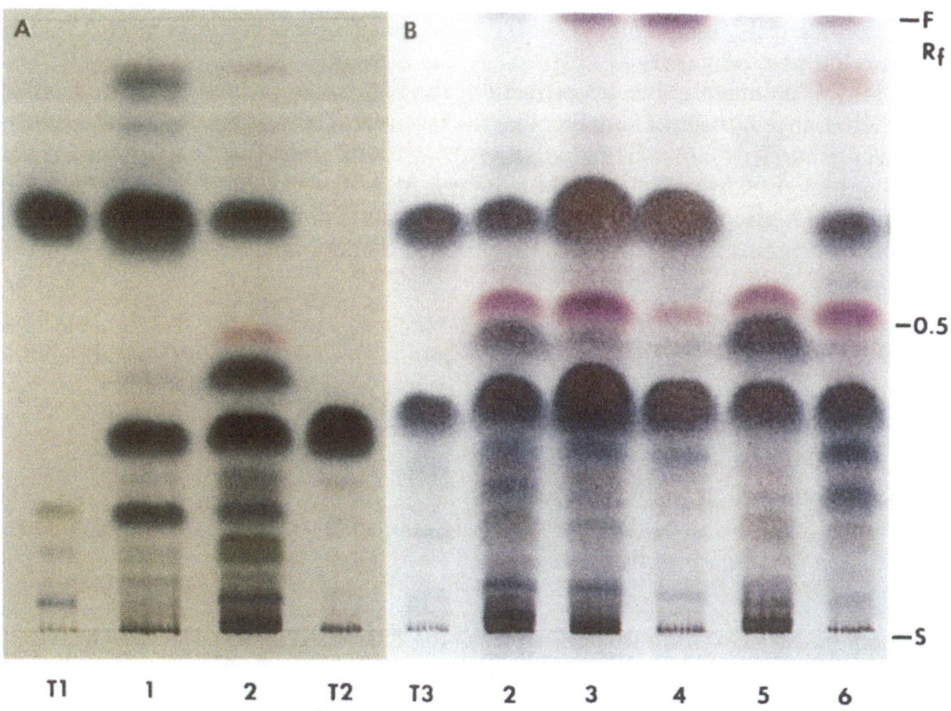

Fig. 16

Aurantii and Citri pericarpium

Drug sample (essential oils)

1 Aurantii peric. (steam distillate)
2 Aurantii peric. (oil, bitter)
3 Aurantii peric. (oil, sweet)
4 Citri peric. (steam distillate)
5 Citri aeth. (oil, squeezed)
6 Citri aeth. (messina oil)
7 Aurantii flos aeth. (neroli oil)
8 Citri var. bergamiae aeth. (bergamot)
9 Citri var. bergamiae aeth. (petit grain)
10 Aurantii pericarpium (MeOH extract 1 g/10 ml, 20 µl)
11 Citri pericarpium (MeOH extract 1 g/10 ml, 20 µl)

Reference T1 citral

Solvent system
Fig. 17 toluene-ethyl acetate (93:7)
Fig. 18A toluene-ethyl acetate (93:7)
B ethyl acetate-formic acid-water (67:7:26/upper phase) - polar system

Detection
Fig. 17 Vanillin-sulphuric acid reagent (VS No. 42) →vis
Fig. 18A UV-365 nm
B Natural products reagent (NP/PEG No. 28)→ UV-365 nm

Fig. 17A **Aurantii pericarpium** (2,3) and **Citri pericarpium** sample (5) are volatile oils squeezed from fresh peels. They contain a higher amount of limonene, seen as a grey-violet zone at the solvent front, than their steam distillates (1,4).
The oil samples 1–3 show up to ten minor grey and red-violet zones of terpene alcohols (R_f 0.1–0.4) and terpene aldehydes (R_f 0.5–0.65).
Citri oil (4) has four prominent greyish-blue zones (R_f 0.2/0.3/0.45/0.6), while in Citri sample(5) citral (R_f ~ 0.45/T1) and limonene at the solvent front are equally concentrated. Commercial **Messina oil** (6) shows a deviating TLC pattern with approximately ten zones in the R_f range 0.1–0.6.

B **Neroli oil** (7), obtained either by extraction, the enfleurage process or by distillation from fresh orange blossoms, contains like the lavender oils (see Fig. 16) the blue zones of linalyl acetate (R_f ~ 0.6) and linalool (R_f ~ 0.3) as main constituents, a further terpene alcohol at R_f ~ 0.15 and a yellow-red pigment zone at R_f ~ 0.45.
Bergamot oil (8) also has linalyl acetate and linalool as major compounds, whereas **petit grain oil** (9) contains mainly linalyl acetate besides a minor terpene alcohol (R_f ~ 0.15).

Fig. 18 **Aurantii pericarpium** (1–3)
A For essential oils squeezed from fresh peels, such as samples (2) and (3), the blue fluorescent zones of methyl anthranilates, coumarins and methoxylated lipophilic flavonoids (e.g. sinensetin) are characteristic. Sample 2 has up to six, sample 3 shows two to three blue fluorescent zones, while in distillate 1 only one weak zone at R_f ~ 0.4 is seen.

Citri pericarpium (5–6)
The samples 5 and 6 show the coumarins bergamottin (a), geranyl methoxy coumarin (b), citropten (c) and a psoralen derivative (d) in the R_f range 0.1–0.5.

B **Flavonoids**
A methanolic extract of Aurantii pericarpium (10), developed in the polar solvent system shows the blue fluorescent anthranilate and coumarin zones in the R_f range 0.8–0.99. Additional blue and orange-yellow fluorescent zones of flavanon and flavanonol glycosides are seen at R_f 0.05–0.25. (For separation of flavonoid glycosides, see Section 7.1.7, Fig. 23) The flavonoid zones of the Citri pericarp methanolic extract 11 are less prominent than those in extract 10. The blue coumarin zones are found at the solvent front.

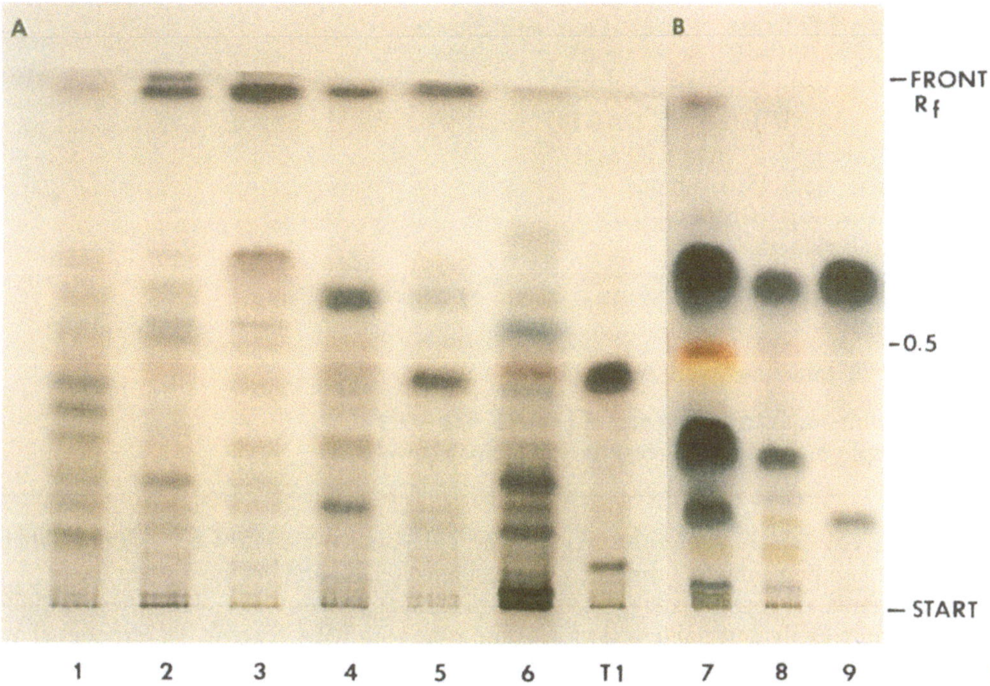
A
B
FRONT
Rf
0.5
START
1
2
3
4
5
6
T1
7
8
9

Fig. 17

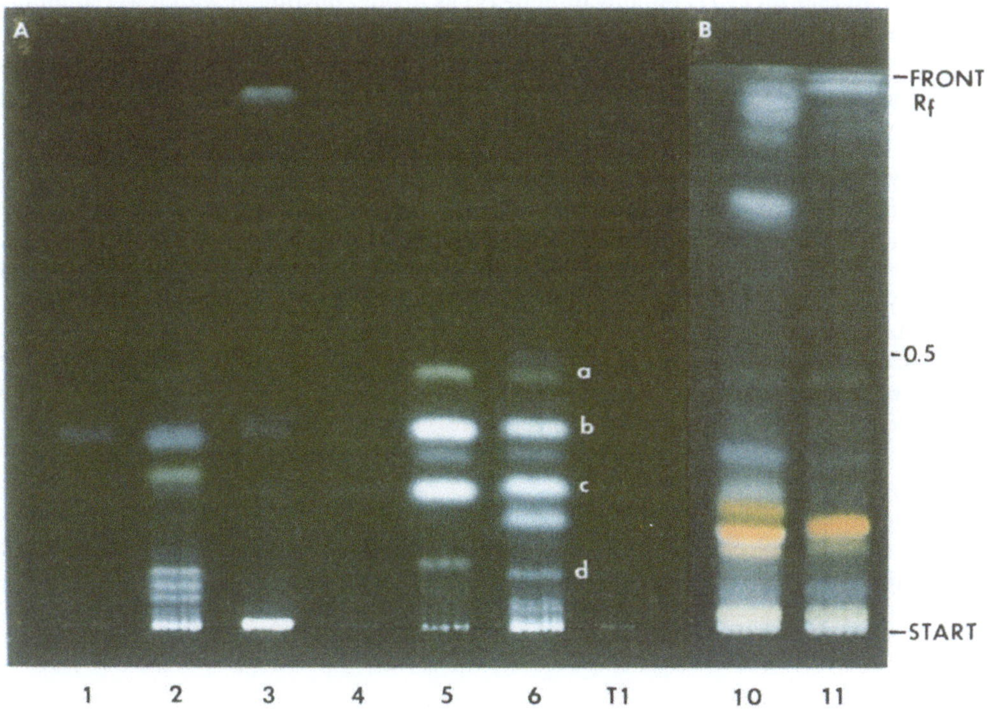
A
B
FRONT
Rf
0.5
a
b
c
d
START
1
2
3
4
5
6
T1
10
11

Fig. 18

Salviae folium
Eucalypti folium Essential oils with cineole

Drug sample (essential oil)
1,2 Salviae aeth. (Dalmatian oil I/II)
3 Salviae aeth. (Greek oil, DAB 10)
4 Salviae aeth. (commercial sage oil)
5 Salviae aeth. (Spanish oil)
6,7 Salviae aeth. (Greek oil I/II)
8 Eucalypti aeth.

Reference compound
T1 α-β-thujone = ((−)-thujone > 35%, (+)-thujone > 65%)
T2 cineole

Solvent system Fig. 19, 20 toluene-ethyl acetate (93:7)

Detection
V Vanillin-sulphuric acid reagent (VS No. 42) →vis
P Phosphomolybdic acid reagent (PMA No. 34) →vis

Fig. 19 Commercial **Salviae aetherolea** (1–3) (sage oils) can be classified according to their content and percentage of thujone (T1/a), cineole (T2/c) and bornyl acetate (b).
The essential oil constituents react with VS reagent as blue or violet-blue zones. Thujone (a) is more easily detectable as a violet-blue zone with PMA reagent. All terpenes show a blue to violet-blue colour in vis.

VS reagent in combination with PMA reagent, vis
Dalmatian sage oil (1,2) contains thujone (a) as major constituent with lower amounts of cineole (c), two terpene alcoholes (R_f 0.2–0.4) and THC at the solvent front.
Greek sage oil (3) contains mainly cineole (c), only traces of thujone (a), two to three terpene alcohols (R_f 0.2–0.4) and THC at the solvent front. Bornyl acetate (b) moves directly ahead of the thujone zone (a).

Fig. 20 **TLC synopsis of sage oils** (PMA reagent, vis)
In many commercial salvia drug preparations or essential oils, thujone and cineole are present in approximately equal concentrations (4).
Spanish oil (5) can be differentiated from the Greek oil 6 by a lower content of cineole (c) and by the absence of thujone (a). Bornyl acetate (b) and four terpene alcohols are detectable in the R_f range 0.2–0.4.
Greek oil (6) shows cineole (c) as major zone, traces of thujone (a) and bornyl acetate (b), two to three terpene alcohols (R_f 0.2–0.4) and THC at the solvent front. In Greek oil sample (7) thujone is missing and cineole is less concentrated than in (6).

Eucalypti folium aeth.
Sample (8) is characterized by the major zone of cineole at $R_f \sim 0.5$ (T2/a), two minor zones of terpene alcohols ($R_f \sim 0.25$–0.35) and THC at the solvent front. In the R_f range of thujone and bornyl acetate no prominent zones are found.

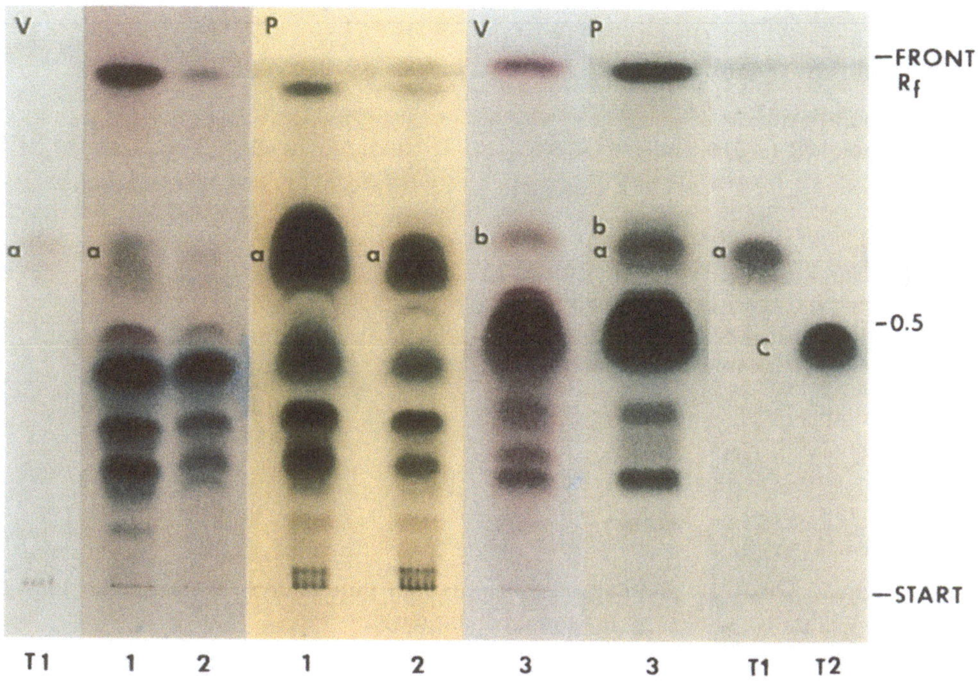
V
P
V
P
—FRONT
Rf
a
a
a
a
b
b
a
a
-0.5
c
—START
T1
1
2
1
2
3
3
T1
T2

Fig. 19

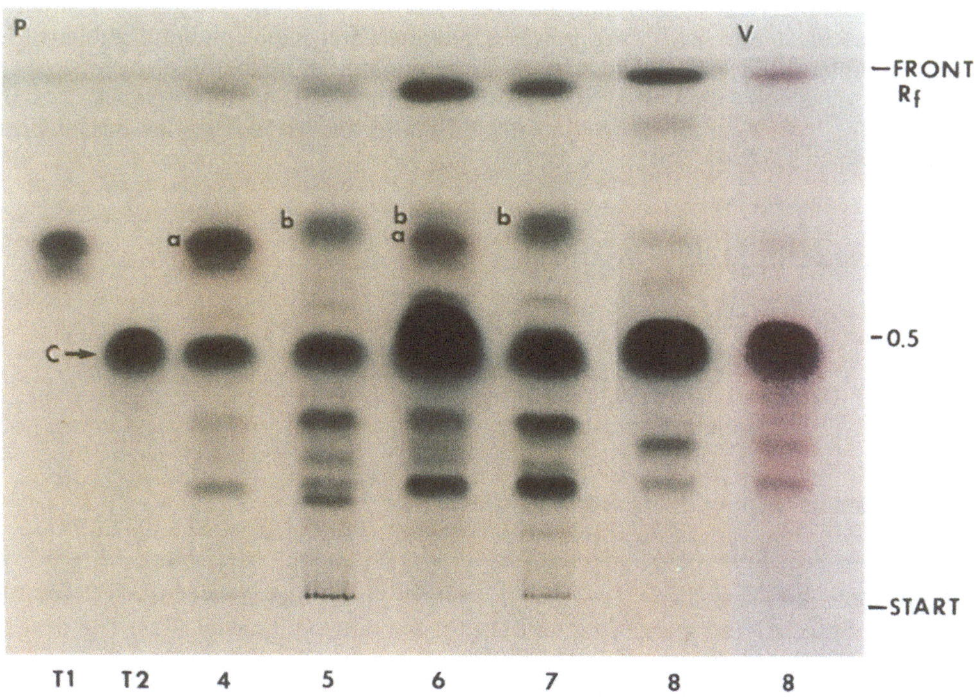
P
V
—FRONT
Rf
a
b
b
a
b
c
-0.5
—START
T1
T2
4
5
6
7
8
8

Fig. 20

Matricariae flos Essential oils with sesquiterpenes

Drug sample (essential oil)
1 Matricariae flos (steam distillate/5 µl)
2–13 Matricariae flos (steam distillates/5 µl various origin of drugs)

Reference compound
T1 bisabolol oxide A
T2 bisabolol (R_f 0.35) ▶ azulene (R_f 0.85)
T3 bisabolol oxide A (I) ▶ bisabolol (III) (R_f ~ 0.35)

Solvent system Fig. 21 toluene-ethyl acetate (93:7)

Detection Vanillin-sulphuric acid reagent (VS No. 42) →vis

Fig. 21 **Official Matricariae flos aetheroleum** (1) is characterized by the following zones:

I	bisabolol oxide A/B	R_f ~ 0.2	yellow-green
II	spathulenol	R_f ~ 0.25	violet
III	bisabolol	R_f ~ 0.35	violet
IV	polyines	R_f ~ 0.5 - 0.6	brown
V	azulene	R_f ~ 0.95	red-violet
VI	THC, farnesene	R_f ~ 0.99	blue-violet

TLC synopsis: The steam distillates of 13 chamomile flowers of the trade market show a different qualitative pattern of the main constituents.
All oils of good quality, according to most pharmacopoeias, contains the compounds I–VI in high concentration, e.g. oils 1 and 6.
The oils 8–10 have less concentrated zones in the R_f range 0.2–0.5, but prominent zones of azulene and THC at the solvent front. Oils 5 and 13 show a high polyine (IV) content, while oils 8 and 9 have hardly any polyines, but a relatively high amount of azulene (V) and bisabolol oxides A/B (I). Oil 12 has a higher amount of bisabolol.
Oils with a generally low concentration of the constituents II, IV and V (e.g. oils 2–4) or azulene free (e.g. 7) are considered as oils of inferior quality and are not accepted by most of the pharmacopoeias.

Anthemidis and Cinae flos

Drug sample 1 Anthemidis flos (DCM extract) 2 Cinae flos (DCM extract)

Reference compound T1 linalool T2 cineole T3 α-santonin

Solvent system Fig. 22 A–C toluene-ethyl acetate (93:7)
D dichloromethane

Detection A+B Vanillin-sulphuric acid reagent (VS No. 42) →vis
C+D Phoshormolybdic acid reagent (PMA No. 27) →vis

Fig. 22A **Anthemidis flos** (1) is characterized by prominent grey-violet ester zones at R_f 0.8–0.9 (e.g. butylangelat) and a blue zone at R_f 0.2 in the range of linalool (T1). The drug is sometimes used as a substitute for Matricariae flos.

B **Cinae flos** (2) shows cineole as the major blue zone at R_f 0.45 (T2) and α-santonin at R_f 0.1 (T3) (VS reagent).

C With PMA reagent thujone is detectable at R_f ~ 0.55 as a violet-blue zone; α-santonin and cineole (T2) get dark blue.

D Separation in dichloromethane shows α-santonin at R_f ~ 0.3 (T2), cineole at R_f ~ 0.7 and thujone at R_f ~ 0.85.

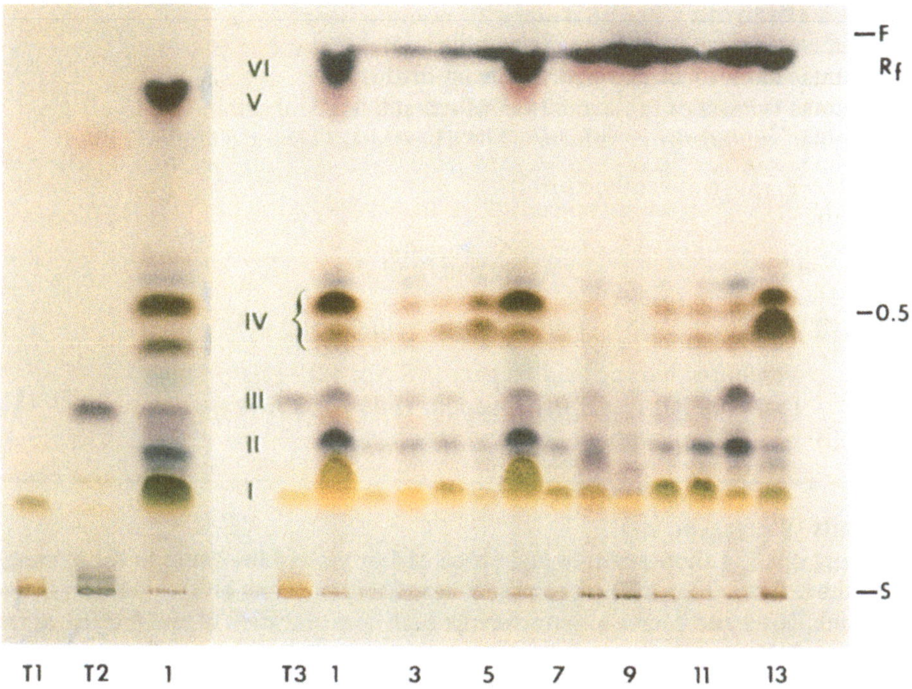
VI
V
IV
III
II
I
—F
Rf
—0.5
—S
T1 T2 1 T3 1 3 5 7 9 11 13

Fig. 21

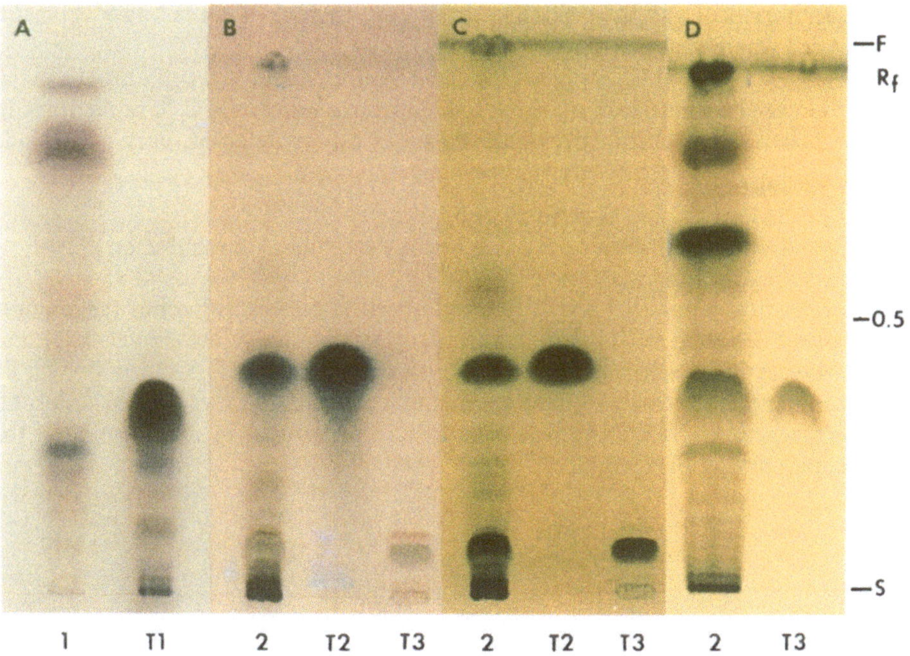
A
B
C
D
—F
Rf
—0.5
—S
1 T1 2 T2 T3 2 T2 T3 2 T3

Fig. 22

Curcumae rhizoma Essential oils with sesquiterpenes

Drug sample (essential oil)
1 Curcumae domesticae rhizoma (steam distillate)
2–4 Curcumae xanthorrhizae rhizoma (steam distillate)
5 Curcumae domesticae rhizoma (MeOH extract, 1 g/5 ml; 5 min/60°, 15 µl)
6 Curcumae xanthorrhizae rhizoma (MeOH extract, 1 g/5 ml; 5 min/60°, 15 µl)

Reference compound
T1 thymol (~R_f range of xanthorrhizol)
T2 curcumin
T3 fluorescein (~R_f range of bisdemethoxycurcumin)

Solvent system
Fig. 23A, 24A toluene-ethyl acetate (93:7)
Fig. 23B, 24B chloroform-ethanol-glacial acetic acid (95:5:1)

Detection
Fig. 23A Vanillin-sulphuric acid reagent (VS No. 42) →vis
Fig. 24A Fast blue salt reagent/NH_3 vapour (FBS No. 15) →vis
Fig. 23B, 24B UV-365 nm (without chemical treatment)

Fig. 23A **Essential oils** (VS reagent, vis)
The curcuma oils 1–4 show seven to eight blue, red or violet-blue zones in the R_f range 0.3 up to the solvent front with a prominent sesquiterpene zone at R_f ~ 0.8 and at the solvent front. Oils 2 and 4 have a characteristic high concentration of zingiberene at the solvent front. THC is present at a low concentration in oils 1 and 3.
The phenolic sesquiterpene xanthorrhizol is found as a blue-violet zone at R_f ~ 0.55, directly above the reference compound thymol (T1).

Fig. 24A **Essential oils** (FBS reagent, vis)
Xanthorrhizol is a characteristic constituent of C. zanthorrhiza. Due to the phenolic structure xanthorrhizol and the reference compound thymol (T1) react to give an intense violet-red when treated with the FBS reagent.
The distillates 2 and 4 from C. zanthorrhiza show xanthorrhizol as a prominent zone at R_f ~ 0.55, in lower concentration in oil 3, distilled from commercial C. xanthorrhiza. In oil 1 from Curcuma domestica, only weak red zones can be detected. Very often trade samples are mixtures of both turmeric rhizomes.

Fig. 23, 24B **Pigments** (UV-365 nm)
Another identification method of turmeric is by the detection of the characteristic yellow pigments in methanolic extracts.
Curcuma domestica extract (5) shows five yellow-white fluorescent zones (yellow/vis) with curcumin (T2) at R_f ~ 0.6, demethoxycurcumin directly below (R_f 0.5–0.55) and bisdemethoxycurcumin at R_f ~ 0.3 (T3).
Curcuma zanthorrhiza extracts (6) contain mainly curcumin (T2) with a small amount of demethoxycurcumin. No prominent zone should be present in the R_f range of the reference compound fluorescein (T3).

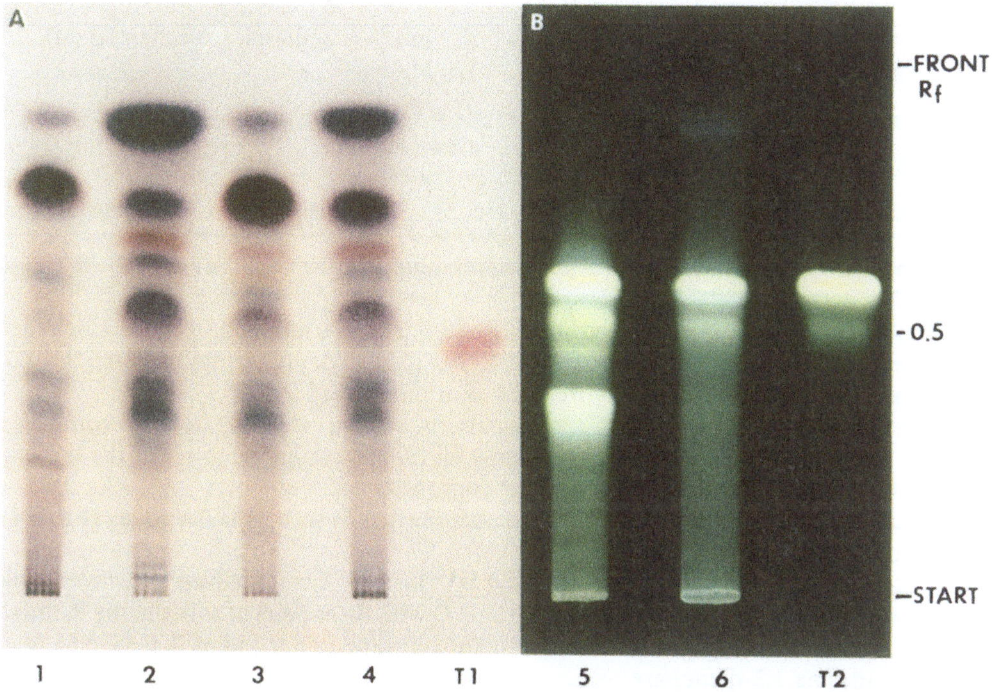
A
B
–FRONT
Rf
-0.5
–START
1
2
3
4
T1
5
6
T2

Fig. 23

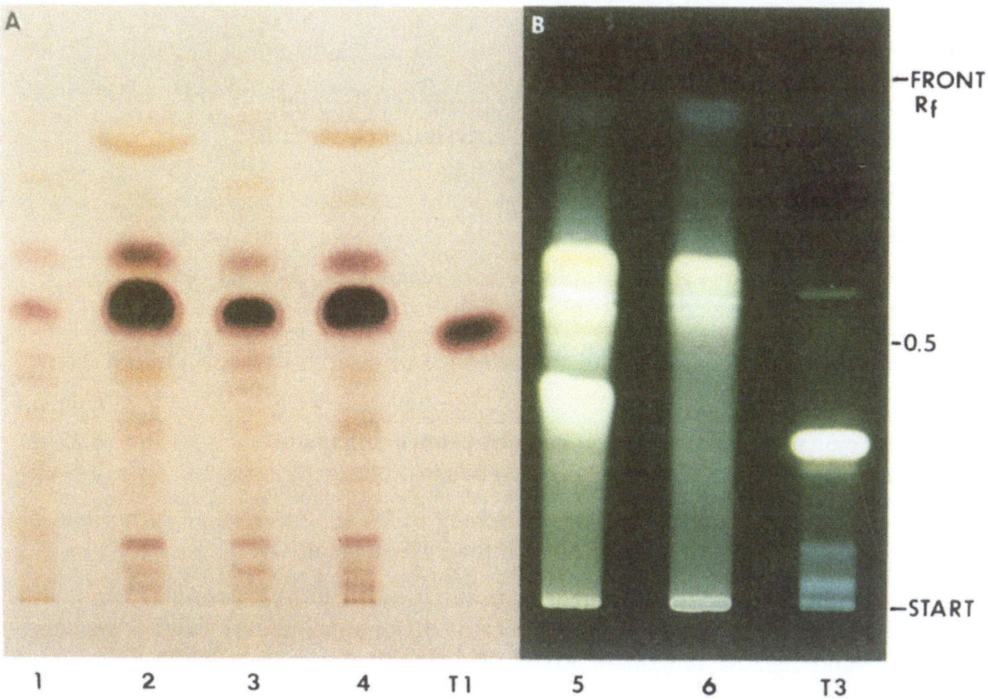
A
B
–FRONT
Rf
-0.5
–START
1
2
3
4
T1
5
6
T3

Fig. 24

Juniperi aetherolea, Myrrha

Drug sample (essential oil)	1 Juniperi aetherol. (ex fructu) 2 Juniperi aetherol. (ex ligno)	3 Juniperi aetherol. (commercial oil) 4,5 Myrrha
Test	T1 linalool ► carvon ► thymol ►linalyl acetate ► anethole	
Solvent system	Fig. 25 toluene-ethyl acetate (93:7)	
Detection	Anisaldehyde-sulphuric acid reagent (AS No. 3) A vis	B UV-365 nm

Fig. 25A **Juniperi fructus aeth.** (1) generates six to seven blue, grey or violet zones in the R_f range 0.2–0.7:

terpene alcohols (e.g. borneol, terpineol; R_f 0.15–0.25; T1/linalool), terpene aldehydes and ketones R_f 0.45; T1/carvon, terpene esters (e.g. bornyl and terpinyl acetate; R_f 0.65; T1/linalyl acetate) and terpene hydrocarbons at the solvent front.

Juniperi lignum aeth. (2) shows a similar pattern, but the esterzone (R_f 0.65) compared to 1 is missing. Commercial **Juniperi fructus** oil (3) is comparable to 1, but the terpene compounds are present in a slightly lower concentration.

Myrrha (4, 5) are characterized by furano sesquiterpenes seen as violet zones at R_f 0.6–0.7 and at R_f 0.25 (vis).

B All zones can be more easily detected under UV-365 nm. A band of blue and violet-pink fluorescent zones, mainly in the R_f range 0.2–0.75 with three pairs of zones in the R_f range 0.1–0.15 and 0.2–0.25 (e.g. curzerenone, methoxyfuranodiene) and at R_f 0.55–0.65 (e.g. furanoeudesma-1,3-diene) are seen.

Benzoin and Balms

Drug sample (essential oil)	1 Banzoe Sumatra 2 Benzoe tonkinemsis (Siam)	3 Tolutanum balsamum 4 Peruvianum balsamum
Reference compound	T1 eugenol	
Solvent system	Fig. 26 toluene-ethyl acetate (93:7)	
Detection	A UV-254 nm B Vanillin-sulphuric acid reagent (VS No. 42) →vis C Phosphomolybdic acid reag. (PMS No. 34) → vis	

Fig. 26 **Benzoins** (1,2) and **balms** (3,4) are characterized by a series of free acids and esters:

benzoic acid, cinnamic acid	R_f ~ 0.05–0.1
coniferyl cinnamate, cinnamoyl cinnamate, propyl cinnamate	R_f ~ 0.25–0.3
cinnamoyl benzoate, coumaroyl benzoate, benzoyl benzoate	R_f ~ 0.7–0.8

These compounds show prominent quenching in UV-254nm (A), all turn violet blue with VS-reagent (B) or get dark blue with the PMS-reagent (C)

In the samples 1, 3 and 4, the benzoates in the R_f range 0.7–0.8 dominate, while in (2) coniferyl benzoate in the R_f ~ range 0.35 is the major zone. In peru balm 4, the benzoyl benzoate and benzoyl cinnamate mixture (= cinnamein) is more highly concentrated and, in addition, nerolidol at R_f ~ 0.35 is detectable (→C).

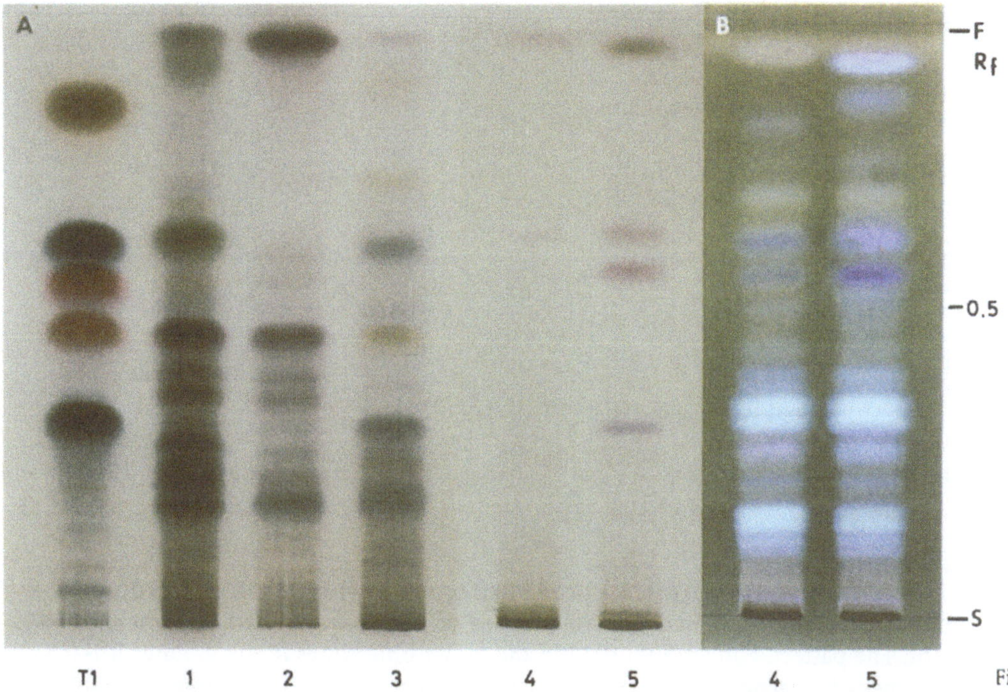

Fig. 25

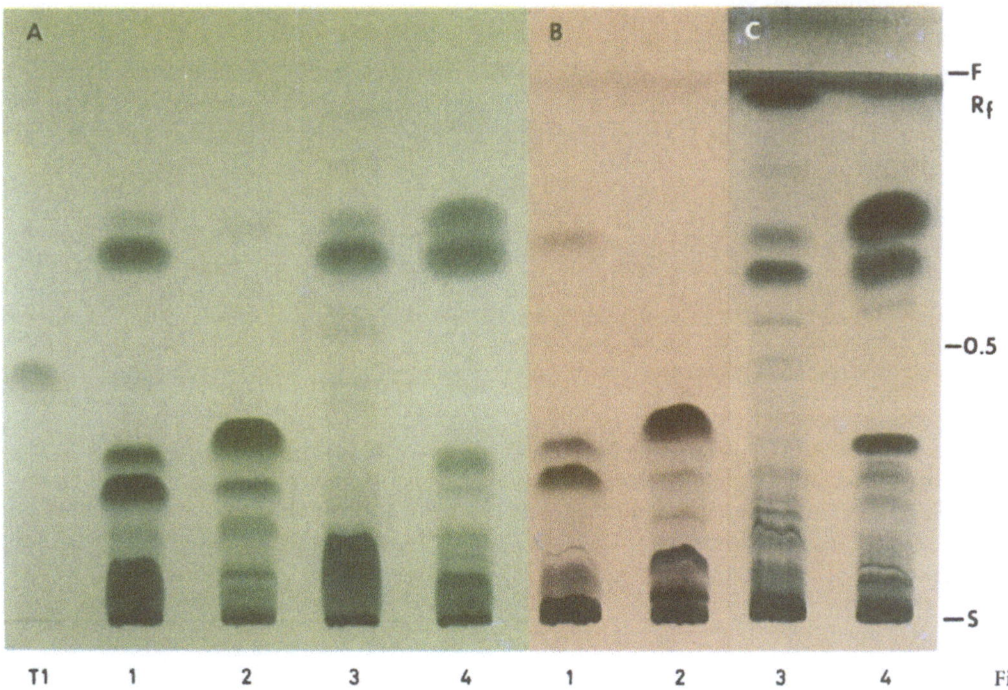

Fig. 26

Pini aetherolea
Terebinthinae aetherolea

Commercial oil
1 Pini sibirici aetheroleum
2 Pini pumilonis aetheroleum
3–5 Pini aetheroleum
6,7 Terebinthinae aetheroleum

Reference compound
T1 bornyl acetate
T2 borneol
T3 linalool ($R_f \sim 0.35$) ▶ carvon ($R_f \sim 0.5$) ▶ thymol ($R_f \sim 0.55$) ▶ linalyl acetate ($R_f \sim 0.7$) ▶ anethole ($R_f \sim 0.9$)

Solvent system
Fig. 27, 28 toluene-ethyl acetate (93:7)

Detection
Anisaldehyde-sulphuric acid reagent (AS No. 3)
Fig. 27 ▶ vis.
Fig. 28 ▶ UV-365 nm

Fig. 27

Pini aetherolea (1–5)
All samples are characterized by a prominent brown ester zone at $R_f \sim 0.75$ due to bornyl and/or terpinyl acetate and the violet zones of terpenes (e.g. cadinene) at the solvent front. The pattern and amount of blue and violet-blue zones in the R_f range 0.4–0.6 and the zones of terpene alcohols (e.g. borneol T2, terpineol) in the R_f range 0.25–0.4 varies in the commercial oil samples 1–5.

Therebinthinae aetherolea (6–7)
The commercial oil sample 7 shows three blue to red-violet terpene alcohols at R_f 0.2–0.3; a prominent violet-brown zone at $R_f \sim 0.5$ in the R_f range of the carvon test (T3), two minor grey zones in the R_f range of terpene esters (T3/linalyl acetate) and terpene zones at the solvent front.

Fig. 28

Pini aetherolea (1–5)
The THC zone at the solvent front, the prominent ester zone at $R_f \sim 0.75$ and some terpene alcohols in the lower R_f range show a red-brown fluorescence. In addition red, violet, blue and green-blue fluorescent zones in the lower R_f range are seen.
The fluorescence of the zones changes after spraying with the AS reagent, but reaches stable fluorescence after 30–60 min.

Therebinthinae aetherolea (6–7)
The zones fluoresce in UV-365 nm mostly light yellow-brown in the upper R_f range and more red or blue-violet and red-brown in the lower R_f range.
Oils of good quality are characterized by a relatively prominent THC zone, e.g. α-pinene, α-/β-phellandren, limonene and a lower content of terpene alcohols in the R_f range 0.2–0.4.

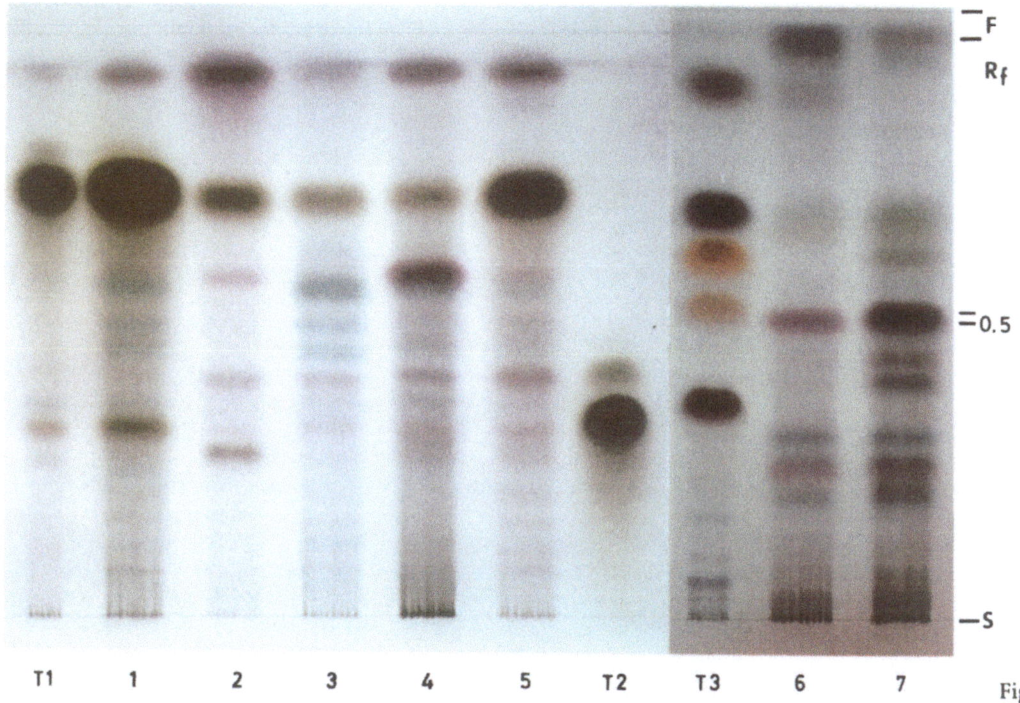

Fig. 27

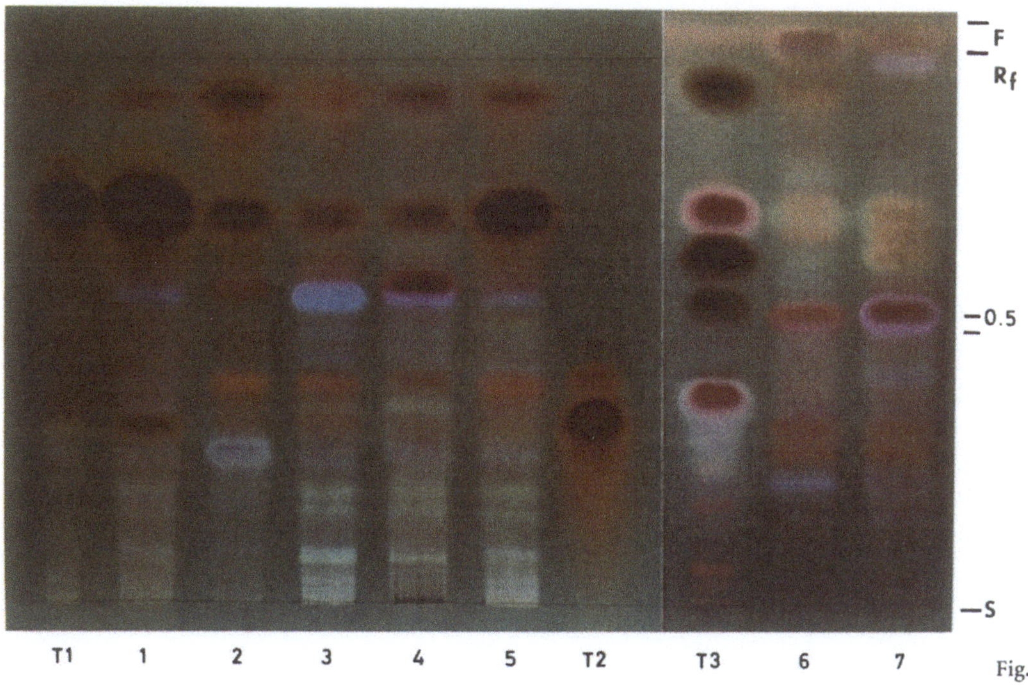

Fig. 28

7 Flavonoid Drugs Including Ginkgo Biloba and Echinaceae Species

7.1 Flavonoids

The main constituents of flavonoid drugs are 2-phenyl-γ-benzopyrones (2-phenylchromones) or structurally related, mostly phenolic, compounds.
The various structure types of flavonoids differ in the degree of oxidation of the C ring and in the substitution pattern in the A and/or B rings (see 7.1.5 Formulae). Most of these compounds are present in the drugs as mono- or diglycosides.

7.1.1 Preparation of Extracts

General Method

Powdered drug (1 g) is extracted with 10 ml methanol for 5 min on a water bath at about 60°C and then filtered; 20–30 µl is used for chromatography (flavonoid content, 0.5%–1.5%). This rapid method extracts both lipophilic and hydrophilic flavonoids.

Enrichment with ethyl acetate

A total of 5 ml of the methanolic extract (see "General method") is concentrated to about 2 ml; 1 ml water and 10 ml ethyl acetate are added and shaken several times. The ethyl acetate phase is separated and reduced to a volume of 1 ml, and 10 µl is used for TLC investigation.

Cardui mariae fructus

Powdered drug (1 g) is first defatted by heating under reflux for 30 min with 50 ml light petroleum. The petroleum extract is discarded and the drug residue is heated under reflux for 15 min with 10 ml methanol. The filtrate is concentrated to 5 ml, and 30 µl is used for chromatography.

Orthosiphonidis folium

Powdered drug (1 g) is extracted by shaking for 15 min with 10 ml dichloromethane; 30 µl of the filtrate is used for chromatography.

Farfarae folium, Petasitidis folium (test for petasins)

Powdered drug (2 g) is extracted by heating under reflux for about 20 min with about 40 ml light petroleum on a water bath. The clear filtrate is concentrated to about 1 ml, and 30 µl is used for chromatography.

Arnicae flos (test for sesquiterpene lactones)

A total of 30 ml hot water is added to 2.5 g powdered drug. After 5 min, the mixture is filtered through a wet filter with additional washing of the filter with 10 ml water; 15 ml $CHCl_3$ is then added to the water extract and shaken carefully several times. The $CHCl_3$ phase is separated and reduced to dryness. The residue is dissolved in 0.5 ml $CHCl_3$ and 10–30 µl is used for TLC.

Crataegi folium Lespedezae herba (Procyanidines) Powdered drug (5 g) is extracted with 75 ml ethanol (45%) for 1 h under reflux. The filtrate is evaporated to approximately 20 ml and transferred into a separation funnel; 30 ml dichloromethane and 2 ml ethanol are added and shaken for 5 min, and the lower phase is discarded. Another 20 ml dichloromethane is added, and after shaking the lower phase is removed. This is repeated twice. The resulting extract is evaporated to approximately 10 ml.
A total of 5 g polyamide powder (trade quality) is added to the extract and thoroughly mixed, and the mixture is filled in a glass column (diameter, 1 cm; length, 15 cm) and eluted in three steps:

fraction 1: elution with 300 ml ethanol → contains mostly flavonoids.
fraction 2: elution with 100 ml ethanol-acetone-water (80:16:4) → contains mostly dimeric and oligomeric procyanidines.
fraction 3: elution with 120 ml of acetone-water (7:3) → contains polymeric procyanidines.

Each fraction is evaporated to dryness and dissolved in 5 ml ethanol; 10–30 μl is used for TLC comparison.

7.1.2 Thin-Layer Chromatography

Reference compounds Standard compounds are prepared as 0.05% solutions in methanol, and 10 μl is used for chromatography. The average detection limit for flavonoids is 5–10 μg.
For a general description of the flavonoid pattern of a drug, 10 μl of a mixture of the compounds rutin, chlorogenic acid and hyperoside is used for TLC (test mixture T1).

Adsorbent Silica gel 60 F_{254}-precoated TLC plates (Merck, Germany).

Chromatography solvents

- **Ethyl acetate-formic acid-glacial acetic acid-water (100:11:11:26)**
  ▶ suitable as a screening system for the TLC investigation of flavonoid glycosides.
- Ethyl acetate-formic acid-glacial acetic acid-ethylmethyl ketone-water (50:7:3:30:10)
  ▶ by addition of ethylmethyl ketone rutin and vitexin-2″-O-rhamnoside can be separated.
- Chloroform-acetone-formic acid (75:16.5:8.5)
  ▶ separation of flavanolignans of Cardui mariae fructus and amentoflavone, scopoletin and catechin of Viburni cortex.
- Chloroform-ethyl acetate (60:40)
  ▶ separation of flavonoid aglycones of Orthosiphonidis folium or Aurantii pericarpium.
- Chloroform (100)
  ▶ separation of petasines in Petasitidis species, adulterants of Farfarae folium.
- benzene-pyridine-formic acid (72:18:10)
  toluene-ethyl formiate-formic acid (50:40:10)
  toluene-dioxan-glacial acetic acid (90:25:4)
  ▶ separation of flavonoid aglycones.

7.1.3 Detection

The solvent (acids) must be thoroughly removed from the silica gel layer before detection.

- UV-254 nm All flavonoids cause fluorescence quenching.
- UV-365 nm Depending on the structural type, flavonoids show dark yellow, green or blue fluorescence, which is intensified and changed by the use of various spray reagents.

Flavonoid extracts often contain phenol carboxylic acids (e.g. caffeic acid, chlorogenic acids) and coumarins (e.g. scopoletin), which form blue fluorescent zones.

- Spray reagents (see Appendix A)

- Natural products reagent (NP/PEG No. 28)
  - Typical intense fluorescence in UV-365 nm is produced immediately on spraying. Addition of polyethylene glycol solution lowers the detection limit and intensifies the fluorescence behaviour, which is structure dependent.

Flavonols:	quercetin, myricetin and their glycosides	orange-yellow
	kaempferol, isorhamnetin and their glycosides	yellow-green
Flavones:	luteolin and their glycosides	orange
	apigenin and their glycosides	yellow-green

  - Fast blue salt B (FBS No. 15)
    Blue or blue-violet (vis) azo-dyes are formed. The colour can be intensified by further spraying with 10% sodium hydroxide or potassium hydroxide solution.

7.1.4 Drug List

Grouping of drug chromatograms according to plant parts and in alphabetical order:

Flos:	Figs. 3–11
Folium, Herba:	Figs. 11–22
Gemma, Pericarpium:	Figs. 21–24
Drugs with aglycones:	Figs. 24–26

For explanation of trivial names see 7.1.5 Formulae.

Drug/plant source Family/pharmacopoeia	Main flavonoids and other specific constituents	
Arnicae flos Arnica, celtic bane Arnica montana L. Arnica chamissonis LESS ssp. foliosa ssp. chamissonis Asteraceae DAB 10, ÖAB 90, Helv VII, MD	0.4%–0.6% total flavonoids Quercetin-3-O-glucoside and -3-O-glucogalacturonide, luteolin-7-O-glucoside, kaempferol-3-O-glucoside 0.2%–1.5% sesquiterpene lactones (pseudoguainolide type) helenaline, 11α, 13-dihydrohelenaline and esters Adulterants: e.g. Calendulae, Farfarae flos Heterothecae inuloidis flos, (see Figs. 5,6)	**Fig. 3,5,6**
Acaciae robiniae flos Acacia flowers Robinia pseudoacacia L. Fabaceae	Kaempferol-3-O-rhamnosylgalactosyl-7- rhamnoside (=robinin), acacetin-7-O- rutinoside, acaciin (Acaciae farnesinae flos, true Acaciae flos) Adulterant: Pruni spinosae flos (see Fig. 9)	**Fig. 9**

	Drug/plant source Family/pharmacopoeia	Main flavonoids and other specific constituents
Fig. 4	**Anthemidis flos** Chamomile (Roman) Chamaemelum nobile (L.) ALL. (syn. Anthemis nobilis L.) Asteraceae DAB 10, ÖAB 90, Helv VII, BP 88, MD	0.5%–1% total flavonoids Apigenin-7-O-glucoside and-7-apiosyl-glucoside (=apiin) Quercetin-3-O-rhamnoside (=quercitrin), luteolin-7-O-glucoside, caffeic and ferulic acid (free acids and as glucosides) ► Coumarins: scopoletin-7-o-glucoside ► Essential oil (see Chap. 6)
Fig. 7,8	**Cacti flos** Night-blooming Cereus Selenicereus grandiflorus (L.) BRITT. et ROSE Cactaceae	1%–1.5% total flavonoids Isorhamnetin-3-O-galactoside (=cacticin), -3-O-galactosyl-rutinoside, -3-O-rutinoside (=narcissin), -3-O-xylosyl-rutinoside Rutin
Fig. 7,8	**Calendulae flos** Marigold flowers Calendula officinalis L. Asteraceae	0.3%–0.6% Isorhamnetin glycosides I-3-O-glucoside, I-3-O-rutinoside (=narcissin), I-3-O-rutinosyl-rhamnoside Quercetin-3-O-glucoside and 3-O-gluco-rhamnoside (<0.2%) ► Saponins: oleanolic acid glycosides (=calendulosides)
Fig. 15	**Crataegi flos, C. folium** Hawthorn flowers DAC 86 Hawthorn leaves, Helv VII Crataegi folium C. flore Hawthorn herb DAB 10 **Crataegi fructus** Hawthorn fruits MD Crataegus species e.g. Crataegus laevigata DC Crataegus azarolus L. Crataegus pentagyna, C. nigra WALDST. et KIT. Rosaceae	1%–2% total flavonoids 0.25% quercetin glycosides: hyperoside, rutin, quercetin-rhamnogalactoside and-4′-glucoside (=spiraeoside); methoxykaempferol-3-O-glucoside Flavon-C-glycosides: vitexin, vitexin-2″-O-rhamnoside, monoacetyl-vitexin-rhamnoside, isovitexin-rhamnoside, vincenin-2, schaftoside, isoschaftoside 1%–3% procyanidines: e.g. dimeric procyanidine B-2 (0.05%–0.25%, leaves)
Fig. 11,12	**Farfarae flos** Coltsfoot flowers Helv VII, China **Farfarae folium** Coltsfoot leaves DAB 10 Tussilago farfara L. Asteraceae	0.05%–0.2% Quercetin glycosides: rutin, hyperoside and isoquercetin in varying concentrations in both drug parts Phenol carboxylic acids Adulterant: Petasitidis folium (see Fig. 12, 7.1.4 Drug List)

Drug/plant source Family/pharmacopoeia	Main flavonoids and other specific constituents	
Heterothecae flos	► see Arnicae flos	**Fig. 5,6**
Matricariae flos Chamomillae flos German chamomile flowers Chamomilla recutita (L.) S. RAUSCHERT (syn. Matricaria chamomilla L.) Asteraceae DAB 10, ÖAB 90, Helv VII, BP 88, MD	0.5%–3% total flavonoids Apigenin-7-O-glucoside (~ 0.45%), quercimeritrin, luteolin-7-O-glucoside, patuletin-7-O-glucoside, and seven flavonoid aglycones Adulterant: Anthemidis flos ► Essential oil (see Chap. 6)	**Fig. 4**
Primulae flos Primrose flowers, cowslip Primula veris L. Primula elatior (L.) HILL Primulaceae	Quercetin and kaempferol glycosides (0.05%): kaempferol-O-dirhamnoside, k-3-O-gentiotrioside, k-triglucoside; gossypetin-dimethylether	**Fig. 7,8**
Primulae radix	Saponins (see Chap. 14, Fig. 3)	
Pruni spinosae flos Acaciae germanicae flos Blackthorn flowers Prunus spinosa L. Rosaceae DAC 86	Quercetin glycosides: rutin, hyperoside, quercitrin, quercetin-3-O-arabinoside Kaempferol -3,7-O-dirhamnoside, k-3-O-rhamnoside and -3-O-arabinoside	**Fig. 9**
Robiniae flos	► see Acaciae flos	**Fig. 9**
Sambuci flos Elder flowers Sambucus nigra L. Sambucaceae (Caprifoliaceae) ÖAB 90, Helv VII, DAC 86, MD (fruit), BHP 83	1.5%–2% total flavonoids Quercetin glycosides: hyperoside, rutin, quercitrin, isoquercitrin Kaempferol-7-O-rhamnoside ► 3% phenol carboxylic acids: chlorogenic, caffeic and ferulic acid and their esters	**Fig. 9**
Spiraeae flos Meadow-sweet flowers Filipendula ulmaria (L.) MAXIM Rosaceae Helv V	3%–5% total flavonoids Quercetin-4'-O-glucoside (=spiraeoside 3%), hyperoside, quercetin-3-O-arabinoside, -3-O-glucuronide, rutin Kaempferol glycosides ► 0.6%–0.8% salicylic acid and its methylester (0.14%)	**Fig. 9**

	Drug/plant source Family/pharmacopoeia	Main flavonoids and other specific constituents
Fig. 3,4	**Stoechados flos** (syn. Helichrysi flos) Yellow chaste weed Everlasting Cats foot, Helichrysum arenarium (L.) MOENCH Asteraceae	>0.4% total flavonoids Kaempferol-3-O-glucoside and -3-0-diglucoside; quercetin-3-O-glucoside; luteolin-7-O- and apigenin-7-O-glucoside Helichrysin A, B: A = (+)-naringenin-5-β-O-D-glucoside B = (−)-naringenin-5-β-O-D-glucoside (syn. salipurposide) 2′,4,4′,6′-tetrahydroxychalcon-6′-O-glucoside (=isosalipurposide)
Fig. 10	**Tiliae flos** Lime flowers Tilia cordata MILL. Tilia platyphyllos SCOP. Tiliaceae DAB 10, ÖAB 90, Helv VII, MD	~1% total flavonoids Quercetin glycosides: quercitrin, isoquercitrin, q-3-O-glucosyl-7-O-rhamnoside Kaempferol glycosides: k-3-O-glucoside, -3-O-rhamnoside, -3-O-glucosyl-7-O-rhamnoside, -3,7-O-dirhamnoside, k-3-O-[6-(p-coumaroyl)]-glucoside (=tiliroside) Myricetin -3-O-glucoside, -3-O-rhamnoside Adulterant: Tilia argentea
Fig. 3,4	**Verbasci flos** Mullein, torch weed flowers Verbascum densiflorum BERTOL. Scrophulariaceae DAC 86, ÖAB 90, Helv VII	1.5%–4% total flavonoids Rutin, hesperidin, apigenin-, luteolin-7-O-glucoside, kaempferol; Phenol carboxylic acids Adulterant: Primulae and Genistae flos ▶ Bitter principle: Aucubin (see Fig. 5,6, Sect. 3.7)

Folium

Fig. 13	**Betulae folium** Birch leaves Betula pendula ROTH B. pubescens EHRHART Betulaceae DAB 10, ÖAB 90, Helv VII	1.5%–3% total flavonoids >1.5% Quercetin glycosides: Quercitrin, isoquercitrin, hyperoside, rutin, quercetin-3-O-arabinoside (=avicularin) Myricetin-3-O-galactoside and -digalactoside Kaempferol-3-O-glucoside and rhamnoside Isorhamnetin-3-O-galactoside, hesperidin Chlorogenic and caffeic acid

Drug/plant source Family/pharmacopoeia	Main flavonoids and other specific constituents	
Castaneae folium Chestnut leaves Castanea sativa MILL. Fagaceae (Cupuliferae)	>1% total flavonoids Quercetin glycosides: isoquercitrin, rutin, q-3-O-glucuronide (=miquelianin), q-3-O-galactopyranoside (=hyperin) Kaempferol-glycosides: astragalin, k-3-O-[6-(p-coumaroyl)]-glucopyranoside (=tiliroside), k-3-O-[6-(p-coumaroyl)]-rhamnoglucoside; 3-O-p-coumaroylquinic acid 6%–8% Tannins	**Fig. 14**
Crataegi folium	► see Crataegi flos	**Fig. 15**
Farfarae folium	► see Farfarae flos	**Fig. 11,12**
Juglandis folium Walnut leaves Juglans regia L. Juglandaceae DAC 86, MD (oil)	2%–3% total flavonoids Quercetin glycosides: hyperoside (>0.2%), quercitrin, avicularin Kaempferol-3-O-arabinoside Neochlorogenic, caffeic and gallic acid	**Fig. 13**
Petasitidis folium Butter bur or umbrella leaves Petasites hybridus (L.) GAERTN., MEYER et SCHERB. Asteraceae **Petasitidis radix**	Flavonol glycosides: isoquercitrin, astragalin Ester of sesquiterpene alkohols (eremophilans), petasol, neo- and isopetasol, methacrylpetasol, angeloylneopetasol, petasin, isopetasin Petasin-free race (=furan-race) contains furanoeremophilanes ~20 petasins (e.g. petasin, iso and S-petasin)	**Fig. 12**
Rubi fruticosi folium Bramble (Blackberry) leaves Rubus fruticosus L. Rosaceae DAC 86	Flavonol glycosides Phenol carboxylic acids Gallotannins (>10%)	**Fig. 13**
Rubi idaei folium Raspberry leaves Rubus idaeus L. Rosaceae	~0.2% total flavonoids Quercetin glycosides Gallo-, ellag tannins	**Fig. 13**
Ribis nigri folium Black current leaves Ribes nigrum L. Grossulariaceae	1%–1.5% total flavonoids Quercetin-, kaempferol-, myricetin and isorhamnetin glycosides Procyanidines (dimeric, trimeric)	**Fig. 13**

	Drug/plant source Family/pharmacopoeia	Main flavonoids and other specific constituents

Herba

Fig. 21	**Anserinae herba** Silverweed Potentilla anserina L. Rosaceae DAC 86	Quercetin-3-O-glucoside and -3-O-rhamnoside Myricetin and myricetin-rhamnoside Ellagtannins (6%–10%)
Fig. 17,18	**Equiseti herba** Common horsetail Equisetum arvense L. Equisetaceae DAB 10, Helv VII, MD	Flavonoids: luteolin-5-O-glucoside (=galuteolin), kaempferol-3-O- and 7-O-diglucoside (=equisetrin), k-3,7-diglucoside, quercetin-3-O-glucoside (=isoquercitrin) Adulterant: E. palustre: 0.1%–0.3% palustrine (alkaloid)

Fig. 18 Flavonoid pattern of **Equisetum species** (see Fig. 18, samples 1–7)

E. arvense L. (1,2):	Kaempferol-3-glucoside, k-7-glucoside, k-3,7-diglucoside; quercetin-3-glucoside
E. palustre L. (3,4):	Kaempferol-3,7-diglucoside, k-3-diglucosyl-7-glucoside, k-3-rutinosyl-7-glucoside
E. fluvatile L. (5):	Kaempferol-3-glucoside, k-7-glucoside, k-3,7-diglucoside, k-3-diglucosyl-7-glucoside; apigenin-4'-glucoside, herbacetin-7-glucoside (=herbacitrin); quercetin-7-glucoside, gossypetin-7-glucoside (=gossypitrin)
E. sylvaticum L. (6):	Kaempferol-3-glucoside, k-7-glucoside, k-3-diglucoside, k-3,7-diglucoside; herbacitrin, quercetin-3,7-diglucoside
E. telmateia EHRH. (7):	Kaempferol-3-glucoside, k-7-glucoside, k-3-rutinoside, k-3,7-diglucoside, k-3-rutinosyl-7-glucoside

Fig. 16	**Lespedezae herba** Round-headed bush clover Lespedeza capitata MICHX. Fabaceae	~1% total flavonoids Flavon-C-glycosides: orientin, iso-orientin, vitexin, isovitexin, schaftoside Flavonol-O-glycosides: rutin, hyperoside, isoquercitrin, isorhamnetin-, kaempferol-3-rhamnoglucoside, kaempferol-3,7-dirhamnoside (=lespedin), astragalin Procyanidines di-, trimeric
Fig. 21	**Passiflorae herba** Passion flower, Maypop Passiflora incarnata L. Passifloraceae DAB 10, Helv VII, MD	0.4%–1.2% total flavon-C-glycosides isovitexin and -glucoside (25%), vitexin, orientin, iso-orientin, iso-schaftoside, schaftoside

Drug/plant source Family/pharmacopoeia	Main flavonoids and other specific constituents	
Virgaureae herba (Solidaginis virgaureae herba) Golden-rod BHP 83 Solidago virgaurea L. **Solidaginis (giganteae) herba** Solidago gigantea AIT. Asteraceae	1%–3.85% Flavonolglycosides 1%–1.5% (rutin 0.8%): S. virgaurea 3%–3.85% (quercitrin 1.3%): S. gigantea Isoquercitrin, hyperoside Kaempferol glycosides: k-3-O-glucoside and -galactoside, k-rutinosid (=nicotiflorin) rhamnetin-3-O-glucoside and -galactoside >0.4% chlorogenic acid, caffeic acid glucosylester Estersaponins (>2.4%), Virgaurea saponin 1–3	**Fig. 19**
Violae tricoloris herba Whild pansy, heart sease herb Viola tricolor L. ssp. tricolor OBORNY ssp. arvensis GAUDIN Violaceae DAC 86, ÖAB 90	0.4%–0.6% total flavonoids Quercetin, kaempferol or isorhamnetin glycosides; Luteolin -7-O-glucoside, violanthin, saponarin; rutin (0.15% white-yellow flowers) Salicylic acid (0.06%–0.3%), methylester and glucosides	**Fig. 20**
Sophorae gemma Sophora buds Sophora japonica L. Fabaceae MD, China (flos, fructus)	Flavonol glycosides Rutin (about 20%)	**Fig. 21**
Aurantii pericarpium Seville orange peel Citrus aurantium L. ssp. aurantium Rutaceae DAB 10, MD, Japan, China	Flavanon glycosides: eriodictyol-7-O-rutinoside (=eriocitrin), naringenin-7-O-neohesperidoside (=naringin), hesperetin-7-O-neohesperidoside (=neohesperidin), hesperetin-7-O-rutinoside (=hesperidin) Flavonol glycoside: rutin. Sinensetin Bitter principles: see Fig. 1,2, 3.5 Formulae Essential oils: see Fig. 17,18, 6.7	**Fig. 23,24**
Citri pericarpium Lemon peel Citrus limon (L.) BURMAN fil. Rutaceae	Flavanon glycosides: eriocitrin, naringin, hesperidin (see Aurantii pericarpium) Flavonoid glycosides: luteolin-7-O-rutinoside, isorhamnetin-3-arabino-glucoside, apigenin-C-glucoside; limocitrin glycosides Essential oil: see Fig. 17,18, 6.7	**Fig. 23**

Drug/plant source Family/pharmacopoeia		Main flavonoids and other specific constituents
	Drugs containing predominantly flavonoid aglycones	
Fig. 24	**Eriodictyonidis herba** Yerba Santa Eriodictyon californicum (HOOK et ARNTT.) J. TORREY Hydrophyllaceae MD	Flavanones: homoeriodictyol (=eriodictyone), eriodictyol, chrysoeriodictyol, xanthoeriodictyol Adulterant: Eriodictyon crassifolium BENTH.
Fig. 24	**Orthosiphonis folium** Orthosiphon leaves Orthosiphon aristatus (BLUME) MIQUEL Lamiaceae DAB 10, Helv VII	0.19%–0.22% total flavonoids sinensetin (3′,4′,5,6,7-pentamethoxy-flavone), scutellarein tetramethyl ether, eupatorin (3′,5-dihydroxy-4′,6,7-trimethoxyflavone)
Fig. 25	**Cardui mariae fructus** Milk-thistle fruits Silybum marianum GAERTNER Asteraceae DAB 10, MD	1.5%–3% Flavanolignans: silybin, silychristin, silydianin and 2,3- dehydroderivatives Flavanonol taxifolin
Fig. 26	**Viburni prunifolii cortex** Black haw bark Viburnum prunifolium L. Caprifoliaceae	Amentoflavone, bi-apigenin, scopoletin, hydroquinone (<0.5%) Adulterant: Viburni opuli cortex

7.1.5 Formulae

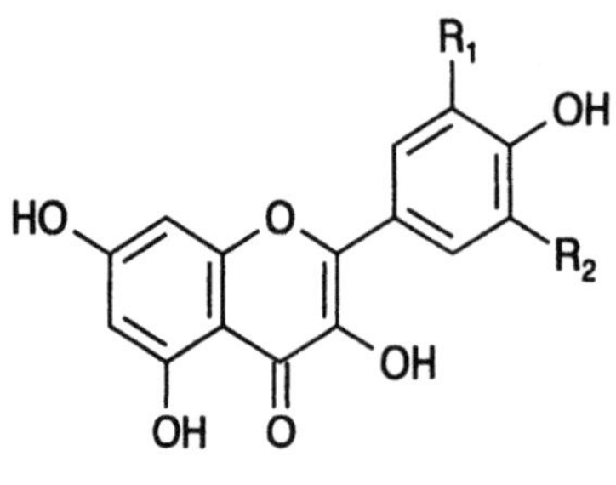

Flavonols	R_1	R_2	Aglycone
	OH	H	Quercetin
	H	H	Kaempferol
	OH	OH	Myricetin
	OCH_3	H	Isorhamnetin

Common glycosides:

Quercetin	**Kaempferol**	**Myricetin**
Q-3-O-glucoside (isoquercitrin) Q-3-O-rhamnoside (quercitrin) Q-3-O-arabinofuranoside (avicularin) Q-3-O-galactoside (hyperoside) Q-3-O-glucuronide (miquelianin) Q-3-O-rutinoside (rutin) Q-4'-O-glucoside (spiraeoside) Q-7-O-glucoside (quercimeritrin)	K-3-O-galactoside (trifoliin) K-3-O-glucoside (astragalin) K-3-O-rhamnoside (afzelin) K-3-O-arabinofuranoside (juglanin) K-3-O-diglucoside K-7-O-rhamnoside K-7-O-diglucoside (equisetrin) K-3,7-O-dirhamnoside (lespedin) K-3-O-rutinoside (nicotiflorin) K-3-(6″-p-coumaroyl-glucoside (tiliroside)	M-3-O-glucoside M-3-O-galactoside M-3-O-rhamnoside (myricitrin) **Isorhamnetin** I-3-O-galactoside (cacticin) I-3-O-glucoside I-3-O-galactosyl-rutinoside I-3-O-rutinoside (narcissin) I-3-O-rutino-rhamnoside

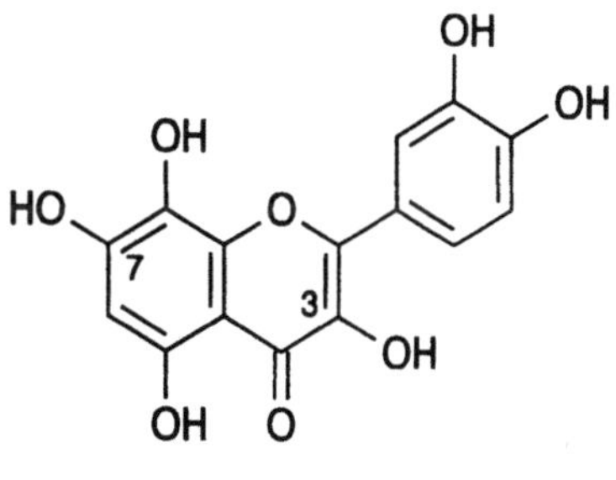

Gossypetin

8-Hydroxy-quercetin = gossypetin
6-Hydroxy-quercetin = quercetagetin
Quercetagetin-6-methylether = patuletin
Kaempferol-7-O-methylether = rhamnocitrin
Rhamnocitrin-4′-rhamnosyl (1 → 4) rhamnosyl (1 → 6) galactoside = catharticin

Flavones

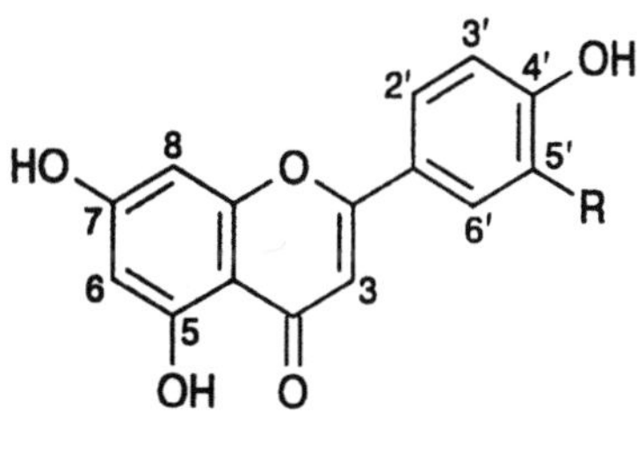

Aglycone	Glycoside
Apigenin R = H	A-8-C-glucoside (vitexin) A-6-C-glucoside (isovitexin) A-7-O-apiosyl-glucoside (apiin) A-6-α-L-arabinopyranoside-8-C-glucoside (schaftoside)
Luteolin R = OH	L-5-O-glucoside (galuteolin) L-8-C-glucoside (orientin) L-6-C-glucoside (iso-orientin)

Flavones

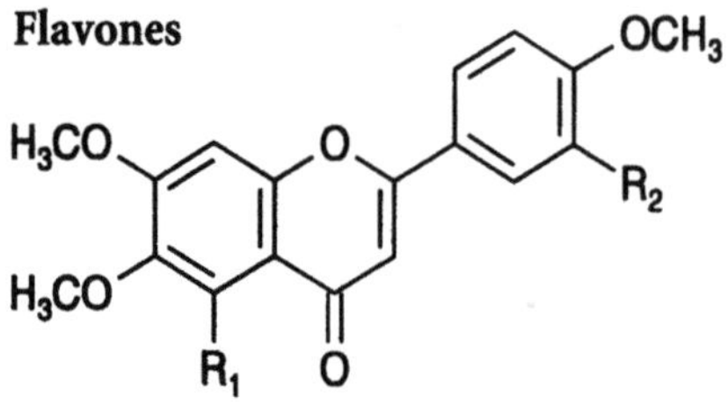

R₁	R₂	Aglycone
OCH_3	OCH_3	Sinensetin
OCH_3	H	Scutellarein tetramethylether
OH	OH	Eupatorin

Flavanon(ol)s

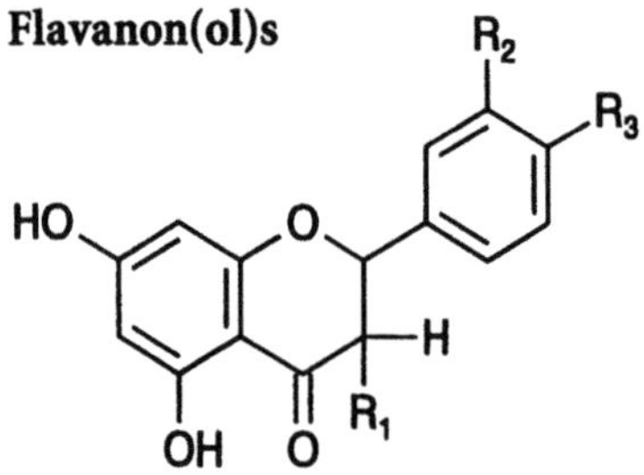

R_1	R_2	R_3	Aglycone -7-O-Glycoside
H	H	OH	Naringenin Naringin (a)
H	OH	OH	Eriodyctiol Eriocitrin (b)
H	OCH_3	OH	Homoeriodyctiol
H	OH	OCH_3	Hesperetin Neohesperidin (a) Hesperidin (b)
OH	OH	OH	Taxifolin

(a) = Neohesperidose
(b) = Rutinose

Flavanolignan

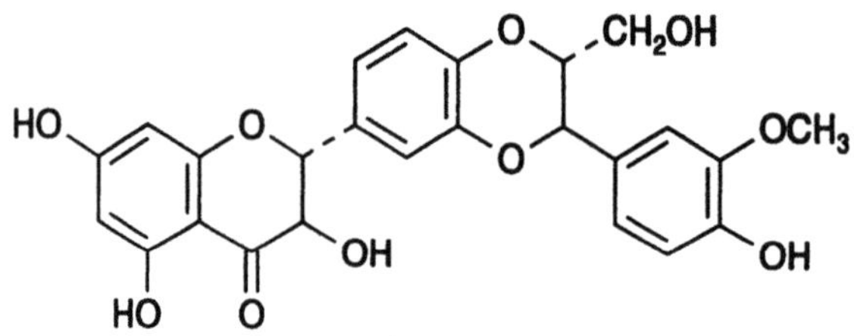

Silybin (=Silybinin)

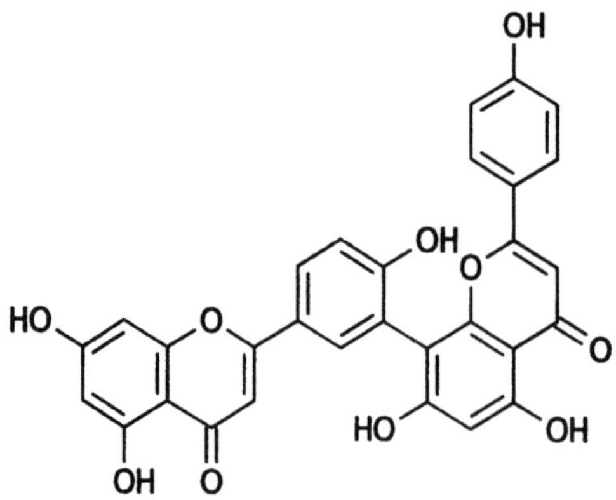

Amentoflavone

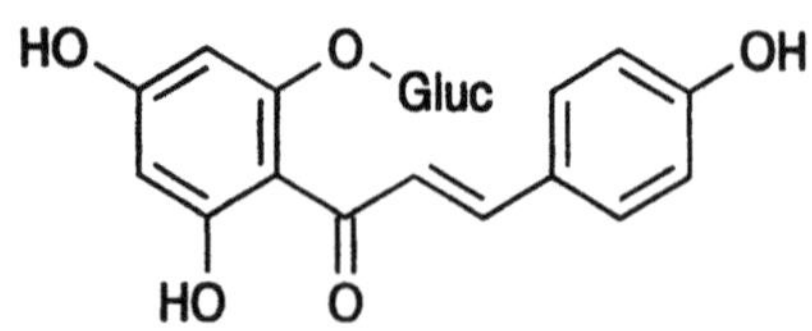

Isosalipurposide

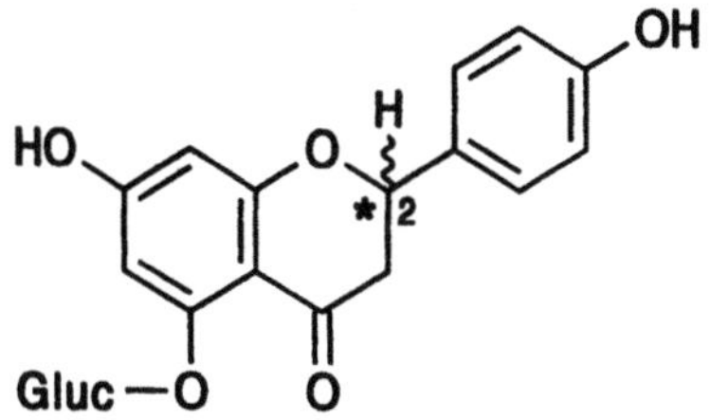

Helichrysin A (2R), B (±form)

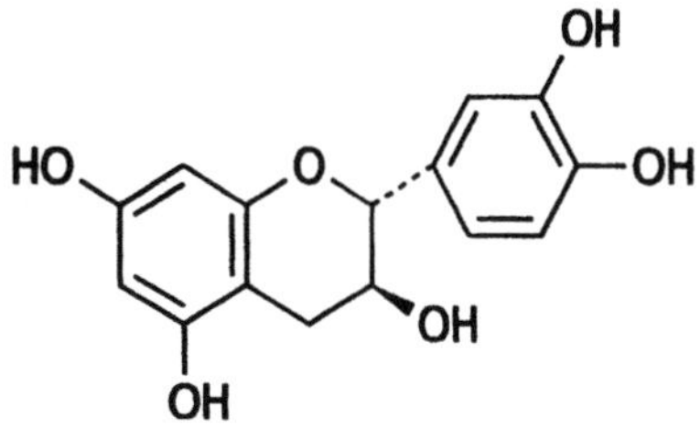

(+) Catechin

Miscellaneous compounds

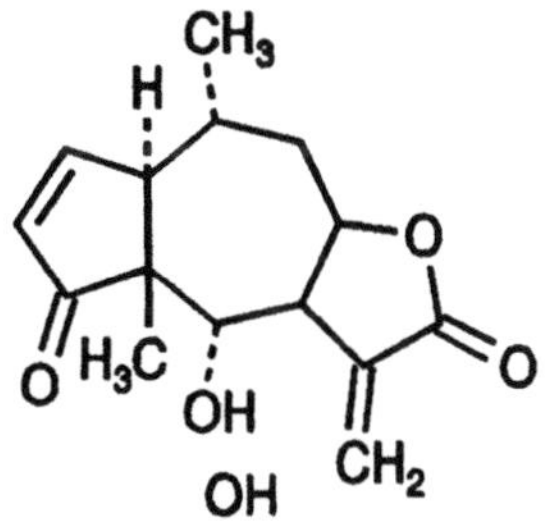

Helenalin (sesquiterpene lacton)

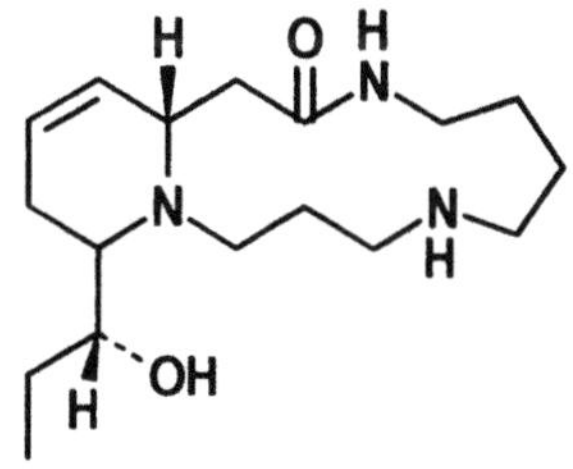

Palustrin (alkaloid)

R:

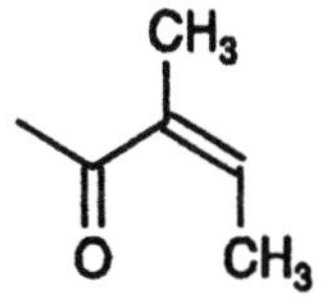

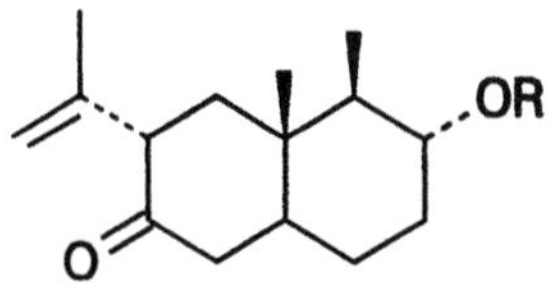

Petasin

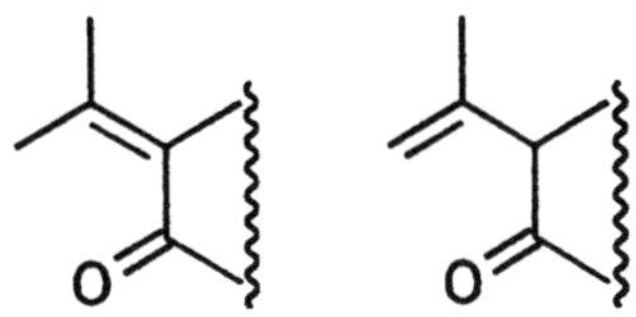

Iso- Neo-

Petasin

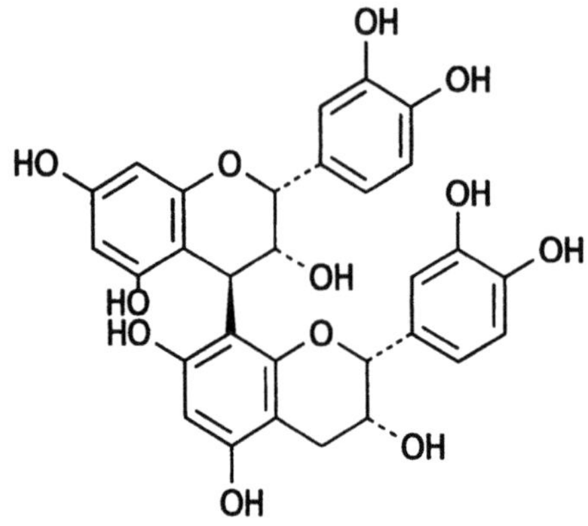

Procyanidin B-2

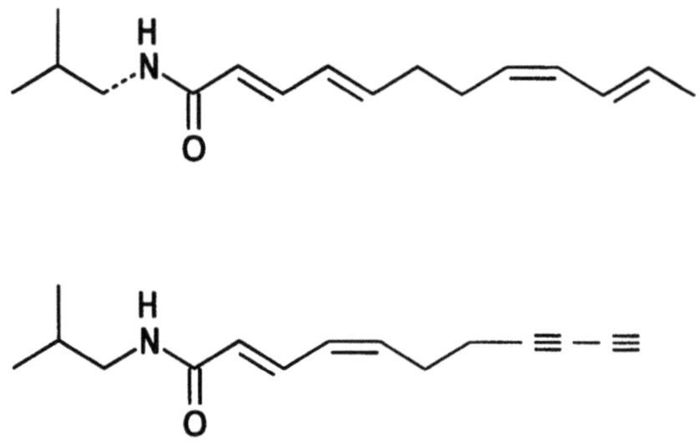

Alkylamides

Phenol carboxylic acids (PCA's)

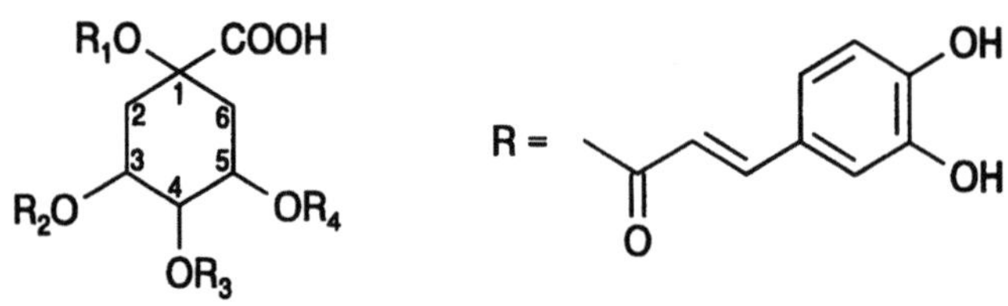

R_1	R_2	R_3	R_4	caffeoyl quinic acids:	
R	H	H	H	pseudo chlorogenic acid	(1-O-caffeoyl quinic acid)
H	R	H	H	chlorogenic acid	(3-O-caffeoyl quinic acid)
H	H	R	H	cryptochlorogenic acid	(4-O-caffeoyl quinic acid)
H	H	H	R	neochlorogenic acid	(5-O-caffeoyl quinic acid)

				dicaffeoyl quinic acids	
R	R	H	H	isochlorogenic acids	1,3-dicaffeoyl quinic acid
H	R	R	H	isochlorogenic acids	3,4-dicaffeoyl quinic acid
H	R	H	R	isochlorogenic acids	3,5-dicaffeoyl quinic acid
H	H	R	R	isochlorogenic acids	4,5-dicaffeoyl quinic acid
R	H	H	R	cynarin (isolated)	1,5-dicaffeoyl quinic acid
				cynarin (native)	1,3-dicaffeoyl quinic acid

Caffeoyl tartaric acids

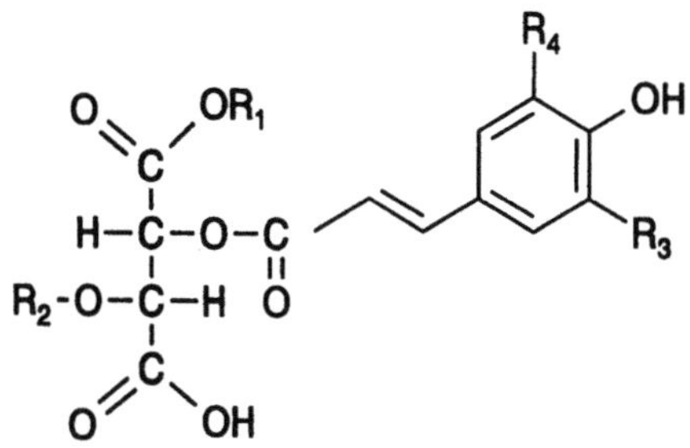

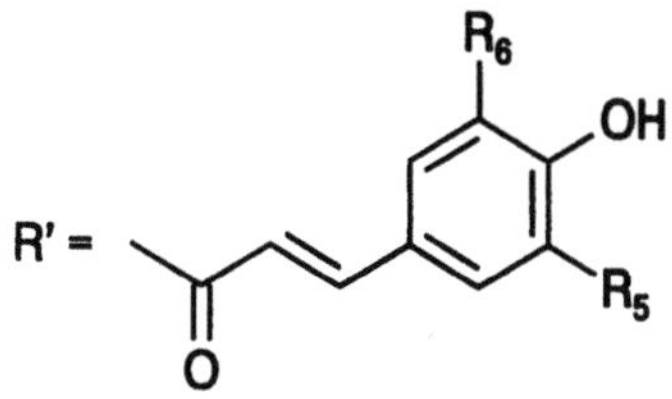

R_1	R_2	R_3	R_4	R_5	R_6	
H	H	OH	H	–	–	2-O-caffeoyl tartaric acid
H	R'	OH	H	OH	H	2,3-O-di-caffeoyl tartaric acid-cichoric acid
CH_3	R'	OH	H	OH	H	2,3-O-di-caffeoyl tartaric acid methyl ester

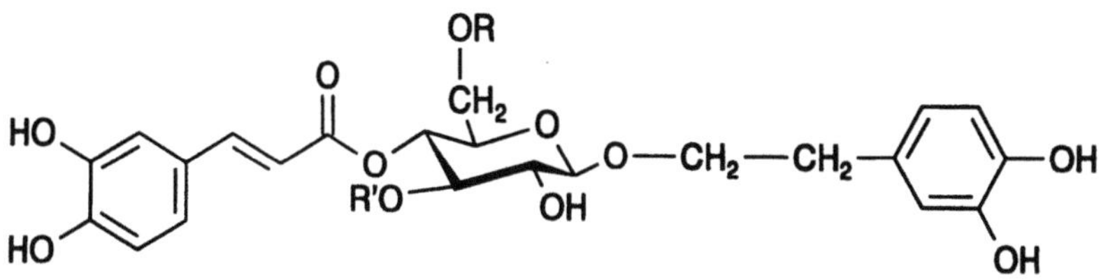

R	R'	dicaffeoyl glycosides
glucose (1,6-)	rhamnose (1,3-)	echinacoside
6-O-caffeoyl-glucoside (1,6-)	rhamnose (1,3-)	6-O-caffeoyl-echinacoside
H	rhamnose (1,3-)	verbascoside
H	H	desrhamnosyl-verbascoside

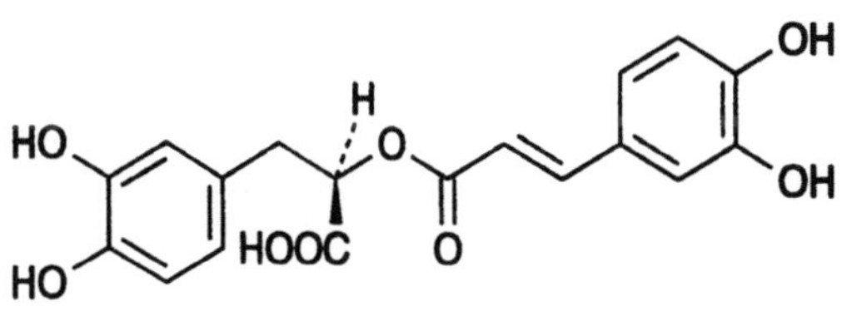

rosmarinic acid

7.1.6 Reference Compounds

Fig. 1 Reference compound series A

1 = quercetin-3-O-gentiobioside
2 = kaempferol-3-O-gentiobioside
3 = quercetin-3-O-rutinoside (rutin)
4 = vitexin-2″-O-rhamnoside
5 = naringin and neohesperidin
6 = chlorogenic acid (R_f ~ 0.45)
7 = luteolin-8-C-glucoside (orientin)
8 = apigenin-8-C-glucoside (vitexin)
9 = isorhamnetin-3-O-glucoside (with isoquercitrin, see Fig. 2)
10 = chlorogenic acid ▶ isochlorogenic acid (R_f ~ 0.8) ▶ caffeic acid (R_f ~ 0.9)
11 = isorhamnetin-3-O-galactoside (cacticin)
12 = quercetin-3-O-rhamnoside (quercitrin, traces of kaempferol-3-O-rhamnoside)
13 = kaempferol-3-O-arabinofuranoside (juglanin)
14 = caffeic acid and ferulic acid (R_f 0.9–0.95)
15 = rutin (R_f ~ 0.4) ▶ chlorogenic acid (R_f ~ 0.45) ▶ hyperoside (R_f ~ 0.6) test mixture T1: these three commercially available compounds are used to characterize the chromatograms of flavonoid drugs

---

Fig. 2 Reference compound series B

1 = quercetin-3-O-gentiobioside
2 = quercetin-3-O-sophoroside
3 = quercetin-3-O-galactosyl-7-O-rhamnoside
4 = kaempferol-3-O-gentiobioside
5 = quercetin-3-O-rutinoside (rutin)
6 = kaempferol-3-O-rhamnoglucoside
7 = quercetin-3-O-glucuronide
8 = quercetin-3-O-galactoside (hyperoside)
9 = quercetin-3-O-glucoside (isoquercitrin)
10 = kaempferol-3,7-O-dirhamnoside (lespedin)
11 = quercetin-3-O-rhamnoside (quercitrin)
12 = kaempferol-3-O-arabinoside
13 = quercetin
14 = kaempferol
15 = mixture of 1–14

Solvent system Fig. 1,2 ethyl acetate-formic acid-glacial acetic acid-water (100:11:11:26)

Detection Natural products-polyethylene glycol reagent (NP/PEG No.28) → UV-365 nm

---

Fig. 1 **Glycosides of the flavone, flavonol and flavanone type**
Treatment with NP/PEG reagent generates in UV-365 nm predominantly orange and yellow-green fluorescences for the flavone and flavanol type and a dark-green one for the flavanone type. Phenol carboxylic acids, which frequently occur in flavonoid drugs, appear as intense, light-blue zones.

Fig. 2 **Various quercetin- and kaempferol-O-glycosides**
Orange or yellow-green fluorescences in UV-365 nm, following NP/PEG treatment, are related to the specific substitution pattern in ring B: two adjacent hydroxyl groups in ring B (e.g. quercetin) give rise to orange fluorescence, whereas a single free hydroxyl group (e.g. kaempferol) results in yellow-green fluorescence.

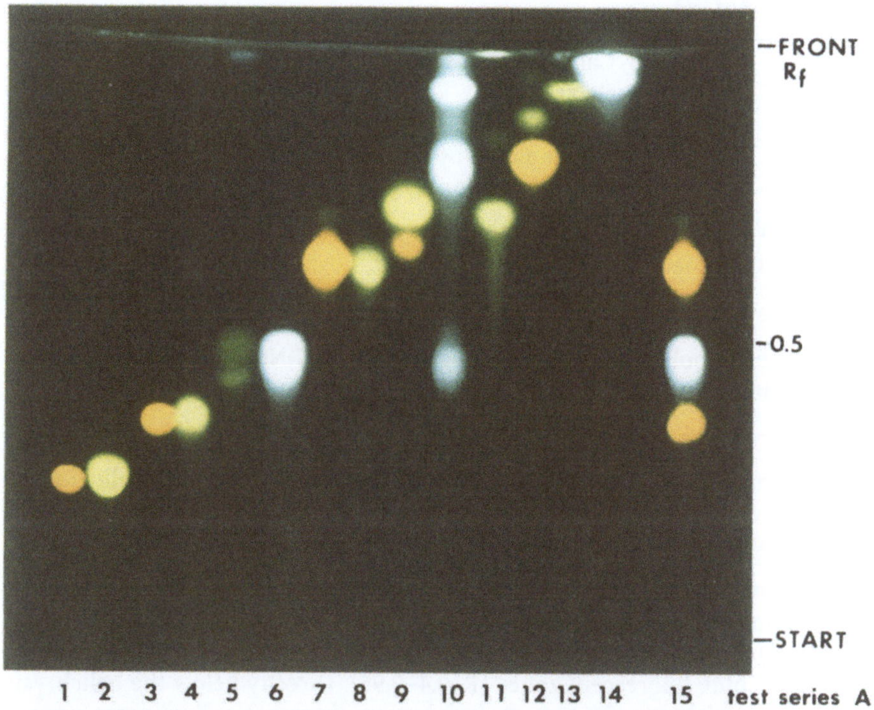

—FRONT
R_f
-0.5
—START
1 2 3 4 5 6 7 8 9 10 11 12 13 14 15 test series A

Fig. 1

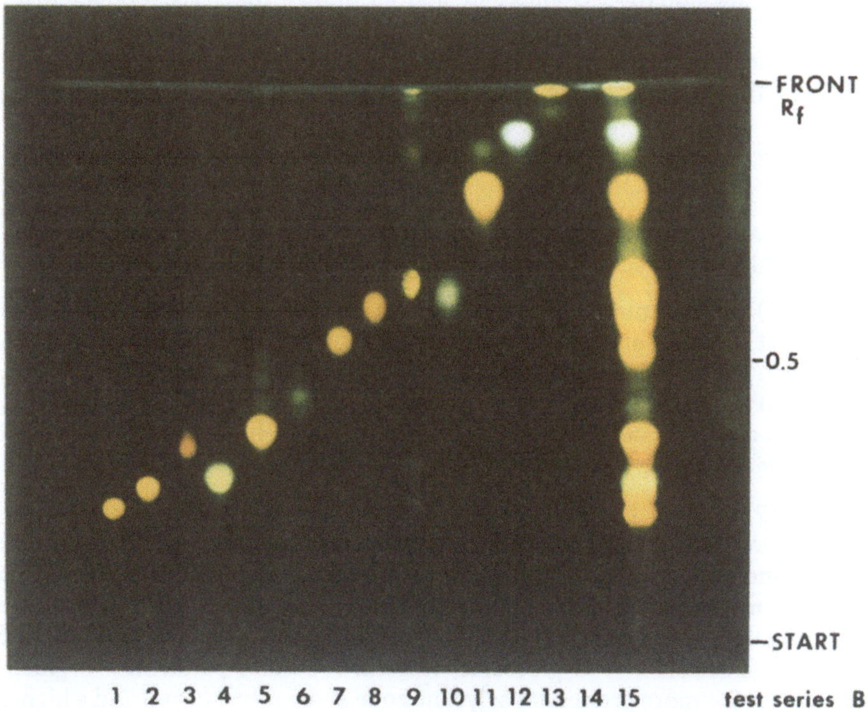

—FRONT
R_f
-0.5
—START
1 2 3 4 5 6 7 8 9 10 11 12 13 14 15 test series B

Fig. 2

7.1.7 TLC Synopsis "Flos"

Drug sample

1	Tiliae flos	6	Calendulae flos
2	Arnicae flos	7	Cacti flos
3	Stoechados flos	8	Primulae flos
4	Sambuci flos	9	Anthemidis flos
5	Verbasci flos	10	Matricariae flos

(methanolic extracts, 20–30 μl)

Reference compound T1 rutin ($R_f \sim 0.4$) ▶ chlorogenic acid ($R_f \sim 0.5$) ▶ hyperoside ($R_f \sim 0.6$)
T2 apigenin-7-O-glucoside

Solvent system Fig. 3,4 ethyl acetate-formic acid-glacial acetic acid-water (100:11:11:26)

Detection A, C Natural products/polyethylene glycol reagent (NP/PEG No. 28) → UV-365 nm
B Natural products reagent (NP No. 28) → UV 365 nm

Description Each extract shows a characteristic TLC fingerprint of yellow-orange or yellow-green flavonoid glycosides (fl.gl) and blue fluorescent phenol carboxylic acids (PCA). The major flavonoids of the individual drugs are identified in Figs. 4–10.

Fig. 3A **Tiliae flos** (1): six orange fl.gl zones (R_f 0.4–0.8) (see Fig. 10)
Arnicae flos (2): three orange fl.gl zones (R_f 0.5–0.7) (see Fig. 5,6)
Stoechados flos (3): three prominent yellow or orange fl.gl zones (R_f 0.6–0.95) accompanied by four major blue fluorescent zones (R_f 0.4–0.95). Directly below the helichrysin (>) zone (R_f 0.85), which appears almost dark brown, there is a yellow-green zone of apigenin-7-O- and kaempferol-3-O-glucoside and an orange zone (e.g. isoquercitrin) (see Fig. 4)
Sambuci flos (4): one major orange fl.gl zone above and below chlorogenic acid (Fig. 9)
Verbasci flos (5): three almost equally concentrated orange fl.gl zones (R_f 0.4/0.5/0.6) (see Fig. 4)
Calendulae (6), **Cacti** (7) and **Primulae flos** (8): characteristic pairs of fl.gl zones (R_f 0.1–0.4) (see Fig. 7,8)

Phenol carboxylic acids: absent in samples 7,8; small amounts in 1,5,6; high concentration in 2–4 (e.g. chlorogenic acid, $R_f \sim 0.5$, or caffeic acid, $R_f \sim 0.9$).

Fig. 4B **Stoechados flos** (3). Treatment with NP reagent reveals helichrysin A/B (see fig. 3) as an olive-green zone at $R_f \sim 0.85$ and apigenin-7-O- and kaempferol-3-O-glucoside as green zones below.
Verbasci flos (5). Two almost equally concentrated flavonoid glycosides are found at R_f 0.60 and 0.75 (e.g. apigenin and luteolin glucoside). The green hesperidin zone at $R_f \sim$ 0.45 is more easily detectable with NP than with NP/PEG reagent (see Fig. 3).

C **Anthemidis flos** (9) is characterized by two and **Matricariae flos** (10) by three yellow-orange or yellow-green major zones in the R_f range 0.55–0.75 and four to six almost white fluorescent PCB and/or coumarins (R_f 0.45–0.95). Apigenin-7-O-glucoside (R_f 0.75/T2) is present in both samples, but is more concentrated in (9). The zones directly below are due to luteolin-7-O-glucoside (sample 9/$R_f \sim 0.7$) and in the chamomile sample (10) due to quercetin-3-O-galactoside, -7-O-glucoside, luteolin-7-O- and patuletin-7-O-glucoside. Sample (9) has the more prominent aglycone zone at the solvent front and a higher concentration and variety of blue fluorescent PCA zones (glucosides of caffeic and ferulic acid) and the coumarin scopoletin-7-O-glucosid at $R_f \sim 0.45$.

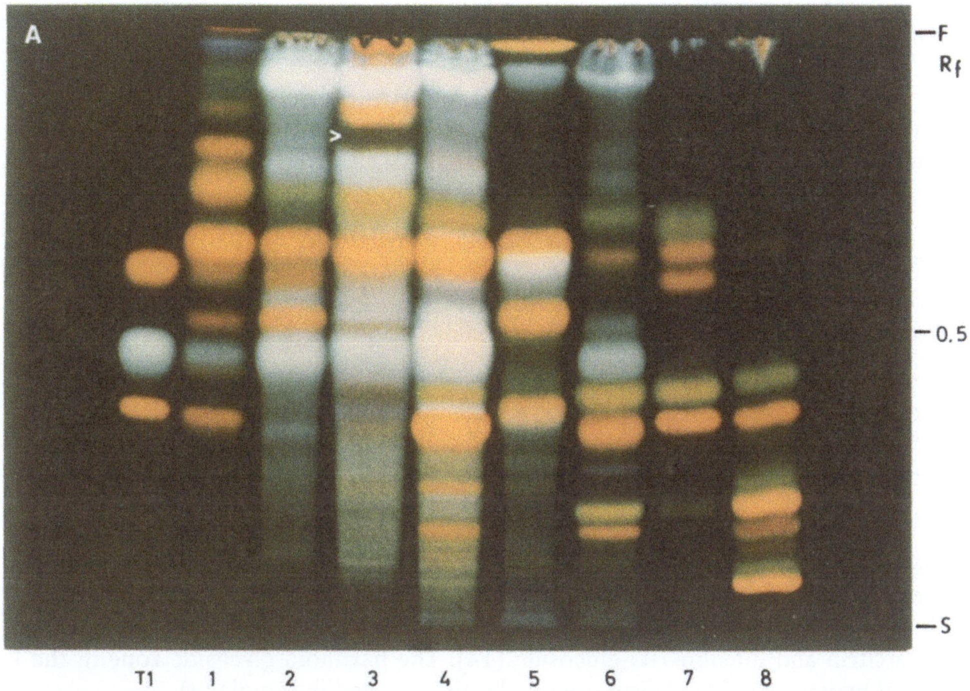

Fig. 3

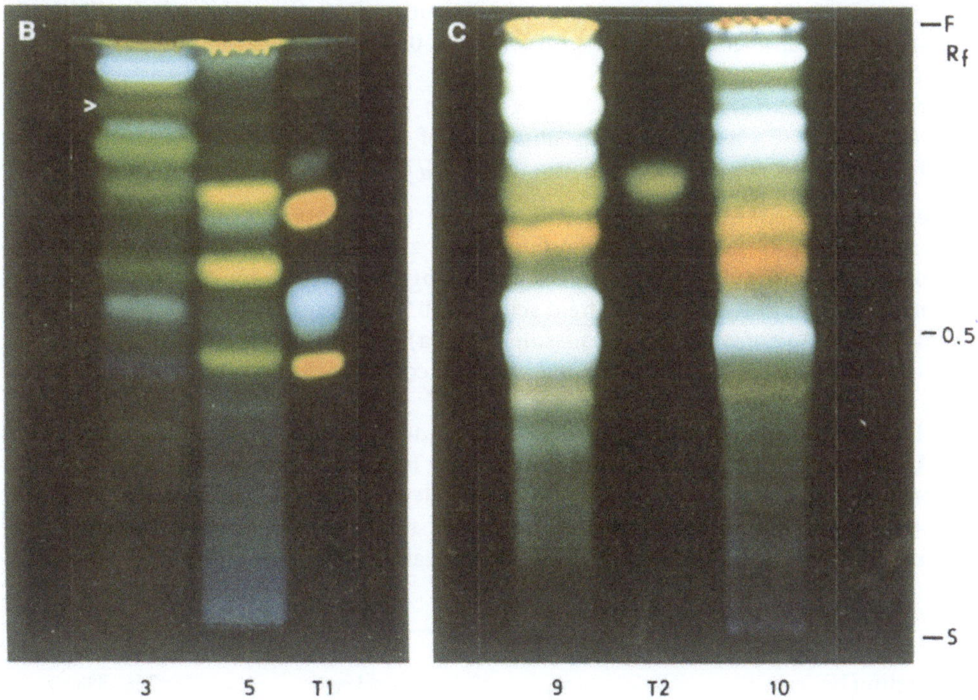

Fig. 4

7.1.8 Chromatograms

Arnicae flos and adulterants

Drug sample

1 Arnicae flos (official drug)
2 Calendulae flos
3 Heterothecae flos (H. inuloides)
4 Arnicae flos (poor in astragalin)
5–8 Arnicae flos (rich in astragalin)
(1.8: MeOH extracts, 30 µl)
5a–8a Arnicae flos ($CHCl_3$ extract, 30 µl)

Reference compound

T1 rutin (R_f 0.35) ▶ chlorogenic acid (R_f 0.45) ▶ hyperoside (R_f 0.6) ▶ isochlorogenic acids (R_f 0.75–0.95)
T2 rutin ▶ chlorogenic acid ▶ hyperoside
T3 quercetin T4 luteolin-7-O-glucoside T5 astragalin

Solvent system

Fig. 5, 6 Aethyl acetate - formic acid - glacial acetic acid - water (100:11:11:26)
Fig. 6B n - pentane - ether (25:75) → system PE

Detection

Fig. 5, 6A Natural products-(polyethylene glycol) reag. (NP/PEG No.28) UV-365 nm
Fig. 6B Zimmermann reagent (ZM No. 44) → vis

Fig. 5A,B **Arnicae flos** (1,4). Arnica montana (1) and Arnica chamissonis (4) show a similar TLC pattern of three orange-yellow flavonoid zones between the blue zones of chlorogenic acid (R_f ~ 0.45/T1) and isochlorogenic acids (R_f 0.7–0.95/T1). The upper zone is due to isoquercitrin and luteolin-7-O-glucoside (T4). The flavonoid glycoside zone in the R_f range of hyperoside (T1/T2) is more highly concentrated in sample (4).
Calendulae flos (2) is characterized by pairs of yellow-orange isorhamnetin and quercetin glycosides. The major zones are due to rutin (R_f 0.4) and narcissin (R_f 0.45), isorhamnetin- and quercetin-3-O-glucoside (R_f 0.6–0.7) and isorhamnetin-rutinosyl-glucoside at R_f ~ 0.2.
Heterothecae flos (3) has a similar TLC pattern with Arnicae flos (4). Together with Calendulae flos (2), it counts as an adulterant of Arnicae flos. The adulterants can be easily detected by the presence of rutin (R_f ~ 0.4/T1/T2) (▶ see also Fig. 7).

Fig. 6A Detection with NP reagent only: **Arnicae flos** (A. montana, A. chamissonis; 5–8) Astragalin is found as a bright green fluorescent zone at R_f ~ 0.8 (T5), while the other flavonoid glycosides below only appear pale orange-brown (see fig. 5). Blue fluorescent chlorogenic acid at R_f 0.45 and caffeic acid at R_f 0.9 are detectable.

B **Arnicae flos** (A. montana, A. chamissonis; 5a–8a). The $CHCl_3$ extracts of Arnicae flos (method see Sect. 7.1.1), developed in system PE, contain sesquiterpene lactones, detectable as violet-grey zones with ZM reagent (vis).
11, 13-Dihydrohelenalin (DH), helenaline (H) and their esters are major compounds in A. montana and A. chamissonis. DH and Higrate in the lower R_f range. The isobutyryl-, methacryl-, tigloyl- and isovaleryl-helenaline and -11, 13-dihydrohelenaline, respectively, are found in the upper R_f range. DH and H are always present, whereas the amount of their esters varies. No sesquiterpenes are found in Heterotheca.

Sesquiterpene lactones	A. montana	A. chamissonis
11, 13-dihydrohelenalin (DH)	<3.6%	2%–7%
helenalin (H)	<1%	2%–8%
6-O-tigloyl-11, 13-dihydrohelenalin	<11%	20%–26%
6-O-tigloyl-helenalin	<37%	17%–30%
chamissonolide/6-O-acetyl-chamissonolide	-	0.8%–9.5%
arnifoline/ dihydroarnifoline	-	3%–12%

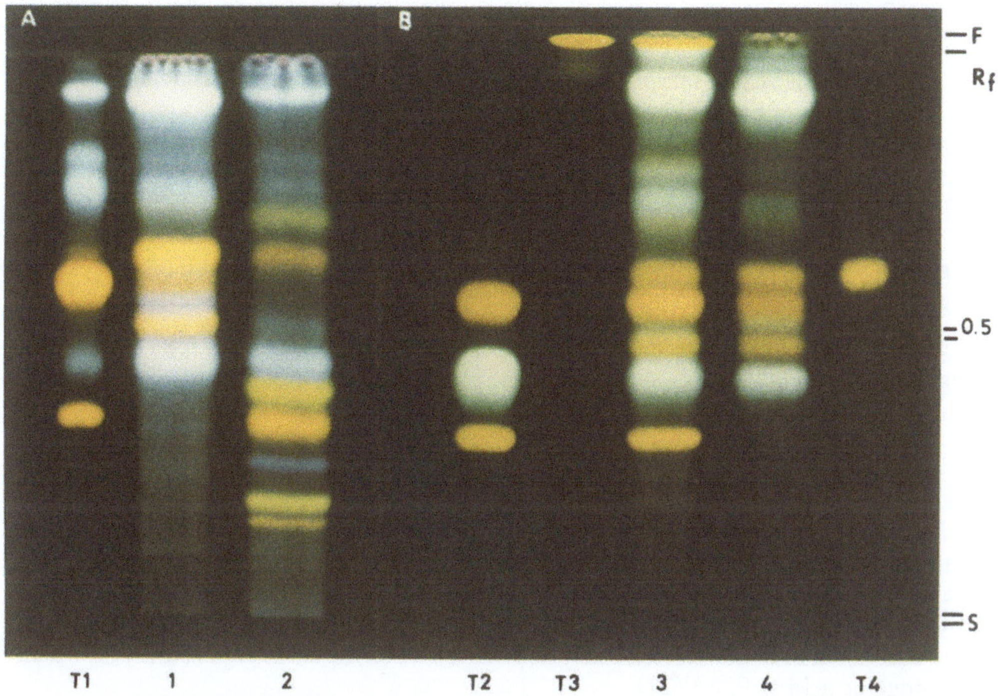

Fig. 5

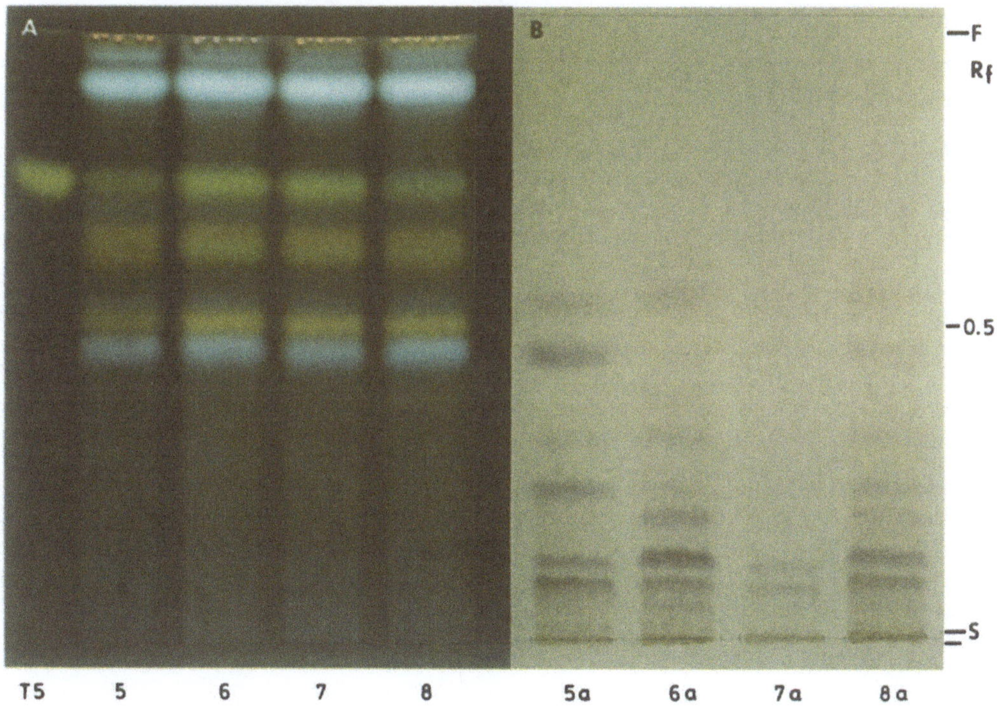

Fig. 6

Calendulae, Cacti, Primulae flos

Drug sample	1	Calendulae flos
	2–2b	Cacti flos (trade samples) (methanolic extracts, 20 μl)
	3–3b	Primulae flos (trade samples)
	3c	Primulae flos (high amount of calycibus)

Reference compound T1 rutin ($R_f \sim 0.4$) ▶ chlorogenic acid ($R_f \sim 0.5$) ▶ hyperoside ($R_f \sim 0.6$)
T2 narcissin (=isorhamnetin-3-O-rutinoside)

Solvent system Fig. 7,8A ethyl acetate-formic acid-glacial acetic acid-water (100:11:11:26) → system 1
Fig. 8B,C chloroform-acid-glacial acetic acid-methanol-water (60:32:12:8) → system 2

Detection Fig. 7A,B Natural products-polyethylene glycol reagent (NP/PEG No. 28) → UV-365 nm
Fig. 8A,B (NP/PEG No. 28) → UV-365 nm
C Anisaldehyde-H_2SO_4 reagent (AS No. 3) vis

Fig. 7A **Calendulae** (1) and **Cacti flos** (2) are both characterized by orange-yellow quercetin and yellow-(green) fluorescent isorhamnetin glycosides in the R_f range 0.2/0.4/0.7 with rutin and narcissin as major compounds (R_f 0.4–0.45 /T1/T2).
Cacti flos (2) contains more monoglycosides (R_f 0.6–0.7), e.g. cacticin an isorhamnetin-3-O-glucoside at $R_f \sim 0.7$, while Calendulae flos (1) has more triglycosides (R_f 0.2–0.25) and additional blue fluorescent zones of phenol carboxylic acids at the solvent front and chlorogenic acid at R_f 0.5 (T1).
Primulae flos (3) shows predominantly di- and triglycosides of quercetin and kaempferol in the R_f range 0.1–0.5 (e.g. kaempferol-gentiotrioside).

B The quantitative distribution of individual flavonoid compounds in Primulae flos are due to the varying quantities of flower and calicybes in trade samples (3a, 3b).

Fig. 8A **Cacti flos.** Variations in the flavonoid pattern are demonstrated (2a, 2b).
Cacti flos trade samples (e.g. 2a) normally show the orange zone of rutin and the green one of narcissin in the R_f range 0.4–0.45 (T2 see also Fig. 7).
Sample (2b) is freshly collected material of Selenicereus grandiflorus, which contains mainly narcissin and only traces of rutin and monoglycosides in the upper R_f range.

B **Saponins** in **Calendulae** and **Primulae flos**: For the separation of saponin glycosides, the more polar solvent system 2 is recommended. In this system, the flavonoids of **Primulae flos** (3c) show a similar separation pattern compared to system 1, while the separation of flavonoid and phenol carboxylic acids of Calendulae flos (1) is different (detection NP/PEG reagent. UV-365 nm).

C Detection with the AS reagent reveals three prominent grey-blue oleanolic acid glycosides ($R_f \sim 0.2/0.4/0.6$) which characterize the saponin pattern of Calendulae flos (1). Primulae flos (3c) shows weak yellow-brown zones in the R_f range 0.05–0.4, mostly due to flavonoid glycosides. According to the literature, saponins might be present preferrably in the calicybes of the drug.

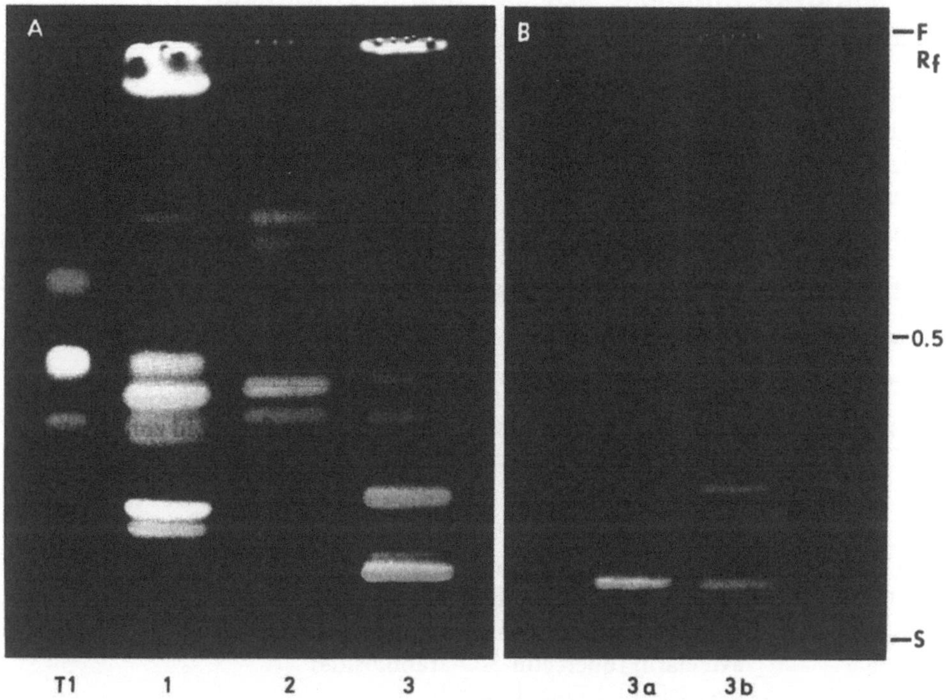

Fig. 7

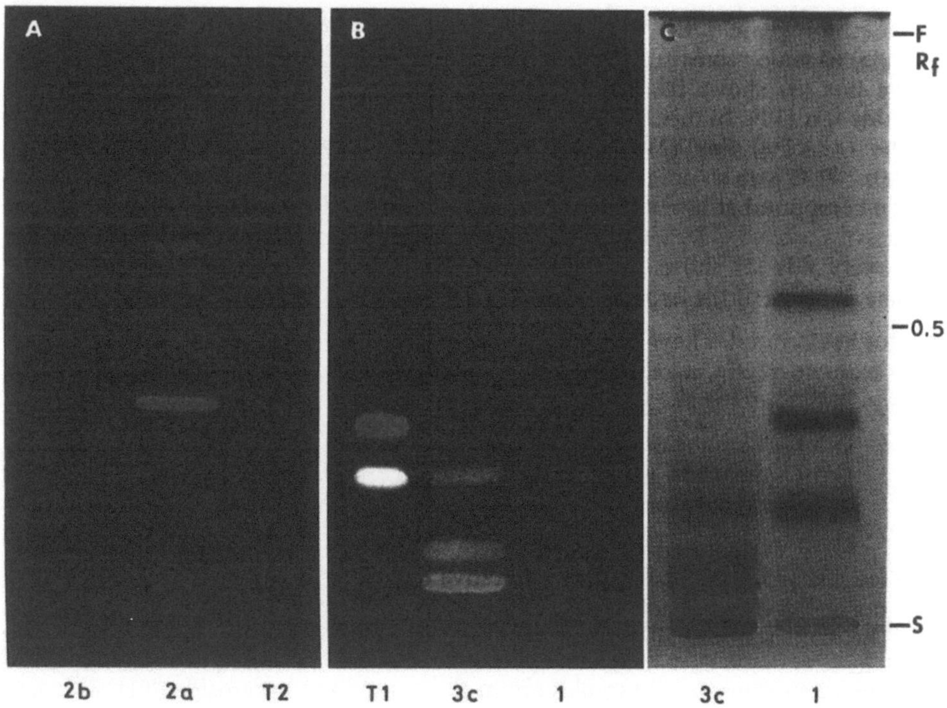

Fig. 8

Pruni spinosae, Robiniae, Acaciae, Sambuci, Spiraeae and Tiliae flos

Drug sample	1 Pruni spinosae flos 2 Sambuci flos 3 Spiraeae flos (methanolic extracts, 20 µl)	4 Robiniae (Acaciae) flos 5 Acaciae verticil. flos 6–10 Tiliae flos (commercial drugs)
Test mixture	T1 rutin ($R_f \sim 0.4$) ► chlorogenic acid ($R_f \sim 0.5$) ► hyperoside ($R_f \sim 0.6$)	
Solvent system	Fig. 9,10 ethyl acetate-formic acid-glacial acetic acid-water (100:11:11:26)	
Detection	Natural products-polyethylene glycol reagent (NP/PEG No. 28) → UV-365 nm	

Fig. 9 **Pruni spinosae flos** (1) shows eight prominent orange or green flavanoid zones between $R_f \sim 0.35$ and the solvent front and two blue zones in the R_f range of chlorogenic acid (T1):

$R_f \sim 0.35$	rutin
$R_f \sim 0.4$	kaempferol-diglycoside
R_f 0.6–0.7	isoquercitrin, kaempferol-3, 7-dirhamnoside
R_f 0.75–0.8	quercetin-3-O- and kaempferol-3-O-rhamnoside
$R_f \sim 0.85$	avicularin (quercetin-3-O-arabinoside)
$R_f \sim 0.9$	kaempferol-3-O-arabinoside
front	quercetin

Sambuci flos (2) is characterized by a pair of orange and green zones above and below the blue chlorogenic acid of almost equal intensity: rutin (T1) and isoquercitrin ($R_f \sim 0.65$) as major constituents, caffeic acid at $R_f \sim 0.9$.

Spiraeae flos (3) shows its main constituents as blue fluorescent zones above the hyperoside test (T1). In this R_f range spiraeoside (quercetin-4'-O-glucoside) is found.

Robiniae (Acaciae) flos (4) has predominantly green-yellow zones in the lower R_f range 0.2–0.45 with robinin (kaempferol-3-O-rhamnosyl-galactosyl-7-rhamnoside) as the main compound at $R_f \sim 0.2$. Acacetin-7-O-rutinoside migrates directly above rutin (T1).

Acaciae vert. flos (5) shows additional high amounts of flavonoid glycosides in and above the R_f range of the hyperoside test.

Fig. 10 **Tiliae flos** has a complex flavonoid pattern consisting of at least eight different glycosides derived from quercetin, myricetin and kaempferol:

$R_f \sim 0.9$	tilirosid		
$R_f \sim 0.8$	Q-3-O-rhamnoside	M-3-O-rhamnoside	K-3-O-rhamnoside
$R_f \sim 0.7$	Q-3-O-glucoside	M-3-O-glucoside	K-3-O-glucoside
$R_f \sim 0.7$	Q-3, 7-dirhamnoside		K-3, 7-dirhamnoside
$R_f \sim 0.4$	rutin (T1)		

The Tiliae flos samples 6–10 show quantitative differences in their flavonoid content, depending on the corresponding Tilia species (T. cordata, T. platyphyllos or a mixture of both). The main zones in all samples are in the R_f range of hyperoside (T1). Blue and orange zones in the R_f range of chlorogenic acid (T1) can be absent.

Note: The adulterant Tilia argentea contains, instead of rutin, a prominent flavonoid glycoside zone in the R_f range 0.2–0.3.

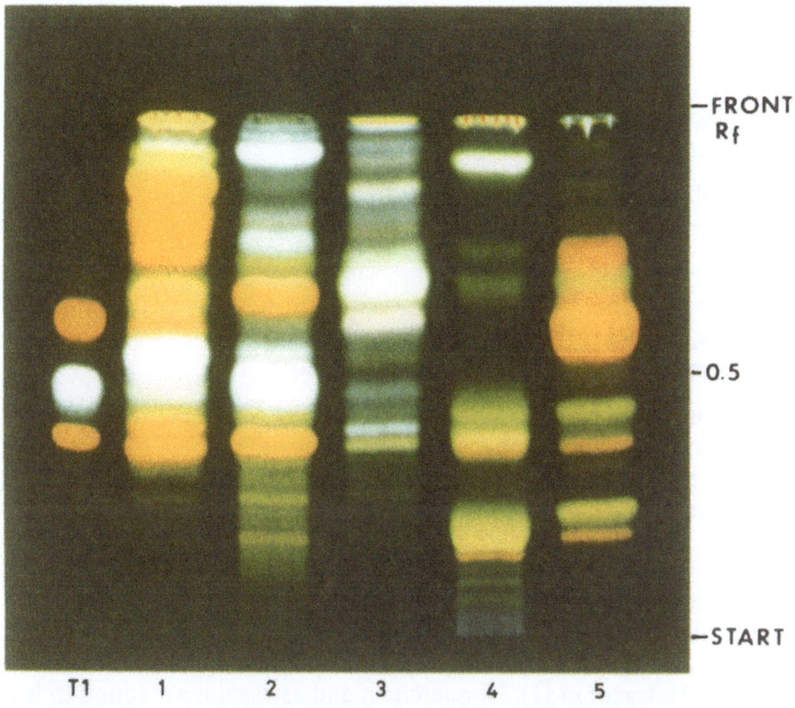

Fig. 9

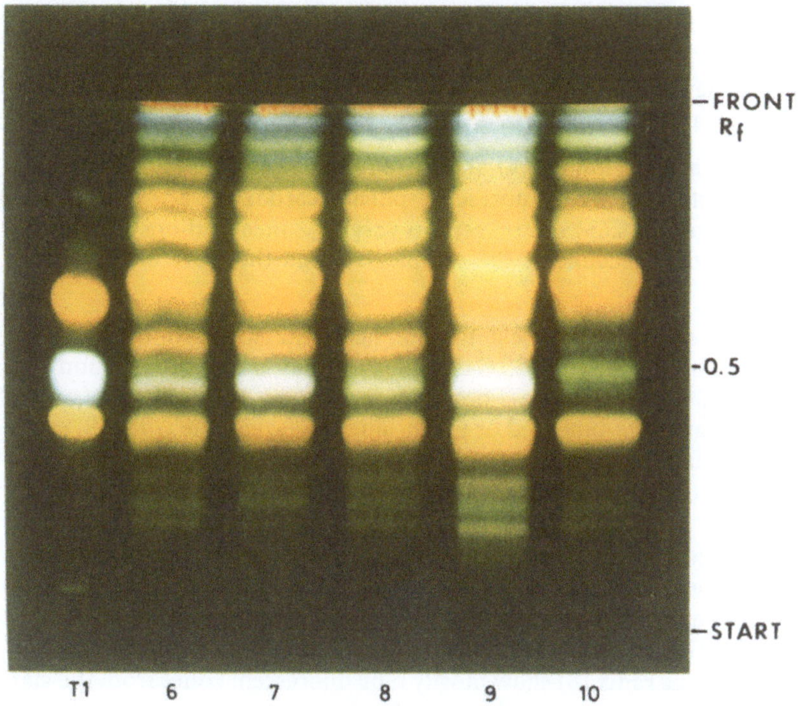

Fig. 10

Farfarae folium, flos; Petasitidis folium, radix

Drug sample
1,2 Farfarae folium
3 Farfarae flos
4 Farfarae folium (trade sample)
5–7 Petasitidis folium (P. hybridus different origin)
8 Petasitidis radix
(extracts, 20 µl, preparation see 7.1.1)

Reference compound
T1 rutin ($R_f \sim 0.4$) ► chlorogenic acid ($R_f \sim 0.5$) ► hyperoside ($R_f \sim 0.6$)
T2 eugenol
T3 petasin

Solvent system
Fig. 11A, 12C ethyl acetate-formic acid-glacial acetic acid-water (100:11:11:26) ► flavonoids, phenol carboxylic acids
Fig. 11B, 12D chloroform ► lipophilic compounds (e.g. petasin)

Detection
Fig. 11A, 12C Natural products-polyethylene glycol reagent (NP/PEG No. 38) → UV-365 nm ► flavonoids, phenol carboxylic acids
Fig. 11B Anisaldehyde-H_2SO_4 reagent (AS No. 3) → vis ► petasin
Fig. 12D Concentrated H_2SO_4 → vis ► petasin

Fig. 11A **Farfarae folium** (1,2) and **flos** (3). Methanolic extracts show mainly blue fluorescent zones of phenol carboxylic acids e.g. chlorogenic acid ($R_f \sim 0.5$/T1), isochlorogenic acids (R_f 0.7–0.75) and caffeic acid ($R_f \sim 0.9$). Rutin is detectable as an orange fluorescent zone in (2) and (3) ($R_f \sim 0.4$/T1); traces in (1). Isoquercitrin and astragalin are found in low concentrations above the hyperoside test (T1) in samples 1 and 3.

B Petrol ether extracts of Farfara samples 1–3, developed in $CHCl_3$ and detected with AS reagent, do not show prominent zones (vis) below the R_f range of the reference compound eugenol (T2). The German pharmacopoeia DAB 10 requires this TLC characterization of authentic Farfarae folium (1,2).

Fig. 12C Methanolic extracts of **Farfarae folium** (4) and the adulterant **Petasitidis folium** (5–7) have a very similar pattern of blue fluorescent phenol carboxylic acids. Flavonoid monoglycosides (R_f range 0.5–0.65) are present in varying concentrations in Petasites species, while rutin is not detectable in the samples 5,6 and there are only traces in sample 7.

D For the detection of the sesquiterpenes e.g. petasol, neo- and isopetasol derivatives, a petrol ether extract has to be prepared. TLC development in $CHCl_3$ and detection with concentrated H_2SO_4 (98%) reveals the lipophilic compounds of Farfarae (4) and Petasitidis folium (5–7) as white-blue, red or green-blue zones over the whole R_f range. The sesquiterpene petasin/isopetasin are found in the R_f range 0.4–0.45 as green-blue fluorescent zones. The concentration of the esters (e.g. methacryl petasol) varies, depending on the Petasites species.
Petasites sample (7) shows the petasin/isopetasin as prominent zones, while in Petasitidis folium samples 5 and 6 the petasin is present in low concentration. Petasin-free chemical races also exist. In these cases Patasitidis and Farfarae folium extracts are hardly distinguishable. The red zones are due to furanoeremophilanes.
$CHCl_3$ extracts of **Petasitidis radix** (8) show mostly blue fluorescent zones from the start up to the solvent front, due to more than 20 sesquiterpenes.

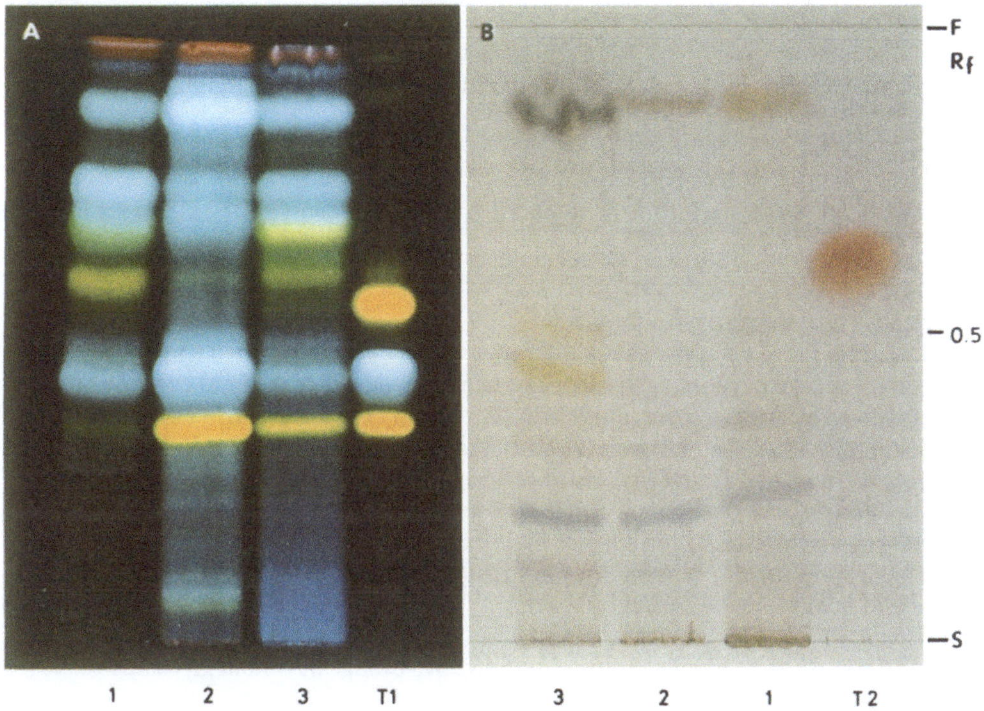

Fig. 11

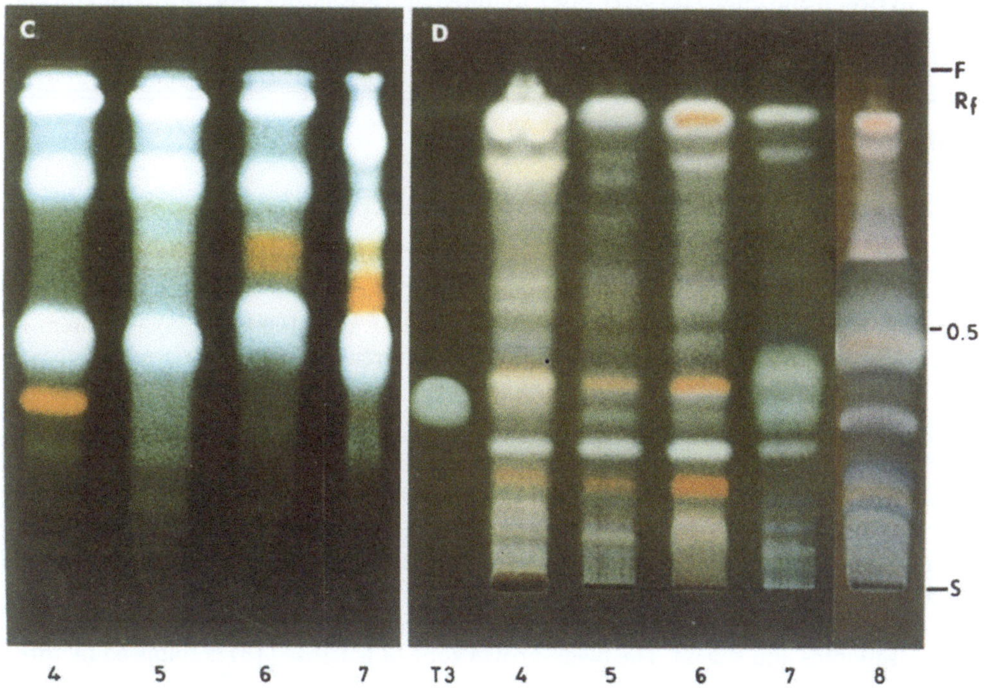

Fig. 12

Betulae, Juglandis, Rubi and Ribis folium

Drug sample 1 Betulae folium 2 Juglandis folium 3 Rubi fruticosi folium 4 Rubi idaei folium 5 Ribis nigri folium (methanolic extracts, 20 µl)

Reference compound T1 quercitrin
T2 rutin (R_f ~ 0.4) ▸ chlorogenic acid (R_f ~ 0.5) ▸ hyperoside (R_f ~ 0.6) → test mixture

Solvent system Fig. 13A,B ethyl acetate-formic acid-glacial acetic acid-water (100:11:11:26)

Detection Natural products-polyethylene glycol reagent (NP/PEG No. 28) → UV-365 nm

---

Fig. 13A **Betulae** (1) and **Juglandis** (2) **folium.**
Both extracts show a similar flavonoid pattern in the R_f range 0.55–0.85 with five to six prominent orange flavonoid glycoside zones: hyperoside (T2) as major compound (R_f ~ 0.6), followed by isoquercitrin, quercitrin (R_f ~ 0.8/T1) and avicularin (R_f ~ 0.85). They are distinguished by the orange myricetin-digalactoside, only present in (1), R_f range of chlorogenic acid, and rutin (T2), which can be present in higher concentrations in other Betula species, and the green zone of kaempferol-3-arabinoside (R_f ~ 0.95) and neochlorogenic acid as a blue zone at R_f ~ 0.55, detectably only in 2.

B **Rubi** (3,4) and **Ribis** (5) **folium.** They are easily distinguishable by their different qualitative and quantitative flavonoid-pattern.
Rubi fruticosi fol. (3) has two prominent blue besides three weak green fluorescent zones (R_f 0.5–0.95), while **Rubi idaei folium** (4) shows five orange flavonoid glycoside zones (R_f range 0.3–0.75).
Ribis nigri folium (5). One major orange flavonoid glycoside above the hyperoside test (e.g. myricetin and quercetin glycoside), followed by a green zone (e.g. isorhamnetin and kaempferol monoglycosides); a minor zone of rutin (T2) is seen at R_f 0.4.

Castaneae folium

Drug sample 1 Castaneae folium (methanolic extract, 20 µl)

Reference compound T1 rutin (R_f ~ 0.4) ▸ chlorogenic acid (R_f ~ 0.5) ▸ hyperoside (R_f ~ 0.6)
T2 fructose T3 rutin

Solvent system Fig. 14A ethyl acetate-formic acid-glacial acetic acid-water (100:11:11:26) → system 1
B,C chloroform-glacial acetic acid-methanol-water (60:32:12:8) → system 2

Detection A,C Natural products-polyethylene glycol reagent (NP/PEG No. 28) → UV-365 nm
B Anisaldehyde-H_2SO_4 reagent (AS No. 3) → vis

Fig. 14A **Castaneae folium** (6) is characterized in system 1 (NP/PEG) by the yellow zone of rutin (R_f ~ 0.4), blue O-p-coumaroyl quinic acid (R_f ~ 0.45) followed by two prominent orange-yellow zones of isoquercitrin and quercetin galacturonide (=miquelianin). Traces of astragalin are found at R_f ~ 0.75.

B Separation in system 2 and treatment with the AS reagent are efficient for the detection of e.g. saponins and sugars. The black-brown zone of fructose (T3) is followed by three brown flavonoid glycoside zones, five additional violet-blue zones (R_f 0.55–0.75; saponin glycosides ?) and the prominent violet zone of lipophilic compounds (e.g. ursolic acid, lupeol) at the solvent front.

C Detection with NP/PEG reagent shows the three flavonoid glycosides as orange-yellow and the phenol carboxylic acids as blue fluorescent zones at lower R_f values compared with system 1 (→ A).

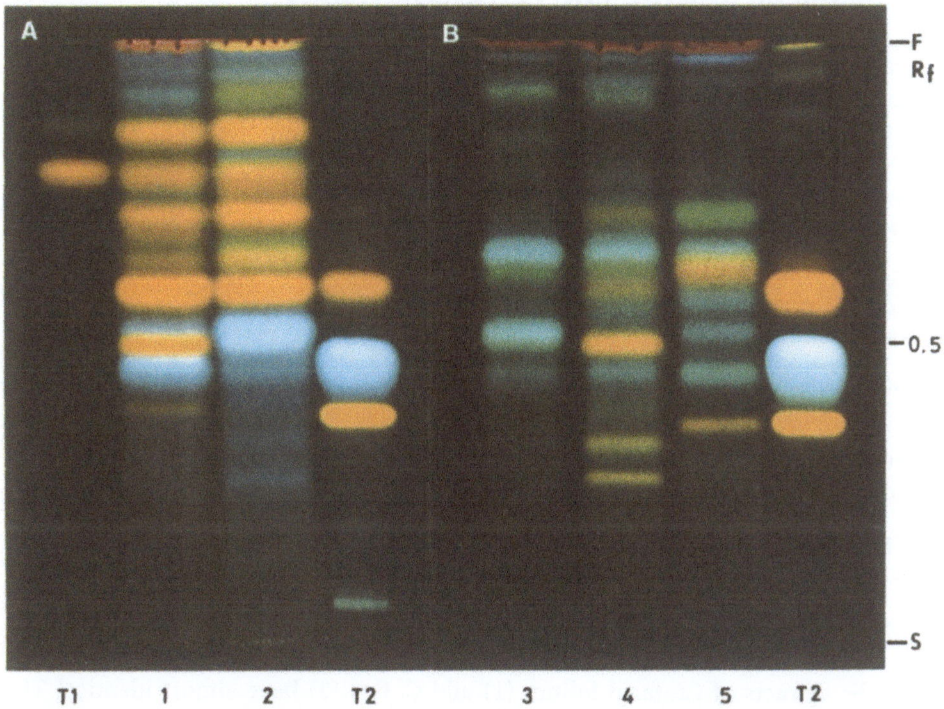

Fig. 13

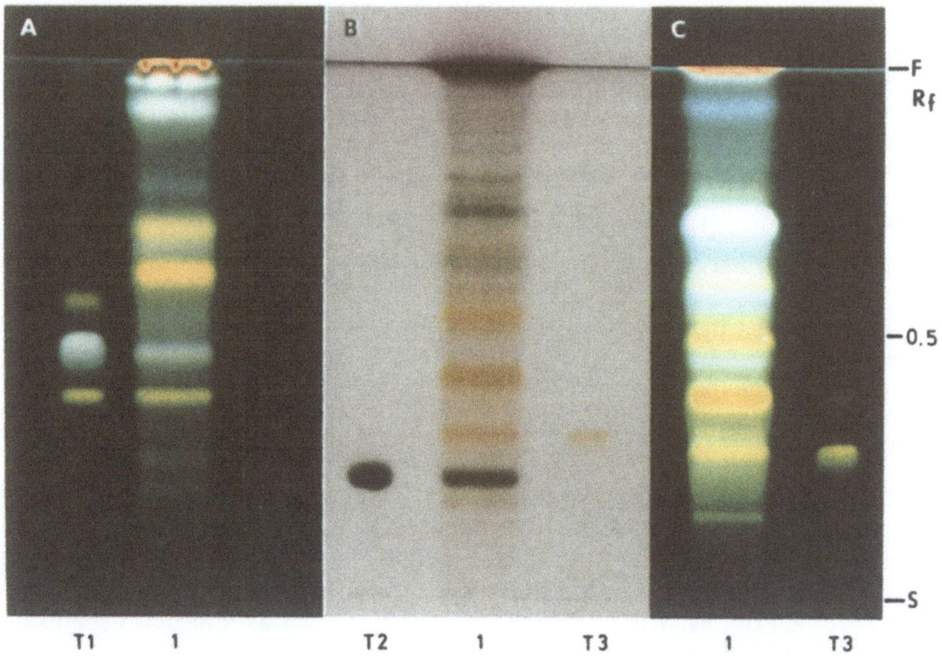

Fig. 14

Crataegi folium, fructus, flos; Lespedezae herba

Drug sample

1 Crataegi folium (MeOH extract 20, µl)
2 Crataegi fructus (MeOH extract 20, µl)
3 Crataegi flos (MeOH extract 20, µl)
4 Lespedezae herba (MeOH extract 20, µl)
5 Lesp.herba (10:1/EtAc enrichment 10, µl)
6 Lesp.herba (5:1/EtAc enrichment 1,0 µl)

1a C. folium (fraction 1)
1b C. folium (fraction 2)
1c C. folium (fractions 3)
5a procyanidin fraction of (5)
6a procyanidin fraction of (6)

Reference compound

T1 rutin ($R_f \sim 0.3$) ► chlorogenic acid ($R_f \sim 0.4$) ► hyperoside ($R_f \sim 0.55$)
T2 vitexin-2″-rhamnoside ($R_f \sim 0.45$) ► vitexin ($R_f \sim 0.7$)
T3 isoorientin ($R_f \sim 0.5$) T4 orientin ($R_f \sim 0.6$)

Solvent system

Fig. 15,16A ethyl acetate-formic acid-glacial acetic acid-water (100:11:11:26) flavonoids
B ethyl acetate-glacial acetic acid-water (100:20:30/upper phase) procyanidins

Detection

A Natural products-polyethylene glycol reagent (NP/PEG No. 28) → UV-365 nm
B Vanillin-phosphoric acid (VP reagent No. 41) → vis

Description

Crataegus and **Lespedeza contain flavonoids** (A) and procyanidins (B). The procyanidins can be separated from flavonoids over a polyamide column (see Sect. 7.1.1).

Fig. 15A Methanolic extracts of **Crataegi folium** (1) and **C. flos** (3) have almost identical TLC flavonoid patterns in UV-365 nm (NP/PEG reagent):

R_f 0.3	orange	rutin (T1)
R_f 0.35	yellow-green	vitexin-2″-O-rhamnoside (T2)
R_f 0.4–0.5	blue	caffeoyl quinic acids (e.g. chlorogenic acid, T1)
R_f 0.5–0.6	orange	hyperoside (T1), luteolin-5-O-glucoside
R_f 0.6–0.65	yellow-green	vitexin (T2), overlapped by blue spiraeoside
R_f 0.9	blue	caffeic acid

C. fructus (2) shows very weak zones of caffeoyl quinic acids and hyperoside only.

B **Procyanidins** in Crataegi folium (1a–1c) can be separated from flavonoids (VP reagent vis.):
► 1a: EtOH eluate with flavonoids (yellow/R_f 0.3–0.5) and red zones of di- and trimeric procyanidines (R_f 0.7–0.8)
► 1b: EtOH/acetone (8:2) eluate with tri- and tetrameric procyanidins (R_f 0.5–0.7).
► 1c: acetone eluate with the enriched tetra- and hexa polymeric procyanidins (R_f range 0.05–0.5).

Fig. 16A A methanolic extract of **Lespedezae herba** (4) contains more highly glycosidated flavonoids (R_f range 0.05–0.35) such as kaempferol- and/or isorhamnetin-rhamno-glucoside and rutin (T1). The EtAc extract (5,6) contains the enriched yellow zones of flavon-C-glycosides such as isoorientin ($R_f \sim 0.5$/T3), orientin ($R_f \sim 0.65$/T3) over-lapped by isovitexin, as well as flavonol-O-glycosides such as lespedin, hyperoside, isoquercitrin and quercitrin (R_f 0.6–0.8).
After treatment with the VP reagent, Lespedezae herba (5a, 6b) shows the phenolic compounds as red-brown zones (vis.). In addition to free catechin, epicatechin ($R_f \sim$ 0.9), the dimeric procyanidins migrate into the upper R_f range and trimeric procyanidins are found in the lower R_f range.

B

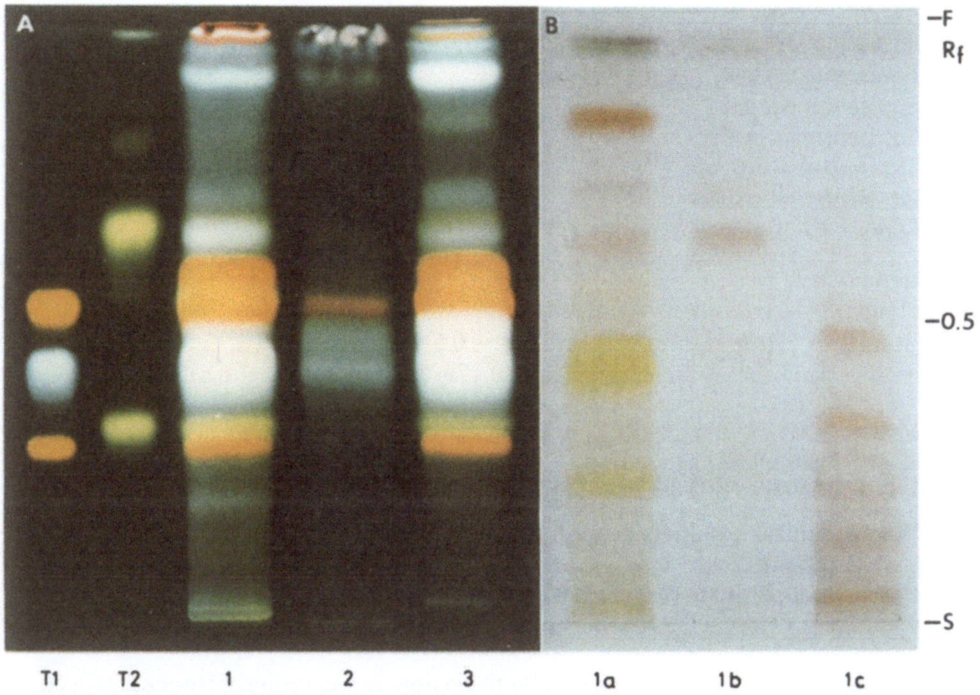

Fig. 15

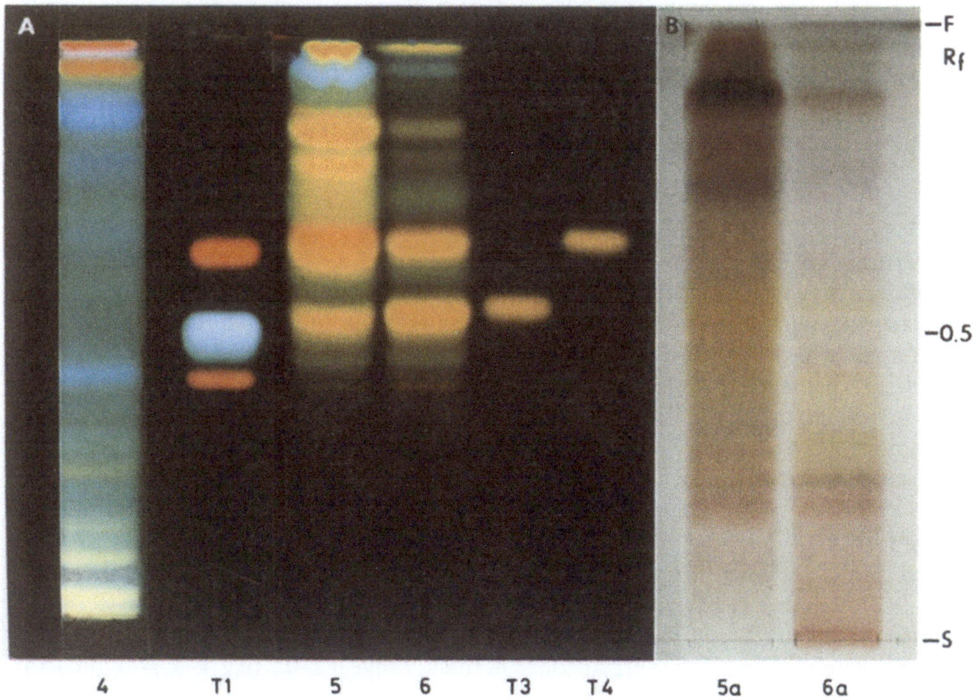

Fig. 16

Equiseti herba

Drug samples "Equiseti herba"

1 Equisetum arvense (origin Germany)
2 Equisetum arvense (origin Sweden)
3,4 Equisetum palustre
5 Equisetum fluvatile
6 Equisetum silvaticum
7 Equisetum telmateia
8,9 Equiseti herba (trade sample) (methanolic extracts, 20 µl)

Reference compound

T1 isoquercitrin
T2 luteolin-5-O-glucoside (=galuteolin)
T3 rutin ($R_f \sim 0.4$) ► chlorogenic acid ($R_f \sim 0.5$) ► hyperoside ($R_f \sim 0.6$)
T4 brucine ($R_f \sim 0.2$) ► strychnine ($R_f \sim 0.4$) ► papaverine ($R_f \sim 0.6$)

Solvent system

Fig. 17A,18 ethyl acetate-formic acid-glacial acetic acid-water (100:11:11:26) → flavonoids
Fig. 17B toluene-ethyl acetate-diethylamine (70:20:10) → alkaloids

Detection

Fig. 17A,18 natural products-polyethylene glycol reagent (NP/PEG No. 28) → UV-365 nm
Fig. 17B iodoplatinate reagent (IP No. 21) → vis

Fig. 17A **Equisetum arvense** (1) is characterized by the yellow-orange zone of isoquercitrin ($R_f \sim$ 0.6/T1), two blue zones above (e.g. caffeic acid) and three weak blue or green zones blow (R_f 0.4–0.55, e.g. galuteolin/T2). The official "horsetail" does not contain green kaempferol glycoside zones in the lower R_f range which indicate one of the adulterants (see Fig. 18). Alkaloids are absent or in extremely low concentrations only.

B **Equisetum palustre** (3,4) is a common adulterant of E. arvense but is easily distinguishable by its alkaloid content. Four to seven zones respond to IP reagent, a major zone (palustrine) directly below the brucine test and very weak zones in the R_f range of the alkaloid test mixture T4.

Fig. 18 **TLC synopsis of Equisetum species** (see Drug list, 7.1.4, "flavonoid pattern").

Equisetum arvense (1,2): the same flavonoid glycoside pattern as shown in Fig. 17A, but with a better separation in the R_f range of isoquercitrin, indicating the presence of an additional flavonoid glycoside (luteolin-7-O-glucoside?).

Equisetum palustre (3,4): six green zones of kaempferol glycosides in the R_f range 0.05–0.5, such as k-3-diglucosyl-7-glucoside, k-3-rutinosyl-7-glucoside and k-3, 7, diglucoside.

Equisetum fluvatile (5): two prominent (R_f 0.3/0.75), three weak green (R_f 0.1/0.35/0.5) zones, no blue zones (R_f 0.75/0.85) and a prominent yellow-orange aglycone zone (solvent front).

Equisetum silvaticum (6): similar pattern of green kaempferol glycoside zones as in sample 5. In addition, the yellow-orange zones of quercetin-3, 7-diglucoside ($R_f \sim 0.4$) and isoquercitrin ($R_f \sim 0.7$) as in E. arvense (1,2).

Equisetum telmateia (7): five green zones in the R_f range 0.2–0.7 (e.g. k-3, 7-diglucoside, k-3-rutinosyl-7-glucoside).

"Equiseti herba" (8,9): Trade samples are often mixtures of various Equisetum species. In the upper R_f range sample 8 shows identical zones with E. arvense, but additional zones in the lower R_f range similar to Equisetum species (6,3,4). Drug sample 9 is almost a 1:1 mixture of 1 and 3.

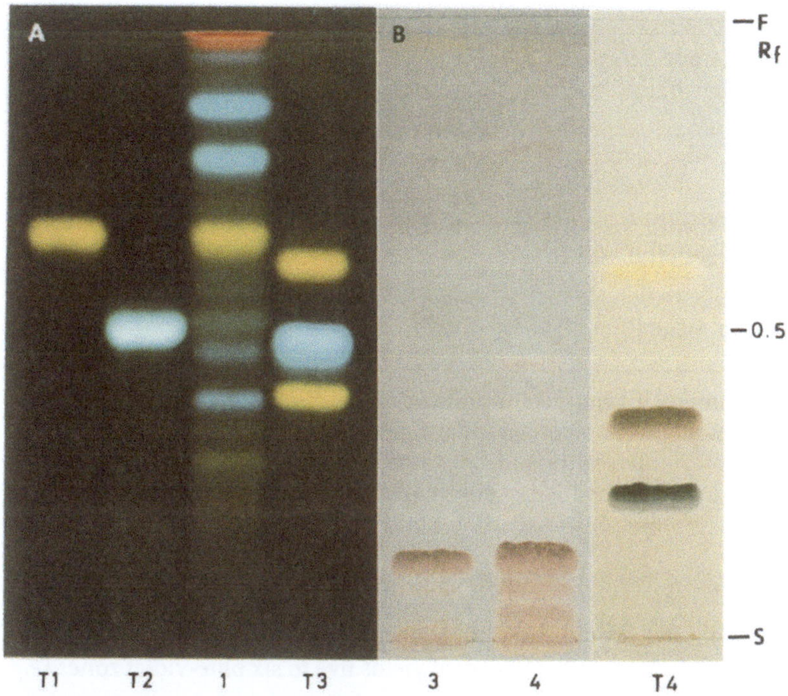

Fig. 17

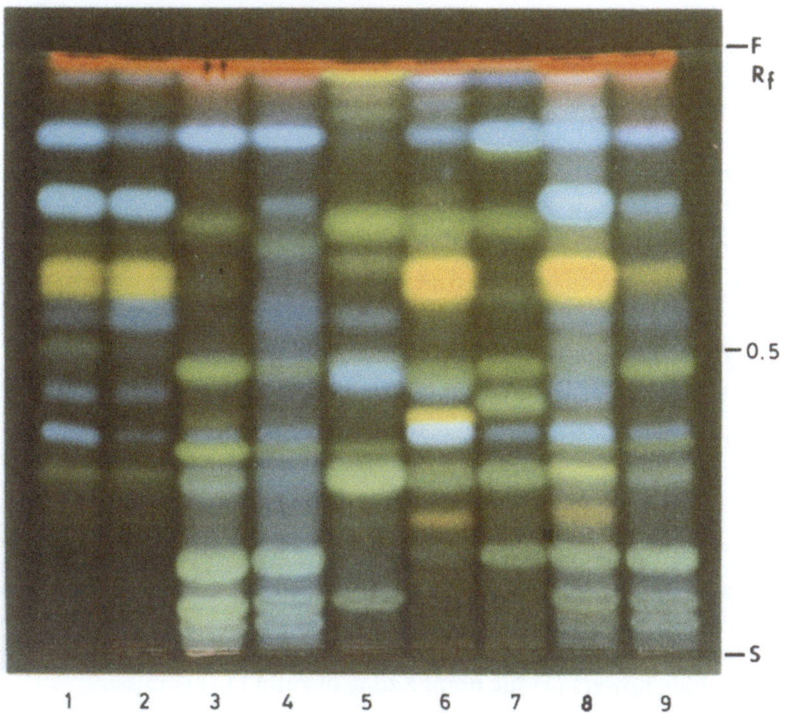

Fig. 18

Virgaureae herba

Drug sample 1,2 Virgaureae herba (trade sample)
3 Solidaginis giganteae herba (methanolic extracts, 20 µl)

Reference compound T1 rutin (R_f ~ 0.4) ▸ chlorogenic acid (R_f ~ 0.45) ▸ hyperoside (R_f ~ 0.55)
T2 isoquercitrin T3 quercitrin T4 oleanolic acid

Solvent system Fig. 19A ethyl acetate-formic acid-glacial acetic acid-water (100:11:11:26)
B chloroform-methanol (90:10)

Detection A Natural products-polyethylene glycol reagent (NP/PEG No. 28) → UV-365 nm
B Anisaldehyde-H_2SO_4 reagent (AS No. 3) → vis

Fig. 19A Extracts from trade samples **Virgaureae or Solidaginis herba** (1–3), show three to four orange or yellow-green quercetin and kaempferol glycosides in varying concentrations in the R_f range 0.4–0.75: quercetin and/or kaempferol rutinoside (R_f ~ 0.4/T1), quercitrin as the main zone (R_f ~ 0.75/T3) and small amounts of isoquercitrin (R_f ~ 0.6/T2) and hyperoside. The blue zones (R_f ~ 0.9 and R_f ~ 0.5/T1) are due to phenol carboxylic acids.
Sample 3 shows the highest flavonoid content and in addition astragalin (R_f ~ 0.85).

B Solidago species contain ester **saponins** (Virgaurea saponin). Alkaline hydrolysis of the methanolic extract and detection with AS reagent yields five to six blue-violet zones (e.g. polygalic acid) in the R_f range 0.4–0.9.

Violae herba

Drug sample 1 Violae herba (V. tricolor, blue flowers)
2,3 Violae herba (V. tricolor, yellow-white flowers)
(methanolic extracts, 20 µl)

Reference compound T1 rutin (R_f ~ 0.4) ▸ chlorogenic acid (R_f ~ 0.45) ▸ hyperoside (R_f ~ 0.55)
T2 salicylic acid

Solvent system Fig. 20A ethyl acetate-formic acid-glacial acetic acid-water (100:11:11:26)
B chloroform-toluene-ether-formic acid (60:60:15:5)

Detection A Natural products-polyethylene glycol reagent (NP/PEG No. 28) → UV-365 nm
B UV-365 nm
C 10% $FeCl_3$ → vis

Fig. 20A **Violae tricoloris herba** (1–3) shows in UV-365 nm mainly green fluorescent glycosides in the R_f range 0.2–0.4. The major zone is due to violanthin, accompanied by lower concentrated zones of saponarin and scoparin below. The yellow-white flowers have rutin (2,3) while in extracts of the blue flowers (1) the orange zone of rutin (T1) is missing.

B,C Violae tricoloris herba contains salicylic acid and methylester. Salicylic acid is detectable in sample 2 at R_f ~ 0.65 (system B) as a weak violet fluorescent zone in UV-365 nm and as a brownish zone after $FeCl_3$ treatment in vis (T2).

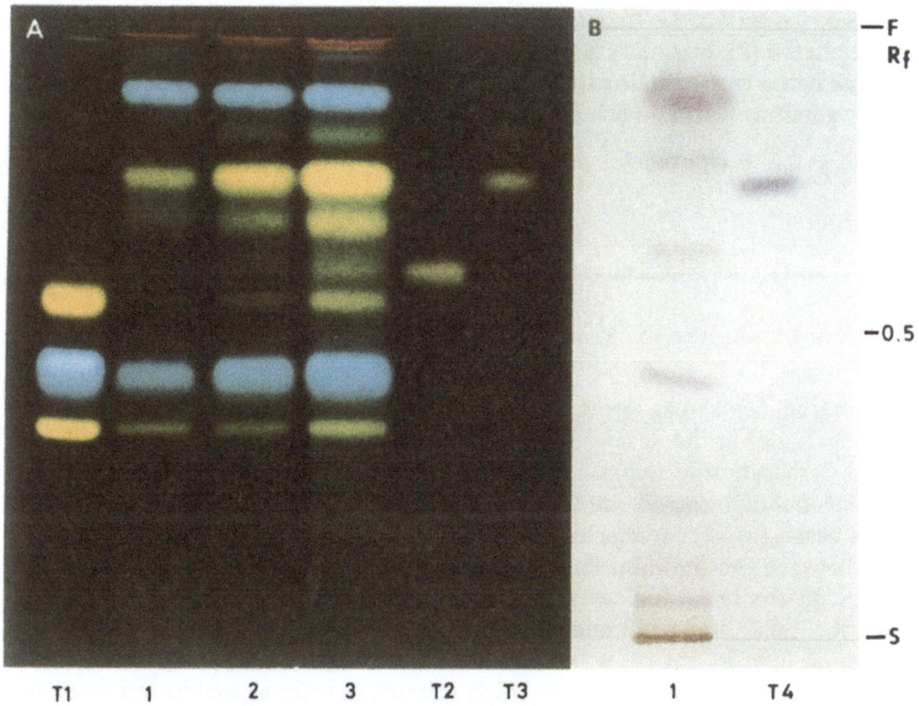

Fig. 19

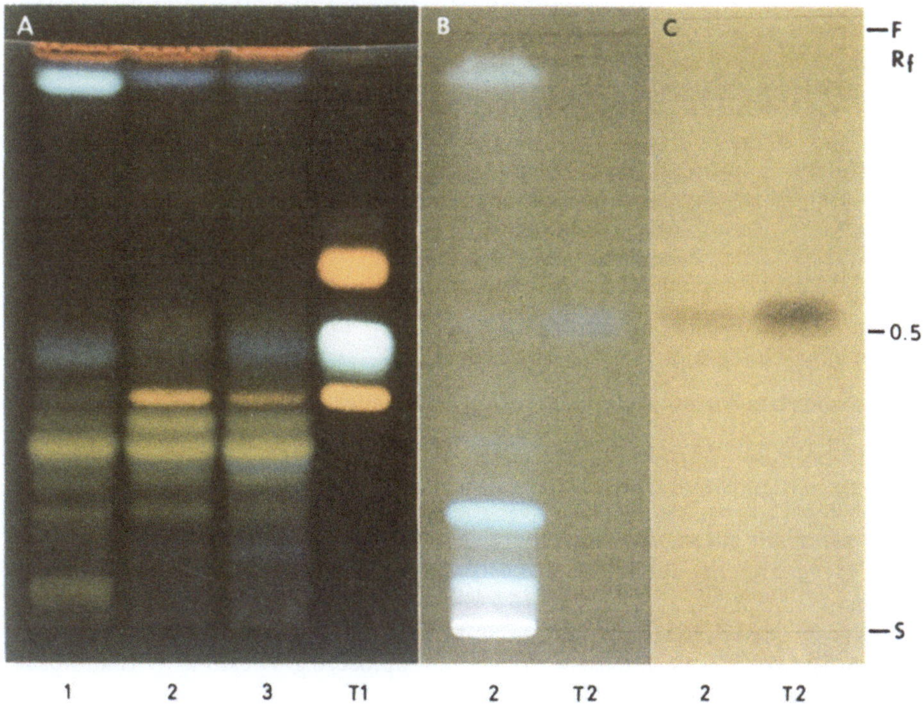

Fig. 20

Anserinae, Passiflorae herba; Sophorae gemmae

Drug sample
1 Anserinae herba (methanolic extract, 30 µl)
2 Passiflorae herba (methanolic extract, 30 µl)
3 Sophorae gemmae (methanolic extract, 5 µl)

Reference compound
T1 rutin ($R_f \sim 0.4$) ► chlorogenic acid ($R_f \sim 0.45$) ► hyperoside ($R_f \sim 0.55$)
T2 vitexin
T3 saponarin

Solvent system Fig. 21 ethyl acetate-formic acid-glacial acetic acid-water (100:11:11:26)

Detection Natural products-polyethylene glycol reagent (NP/PEG No. 28) → UV-365 nm

Fig. 21 **Anserinae herba** (1) shows seven to eight orange-yellow zones of quercetin and myricetin glycosides in the R_f range 0.35–0.75: isoquercitrin ($R_f \sim 0.6$), myricetin- and quercetin-3-O-rhamnoside migrate above the hyperoside test (T1), while the corresponding flavonol diglycosides are found in the R_f range of the rutin test (T1).
Passiflorae herba (2) is characterized by six to eight yellow-green zones of flavon-C-glycosides between the start and $R_f \sim 0.65$: iso-orientin as major zone ($R_f \sim 0.45$), the green zones of isovitexin and vitexin (T2/compound 18,19/Fig. 22), isovitexin-2″-O-glucoside ($R_f \sim 0.2$) and additional zones above and below saponarin-test (T3) e.g. schaftoside.
Sophorae gemmae (3): a charactistically high amount of rutin as well as five flavonoid oligosides in the R_f range 0.05–0.3 and three glycosides in the R_f range 0.45–0.65.

Flavon-C-glycosides as reference compounds

1 saponarin
2 saponaretin
3 schaftoside
4 violanthin
5 isoviolanthin
6 spinosin
7 6″-O- feruloyl-violanthin
8 adonivernith
9 swertiajaponin
10 swertisin
11 aspalathin
12 scoparin
13 orientin
14 isoorientin
15 vitexin-2″-O-glucoside
16 vitexin-2″-O-rhamnoside
17 isovitexin-2″-O-rhamnoside
18 isovitexin
19 vitexin

Solvent system Fig. 22 ethyl acetate-formic acid-glacial acetic acid-water (100:11:11:26)

Detection Natural products-polyethylene glycol reagent (NP/PEG No. 28) → UV-365 nm

Fig. 22 The TLC **synopsis** shows the apigenin-C-glycosides (e.g. vitexin, isovitexin) with generally green and the luteolin-C-glycosides (e.g. orientin) with generally orange fluorescence.
Flavonoid glycosides derived from the same aglycone show ascending R_f values in the following order: galactose ► glucose ► rhamnose ► apiose; e.g. vitexin-2″-O-rhamnoside has a higher R_f value than vitexin-2″-O-glucoside.

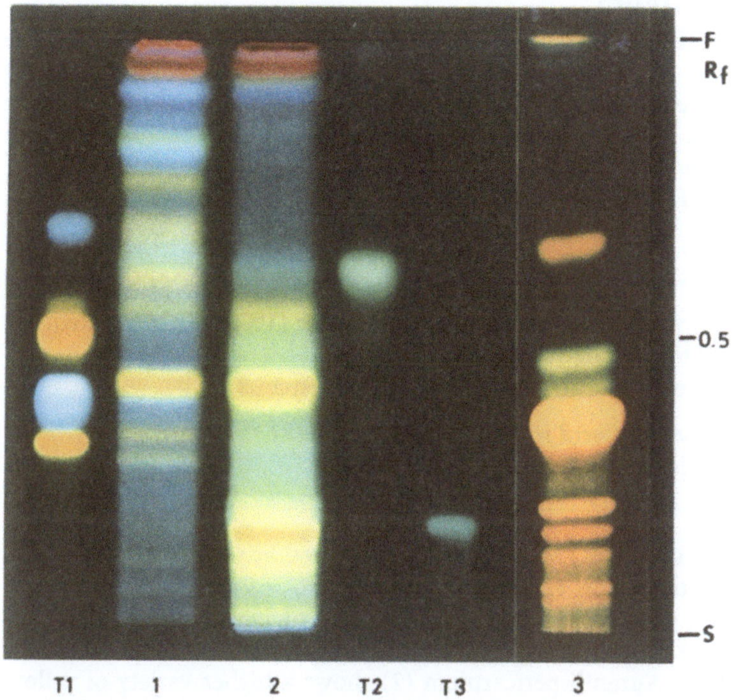
—F
Rf
—0.5
—S
T1
1
2
T2
T3
3

Fig. 21

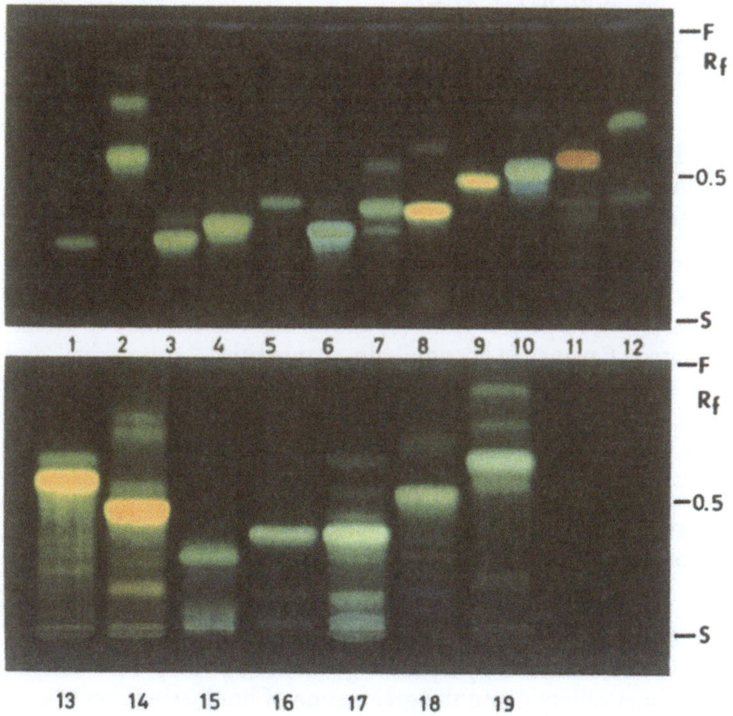
—F
Rf
—0.5
—S
1
2
3
4
5
6
7
8
9
10
11
12
—F
Rf
—0.5
—S
13
14
15
16
17
18
19

Fig. 22

Citri, Aurantii pericarpium
Orthosiphonis, Eriodictyonis folium

Drug sample	1 Citri pericarpium (MeOH extract, 25 µl) 2,3 Aurantii pericarpium (MeOH extract, 25 µl) 4 Orthosiphonis folium (DCM extract, 20 µl) 5 Eriodictyonis herba (DCM extract, 20 µl)
Reference compound	T1 rutin T3 eriodictyol T2 sinensetin T4 homoeriodictyol
Solvent system	Fig. 23A,B ethyl acetate-formic acid-glacial acetic acid-water (100:11:11:26), system 1 Fig. 24C chloroform-ethyl acetate (60:4), system 2 D,E chloroform-acetone-formic acid (75:16.5:8.5), system 3
Detection	Natural products-polyethylene glycol reagent (NP/PEG No. 28) ► UV-365 nm (A,C,E) ► vis (B,D)

Fig. 23A **Citri** (1) and **Aurantii pericarpium** (2) are both characterized by the prominent yellow rutin (R_f ~ 0.35/T1) and the yellow-red eriocitrin zone at R_f ~ 0.45. The broad, dark-green band directly above eriocitrin is due to the bitter-tasting naringin, neohesperidin and the non-bitter hesperidin in sample 2. In (1), only traces of hesperidin and neohesperidin are present. Aurantii pericarpium (2) shows a higher variety of yellow flavonoid glycosides below rutin and blue fluorescent zones in the upper R_f range. These are separeated in system 2, as shown in Fig. 24(C).

Eriocitrin is first seen as a yellow zone in UV-365 nm. After exposure (30–60 min) of the TLC plate to UV-365 nm or daylight the zone turns red (UV-365 nm) and violet (vis.), respectively.

B The higher amount of naringin, neohesperidin and hesperidin in Aurantii pericarpium (2) is seen as a broad yellow band directly above the violet-red eriocitrin. The R_f value depression is caused by other plant products.

Fig. 24C Development of DCM extracts of **Aurantii pericarpium** (3) and **Orthosiphonis folium** (4) in system 2 yields a series of blue fluorescent aglycones (UV-365 nm).
Besides sinensetin (R_f ~ 0.35/T2), A. pericarpium (3) shows eight to ten blue to violet-blue zones of hydroxylated flavans (e.g. nobiletin, tangeritin), coumarins and methylanthranilate in the R_f range 0.05–0.8 (see also Chap. 6, Fig. 17,18).
Orthosiphonis folium (4): sinensetin (T1) is the major compound with scutellarein tetramethyl ether directly above and eupatorin and 3'-hydroxy-5,6,7,4'-tetramethoxy-flavone below (R_f 0.3–0.4).

D **Eriodictyonis herba** (5). Treatment with NP/PEG reagent generates yellow and red zones (vis.) after 40–60 min exposure of the developed TLC plate (system 3) to UV-365 nm or daylight.

E In UV-365 nm six yellow, orange-red or green fluorescent flavonoid aglycones are detectable: eriodictyol (R_f ~ 0.3/T3) followed by a yellow-green zone of chrysoeriodictyol, an orange-red zone of xanthoeriodictyol and the green zone of homoeriodictyol (R_f ~ 0.55/T4).

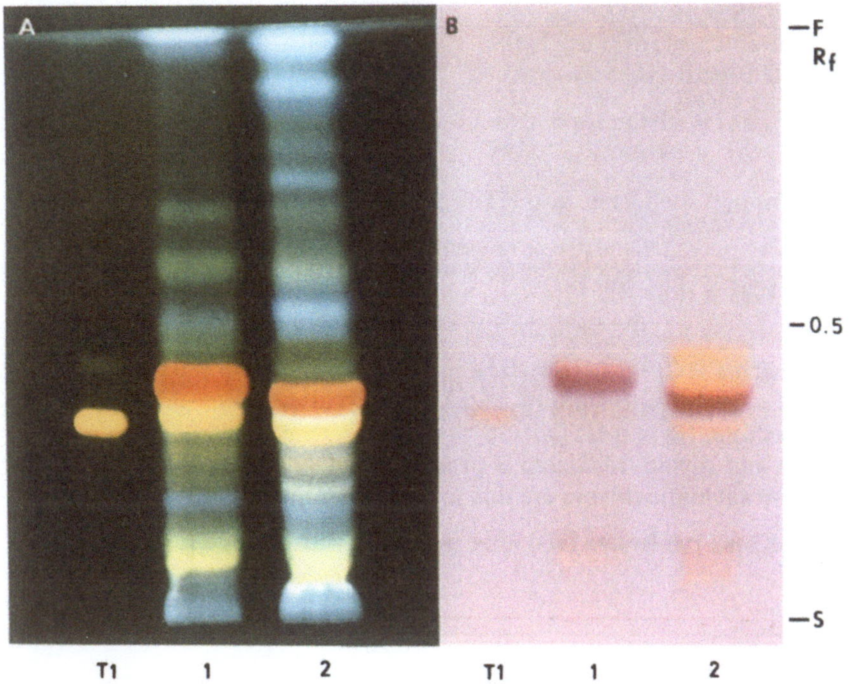

Fig. 23

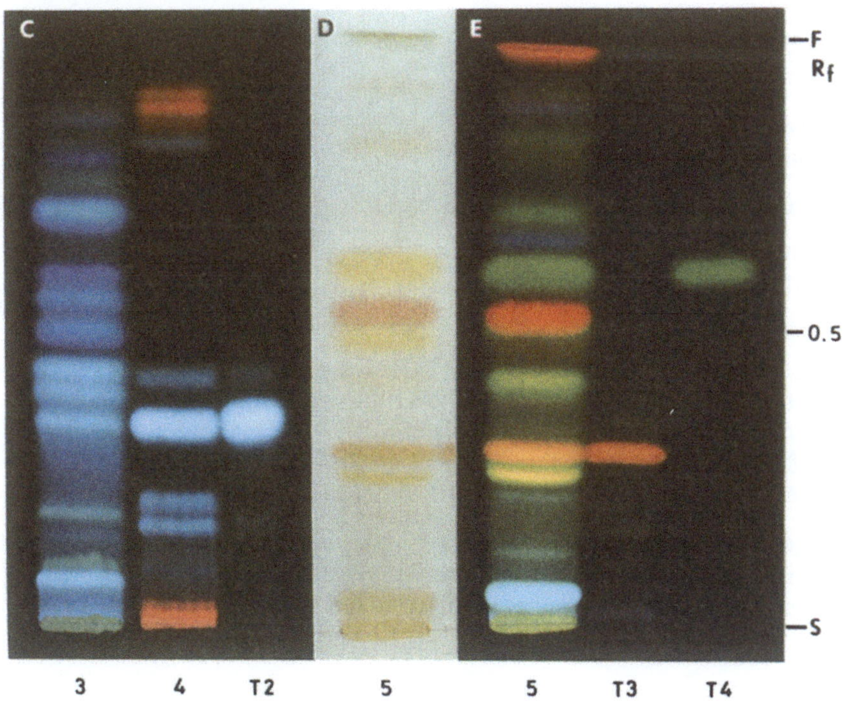

Fig. 24

Cardui mariae (Silybi) fructus

Drug sample 1,2 Cardui mariae fructus (Silybi fructus)
(methanolic extract, 20 µl)

Reference compound T1 taxifolin (R_f ~ 0.4) ▶ silybin (R_f ~ 0.6)
T2 silychristin

Solvent system Fig. 25 chloroform-acetone-formic acid (75:16.5:8.5)

Detection A Natural products-polyethylene glycol reagent (NP/PEG No. 28) → UV-365 nm
B Fast blue salt reagent (FBS No. 15) → vis

Fig. 25A **Cardui mariae fructus** (1,2) is characterized in UV-365 nm by two intense green-blue fluorescent zones of silybin/isosilybin (R_f ~ 0.6/T1), silychristin (R_f ~ 0.35/T2) and the orange zone of taxifolin (R_f ~ 0.4/T1).
Between taxifolin and silybin, silydianin is present in the silydianin race (2) only. The minor zones above silybin/isosilybin are due to their dehydroderivatives.

B All main zones become red-brown (vis) after treatment with the FBS reagent.

Viburni cortex

Drug sample 1 Viburni prunifoli cortex
2 Viburni opuli cortex (methanolic extracts, 30 µl)

Reference compound T1 scopoletin
T2 amentoflavone
T3 catechin/epicatechin mixture

Solvent system Fig. 26 chloroform-acetone-formic acid (75:16.5:8.5)

Detection A KOH reagent (No. 35) → UV-365 nm
B Fast blue salt reagent (FBS No. 15) → vis

Fig. 26A **Viburni cortex.** The samples 1 and 2 show with KOH reagent in UV-365 nm seven to ten blue or greenish fluorescent zones distributed over the whole R_f range.
The presence of the blue fluorescent scopoletin (R_f ~ 0.7/T1) and the dark-green fluorescent biflavonoid amentoflavone (R_f ~ 0.4/T2) is characteristic for **Viburni prunifolii cortex** (1).

B With FBS reagent the **Viburni cortex** samples 1 and 2 develop four to six red-brown zones in the vis. Amentoflavone (T2) is a characteristic main red-brown zone in V. prunifolii cortex (1), while a high amount of catechin/epicatechin (R_f ~ 0.15/T3) characterizes **Viburni opuli cortex** (2).

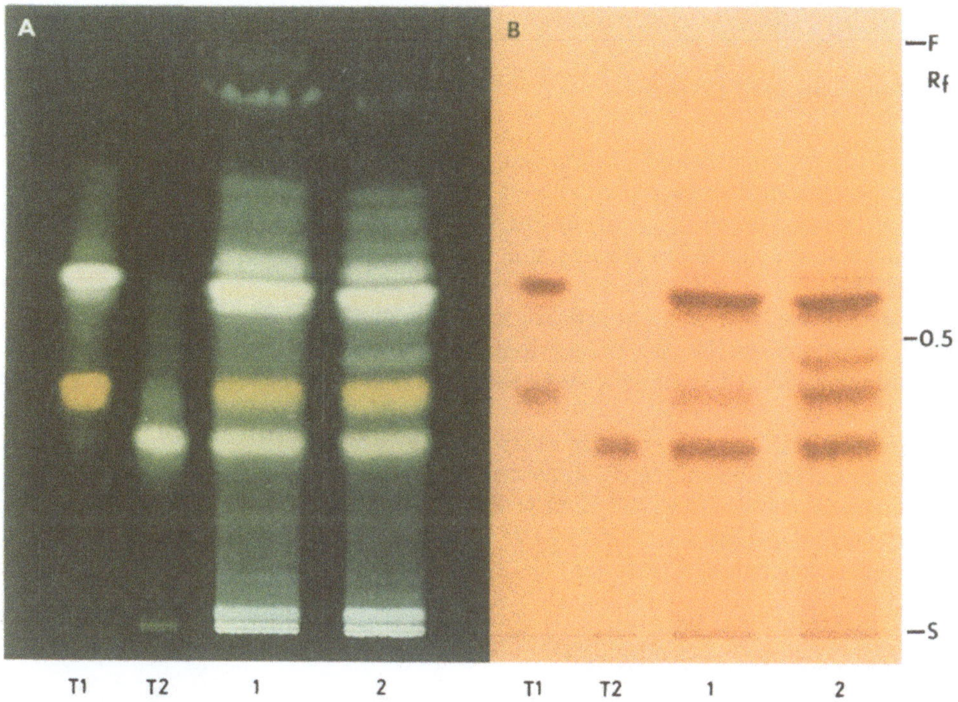

Fig. 25

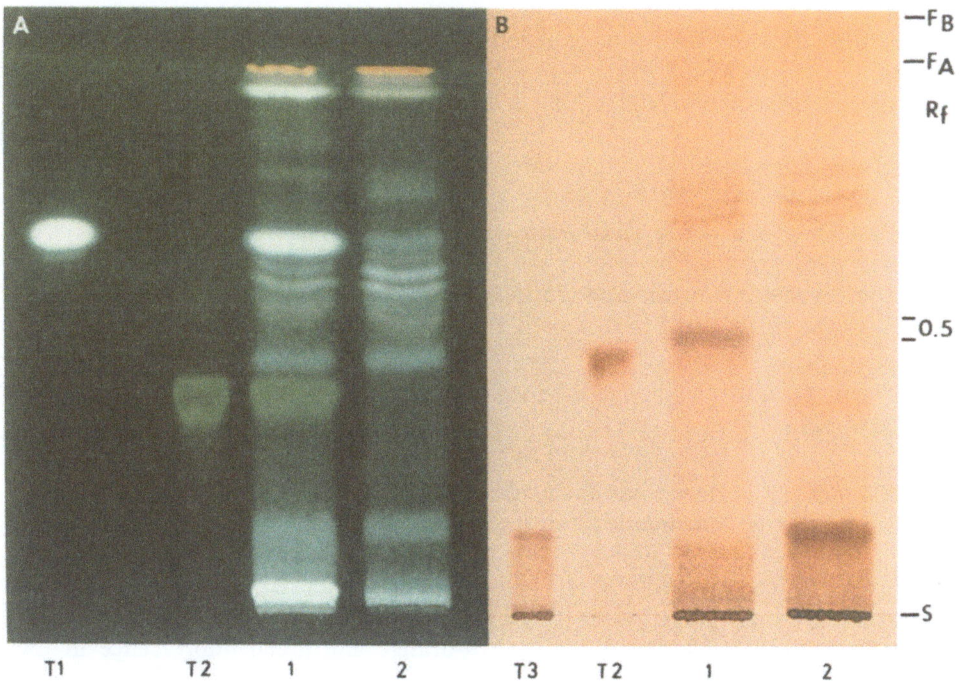

Fig. 26

7.2 Ginkgo biloba

7.2.1 Preparation of Extracts

Flavonoids A total of 10 g fresh leaves and 50 ml methanol are homogenized for 10 min in a Warren blender. The filtrate is evaporated to about 2 ml, and 10 µl–20 µl is used for TLC.
A total of 1 g dried leaves is extracted with 30 ml methanol for 30 min under reflux. The clear filtrate is evaporated to dryness and, dissolved in 2 ml methanol and 10–20µl is used for TLC.
Commercially available pharmaceuticals, such as liquid preparations, are used directly for TLC investigations (10–20 µl) or one to two powdered tablets or dragées are extracted with 5 ml methanol for 5 min on a water bath; 10–20 µl of the filtrate is used for TLC.

Ginkgolides A total of 40 g fresh leaves is boiled in water for 20 min, filtered through Whatman paper followed by Celite (Hyflosupercel). Activated charcoal is added to the filtrate and stirred for 12 h at room temperature. The mixture then is centrifuged, the supernatant discarded and the charcoal residue dissolved in 20 ml acetone after filtration through a glass filter. The filtrate is concentrated to about 1 ml, and 10 µl is used for TLC.

7.2.2 Thin-Layer Chromatography

Reference solutions Test mixture A: 3 mg rutin, 2 mg chlorogenic acid and 3 mg hyperoside in 10 ml methanol.
Test mixture B: 1 mg bilobetin, ginkgetin/isoginkgetin and sciadopitysin in 3 ml methanol.
Ginkgolides A,B,C and bilobalide: 1 mg is dissolved in 1 ml methanol.

Adsorbent Silica gel $60F_{254}$-precoated TLC plates (Merck, Darmstadt).

Chromatography solvents

• Ethyl acetate-glacial acetic acid-formic acid-water (100:11:11:26)	flavonoid glycosides
• Chloroform-acetone-formic acid (75:16.5:8.5)	biflavonoids
• Toluene-acetone (70:30)	ginkgolides

7.2.3 Detection

- UV-254 mm — Flavonoids show quenching
  UV-365 mm — Flavonoids fluoresce brown, dark green

- Spray reagent (see Appendix A)
  - Natural products-polyethylene glycol reagent (NP/PEG No. 28)
    → Flavonoids and biflavonoids: yellow-orange and green fluorescence in UV-365 nm.
  - Water or acetic anhydride → Ginkgolides
    The TLC plate is sprayed either with water or with acetic anhydride and heated for 30–60 min at 120°C. The ginkgolides and bilobalide then develop blue or green-blue fluorescence in UV-365 nm.

7.2.4 Drug Constituents

Drug/plant source Family	Main constituents	
Ginkgo bilobae folium Ginkgo leafs Ginkgo biloba L. Ginkgoaceae	0.5%–1% total flavonoids: (~20 compounds) Quercetin, kaempferol and isorhamnetin glycosides: q-, k-, i-3-O-α-rhamnosyl-(1 → 2)-α-rhamnosyl-(1 → 6)-β-glucoside, quercitrin, isoquercitrin, rutin, k-7-O-glucoside, k-3-O-rutinoside, astragalin, dihydrokaempferol-7-O-glucoside, isorhamnetin-3-O-rutinoside 3'-O-methylmyricetin-3-O-glucoside, luteolin Flavonol acylglycosides: Q-, k-, i- O-α-rhamnopyranosyl-4-O-β-D-(6"'-trans-p-coumaroyl)-glucopyranoside Biflavonoids: amentoflavone, bilobetin, 5'-methoxybilobetin, ginkgetin, isoginkgetin, sciadopitysin 0.01%–0.04% ginkgolides A,B,C; bilobalide Ginkgolic acid, 6-hydroxykynurenic acid, shikimic acid, chlorogenic acid, p-coumaric acid, vanillic acid · (Ginkgol) Catechin, epi-, gallo- and epigallocatechin	Fig. 27,28

7.2.5 Formulae

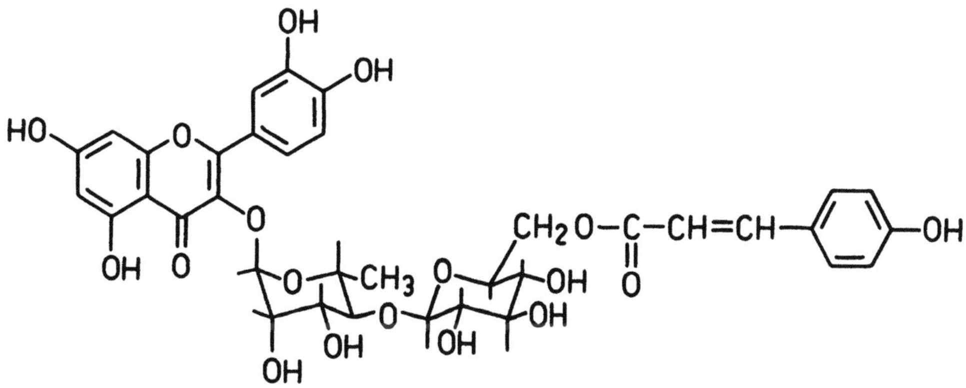

Quercetin-α-rhamnopyranosyl-4"-O-β-D-(6"'-trans-p-coumaroyl)-glucopyranoside

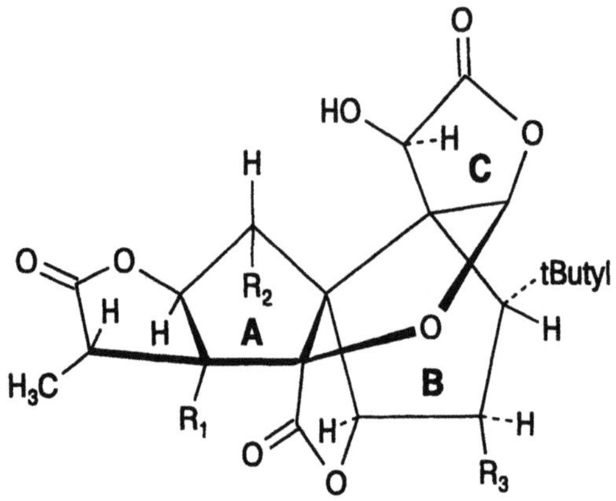

Ginkgolide	R_1	R_2	R_3
A	OH	H	H
B	OH	OH	H
C	OH	OH	OH

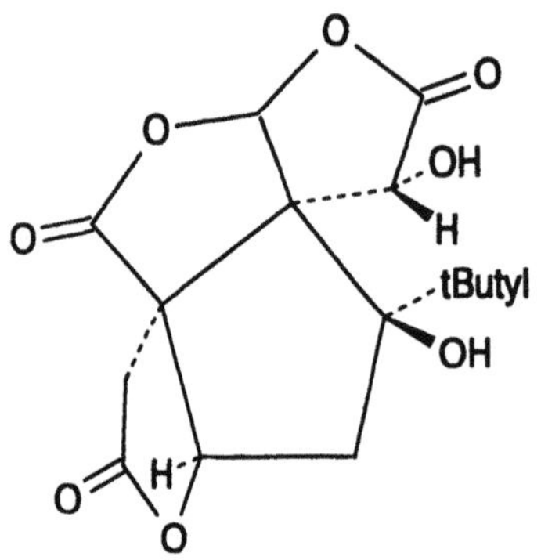

Bilobalide

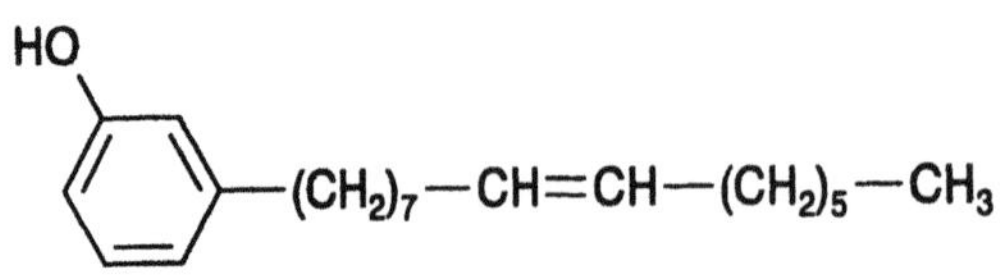

Ginkgol

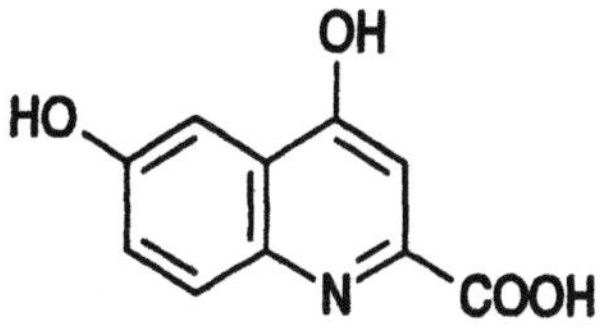

6-Hydroxykynurenic acid

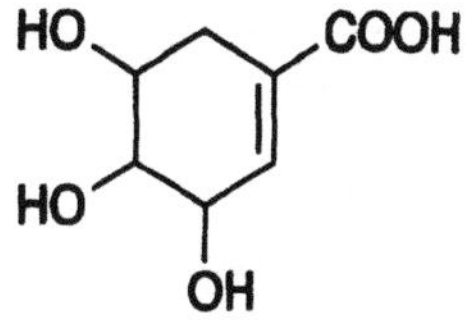

Shikimic acid

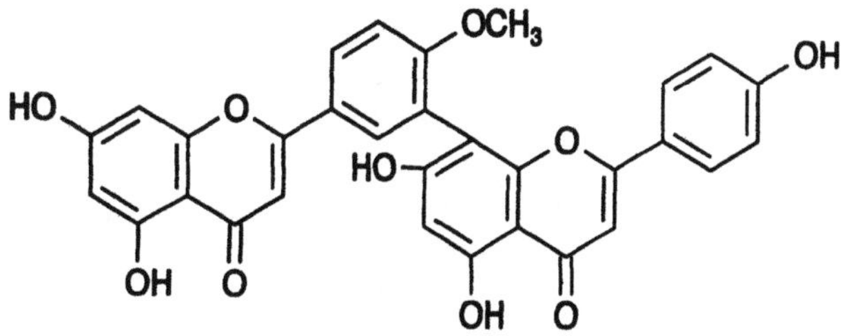

Bilobetin

7.2.6 Chromatogram

Ginkgo bilobae folium

Leaf sample

1 Ginkgo biloba (origin Germany)
2 Ginkgo biloba (origin Korea)
3 Ginkgo biloba (origin Italy)
4 Ginkgo biloba (commercial extract)
5 Ginkgo biloba (pharmaceutical preparation)
6 Ginkgo biloba (green leaf/Germany)
(extract preparation see Sect. 7.2.1)

Reference compound

T1 rutin (R_f ~ 0.45) ▸ chlorogenic acid (R_f ~ 0.5) ▸ hyperoside (R_f ~ 0.6)
T2 bilobetin (R_f ~ 0.45)
T3 bilobetin (R_f ~ 0.45) ▸ ginkgetin (R_f ~ 0.6) ▸ sciadopitysin (R_f ~ 0.8)
T4 bilobalide
T5 ginkgolide A
T6 ginkgolide B
T7 ginkgolide C

Solvent system

Fig. 27A, 28D ethyl acetate-glacial acetic acid-formic acid-water (100:11:11:26)
Fig. 27B chloroform-acetone-formic acid (75:16.5:8.5)
Fig. 28C toluene-acetone (70:30)

Detection

A,B Natural products reagent (NP No. 28) → UV-365 nm
C Acetic anhydride reagent (AA No. 1) → UV-365 nm
D Natural products-polyethylene glycol reagent (NP/PEG No. 28) → UV-365 nm

Fig. 27A **Ginkgo folium** (1–3) is characterized in UV-365 nm (NP reagent) by eight to ten green-yellow or orange-yellow fluorescent flavonol glycoside zones in the R_f range 0.2–0.75 and flavonol aglycones and biflavonoids at the solvent front:

R_f ~ front	flavonol aglycones, biflavonoids	yellow-orange
0.75	isoquercitrin, astragalin,	green-yellow
↑	dihydrokaempferol-7-O-glucoside	green-yellow
0,6	quercitrin	orange-yellow
0,5	6-hydroxykynurenic acid	light blue
↑	kaempferol-, quercetin-3-O-(6"'-trans-p-coumaroyl-4"-glucosyl)-rhamnoside	
0.45	narcissin, isorhamnetin-rutinoside	green-yellow
0.40	rutin (T1), quercetin-, kaempferol-, isorhamnetin-3-O-(2"-6"-di-O-α-L-	yellow-orange/green
↑	rhamnopyranosyl)-β-D-glucopyranoside	yellow-green/orange
0,25	flavonol triosides	yellow-green/orange

B The biflavonoids in **Gingko folium** (2,3) are separated in system B: bilobetin (R_f ~ 0.45/T2), one zone of ginkgetin/isoginkgetin (R_f ~ 0.6) and sciadopitysin (R_f ~ 0.8/T3). The blue fluorescent hydroxykynurenic acid and yellow-orange fluorescent flavonol glycosides remain at the start.

Fig. 28C The reference compounds **ginkgolide A–C** and **bilobalide** are separated in system C and detected in UV-365 nm after spraying with acetanhydride and heating (30 min/120°C) as green and blue fluorescent zones. In Ginkgo leaf preparations, enriched with ginkgolides (see Sect. 7.2.1), these four compounds are present, but overlapped by various blue fluorescent zones and not reliably detectable by TLC methods. Therefore, for an unambigious detection of them the high-performance liquid chromatography (HPLC) method is recommended.

D The enriched and standardized extracts of commercial Ginkgo preparations (4,5) are free of biflavonoids and therefore do not show yellow zones at the solvent front, as seen in extract 6.

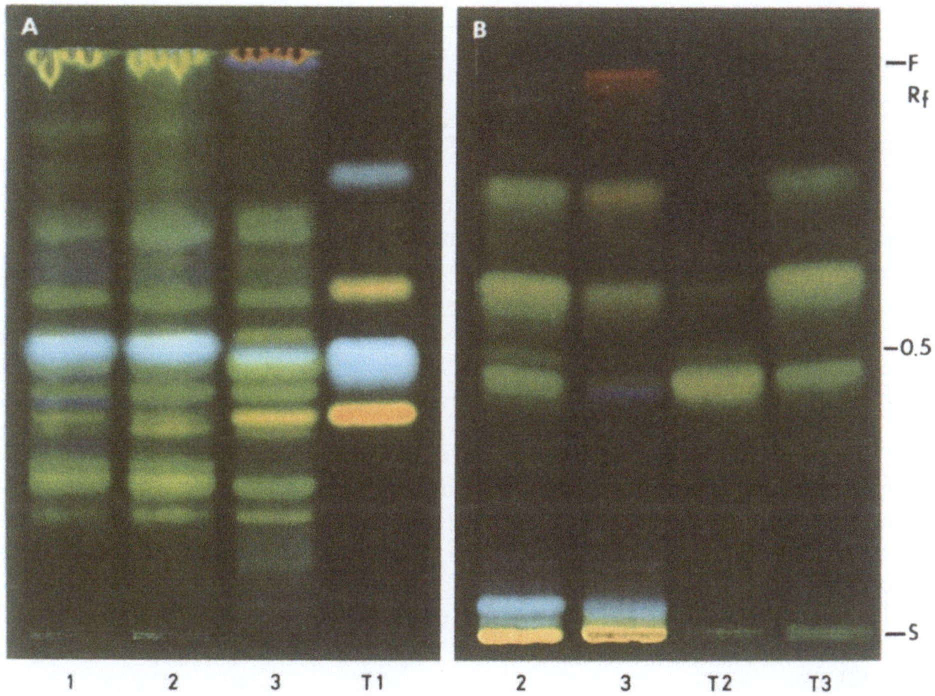

Fig. 27

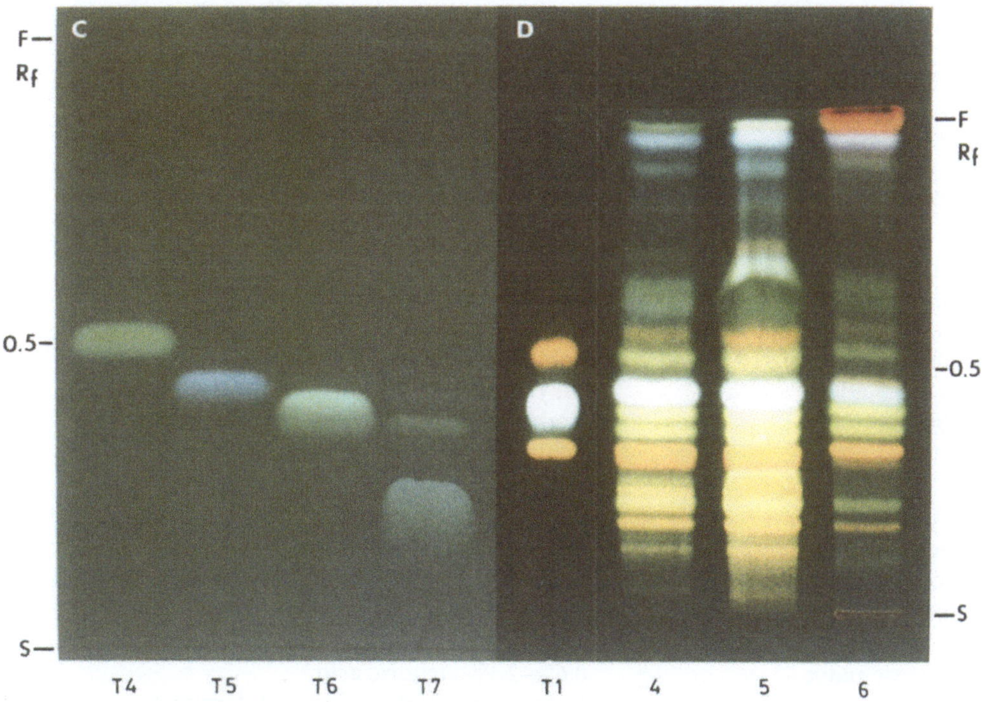

Fig. 28

7.3 Echinaceae radix

7.3.1 Preparation of Extracts

Powdered drug (1 g) is extracted with 75 ml methanol under reflux for 1 h. The filtrate is evaporated to 5 ml, and 30 µl is used for TLC investigations.

7.3.2 Solvent Systems and Detection

Hydrophilic compounds such as caffeic acids derivatives are separated in toluene-ethylformiate-formic acid-water (5:100:10:10) over silica gel 60 F_{254} plates (Merck, Germany) and inspected in UV-254 nm (quenching zones) and, after treatment, with natural products-polyethylene glycol reagent (NP/PEG No. 28) detected in UV-365 nm as blue fluorescent zones.
Lipophilic compounds such as alkyl amides are separated over silica gel in the solvent system toluene-ethyl acetate (70:30) and detected with vanillin-sulphuric acid reagent (VS No. 42), vis.

7.3.3 Drug List

	Drug/plant source Family	Main constituents
Fig. 29,30	**Echinaceae radix** E. angustifoliae radix (narrow-leaved) coneflower root Echinacea angustifolia DC Asteraceae ► Herba ► Flos	 0.3%–1.3% echinacoside Cynarin, traces of cichoric acid, caffeoyl quinic acid derivatives Alkylamides Verbascoside (=desglucosyl-echinacoside) Echinacoside (0.1%–1%), rutin
Fig. 29,30	**E. pallidae radix** E. pallida NUTT. Asteraceae ► Herba	0.4%–1.7% echinacoside 6-O-caffeoyl-verbascoside Caffeic acid derivatives; alkyl amides Desrhamnosyl-verbascoside; rutin
Fig. 29,30	**E. purpureae radix** black sampson root E. purpurea (L.) MOENCH Asteraceae ► Herba	0.6%–2.1% cichoric acid Chlorogenic acid, caffeic acid derivatives, no echinacoside; alkylamides Cichoric acid and methylester, rutin

Drug/plant source Family	Main constituents
Common substitute or adulterant of Echinaceae radix	
Parthenium integrifolium Cutting almond, wild quinine Missouri snake root Parthenium integrifolium L. Asteraceae	Sesquiterpene esters Echinadiol-, epoxyechinadiol-, echinaxanthol- and dihydroxynardol-cinnamate Caffeic acid derivatives

Fig. 29,30

7.3.4 Formulae

See Sect. 7.1.5 Formulae

7.3.5 Chromatogram

Echinaceae radix

Radix samples
1 Echinacea angustifolia
2 Echinacea pallida
3 Echinacea purpurea
4 Parthenium integrifolium (adulterant) (methanolic extracts, 30 µl)

Reference compounds
T1 chlorogenic acid
T2 cichoric acid
T3 caffeic acid
T4 β-sitosterin

Solvent system
Fig. 29 toluene-ethyl formiate-formic acid-water (5:100:10:10) → caffeic acid derivatives
Fig. 30 toluene-ethyl acetate (70:30) → sesquiterpenes, polyacetylenes.

Detection
Fig. 29A Natural products-polyethylene glycol reagent (NP/PEG No. 28) → UV-365 nm
Fig. 30B,C Vanillin-sulphuric acid reagent (VC No. 42) → vis B 100°C/10 min
C 100°C/5 min

Fig. 29A Methanolic extracts of the **Echinaceae radix** samples 1–3 and **Partenium integrifolium** (4) can be differentiated in UV-365 nm by their number, amount and R_f range of blue fluorescent caffeic acid derivatives.
E. angustifoliae radix (1) and **E. pallidae radix** (2) are characterized by echinacoside, seen as main compound at $R_f \sim 0.1$ besides five to six less concentrated zones in the R_f range 0.2–0.8. Desglucosyl-echinacoside (=verbascoside) in (1) and desrhamnosyl-echinacoside in (2) are found in the R_f range of chlorogenic acid (T1) and above. Cynarin ($R_f \sim 0.75$) is found in (1) only. Cichoric acid ($R_f \sim 0.8$/T2) is present in low concentrations in (1) and (2), while in E. purpureae radix (3) cichoric acid is the major compound. Chlorogenic acid is identified at $R_f \sim 0.45$ (T1); echinacoside is not present in sample 3. **Parthenium integrifolium** (4) as a common substitute of E. purpurea shows ten to twelve weaker blue zones of caffeic acid derivatives in the R_f range 0.2–0.8.

Note: In the drug part "Herba" of Echinacea angustifolia verbascoside, of Echinacea pallida desrhamnosyl-verbascoside and of Echinacea purpurea cichoric acid and methylester are found. In addition, rutin is present in E. angustifoliae and E. pallidae herba.
Echinacoside is unstable in solution and is missing in extracts that have been stored for a long time.

Fig 30B The lipophilic compounds separated in solvent 2 and detected with VS reagent allow an easy differentiation of all four radix drugs:
E. angustifolia (1): six to seven blue to violet-blue zones (R_f 0.2–0.55) with alkylamides at R_f 0.35–0.55 (in the R_f range of β-sitosterin-test T4 and below).
E. pallida (2): two prominent grey zones in the R_f range 0.8–0.85 due to ketoalkines, keto alkanes and hydroxylated ketoalkanes, grey zone at $R_f \sim 0.25$.
E. purpurea (3): two weak grey-blue zones at $R_f \sim 0.2$ and $R_f \sim 0.5$ (β-sitosterin/T4)
P. integrifolia (4): a dominating grey zone at R_f 0.6 (see C).

C With VS reagent (3–5 min/100°) sample 4 reveals two blue and one orange zone (vis) due to echinadiol-cinnamate (blue/directly above the β-sitosterin test, T4), epoxi-echinadiol cinnamate (orange) and echinaxanthol cinnamate (blue/ not always present).

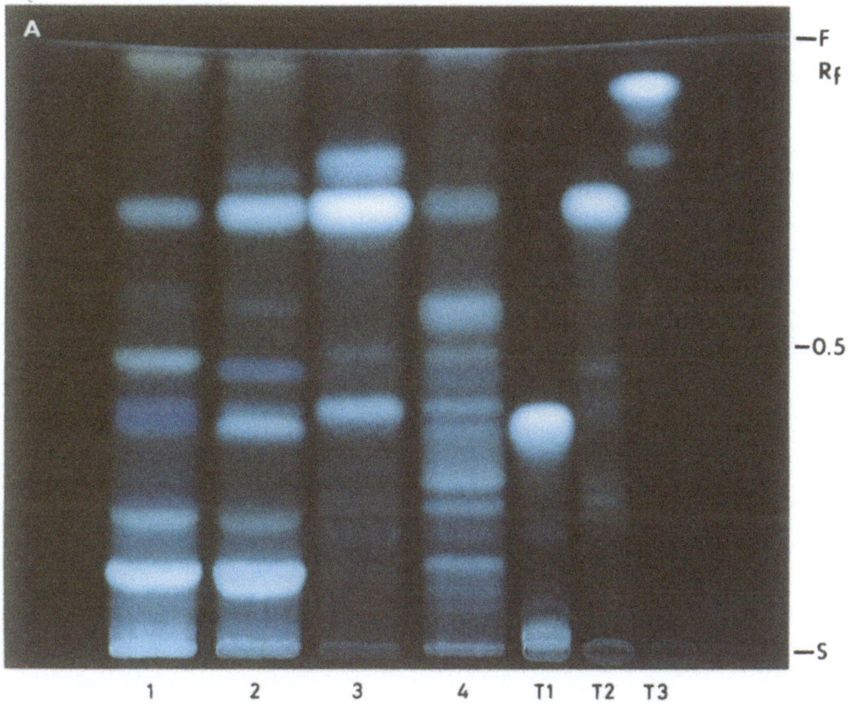

Fig. 29

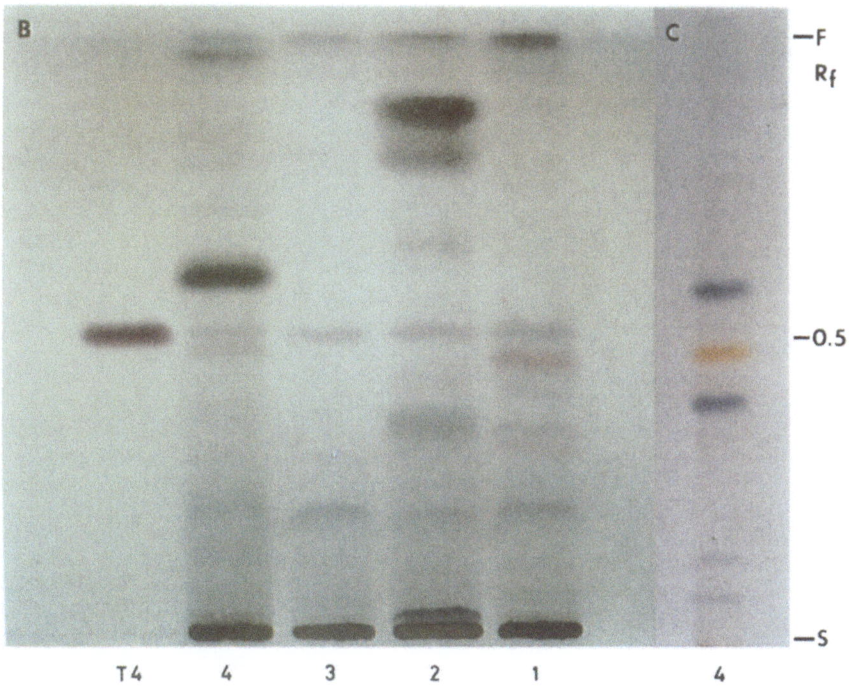

Fig. 30

8 Drugs Containing Arbutin, Salicin and Salicoyl Derivatives

8.1 Drugs with Arbutin (Hydroquinone derivatives)

These drugs contain the hydroquinone-β-O-glucoside arbutin as their major compound, as well as small amounts of methyl-, 2-O-galloyl-arbutin, picein and free hydroquinone. Polyphenols (predominant, >15%), galloyl esters of glucose (e.g. Uvae ursi folium) and ellagtannins are also present. Other plant constituents such as flavonoids, coumarins and phenol carboxylic acids can be used to identify and characterize the "arbutin drugs".

8.1.1 Preparation of Extracts

General method

Powdered drug (0.5–1 g) is extracted under reflux for approximately 15 min with 5 ml 50% methanol. The hot extract is filtered and the filter then washed with methanol up to a total of 5.0 ml; 20–30 µl is used for TLC.
To remove tannins the solution is treated with 0.5 g basic lead acetate, vigorously shaken and then filtered. To remove interfering resins and lipids, the powdered drug can be extracted under reflux for about 15 min with light petroleum before extraction with methanol.

8.1.2 Thin-Layer Chromatography

Reference solutions

25 mg arbutin and 25 mg hydrochinone are dissolved in 10 ml 50% methanol; 10 µl is used for TLC.
A mixture of 1 mg rutin, chlorogenic acid and hyperoside in 5 ml methanol; 10 µl is used for TLC.

Adsorbent

Silica gel 60 F_{254}-precoated TLC plates (Merck, Darmstadt)

Chromatography solvents

Ethyl acetate-methanol-water (100:13.5:10) → arbutin
Ethyl acetate-glacial acetic acid-formic acid-water (100:11:11:26) → flavonoid glycosides

8.1.3 Detection

- UV-254 nm — Arbutin shows prominent quenching.
- UV-365 nm — No fluorescence of arbutin.
- Spray reagents — (see list Appendix A)

- Berlin blue reaction (BB No. 7)
  All phenols appear as blue zones (vis).
- Millons reagent (ML No. 27)
  Hydroquinone derivatives form yellow zones (vis).
- Gibb's reagent (DCC No. 10) Arbutin becomes blue-violet (vis) when the TLC plate is sprayed with a 1% methanolic solution of 2,6-dichloro-*p*-benzoquinone-4-chloroimide and then exposed to ammonia vapour.

8.1.4 Drug List

	Drug	Plant of origin Family	Total hydroquinones
Fig. 1,2	**Uvae ursi folium** Bearberry leaves	Arctostaphylos uva-ursi (L.) SPRENGEL Ericaceae	4%–15%
	Vitis idaeae folium Cowberry leaves	Vaccinum vitis idaea L. Ericaceae	5.5%–7%
	Myrtilli folium Bilberry leaves	Vaccinium myrtillus L. Ericaceae	0.4–1.5%
	Bergeniae folium	Bergenia crassifolia	~12%
	Callunae herba	Calluna vulgaris	~0.65%
	Pyri folium	Pyrus communis	~4.5%
	Viburni cortex	Viburnum prunifolium (see Fig. 26)	~0.5%

8.1.5 Formulae

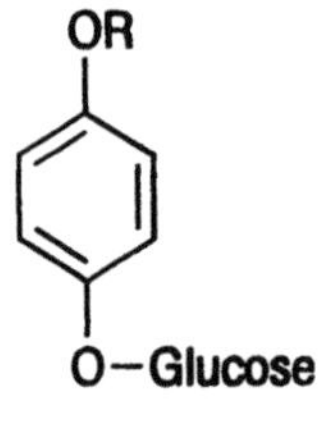

Arbutin R = H
Methylarbutin R = CH_3

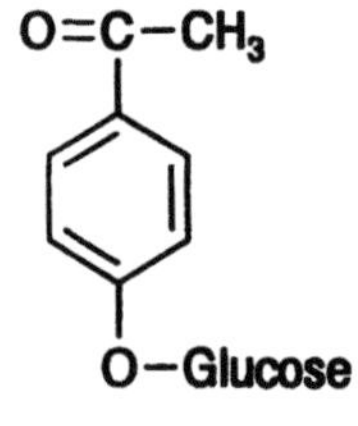

Picein
(Piceoside)

8.2 Drugs Containing Salicin and Its Derivatives

Salicin, a (2-hydroxymethyl)-phenyl-β-D-glucopyranoside, and its derivatives fragilin, salicortin, 2′-O-acetylsalicortin, tremulacin and salireposide are major constituents of various Salix species. Salicin, salicortin and tremulacin are also present in buds of Populus tremula L.
Picein (piceoside), a *p*-hydroxyacetophenone glucoside, has been identified in Salix cinerea, Uvae ursi folium and in sprouts of Pinus picea L. and Picea species.

8.2.1 Preparation of Extracts for TLC

General method
Powdered drug (1 g) is extracted with 50 ml MeOH for 30 min under reflux. The filtrate is evaporated and the residue resolved in 3 ml methanol; 20–40 µl is used for TLC.

Hydrolysis
A total of 1 ml extract (see above) and 0.5 ml 0.1 N NaOH are stirred for 60 min at 60°C; 0.5 ml 1 N HCl is added to stop hydrolysis, 5 ml methanol is added and 20–30 µl is used for TLC.

8.2.2 Thin-Layer Chromatography

Reference compound
2.5 mg salicin or derivatives are dissolved in 1 ml methanol, 20 µl is used for TLC.

Adsorbent
Silica gel 60 F_{254}-precoated TLC plates (Merck, Darmstadt)

Chromatography solvent
Ethyl acetate-methanol-water (77:13:10)

8.2.3 Detection

- UV-254 nm Quenching of salicin and derivatives.
- Spray reagent (see Appendix A)
  - Vanillin-glacial acetic acid reagent (VGA No. 39)
    After spraying, the plate is heated for 3–5 min at 110°C under observation. Salicin and derivatives show grey, violet-grey and brown zones (vis).

8.2.4 Drug List

Salicis cortex Willow bark from various Salix species Salicaceae MD	0.2%–10% phenolic glycosides: depending on the species or season Salicylates calculated as total salicins (after alkaline hydrolysis, for method see section 8.2.1) Salicin, triandrin, fragilin, salicortin, 3′- and 2′-O-acetyl- salicortin, vimalin, salireposide, tremulacin Isosalipurposide; tannins	Fig. 3,4

Salix alba L.	white willow	0.5%–1%	total salicins
Salix cinerea L.	grey willow	~0.4%	total salicins
Salix daphnoides L.	violet willow	4.9%–8.4%	total salicins
Salix fragilis L.	crack willow	4%–10%	total salicins (e.g. 2′-O-acetylsalicortin)
Salix purpurea L.	red willow	3.4%–7.4%	total salicins (e.g. salicortin)
Salix pentandra L.	bay willow	0.9%–1.1%	total salicins
Salix viminalis L.	common osier	~0.2%	total salicins (e.g. triandrin)

8.2.5 Formulae

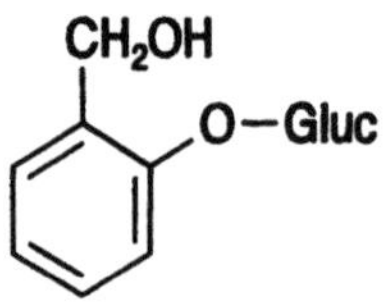

Salicin

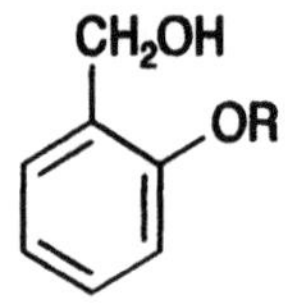

Fragilin R = 6′-O-Acetylglucose
Populin R = 6′-O-Benzoylglucose

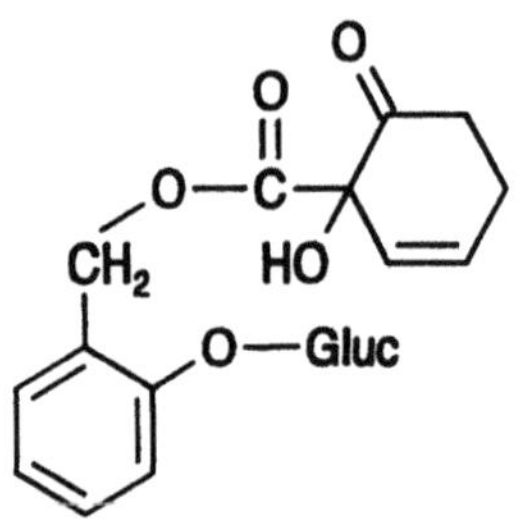

Salicortin
Tremulacin:
2′-O-Benzoyl-Salicortin

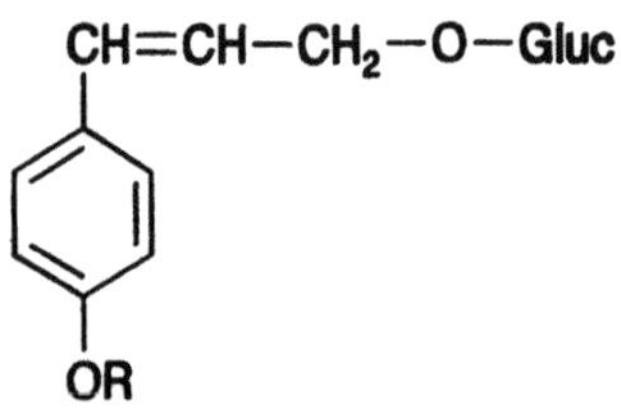

Triandrin R = H
Vimalin R = CH_3

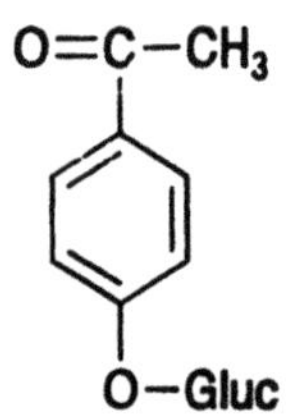

Picein

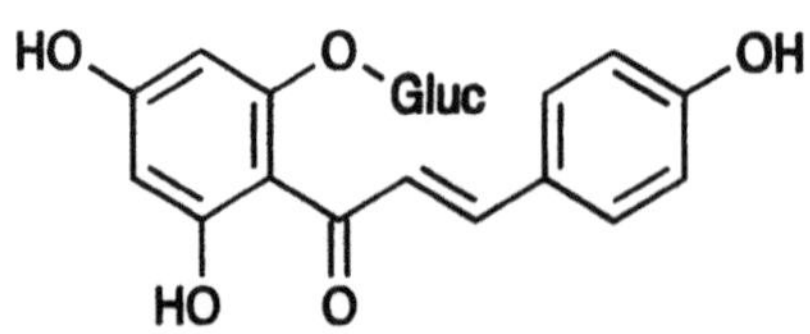

Isosalipurposide

8.3 Chromatograms

Arbutin drugs

Drug sample 1 Vitis idaeae folium
2 Uvae ursi folium
3 Myrtilli folium (methanolic extracts, 20–30 µl)

Test mixture T1 arbutin (R_f ~ 0.4) ▶ hydroquinone (front)
T2 rutin (R_f ~ 0.35) ▶ chlorogenic acid (R_f ~ 0.4) ▶ hyperoside (R_f ~ 0.55)

Solvents system Fig. 1A–C ethyl acetate-methanol-water (100:13.5:10)
Fig. 2D ethyl acetate-glacial acetic acid-formic acid-water (100:11:11:26)

Detection A Gibb's reagent (DCC No. 10) → vis.
B Berlin blue reaction (BB No. 7) → vis
C Millons reagent (ML No. 27) → vis
D Natural products-polyethylene glycol reagent (NP/PEG No.28) → UV-365 nm

---

Fig. 1 **Phenolglucosides**

A **Vitis idaeae folium** (1) and **Uvae ursi folium** (2) are characterized by the prominent distinct blue arbutin zone at R_f ~ 0.4 (T1).
The arbutin content in the extract of **Myrtilli folium** (3) is too low for detection with the DCC reagent. Instead of arbutin, an indistinct grey zone at R_f 0.4 is seen.
Additional grey zones are found below and above arbutin. Hydroquinone (T1) at the solvent front appears brown-violet.

B,C The Berlin blue reaction shows arbutin as a blue (→B), the Millons reagent as a yellow zone (→C) in sample 1. In sample 3, in the R_f range of arbutin two minor blue (→B) and yellow (→C) zones are detectable.

Fig. 2 **Flavonoids**

D The three drug extracts can be distinguished by their different flavonoid and phenol carboxylic acid content when separated in the polar solvent system and detected with NP/PEG reagent in UV-365 nm.
Vitis idaeae folium (1) is characterized by six yellow-orange fluorescent flavonoid glycosides in the R_f range 0.35–0.8. The major zones are found at R_f ~ 0.8 and in the R_f range of the hyperoside test (T2). As a minor zone rutin is detected at R_f ~ 0.35 (T2).
Uvae ursi folium (2) and **Myrtilli folium** (3) also show their principal flavonol glycosides in the R_f range of the hyperoside test (T2) accompanied by minor orange zones directly below, as in sample 2, and above, as in sample 3.
The flavonoid glycosides in Uvae ursi folium are due to quercetin-3-β-D-6′-O-galloyl-galactoside, hyperoside (>1%), isoquercitrin, quercitrin and myricetrin.
The blue fluorescent zones in the R_f range of the chlorogenic acid test (see T2) are prominent in sample 3, while 1 and 2 show more highly concentrated zones in the upper R_f range.

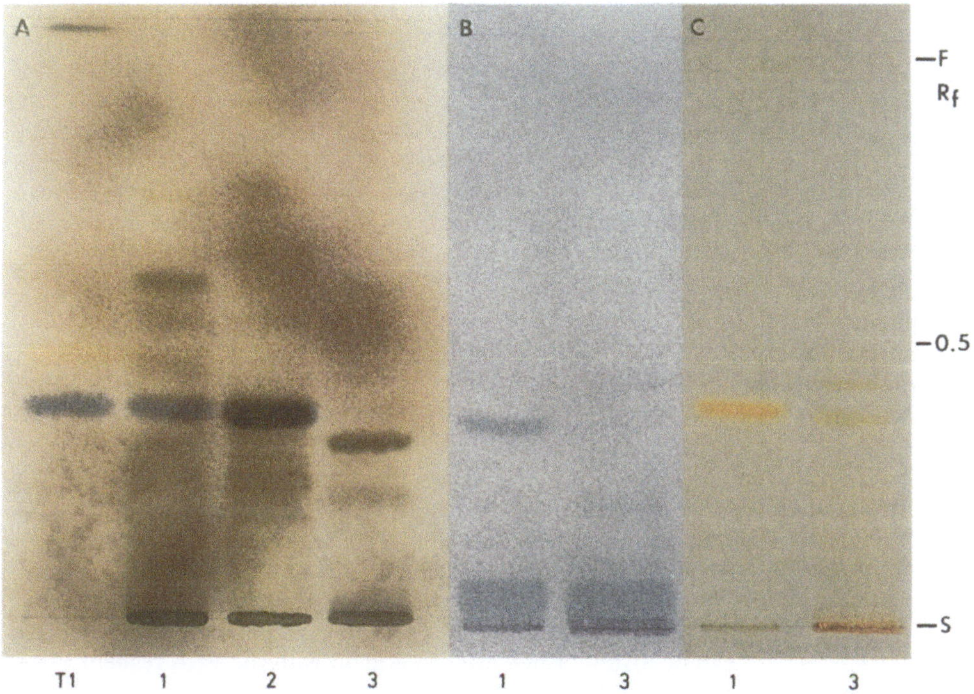

Fig. 1

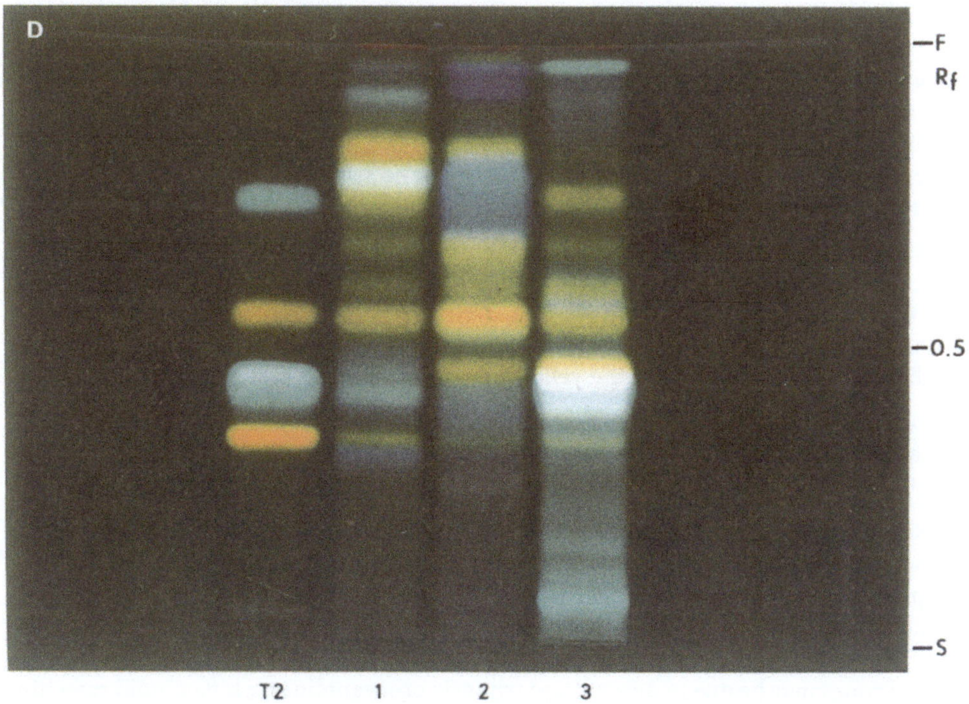

Fig. 2

Salicis cortex

Salicis cortex sample

1 Salix pentandra
2 Salix purpurea
2a Salix purpurea (after alkaline hydrolysis)
3 Salix alba
4 Salix species (trade sample)
5 Salix alba (freshly harvested bark)
5a Salix alba (after alkaline hydrolysis) (methanolic extracts, 20–40 µl)

Reference compound

T1 salicin
T2 salicortin
T3 salireposide
T4 isosalipurposide
T5 triandrin

Solvent system Fig. 3,4 ethyl acetate-methanol-water (77:13:10)

Detection

Fig. 3A,4A Vanillin-glacial acetic acid (VGA No. 39) 5 min/110°C → vis
Fig. 3B Natural products-polyethylene reagent (NP/PEG No.28) → UV-365 nm
Fig. 4C Vanillin-glacial acetic acid (VGA No.39) 10 min/110°C → vis

---

Fig. 3 The Salix species 1–3 can be distinguished after treatment with the VGA reagent (→ A, vis) and the NP/PEG reagent (→ B, UV-365 nm).

A **Salix pentandra** (1) and **S. purpurea** (2) show a similar TLC pattern of four weak grey and violet zones in the R_f range 0.25–0.45 with salicin at $R_f \sim 0.4$ (T1), a red-brown zone at $R_f \sim 0.85$ and two grey zones at the solvent front. Salix purpurea (2) is distinguishable by two additional prominent red zones in the R_f range 0.55–0.6.
In **S. alba** (3), only three weak zones in the R_f range 0.3–0.45 are detectable.
Salicin migrates as a grey-violet zone to $R_f \sim 0.45$. The phenol glycosides salicortin (T2), salireposide (T3), isosalipurposide (T4/chalcone) and triandrin (T5) are found as grey-violet zones in the R_f range 0.45–0.55.

B The Salix species 1–3 differ in their flavonoid glycoside content (1%–4%) and pattern. In the R_f range 0.4–0.6, sample 1 has two yellow-green and sample 3 three weaker green-blue fluorescent zones. Sample 2 is characterized by the green zone of naringenin-7-glucoside directly above the prominent red-orange fluorescent zone (R_f range of eriodictyol-7-O-glucoside) and the yellow isoquercitrin zone at $R_f \sim 0.4$.

Fig. 4A After treatment with the VGA reagent and heating of the TLC plate for 10 min at 110°C, the samples 2,4,5 show up to seven grey, violet or prominent red zones in the R_f range 0.35–0.6. The zone of salicin (T1) can be overlapped by other compounds (e.g. sample 5).

C The salicylates salicortin, tremulacin, 6′-O-acetylsalicin and 2′-O-acetylsalicortin, which are naturally present in the drugs, are easily hydrolyzed to salicin (see Sect. 8.2.1). A comparative TLC analysis of a methanolic **S. purpurea extract** (2) and its hydrolysis product (2a) shows salicin as a prominent grey zone at $R_f \sim 0.4$ (R_f value depression). In a hydrolyzed extract (5a) of a **Salix alba sample** (5), salicin was not detectable.
The red zones may be due to dimeric and trimeric procyanidines, a biflavonoid catechin-taxifolin, catechin and gallotannins.

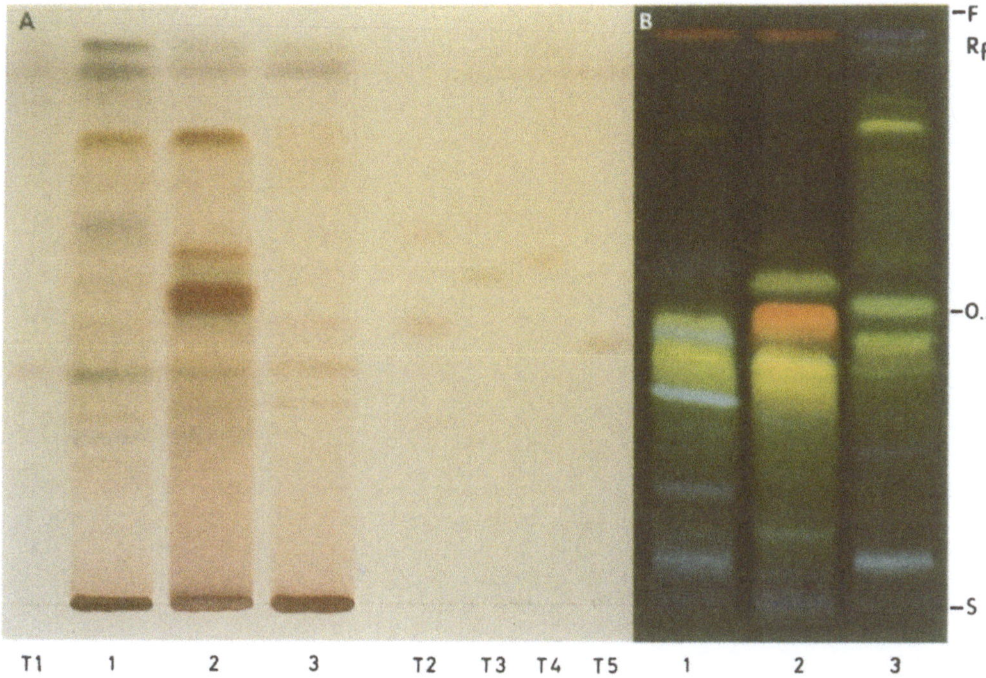

Fig. 3

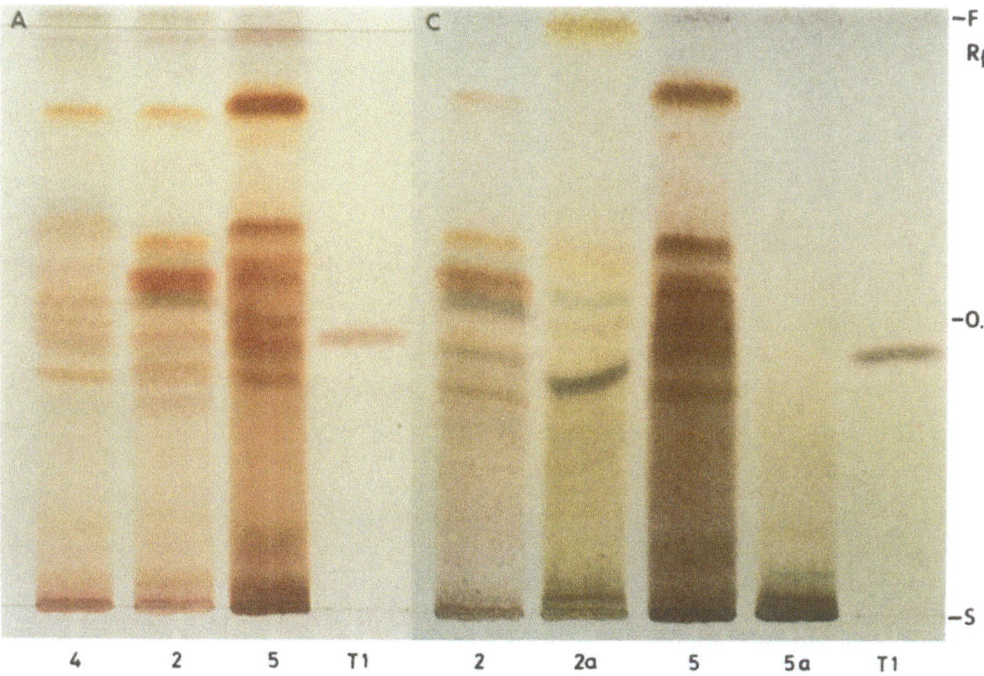

Fig. 4

9 Drugs Containing Cannabinoids and Kavapyrones

9.1 Cannabis Herba, Cannabis sativa var. indica L., Cannabaceae

The cannabinoids are benzopyran derivatives. Only Δ9,10-tetrahydro-cannabinol (THC) shows hallucinogenic activity. The type and quantity of the constituents depend on the geographical origin of the drug, climatic conditions of growth, time of harvesting and storage conditions.
Marihuana: the flowering or seed-carrying, dried branch tips of the female plant.
Hashish: the resin exuded from the leaves and flower stalks of the female plant.

9.1.1 Preparation of Drug Extracts

Powdered drug (1 g) is extracted by shaking at room temperature for 10 min with 10 ml methanol. The filtrate is evaporated and the residue dissolved in 1 ml toluene; depending on the cannabinoid concentration, 5–50 µl is used for TLC.

9.1.2 Thin-Layer Chromatography

Reference solutions

10 mg thymol is dissolved in 10 ml toluene; 5 µl is used for TLC.
1 mg synthetic tetrahydrocannabinol (THC) is dissolved in 5 ml $CHCl_3$; 3 µl is used for TLC.

Adsorbent

Silica gel 60 F_{254}-precoated TLC plates (Merck, Darmstadt)

Chromatography solvent

n-hexane-diethyl ether (80:20) or
n-hexane-dioxane (90:10)

9.1.3 Detection

- UV-254 nm — Prominent quenching of cannabinoids.
- Fast blue salt reagent (FBS No.15) — Cannabinoids appear orange-red or carmine (vis); standard thymol gives an orange colour.

9.1.4 Formulae

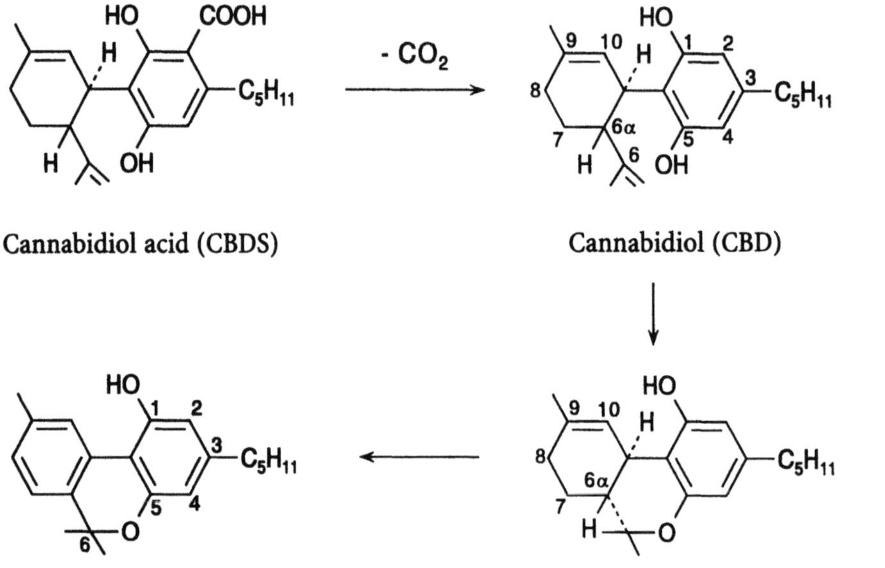

Cannabidiol acid (CBDS)

Cannabidiol (CBD)

Cannabinol (CBN)

Δ9,10-Tetrahydrocannabinol (THC)

9.2 Kava-Kava, Piperis methystici rhizoma, Piper methysticum G. FORST., Piperaceae (MD, DAC 86)

Depending on its geographical origin, the drug contains 5%–9% kavapyrones. These are derivatives of 6-styryl-4-methoxy-α-pyrones with anticonvulsive, muscle-relaxing and generally sedative effects.

9.2.1 Preparation of Drug Extracts for TLC

Powdered drug (0.6 g) is extracted with 10 ml dichloromethane for 10 min under reflux and 0.5 g of a commercial extract is dissolved in 5 ml methanol; 10 µl of each filtrate is used for TLC investigation.

9.2.2 Thin-Layer Chromatography

Reference 1 mg kawain is dissolved in 1 ml MeOH; 10 µl is used for TLC.

Adsorbent Aluminium oxide 60 F_{254} (Merck, Darmstadt)

Chromatography solvent n-hexane-ethyl acetate (70:30) (2 × 15 cm)

9.2.3 Detection

• UV-254 nm	Prominent quenching of all kawapyrones.
• Spray reagent (see Appendix A)	Anisaldehyde sulphuric acid reagent (AS No.3) Red to violet-red zones (vis.).

9.2.4 Formulae

Piperis methystici rhizoma (Kava-Kava)

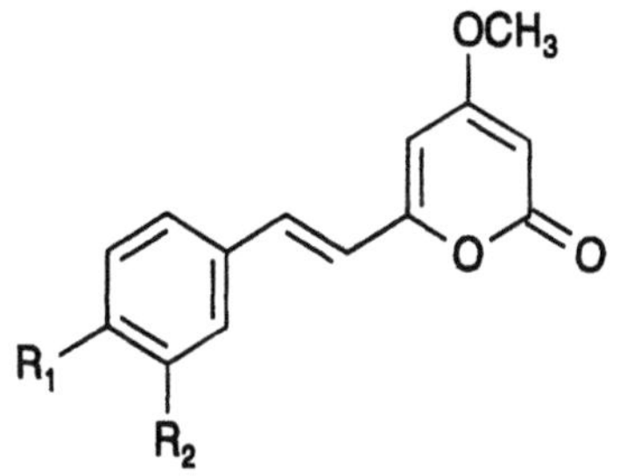

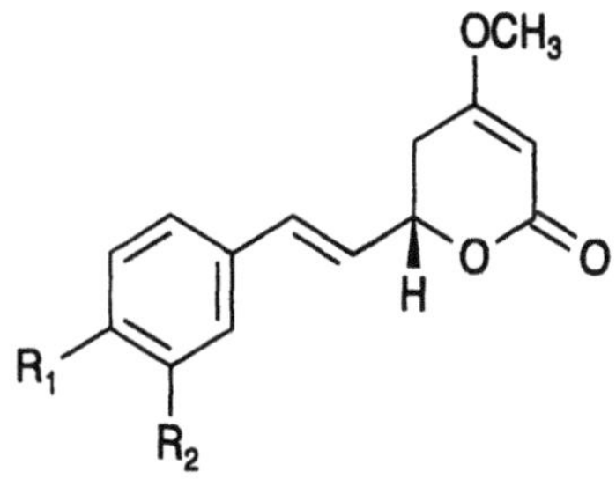

Desmethoxy-yangonin	(0,6–1%)	$R_1 = R_2 = H$
Yangonin	(1–1,7%)	$R_1 = OCH_3$; $R_2 = H$
Kawain	(1,8–2.1%)	$R_1 = R_2 = H$
Methysticin	(1,2–2%)	$R_1, R_2 = -OCH_2O-$

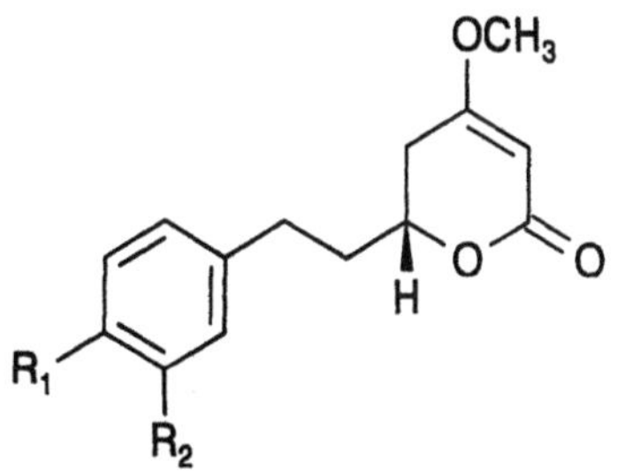

Dihydrokawain	(0,6–1%)	$R_1 = R_2 = H$
Dihydromethysticin	(0,5–0,8%)	$R_1, R_2 = -OCH_2O-$

9.3 Chromatograms

Cannabis herba, Hashish

Drug sample 1 Hashish (Turkish, 1980) 3 Hashish cigarette
2 Hashish (Iranian, 1980) 4–6 Cannabis herba (drug collection)

Reference T thymol
THC tetrahydrocannabinol (synthetic)

Solvent system Fig. 1 n-hexane-diethyl ether (80:20)

Detection Fast blue salt reagent (FBS No. 15) followed by 0.1 *M* NaOH→ vis

Fig. 1 Treatment with FBS-NaOH reagent shows intense red-violet to red-orange zones (vis.). **Hashish** samples 1 and 3 show two prominent red zones in the R_f range 0.45–0.55 due to cannabinol (CBN) and cannabidiol (CBD).
Between CBS and CBN, sample 2 has the additional red-violet zone of tetrahydrocannabinol (THC) at $R_f \sim 0.5$. The three to four red zones from the start up to $R_f \sim 0.15$ are due to cannabidiol acid and other polar cannabinoids.
Cannabis herba samples 4–6 contain cannabinoids in low concentrations only.

Kava-Kava rhizoma, Piper methysticum

Drug sample 1 Kava-kava rhizoma
2 Kava-kava extractum (trade sample)

Reference T1 kawain

Adsorbent Fig. 2 aluminium oxide 60 F_{254} plates

Solvent system n-hexane-ethyl acetate (70:30) (→ 2 × 15 cm)

Detection A without chemical treatment → UV-254 nm
B Anisaldehyde-sulphuric acid reagent (AS No. 3) → vis

Fig. 2A The **Kava-extracts** (1,2) are characterized in UV-254 nm by five lactones in the R_f range 0.4–0.8:

methysticin	$R_f \sim 0.4$	1.2%–2.1%
dihydromethysticin	$R_f \sim 0.45$	0.5%–0.8%
kawain (T1)	$R_f \sim 0.55$	1.8%–2.1%
dihydrokawain	$R_f \sim 0.65$	0.6%–1%
desmethoxykawain	$R_f \sim 0.75$	

B The quenching zones in Fig. 2 A appear red to blue-violet after treatment with the AS reagent. Kawain (T1) is seen as a red-violet zone at $R_f \sim 0.55$, followed by weaker violet zones of dihydrokawain ($R_f \sim 0.65$) and desmethoxykawain ($R_f \sim 0.75$). In the R_f range below kawain, the violet zone of dihydromethysticin ($R_f \sim 0.45$) and methysticin ($R_f \sim 0.4$) are found. A very weak yellow zone of yangonine can overlap the kawain zone at $R_f \sim 0.6$.

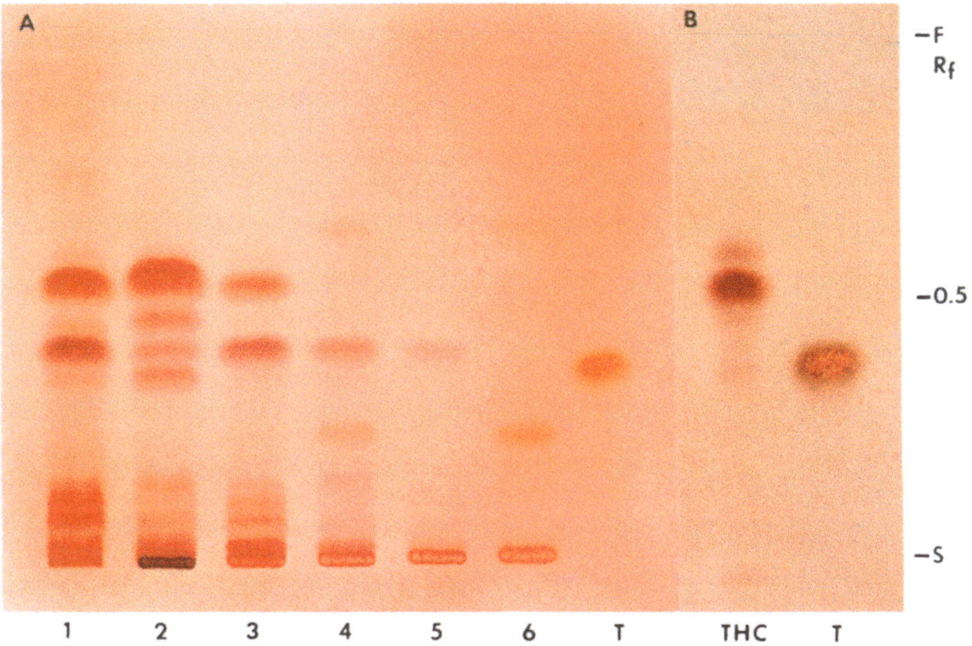

Fig. 1

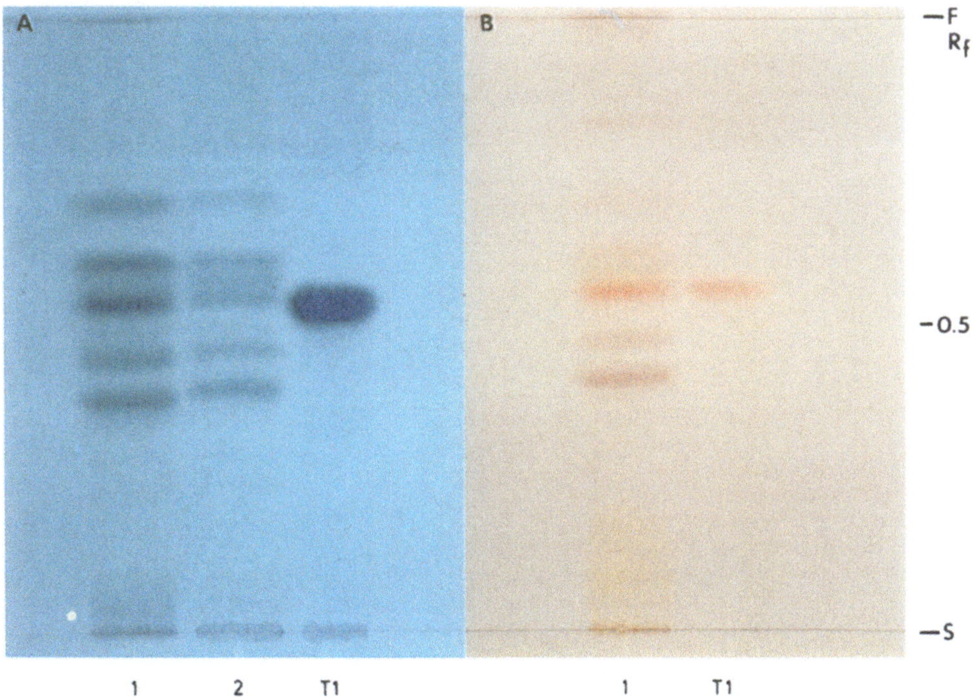

Fig. 2

10 Drugs Containing Lignans

Lignans are formed by oxidative coupling of p-hydroxyphenylpropene units, often linked by an oxygen bridge. They are found in fruits, foliage, heartwood and roots.

10.1 Preparation of Extracts

Powdered drug (1 g) is extracted by heating under reflux for 10 min with 10 ml methanol. The filtrate is evaporated to 3 ml and 20–30 µl is used for TLC.

Cubebae fructus, Podophylli rhiz.

Powdered drug (1 g) is extracted with 10 ml 50% methanol by heating under reflux for 15 min; after cooling, 15 ml water-saturated n-butanol is added. After shaking for 5 min, the butanol phase is separated and evaporated to approximately 1 ml; 20–40 µl is used for TLC.

Eleutherococci radix, Visci albi herba

Powdered drug (10 g) is extracted with 100 ml ethanol by slow percolation. The percolate is concentrated by evaporation until the residue has the consistency of a thin syrup and is then poured, with constant stirring, into 100 ml water containing 1 ml HCl (38%), and precooled to a temperature below 10°C. The precipitate is decanted and washed with two 100-ml portions of cold water; 0.1 g dried resin is dissolved in 2 ml methanol, and 20 µl is used for TLC.

Podophyllin resin

10.2 Thin-Layer Chromatography

Cubebin, podophyllotoxine, eleutherosides B, E, E_1:
1 mg is dissolved in 1 ml methanol; 20 µl is used for TLC investigation.

Reference solution

Silica gel 60 F_{254}-precoated plates (Merck, Darmstadt)

Adsorbent

Chloroform-methanol-water (70:30:4) → Eleutherococci radix, Visci albi herba
Chloroform-methanol (90:10) → 6 cm followed by toluene-acetone (65:35) → Podophylli rhizoma, Podophyllin
Toluene-ethyl acetate (70:30) → Cubebae fructus

Chromatography solvents

10.3 Detection

- UV-254 nm all lignans show prominent quenching.
- UV-365 nm e.g. eleutheroside E_1 gives blue fluorescence.
- Spray reagents (see Appendix A)

50% ethanolic sulphuric acid		→ Cubebae fructus,
Vanillin-phosphoric acid reagent	(VP No. 41)	→ Eleutherococci radix
Fast blue salt reagent	(FBS No. 15)	→ specific for peltatins
Antimony-(III)-chloride reagent	($SbCl_3$ No. 4)	→ syringin/Eleutherococci radix
Vanillin-sulphuric acid reagent	(VS No. 42)	→ essential oil compounds/ Cubebae fructus

10.4 Drug List

	Drug/plant source Family/pharmacopoeia	Main constituents
Fig. 1, 2	**Eleutherococci radix** (Rhizoma) Siberian ginseng Eleutherococcus senticosus MAXIM Araliaceae MD	0.05%–0.1% lignans Eleutheroside E (syringaresinol-4′, 4″-O-di-β-glucopyranoside), eleutheroside E_1 (syringaresinol-4′-O-β-D-monoglucopyranoside), (−)syringaresinol, sesamin Phenylpropane derivatives: eleutheroside B (= syringin, 0%–0.5%). Caffeic acid ethyl ester, coniferylaldehyde, sinapyl alcohol Essential oil (~0.8%) Coumarins: isofraxidin, -7-O-glucoside
Fig. 3, 4	**Visci albi herba** White mistletoe Viscum album L. var. malus Deciduous mistletoe (on practically all European deciduous trees, except beech) var. abies (WIESB.) ABROMEIT Silver fir mistletoe (on Abies species) var. pinus syn. ssp. austriacum (WIESB.) VOLLMANN Scots pine mistletoe (on Pinus spp.; or Picea excelsa LINK) Viscaceae/Loranthaceae DAC 86, MD	Lignans: eleutheroside E, E_1. Phenylpropane derivatives: 0.04%–0.07% syringin (= syringenin-4-O-β-D-glucopyranoside), syringenin-4-O-β-D-apiofuranosyl-1→2-β-D-glucopyranoside Plant acids (~ 15): caffeic, sinapic, syringa, p-coumaric, protocatechuic, chlorogenic, vanillic, ferulic, p-hydroxy, benzoic and shikimic acid Free amino acids <0.4% (leaves) (~18 in fresh leaves): e.g. L-arginine, alanine, proline, L-serin, tyrosine

Drug/plant source Family/pharmacopoeia	Main constituents	

Flavonoid pattern
European mistletoes:
Viscum album L.: quercetin and its methylethers (e.g. rhamnetin, isorhamnetin, rhamnazin)
2′-hydroxy-4′,6′-dimethoxy-chalcon-4-glucoside;
Loranthus europaeus L.: rhamnocitrin-3-O-rhamnoside, rhamnetin-3-O-glucoside, rhamnetin-3-O-rhamnoside
Non-European mistletoe:
Viscum album var. coloratum (Japan):
flavoyadorinin A (= 7,3′-di-O-methylquercetin (rhamnazin)-3-O-glucoside),
flavoyadorinin B (= 7,3′-di-O-methylluteolin-4′-O-mono-glucoside),
homoflavoyadorinin B (= 7,3′-di-O-methylluteolin-4′-O-glucoapioside)
Psittacanthus cuneifolius (Argentinia):
quercetin-3-O-rhamnoside (= quercitrin), quercetin-3-O-xyloside (reynoutrin),
quercetin-3-O-α-arabinofuranoside (avicularin)
Loranthus parasiticus (China): quercetin, quercetin-3-O-arabinoside
Phoradendron tomentosum (Texas): vitexin, 6-C-glucosyl-8-C-arabinosylapigenin (= schaftoside), 6-C-arabinosyl-8-C-glucosylapigenin (= isoschaftoside), apigenin-4′-O-glucoside, apigenin

Podophylli rhizoma Podophyllum May apple, Mandrake root Podophyllum peltatum L. Berberidaceae	3%–6% resin (~16 compounds) with 0.2%–1% podophyllotoxin and the β-D-glucoside; α-, β-peltatine and their β-D-glucosides Picropodophyllin (an artefact due to extraction procedures)	Fig. 5
Podophyllum resin "Podophylline" MD	Only aglycones due to extraction procedures: >20% podophyllotoxin, α-, β-peltatines, desoxy and dehydropodophylline	
Indian Podophyllum Podophyllum emodi WALL. Berberidaceae MD	6%–12% resin with 1%-4% podophyllotoxin, only traces of peltatines, berberine	
Cubebae fructus Cubeb, Java pepper Piper cubeba L. Piperaceae	1.5%–2.5% cubebin 10%–18% essential oil tricyclic sesquiterpene alcohols; 1,4-cineol, terpineol-4, cadinol, cadinene	Fig. 6

10.5 Formulae

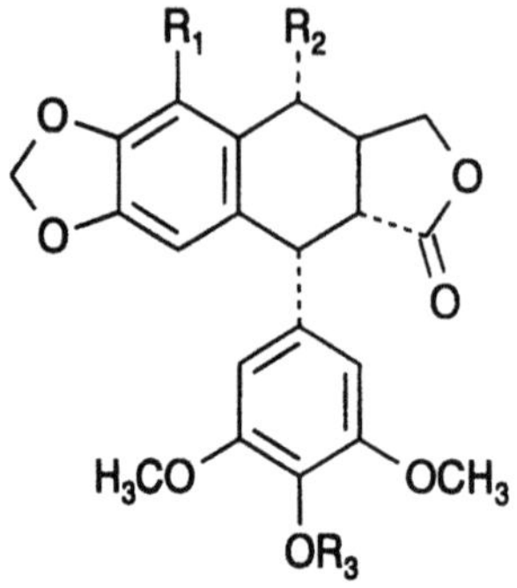

e.g. Podophyllotoxin

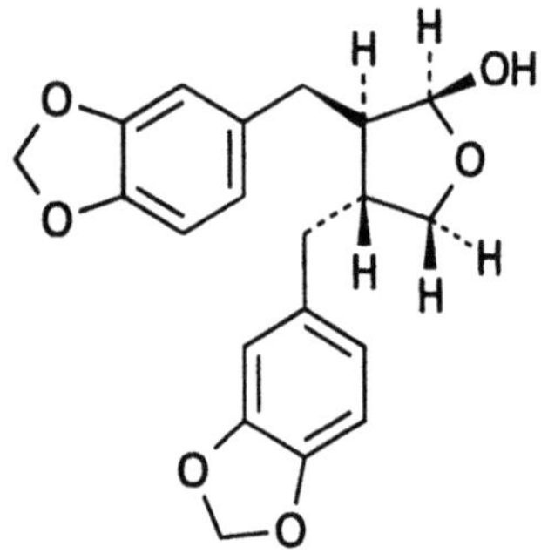

Cubebin

R_1	R_2	R_3	
H	OH	CH_3	Podophyllotoxin
H	O-Gluc	CH_3	Podophyllotoxin glucoside
H	H	CH_3	Desoxypodophyllotoxin
H	OH	H	4′-Desmethylpodophyllotoxin
H	O-Gluc	H	4′-Desmethylpodophyllotoxin glucoside
OH	H	H	α-Peltatin (P1)
O-Gluc	H	H	α-Peltatin glucoside
OH	H	CH_3	β-Peltatin (P2)
O-Gluc	H	CH_3	β-Peltatin glucoside

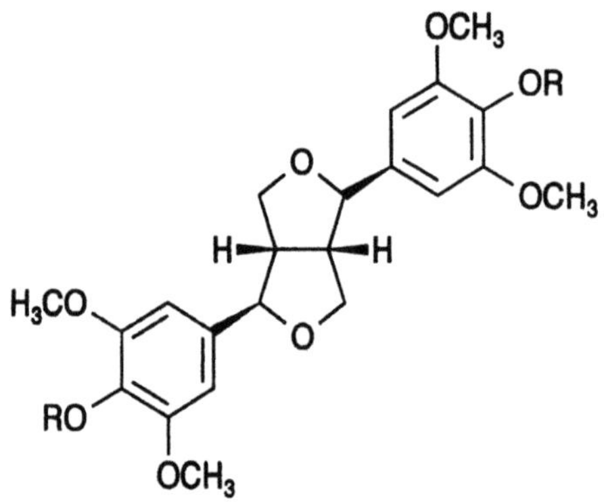

Eleutheroside E R = β-D-Gluc
Syringaresinol R = H

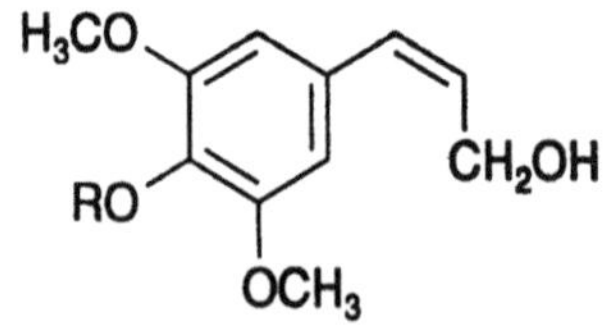

Eleutheroside B R = β-D-Gluc
Sinapyl alcohol R = H

10.6 Chromatograms

Eleutherococci radix (rhizoma)

Drug sample 1 Eleutherococci radix (type A)
2 Eleutherococci radix (type B)
3 Eleutherococci radix (type C)
4–12 Eleutherococci radix (commercial drug samples) (n-butanol extracts, 20 µl)

Reference compound T1 eleutheroside B (syringin)
T2 eleutheroside E
T3 syringaresinol monoglucoside E_1
T4 syringaresinol

Solvent system Fig. 1, 2 chloroform-methanol-water (70:30:4)

Detection A Without chemical treatment → UV-365 nm
B Vanillin-phosphoric acid reagent (VPA No. 41) → vis
C Antimony-(III)-chloride reagent ($SbCl_3$ No. 4) → UV-365 nm

---

Fig. 1A **Eleutherococci radix** samples 1–3 show phenol carboxylic acids and coumarins as blue fluorescent zones: chlorogenic acid at $R_f \sim 0.05$, lipophilic plant acids and coumarins in the R_f range 0.85–0.95. Their presence and amount varies according to plant origin.

B Eleutherococci radix samples 1 and 3 are characterized by the blue to violet-red zones of eleutheroside B (syringin) (T1) at $R_f \sim 0.5$, eleutheroside E (T2) at $R_f \sim 0.35$ and eleutheroside E_1 (T3) at $R_f \sim 0.65$. Syringin (T1) can be absent (e.g. sample 2) or is found in extremely low concentrations only.
The amount of blue aglycone zones in the R_f range 0.8–0.95 varies as do the grey zones in the R_f 0.05–0.15, which are partly due to free sugars.

C The zone of syringin (T1) fluoresces specifically orange-red with $SbCl_3$ reagent. Syringin is accompanied by a blue and yellow fluorescent zone directly above and below, respectively.

Fig. 2 **TLC Synopsis** (VPA reagent, vis)
As demonstrated with the Eleutherococcus samples 4–12, the amount and presence of eleutheroside B at $R_f \sim 0.5$, as well as eleutheroside E at $R_f \sim 0.35$ and its monoglucoside E_1 at $R_f \sim 0.65$, varies depending on origin of the plant and the part of the roots used for investigation. The grey zones in the R_f range 0.05 and 0.2 (e.g. chlorogenic acid, free sugars) and the blue-grey and violet-zones in the upper R_f range are present in varying amounts.

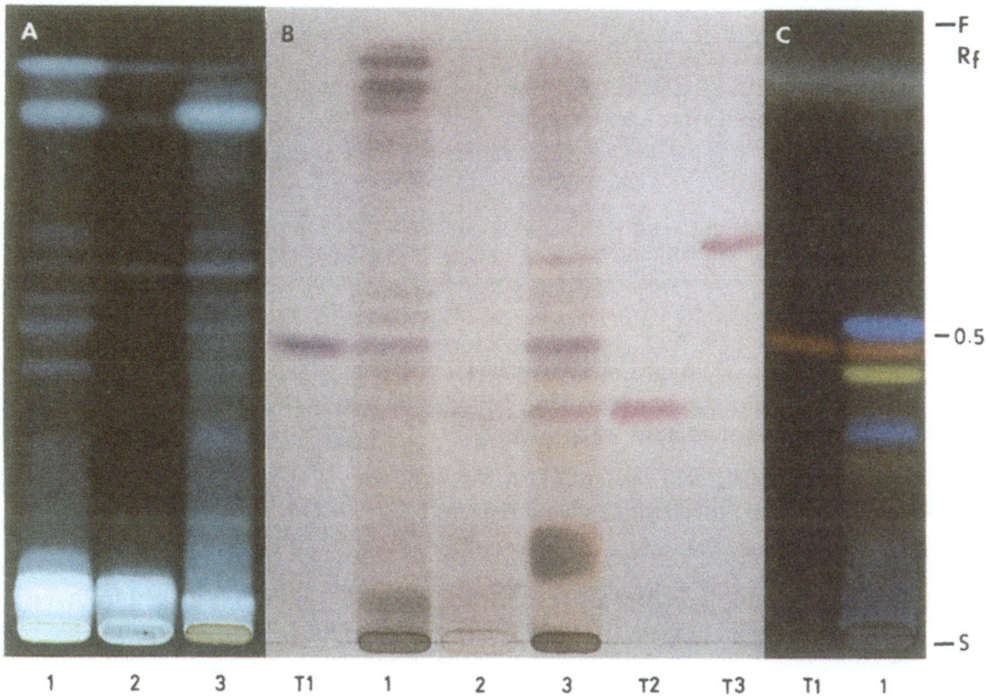
A
B
C
F
Rf
0.5
S
1 2 3 T1 1 2 3 T2 T3 T1 1

Fig. 1

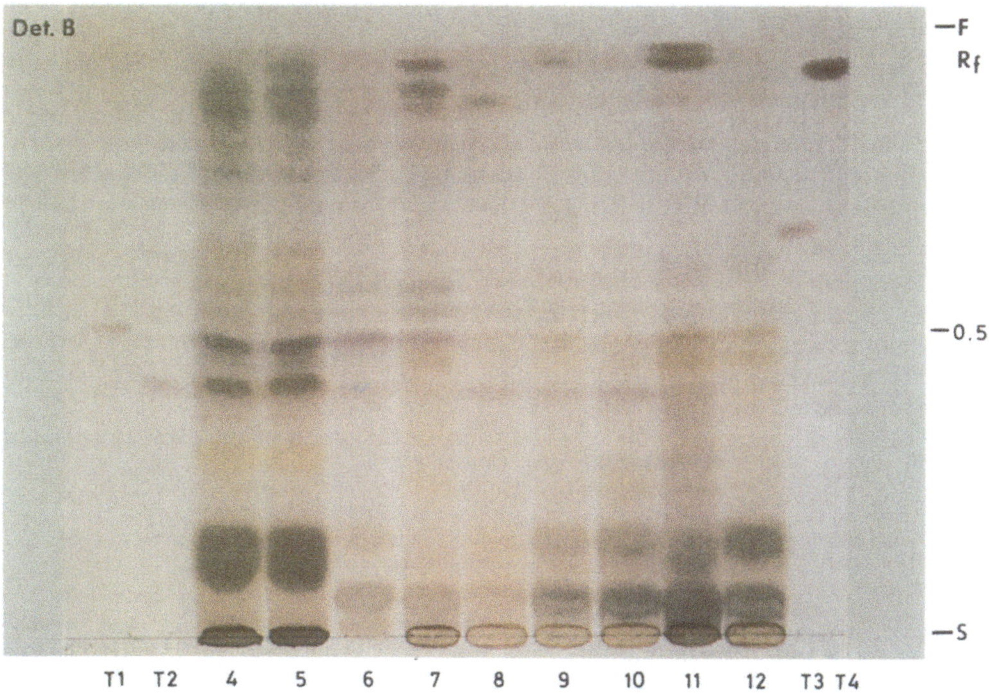
Det. B
F
Rf
0.5
S
T1 T2 4 5 6 7 8 9 10 11 12 T3 T4

Fig. 2

Viscum album

Drug sample
1 Viscum album (n-butanol extract)
2 Viscum album (MeOH extract 1 g/10 ml/for flavonoids, 20µl)
3, 4, 5 Viscum album (n-butanol extracts)
6 Viscum album (pharmaceutical preparation) (n-butanol extracts, 20–40 µl)

Reference compound
T1 eleutheroside E
T2 syringenin-apiosylglucosid ($R_f \sim 0.3$) + syringin ($R_f \sim 0.4$)
T3 rutin (R_f 0.35) ► chlorogenic acid (R_f 0.4) ► hyperoside (R_f 0.55) ► isochlorogenic acid
T4 eleutheroside B (syringin)

Solvent system
Fig. 3 A chloroform-methanol-water (70:30:4)
B ethyl acetate-glacial acetic acid-formic acid-water (100:11:11:26)
Fig. 4 A chloroform-methanol-water (70:30:4)

Detection
A Vanillin-phosphoric acid reagent (VPA No. 41) → vis.
B Natural-products-polyethylene glycol reagent (NP/PEG No. 28) → UV-365 nm

---

Fig. 3A **Viscum album** sample 1 represents the characteristic TLC pattern obtained from Visci albi herba of European origin. After treatment with the VPA reagent, more than ten red or blue-violet and brown zones are found in the R_f range 0.4 up to the solvent front.
Eleutheroside E ($R_f \sim 0.4$/T1) is normally present in most samples as a minor compound.
Eleutheroside B (syringin) ($R_f \sim 0.45$/T4) has a medium concentration in (1) but can be more highly concentrated e.g. sample 3, Fig. 4A.
Syringenin-4-O-β-apiofuranosyl-glucopyranoside can be found at $R_f \sim 0.25$. This very unstable compound (T2) easily forms syringin.

B With the NP/PEG reagent, the methanolic **Viscum album** (2) extract develops a series of blue fluorescent zones from the start till up to $R_f \sim 0.55$, due to various plant acids. The blue-green zones in the higher R_f range might derive from quercetin ethers and chalcon glucosides.

Fig. 4 **TLC Synopsis**
A Three Viscum samples (3–5) collected from different trees show syringin (T4) at $R_f \sim 0.45$, and a variation of blue-violet and yellow zones due to other lignans and various phenol carboxylic acids. The pharmaceutical preparation 6 has an additional prominent blue zone above the syringin test.
The pattern of compounds changes according to the origin of the plant (e.g. Malus, Abies, Pinus).

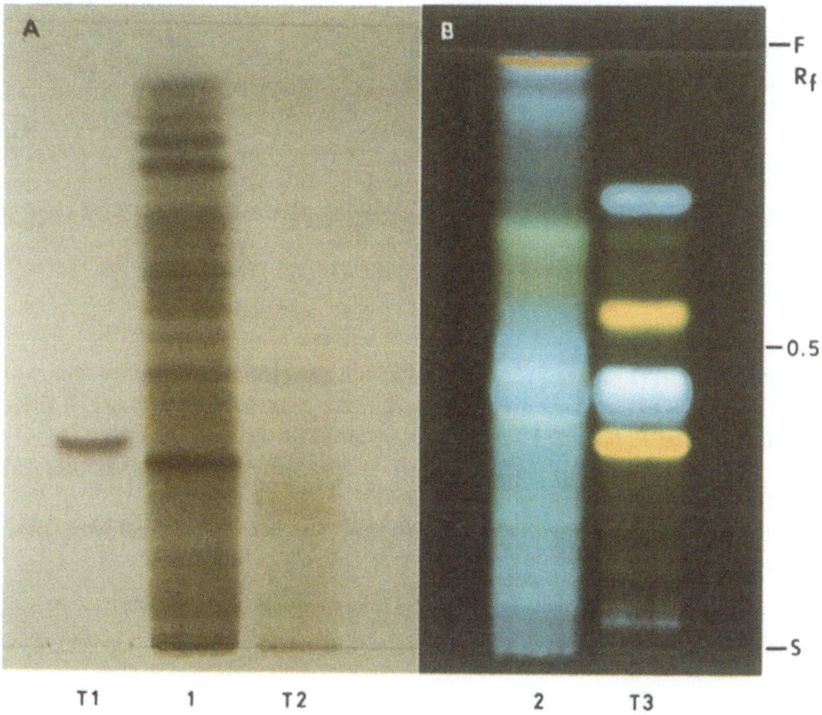

Fig. 3

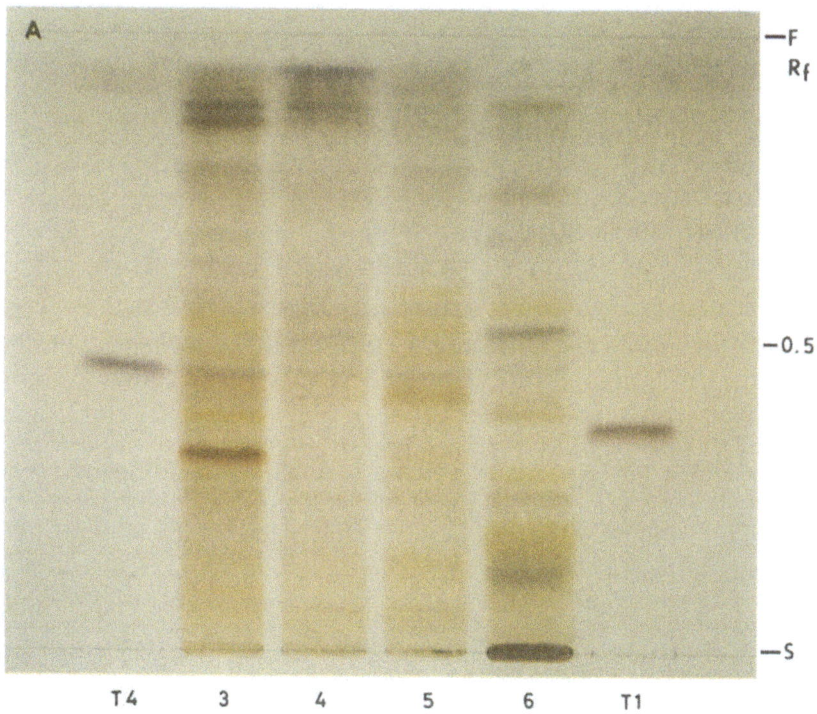

Fig. 4

Podophylli rhizoma

Drug sample 1 Podophylli peltati rhizoma
2 Podophylli emodi rhizoma
3 Resin of Podophylli peltati rhizoma (methanolic extracts, 20–30 µl)

Reference T1 podophyllotoxin

Solvent system Fig. 5 chloroform-methanol (90:10) → 6 cm ▶ then toluene-acetone (65:35) → 15 cm

Detection A Sulphuric acid 50% (H_2SO_4 No. 37) → vis
B Fast blue salt reagent (FBS No. 15) → vis

Fig. 5A With 50% H_2SO_4, **Podophylli rhizoma** (1, 2) shows the blue to violet-blue lignan zones of podophyllotoxin ($R_f \sim 0.7$/T1) and α- and β-peltatin ($R_f \sim 0.65/P_1$; $R_f \sim 0.8/P_2$), their corresponding glucosides at R_f 0.05–0.15 and the aglycones at the solvent front.
Podophyllotoxin (T1) is more highly concentrated in **Podophylli emodi rhizoma** (2) than in **Podophylli peltati rhizoma** (1).
The resin **"podophyllin"** (3) contains podophyllotoxin, α- and β-peltatin, and very small amounts of lignan glucosides.

B With FBS reagent the peltatins (P1, P2) and tannins form red-brown zones in vis (1–3). Podophyllotoxin does not react.

Cubebae fructus

Drug sample 1 Cubebae fructus (methanolic extract, 20 µl)

Reference T1 cubebin

Solvent system Fig. 6 toluene-ethyl acetate (70:30)

Detection A Sulphuric acid, 98% (H_2SO_4 No. 37) → vis
B Vanillin-sulphuric acid reagent (VS No. 42) → vis

Fig. 6A A methanolic extract of **Cubebae fructus** A, B (1) is characterized by the lignan cubebin, which forms with H_2SO_4 reagent a red-violet zone at $R_f \sim 0.45$ (vis) besides diffuse brown zones.

B Cubebin and the essential oil compounds, such as cadinol, a tricyclic sequiterpene alcohol, a mixture of isomer cadinenes, 1,4-cineol and terpineol-4, give prominent blue to violet-blue zones (B) with the VS reagent in the R_f range 0.5 up to the solvent front.

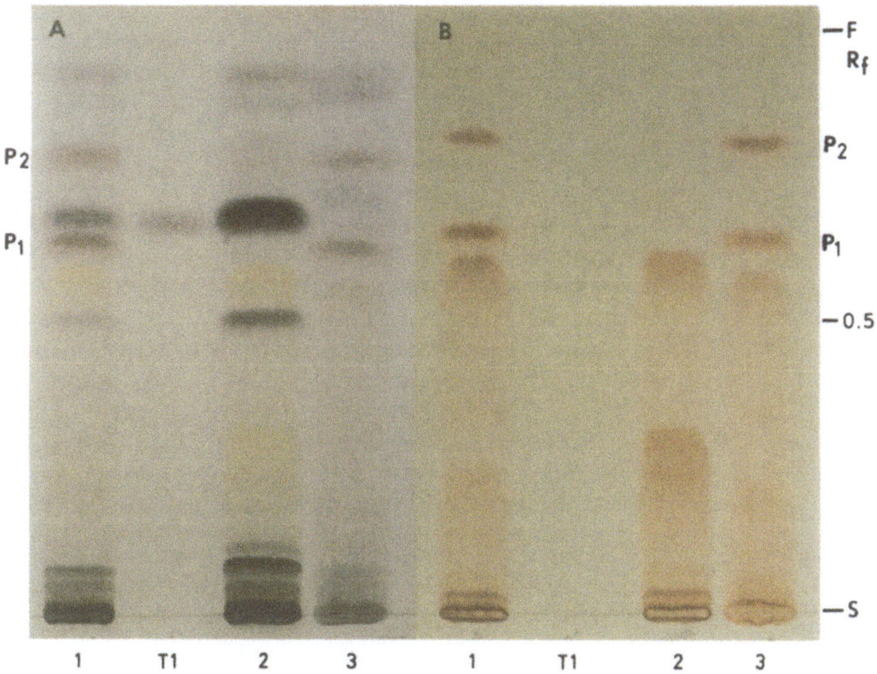

Fig. 5

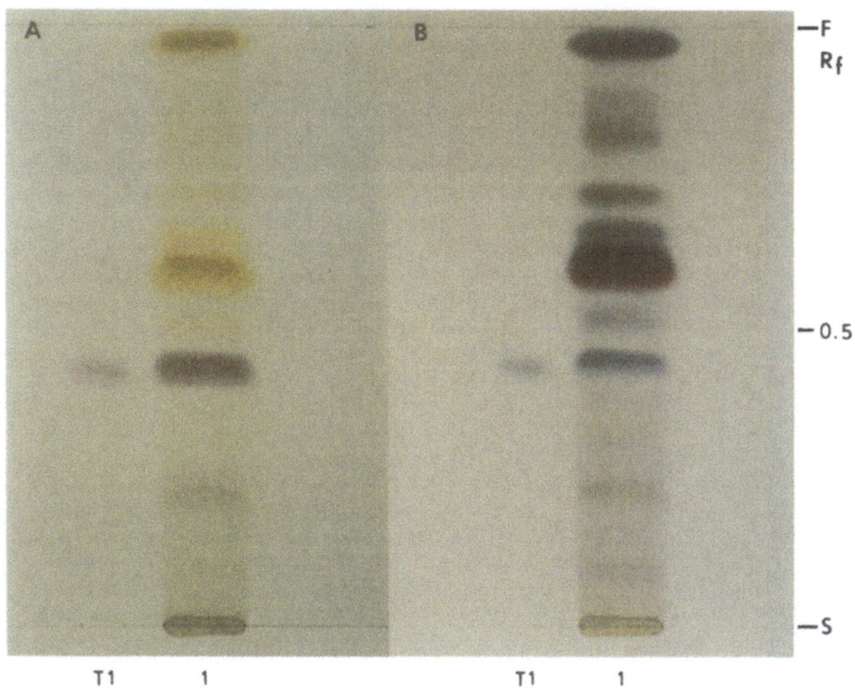

Fig. 6

11 Drugs Containing 1,4-Naphthoquinones Droserae herba, Dionaeae herba

11.1 Preparation of Extract

Droserae herba

1. Powdered drug (1 g) is extracted for 15 min with 10 ml methanol on a water bath; 30 µl of the clear filtrate is used for TLC.
2. Powdered drug (1 g) is distilled with 10 ml water and 1 ml 2 M H_3PO_4 in a 50-ml flask through a glass pipe into a chilled glass tube until 3 ml distillate has been collected (see microdistillation, Sect. 6.1). After cooling, the lipophilic compounds are extracted with 1 ml pentane; 10 µl of this solution is used for TLC.

Dionaeae herba

The whole fresh plant is put through a tincture press until 1 ml plant juice has been collected. The juice is diluted with 9 ml $CHCl_3$ and 20 µl is used for TLC investigations.

11.2 Thin-Layer Chromatography

Reference solutions

10 mg plumbagin and juglone are dissolved in 1 ml methanol and 10 µl are used for TLC investigation.

Adsorbent

Silica gel $60F_{254}$-precoated TLC plates (Merck, Germany)

Solvent systems

Toluene-formic acid (99:1) → naphtoquinone aglycone
Ethyl acetate-formic acid-glacial acetic acid-water (100:11:11:26) → glycosides

11.3 Detection

- All naphthoquinones show quenching in UV-254 nm.
- After spraying with 10% methanolic KOH reagent, naphtoquinones show red fluorescence in UV-365 nm and red to red-brown colour (vis).

11.4 Drug List

	Drug/plant source Family/pharmacopoeia	Main constituents
Fig. 1	**Droserae rotundifoliae herba** Round-leafed sundew Drosera rotundifolia L. ▶ protected plant	>0.5% 1,4-naphthoquinones, plumbagin, 7-methyl-juglone, droserone
Fig. 1	**Droserae longifoliae herba** Long-leafed sundew from various Drosera species e.g.: Drosera ramentacea BURCH. ex HARV. et SOND. Drosera longifolia, D. anglica D. intermedia, D. burmanii Droseraceae MD	>0.25% 1,4-naphthoquinones, plumbagin, ramentaceon and its glucoside rossoliside
Fig. 2	**Dionaeae muscipulae herba** Dionaea muscipula ELLIS (syn. Drosera sessiliflora RAF.) Droseraceae	>0.85% total 1,4-naphthoquinones, plumbagin (~0.2%), hydroplumbagin-4-0-β-glucoside (~0.6%), 3-chloro-plumbagin (~0.01%), droserone (~0.002%)

11.5 Formulae

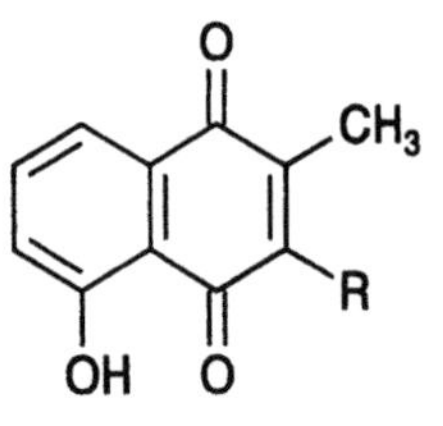

Plumbagin R = H
Droserone R = OH

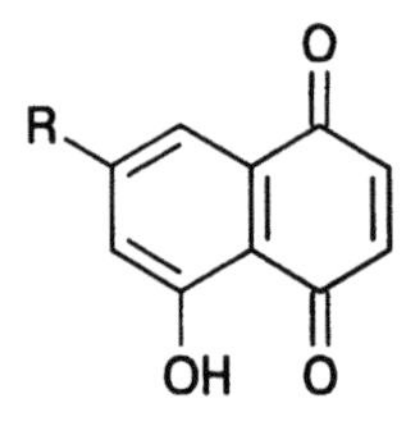

Juglone R = H
Methyljuglone R = CH_3

11.6 Chromatograms

Droserae herba, Dionaeae herba

Drug sample
1 Droserae rotundifoliae herba (MeOH extract, 30 µl)
2 Droserae rotundifoliae herba (distillate)
3 Droserae ramentaceae herba (distillate)
4 Dionaeae muscipulae herba (pressed juice, 20 µl)

Reference compound
T1 plumbagin
T2 juglone
T3 rutin ($R_f \sim 0.35$) ▶ chlorogenic acid ($R_f \sim 0.45$) ▶ hyperoside ($R_f \sim 0.6$)

Solvent system
Fig. 1 A–D toluene-formic acid (99:1) → system I
Fig. 2 A+B toluene-formic acid (99:1) → system I
C ethyl acetate-glacial acetic acid-formic acid-water (100:11:11:26) → system II

Detection
Fig. 1A–D 10% methanolic potassium hydroxide
A,D → vis B,C → UV-365 nm
Fig. 2A,B 10% methanolic potassium hydroxide
A → vis B → UV-365 nm
Fig. 2C Natural products-polyethylene glycol reagent (NP/PEG No. 28) → UV-365 nm

---

Fig. 1 A–D All three **Drosera** samples show the violet-brown (vis) and brown-yellow (UV-365 nm) fluorescent plumbagin (T1) as the main zone at $R_f \sim 0.45$.
Plumbagin is accompanied in sample 1 by 7-methyljuglone (the same R_f value as plumbagin) and juglone ($R_f \sim 0.4$). Juglone is more highly concentrated in sample 3 (D. ramentacea). Droserone ($R_f \sim 0.35$) can be detected in sample 1.

Fig. 2 A,B **Dionaea muscipula** (4) shows the prominent blue (vis/A) and red-brown (UV-365 nm/B) zone of plumbagin (T1) at $R_f \sim 0.5$ in solvent system I. The hydroplumbagin-4-β-D-glucoside remains at the start.

C (NP/PEG reagent, UV-365 nm):
In solvent system II hydroplumbagin glucoside migrates as a blue-green band into the R_f range 0.85–0.9. Further blue to greenish-blue zones are found in low concentrations in the R_f range of hyperoside (T3).

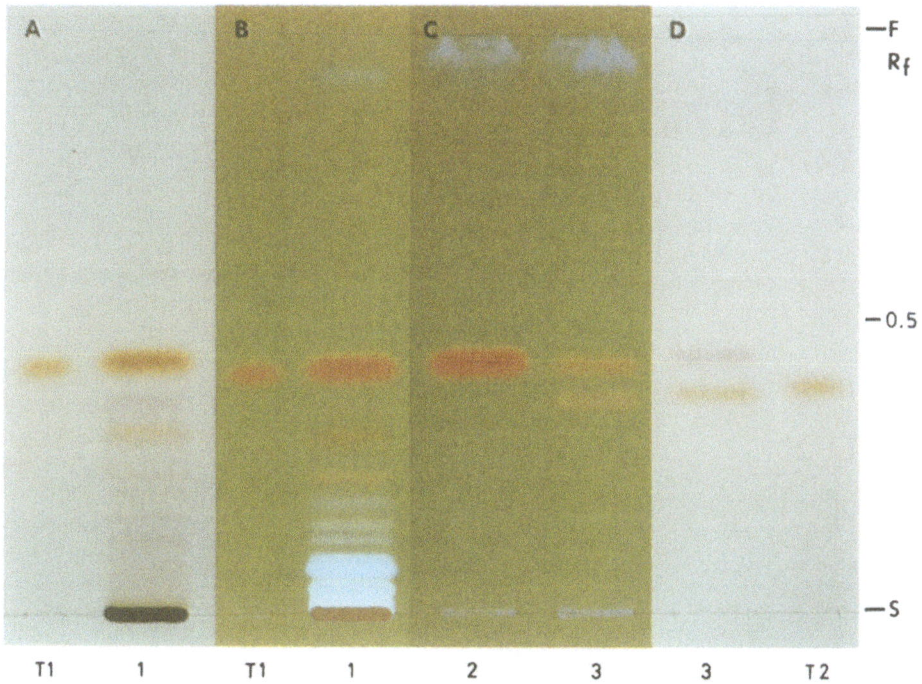

Fig. 1

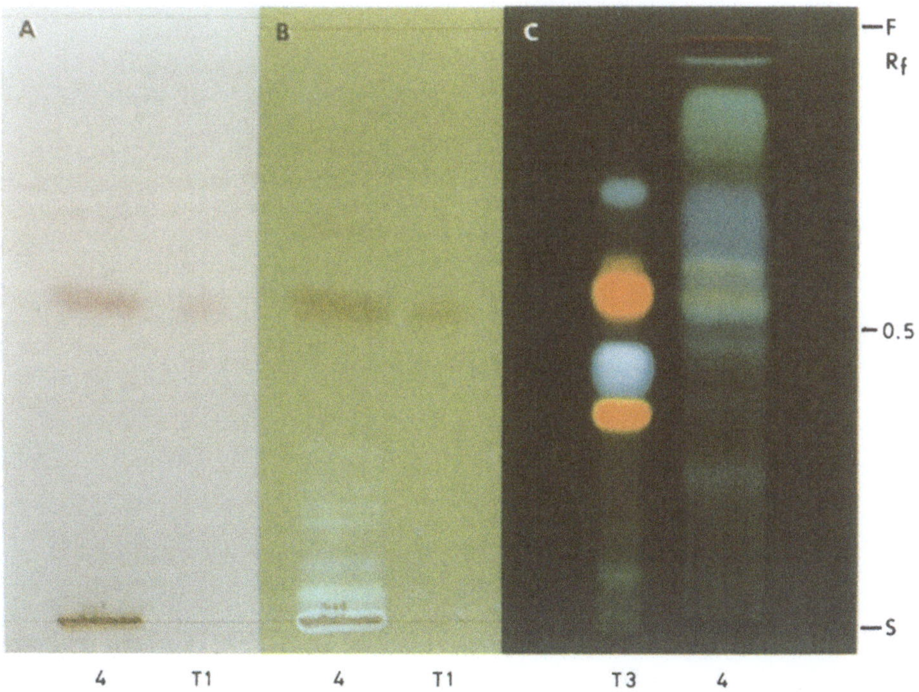

Fig. 2

12 Drugs Containing Pigments

Anthocyanins (Flavylium salts): Anthyocyanidins are responsible for the red, violet and blue colours of flowers and other plant parts. They are mostly present in plants as glycosides of hydroxylated 2-phenylbenzopyrylium salts. Cleavage by acid hydrolysis produces the corresponding free flavylium salts.
Crocus: Croci stigma contains crocetin, a 8,8′-diapocarotenedioic acid. The bright yellow digentiobiosyl ester crocin is water soluble.

12.1 Preparation of Extracts

Anthocyanins

Cyani, Hibisci and Malvae flos, Myrtilli fructus. Powdered drug (1 g) is extracted by shaking for 15 min with 6 ml of a mixture of nine parts methanol and one part 25% HCl; 25 μl of the filtrate is used for TLC investigation.

Croci stigma

Four or five crushed stigma are moistened with one drop of water. After about 3 min, 1 ml methanol is added and the extraction continued for about 20 min in the dark, with occasional shaking; 10 μl of the supernatant or filtrate is used for chromatography.

12.2 Thin-Layer Chromatography

Reference solutions

Anthocyanins:	1 mg standard compound dissolved in 1 ml methanol; TLC sample, 5 μl.
Methylene blue:	5 mg dissolved in 10 ml methanol; TLC sample, 10 μl.
Naphthol yellow:	5 mg dissolved in 5 ml methanol; TLC sample, 5 μl.
Sudan red:	5 mg dissolved in 5 ml chloroform; TLC sample, 5 μl.

Adsorbent

Silica gel 60 F_{254}-precoated TLC plates (Merck, Germany).
Cellulose-precoated TLC plates (Merck, Germany).
Chromatography of flower pigments (anthocyanins) is performed on both silica gel and cellulose plates. Silica gel plates are used for TLC of Croci stigma extracts.

Chromatography solvents

Anthocyanins:	Ethyl acetate-glacial acetic acid-formic acid water (100:11:11:26) n-Butanol-glacial acetic acid-water (40:10:20) or (40:10:50) → upper layer
Croci stigma:	Ethyl acetate-isopropanol-water (65:25:10)

12.3 Detection

- Without chemical treatment
  Anthocyanins show red to blue-violet, Croci stigma constituents yellow colour (vis).
- Anisaldehyde-sulphuric acid reagent (AS No. 3)
  After spraying and heating (8 min/110°C) the picrocrocin appears red-violet, crocin blue-violet (vis).

12.4 Drug List

	Drug/plant source Family/pharmacopoeia	Main compounds Anthocyanins
Fig. 1,2	**Hibisci flos** Hibiscus flowers Hibiscus sabdariffa L. Malvaceae DAB 10	Delphinidin-3-glucosyl-xyloside (hibiscin), delphinidin-3-glucoside, cyanidin-3-glucosyl-xyloside, cyanidin-3-glucoside
Fig. 3	**Cyani flos** Cornflowers Centaurea cyanus L. Asteraceae	Cyanidin-3,5-diglucoside (cyanin), pelargonidin-3,5-diglucoside (pelargonin), pelargonin-3-caffeoylglucoside-5-glucoside
Fig. 3	**Malvae flos** Common mallow flowers Malva sylvestris L. Mauretanian, dark-violet mallow Malva sylvestris L. ssp. mauritania (L.) ASCH. et GRAEBN. Malvaceae ÖAB 90, Helv. VII, MD	6%–7% total anthocyanins Malvidin-3,5-diglucoside (malvin 50%) delphinidin glucosides; petunidin-3-, cyanidin-3- and malvidin-3-O-glucoside
Fig. 3	**Malvae (arboreae) flos** Hollyhock Althaea rosea (L.) CAV. var. nigra HORT. Malvaceae	Delphinidin-3-glucoside, malvidin-3-glucoside, "althaein", the mixture of both glucosides;
Fig. 4	**Myrtilli fructus** Common blue berries Vaccinium myrtillus L. Ericaceae DAC 86, ÖAB 90, Helv. VII, MD	0.5% total anthocyanins Delphinidin-3-glucoside (myrtillin A), -3-galactoside, malvidin-3-glucoside; glycosides of pelargonidin, cyanidin and petunidin
Fig. 4	**Croci stigma** Saffron (crocus)	1.9%–15% crocin (digentiobiosyl ester of crocetin)

Drug/plant source Family/pharmacopoeia	Main compounds Anthocyanins
Crocus sativus L. Iridaceae DAC 86, Ph.Eur.III, ÖAB 90, MD, Japan	2.7%–12.9% picrocrocin (β-hydroxycyclocitral glucoside) β-hydroxycyclocitral and safranal (dehydro-β-cyclocitral) are formed from picrocrocin during storage or steam distillation; carotene glycosides

12.5 Formulae

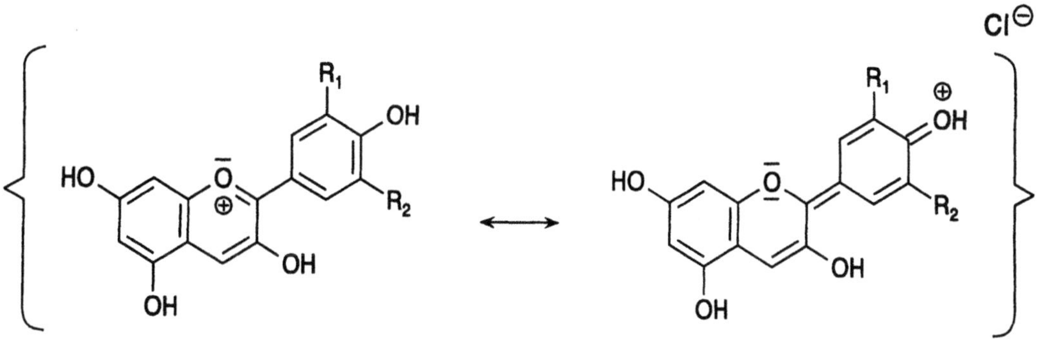

	R_1	R_2
Pelargonidin	H	H
Paeonidin	H	OCH_3
Cyanidin	H	OH
Malvidin	OCH_3	OCH_3
Petunidin	OH	OCH_3
Delphinidin	OH	OH

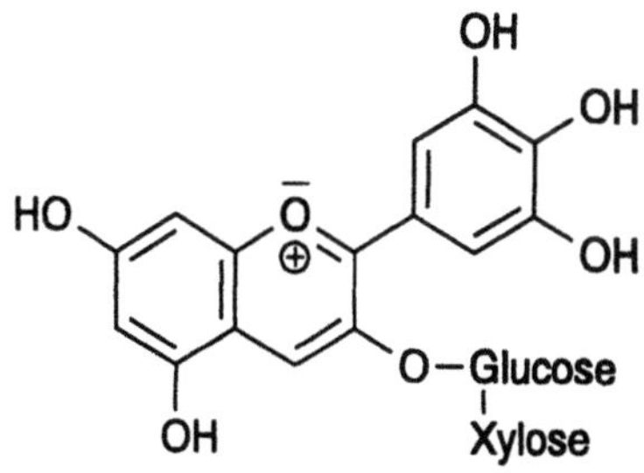

Hibiscin

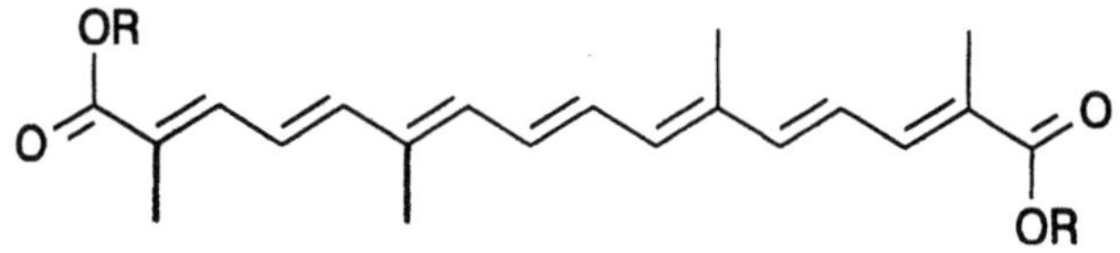

Crocetin R = H
Crocin R = Gentiobiosyl

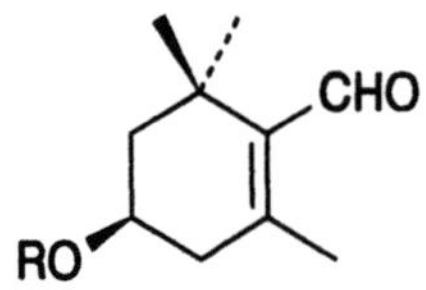

4-Hydroxycyclocitral R = H
Picrocrocin R = Glucosyl

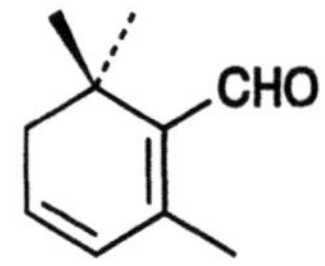

Safranal

12.6 Chromatograms

Hibisci flos

Reference compounds

Drug sample	H Hibisci flos (methanolic extract, 25 µl)	
Reference compound	1 methylene blue 2 delphinidin-3, 5-diglucoside 3 delphinidin-3-glucoside 4 petunidin-3, 5-diglucoside 5 petunidin-3-glucoside 6 cyanidin-3, 5-diglucoside	7 cyanidin-3-glucoside 8 malvidin-3, 5-diglucoside 9 malvidin-3-glucoside 10 paeonidin-3, 5-diglucoside 11 paeonidin-3-glucoside (reference, 5 µl)
Adsorbent	Silicagel 60 F_{254} (Merck, Darmstadt)	
Solvent system	Fig. 1,2 A,B ethyl acetate-glacial acetic acid-formic acid-water (100:11:11:26) → system I C,D n-butanol-glacial acetic acid-water (50:10:20) ▸ upper layer → system II	
Detection	Without chemical treatment → vis	

Fig. 1A The separation of **Hibisci flos** in solvent system 1 reveals three clearly defined blue to violet-blue pigment zones in the R_f range 0.15–0.25. The two major bands at R_f 0.15–0.2 are probably due to delphinidin-3-glucosyl-xyloside (hibiscin) and cyanidin-3-glucosyl-xyloside, reported as major pigments. The monoglucosides delphinidine-3-glucoside (3) and cyanidin-3-glucoside (7) are found in the R_f range 0.2–0.35.

Note: Diglucosides such as delphinidin-3,5-glucoside (3) are found in a lower R_f range than 3-glucosyl-xylosides.

B The 3,5-diglucosides **reference compounds** of delphinidin, petunidin, cyanidin, malvidin and paeonidin (2,4,6,8,10) migrate with low R_f values, slightly increasing in the R_f range 0.05–0.1. The corresponding monoglucosides (3,5,7,9,11) are better separated and show higher R_f values in the R_f range 0.2–0.4.

Fig. 2C,D Development in solvent system 2 shows the pigments of **Hibisci flos** as two major zones. A blue band at R_f 0.2, typical for the delphinidin, petunidin and cyanidin types (1–7), and a violet zone above (R_f 0.35). The reference compounds of the malvidin and paeonidin types (8–11) show clearly defined zones and differences in the R_f values of mono- and diglucosides.
With the TLC technique only a fingerprint of an anthocyanin-containing drug extract can be obtained. Similar pigments often overlap and have to be identified by other techniques.

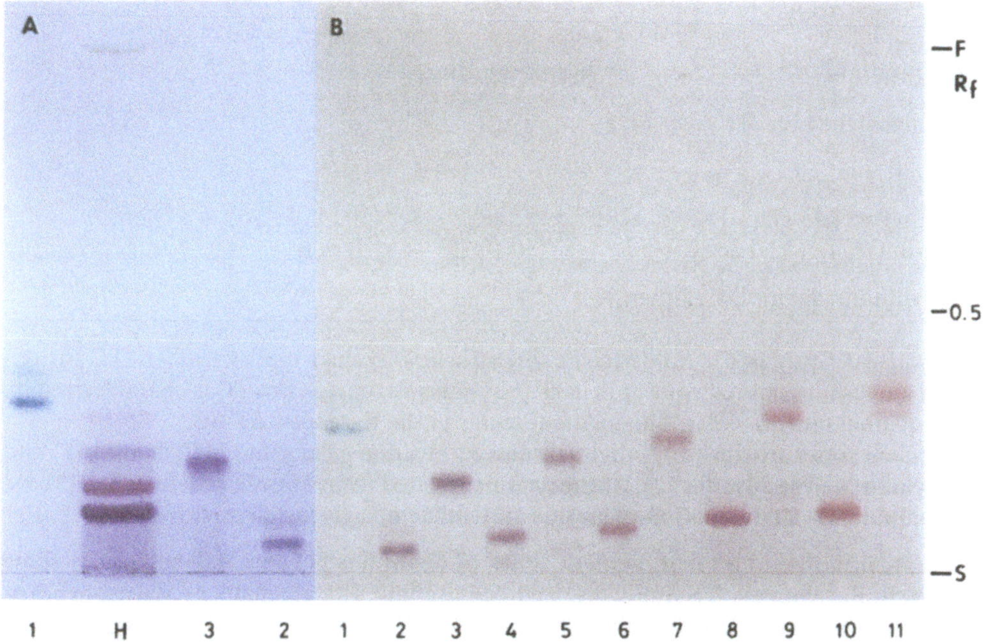
A
B
—F
Rf
—0.5
—S
1
H
3
2
1
2
3
4
5
6
7
8
9
10
11

Fig. 1

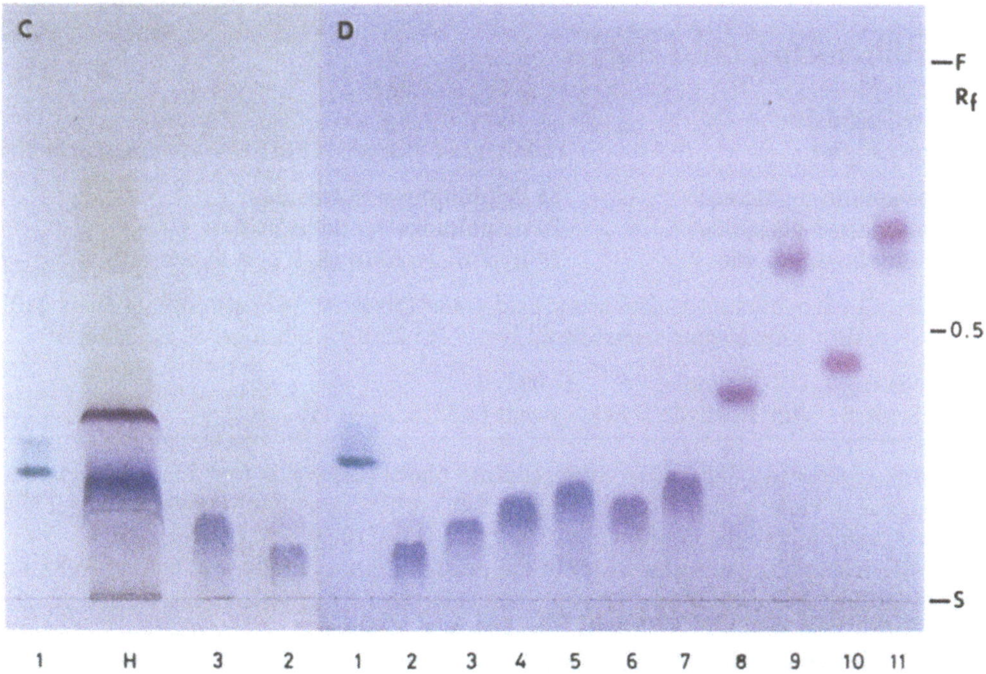
C
D
—F
Rf
—0.5
—S
1
H
3
2
1
2
3
4
5
6
7
8
9
10
11

Fig. 2

TLC Synopsis

Drug sample 1 Cyani flos; 2 Malvae silvestris flos; 3 Malvae arboreae flos (Extracts, 25 µl)

Reference T1 methylene blue

Adsorbent A,B Silica gel 60 F_{254}

Solvent system A ethyl acetate-glacial acetic acid-formic acid-water (100:11:11:26) → system A
B n-butanol-glacial acetic acid-water (50:10:20) → system B

Detection vis (without chemical treatment)

Fig. 3A In system A **Cyani flos** (1) and **Malvae silvestris flos** (2) show quite a similar TLC pattern with one prominent red zone at R_f 0.05–0.1. **Malvae arboreae flos** (3) is characterized by five distinct blue to violet-blue pigment zones in the R_f range 0.05–0.3.
The main zones are due to 3,5-diglucosides, e.g. cyanin, pelargonin in Cyani flos (1) and malvin in Malvae silv. flos (2). Anthocyanins isolated from Malvae arboreae flos (3) are delphinidin-3-, malvidin-3-O-glucoside and althaein, a glycoside mixture.

B In system B, the prominent pigment zones of **Cyani flos** (1) and **Malvae flos** (2) show different R_f values (R_f 0.3 and R_f 0.4) and colouration. The pigments of **Malvae arborea flos** (3) are separated into two violet zones (R_f 0.45/0.6, e.g. malvidin-3-glucoside), which migrate ahead of the broad blue pigment band (R_f 0.1–0.4).

Myrtilli fructus, Croci stigma

Drug sample 4 Myrtilli fructus; 5 Croci stigma (methanolic extracts, 10 µl)

Reference compound

T2 paeonidin-3-glucoside	T5 delphinidin-3-glucoside
T3 malvidin-3-glucoside	T6 delphinidin-3,5-diglucoside
T4 cyanidin-3,5-diglucoside	T7 naphthol-yellow (R_f 0.2) ► Sudan red

Solvent system Fig. 4 Cl+Si n-butanol-glacial acetic acid-water (50:10:20) (Cl Cellulose, Si Silica gel)
C-E ethyl acetate-isopropanol-water (65:25:10) – Silica gel 60 F_{254} (Merck)

Detection A,B vis C UV-254 nm D vis
Cl+Si vis E Anisaldehyde H_2SO_4 reagent (AS No. 3) → vis

Fig. 4 Cl/Si **Myrtilli fructus** (4). Separation over cellulose plates (Cl) yields four major clearly defined blue to violet zones in the R_f range 0.2–0.5, whereas separation over silica gel (Si) shows cyanidin and delphinidin glycosides as a broad blue band between R_f 0.05–0.45. Myrtylli fructus contains glucosides of the pelargonidin, cyanidin and petunidin types, delphinidin-3-galactoside and -3-glucoside (myrtillin A) as well as malvidin-3-glucoside.
The identification of a specific pigment by TLC only is limited. The separation with two adsorbents or different solvent systems, however, can give a helpful TLC fingerprint.

C,D,E **Croci stigma** (5) is characterized by yellow-coloured crocin and crocetin (R_f 0.15–0.25) in vis. Both show fluorescence-quenching in UV-254 nm, as well as picrocrocine (R_f ~ 0.55), and become dark violet-blue with AS reagent (vis). Weak zones in the R_f range of Sudan red (6), e.g. 4-hydroxy-cyclocitral or safranal can be present.

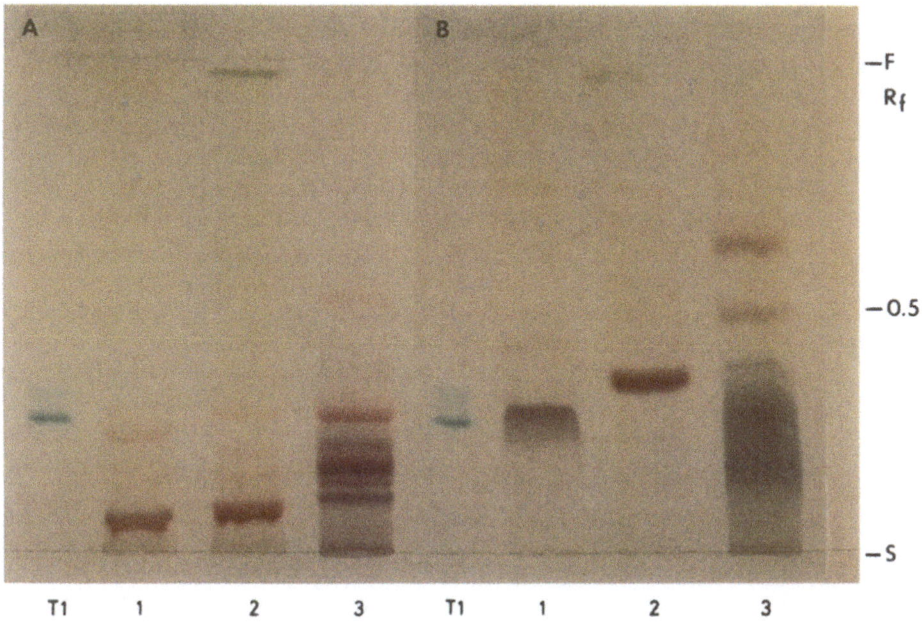

Fig. 3

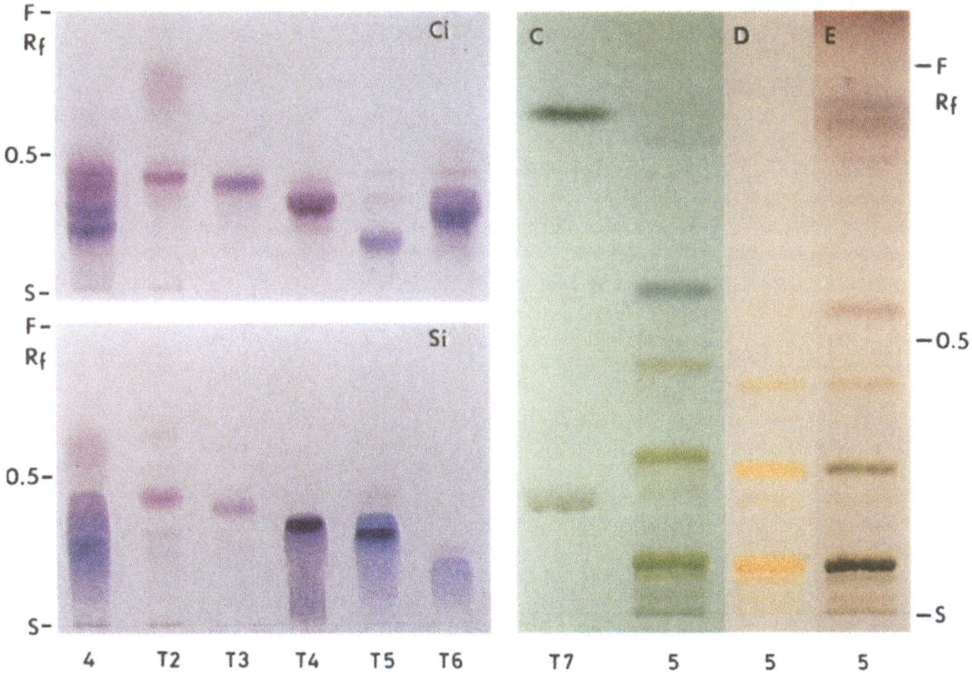

Fig. 4

13 Drugs with Pungent-Tasting Principles

13.1 Pungent-Tasting Constituents

These constituents belong mainly to one of the following types:

- Amides: piperines (Piperis fructus) or capsaicin (Capsici fructus).
- O-Methoxyphenols and propylphenols: gingerols (Zingiberis and Galangae rhizoma), eugenol (Caryophylli flos and Myristicae semen[1]), elemicin and asarone (Calami and Asari rhizoma[1]).
- Phenolic sesquiterpenes: xanthorrhizol in Curcumae rhizoma[1].

13.1.1 Preparation of Extracts

Piperis fructus

Powdered drug (1 g) is extracted by heating under reflux for 10 min with 10 ml methanol. The filtrate is evaporated to 3 ml, and 10 µl is used for chromatography.

Capsici fructus, Galangae and Zingiberis rhiz.

Powdered drug (1 g) is extracted by heating under reflux for 10 min with 10 ml $CHCl_3$ or dichloromethane. The filtrate is evaporated to 3 ml, and 20 µl is used for TLC.

13.1.2 Thin-Layer Chromatography

Reference solutions

1 mg standard compound (capsaicin, piperine, vanillin) is dissolved in 1 ml MeOH; 10 µl is used for TLC.

Adsorbent

Silica gel 60 F_{254}-precoated TLC plates (Merck, Germany).

Chromatography solvents

toluene-ethyl acetate (70:30)	→ Piperis and Capsici fructus, Galangae and Zingiberis rhizoma
toluene-diethyl ether-dioxane (62.5:21.5:16)	→ Piperis fructus
diethyl ether (100)	→ Capsici fructus
hexane-diethyl ether (40:60)	→ Galangae and Zingiberis rhizoma.

[1] For volatile compounds, TLC separation, description of the drugs and formulae see Chap. 6.

13.1.3 Detection

- UV-254 nm — Capsaicin shows fluorescence quenching only at high concentrations. Piperine and gingeroles cause distinct fluorescence quenching.
- UV 365 nm — Piperine gives dark blue, piperyline light blue fluorescence.
- Spray reagents (see Appendix A)
  - Vanillin-sulphuric acid reagent (VS No. 42)
    After spraying, the plate is heated for 10 min at 100°C, evaluation in vis.: piperine lemon yellow; gingeroles blue to violet.
  - Barton reagent (No. 5)
    After spraying and heating for 2–5 min at 100°C, evaluation in vis.: gingeroles, shogaoles, galangol bright blue (vis).
  - Dichloroquinone-chloroimide reagent (DCC No. 10)
    Immediately after spraying, spontaneous reaction as blue-violet (vis) zones, evaluation in vis.: capsaicin and capsaicinoides, detection limit 0.1 μg.

13.1.4 Drug List

	Drug/plant source Family/pharmacopoeia	Pungent principles lipophilic, non-volatile
Fig. 1	**Piperis fructus** (Black) pepper Piper nigrum L. Piperaceae ÖAB	4%–10% amides 2%–5% trans-piperine (pungency index 1:2 000 000) Piperettin, piperanin, piperaestin A, piperyline (about 5%) ► Essential oil: 1%–2.5% (black pepper), >98% terpene hydrocarbons
Fig. 1,2	**Capsici fructus** Capsicums Capsicum annum L. var. longum SENDTN. ÖAB, MD, Japan **Capsici acris fructus** Cayenne pepper, Chillies Capsicum frutescens L. Solanaceae DAB 10, Helv VII, MD, DAC 86 (tincture)	0.1%–0.5% capsaicinoids (C. annum) 0.6%–0.9% (C. frutescens) >30% capsaicin (= vanillylamide of 8-methyl-(trans)-non-6-enoic acid; pungency index 1:2 million) Homo-, dihydro-, homodihydro- and nor-dihydrocapsaicin (50%) ► 0.1% essential oil ► 0.8% carotinoids ► Steroids
Fig. 3,4	**Galangae rhizoma** Chinese ginger Alpinia officinarum HANCE Zingiberaceae Helv VII	Diarylheptanoids, gingerols Galangol (= complex mixture of diarylheptanoids), (8)-gingerol ► 0.3%–1.5% essential oil with sesquiterpene hydrocarbons, 1,8 cineole, eugenol

Drug/plant source Family/pharmacopoeia	Pungent principles lipophilic, non-volatile	
Zingiberis rhizoma/radix Ginger (root) Zingiber officinale ROSCOE Zingiberaceae ÖAB 90, Helv VII, BP, MD Japan, China	1%–2.5% gingerols, shogaols: 5-hydroxy-1-(4-hydroxy-3-methoxy-phenyl)-3-decanone and homologues (= gingerols); (6)-gingerol; the corresponding anhydro-compounds (= shogaols) and vanillyl-acetone (= zingerone) ▶ 1%–3% essential oil sesquiterpenes: (−) zingiberene (30%), β-bisabolene (>10%) sesquiphellandrene (15%–20%) citral, citronellyl acetate	**Fig. 3,4**

▶ Pungent principles present in the essential oil ▶ volatile (O-methoxyphenols or phenols):

Calami rhizoma:	3%–5% essential oil with asarone (0%–95%)
Caryophylli flos:	14%–20% essential oil with eugenol (90%)
Myristicae semen:	12%–16% essential oil with myristicin (6%)
Curcumae xanth. rhizoma:	6%–11% essential oil with xanthorrrhizol (5%)

▶ For TLC separation, description, constituents, formulae see Chap. 6

13.2 Drugs with Glucosinolates (Mustard Oils)

Glucosinolates are β-S-glucosides of isothiocyanates (ITC). They are non-volatile, water-soluble compounds, cleaved by the enzyme myrosinase, a β-thioglucosidase, when plant tissues are damaged to form isothiocyanates (mustard oils).

13.2.1 Preparation of Extracts

General method

Ground seeds (10 g) are added to 50 ml boiling methanol, boiled for 5 min and then allowed to stand for 1 h with occasional shaking. The filtrate is evaporated to 5 ml and then applied to a column (length, about 20 cm; diameter, about 1 cm) containing 5 g cellulose powder (cellulose MN 100, Machery and Nagel, Düren). The column is eluted with methanol and the first 20 ml eluate is discarded. The next 100 ml is collected and evaporated to about 1 ml at 20°–30°C under reduced pressure; 25 μl is used for chromatography.

13.2.2 Thin-Layer Chromatography and Detection Methods

Separation over silica gel 60 F_{254}-precoated TLC plates (Merck, Darmstadt) in the solvent system n-butanol-n-propanol-glacial acetic acid-water (30:10:10:10). The developed TLC plate is dried and sprayed with 25% trichloracetic acid in chloroform. After heating for 10 min at 140°C, the plate is sprayed with a 1:1 mixture of 1% aqueous potassium hexacyanoferrate and 5% aqueous $FeCl_3$ (TPF No. 38). Sinigrin and sinalbin turn blue (vis).

13.2.3 Drug List

Drug/plant source Family	Glucosinolates	Mustard oils
Sinapis nigrae semen Black mustard seeds Brassica nigra (L.) KOCH Brassicaceae DAC 86, ÖAB 90, Helv. VII, MD	1%–2% sinigrin (sinigroside/potassium myronate/potassium allyl glucosinolate) ► Sinapin (choline ester of 3,5-dimethoxy-4-hydroxycinnamic acid	Allylisothiocyanate
Sinapis albae semen (Erucae semen) White mustard seeds Sinapis alba L. Brassicaceae MD	2.5% Sinalbin (p-hydroxybenzoyl-glucosinolate) ► Sinapin (1.2%)	p-Hydroxybenzyliso-thiocyanate

13.3 Drugs with Cysteine sulphoxides and Thiosulphinates Allium sativum L., Allium ursinum L., Allium cepa L. – Alliaceae

Allium sativum and Allium ursinum preparations show a very similar qualitative composition of sulphur-containing compounds. Quantitative differences are known for alliin/allicin and other cysteine sulphoxides and thiosulphinates, respectively.
Alliin, the major compound in Allium sativum and A. ursinum, is unstable in water extracts. The enzyme alkylsulphinate lyase (allinase = allinlyase) splits alliin to allicin, which itself generates further sulphur-containing degradation or transformation products (e.g. ajoenes).
Allicin is absent in Allium cepa preparations. Onions contain cepaenes and different thiosulphinates in comparison to garlic and wild garlic.

13.3.1 Preparation of Extracts for TLC

1. Fresh plant bulbs are cut into small pieces. The fresh plant juice is obtained by pressing the pieces under pressure. The resulting juice is diluted with dichloromethane (1:10) and 20–40 µl is used for TLC investigations.
2. Freshly cut drug (5 g) is extracted with 20 ml distilled water by standing at room temperature for about 30 min; alternatively, one part of drug and 3.5 parts water can be homogenized for 5 min in a blender (e.g. Warring blender).

After 30 min, the extract is filtered and the clear solution extracted with 100 ml dichloromethane. The dichloromethane phase is separated, dried over Na_2SO_4 and evaporated (<40°C) to dryness. The residue is dissolved in 2 ml methanol, and 20–40 µl is used for TLC.

13.3.2 Thin-Layer Chromatography and Detection

Reference solutions

Standard compound: allicin, diallylsulphide, dipropyl- and dimethylsulphinate; 10 mg is dissolved in 2 ml dichloromethane; 15 µl is used for TLC.

Adsorbent

Silica gel 60 F_{254}-precoated TLC plates (Merck, Germany).

Chromatography solvent

- toluene-ethyl acetate (100:30)

Detection

- UV-254 nm: thiosulphinates and diallylsulphide show quenching
- Spray reagent (see Appendix A)
  - Palladium-II-chloride (PC No.32) → evaluation in vis.: yellow-brown zones
  - Vanillin-glacial acid reagent (VGA No.39) → evaluation in vis.: yellow, brown, blue and red zones

13.4 Formulae of Pungent Principles

Capsici fructus

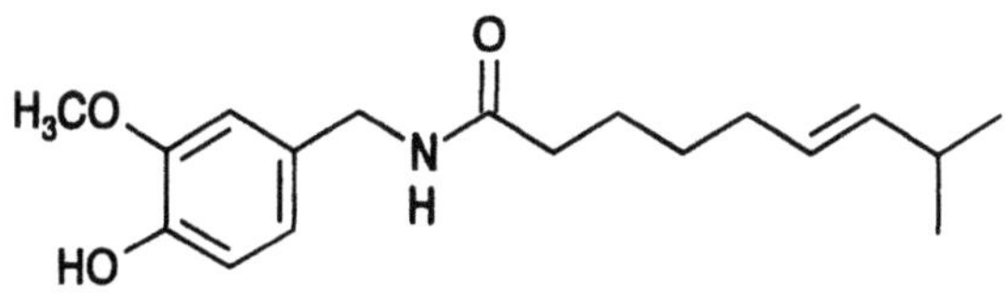

trans-Capsaicin

Piperis fructus

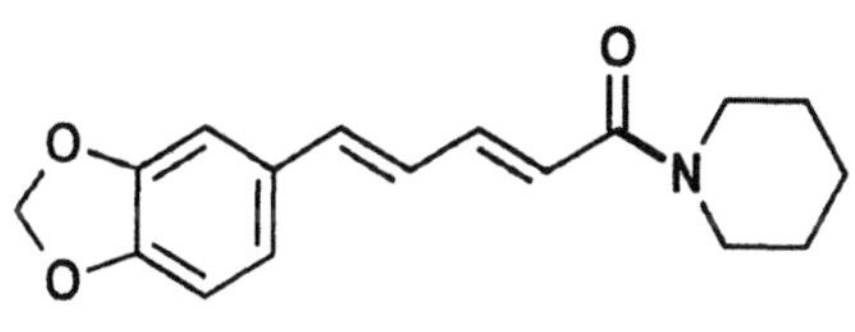

Piperin

Galangae rhizoma

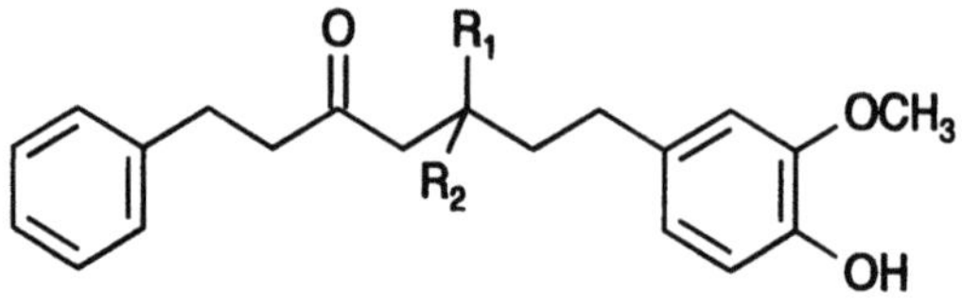

R_1 = H R_2 = OCH_3
R_1 = H R_2 = OH
R_1, R_2 } = O

Diarylheptanoids

Zingiberis rhizoma

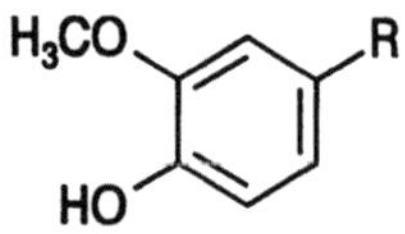

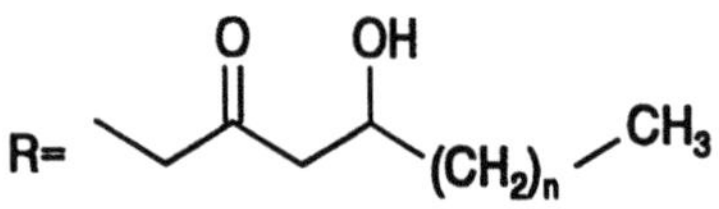

Gingerols

[6]-Gingerol	n = 4
[8]-Gingerol	n = 6
[10]-Gingerol	n = 8

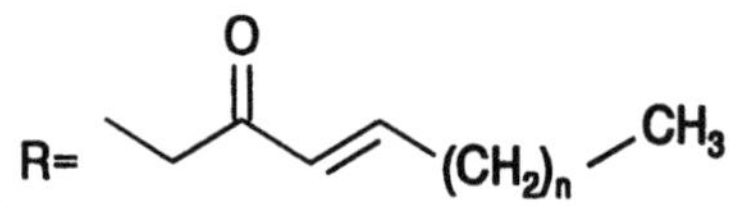

Shogaols

[6]-Shogaol	n = 4
[8]-Shogaol	n = 6
[10]-Shogaol	n = 8

Allium species

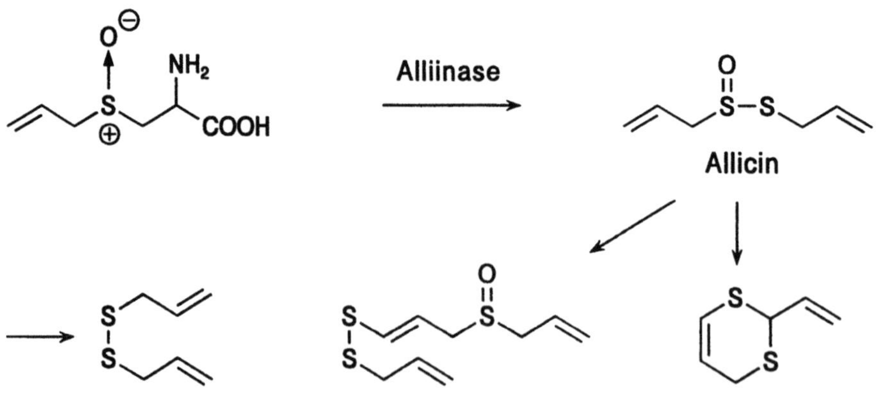

Diallyldisulfid   Ajoene   2-Vinyl-1,3-dithiin

Allium cepa

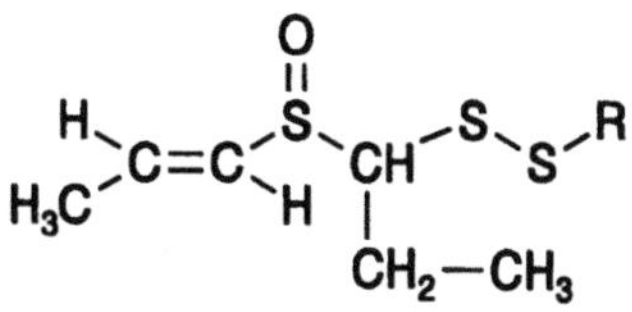

Cepaenes

R : $-CH_2-CH_2-CH_3$

R : $-CH=CH-CH_3$

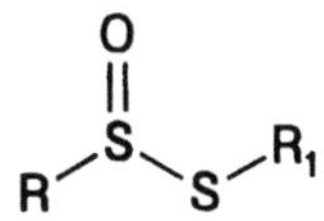

Thiosulphinates (TS)

R = R_1 : Dimethyl (TS)

R : 1-Propenyl R : 1-Propenyl

R_1 : Methyl R_1 : Propyl

Sinapis semen

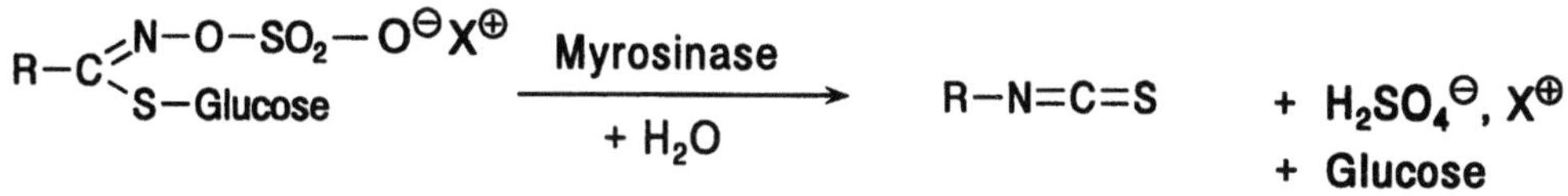

Glucosinolate Alkylisothiocyanate

Sinigrin : ($R = -CH_2-CH=CH_2$) Allyl mustard oil

Sinalbin : 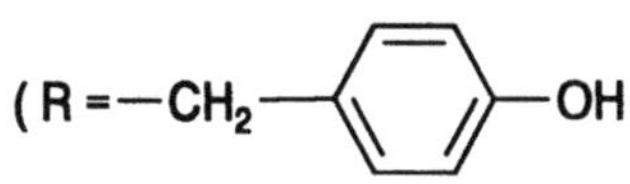 p-Hydroxybenzoyl mustard oil

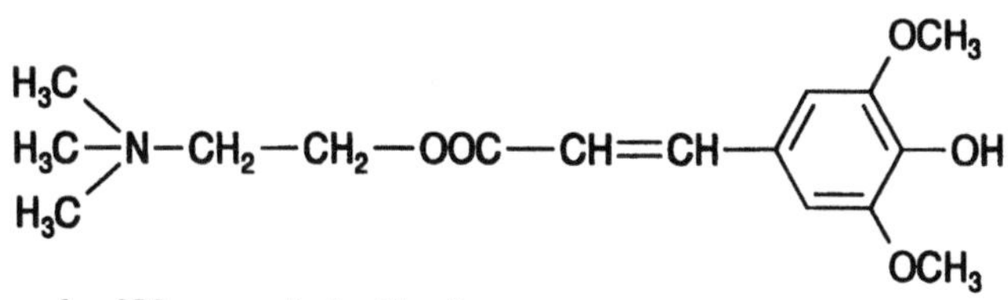

Sinapin (Sinapoylcholine)

13.5 Chromatograms

Capsici and Piperis fructus

Drug sample
1 Capsici acris fructus
2 Piperis nigri fructus
3 Piperis albi fructus (extracts, 10–20 µl)

Reference compound
T1 capsaicin
T2 piperine
T3 piperine-diperine mixture

Solvent system
Fig. 1 A,B toluene-ethyl acetate (70:3)
C toluene-diethyl ether-dioxane (62.5:21.5:16)

Detection
A; C Vanillin sulphuric acid reagent (VS No. 42) → vis.

Fig. 1A **Capsici fructus** (1) shows three weak blue-violet zones in the R_f 0.01–0.2 with capsaicin at R_f ~ 0.2 (T1), a violet-tailed band from R_f ~ 0.25–0.5 (triglycerides?) and two weak blue-violet zones of capsacinoides at R_f ~ 0.7–0.8 (see also Fig. 2A).

B **Piperis nigri fructus** (2) is characterized by the yellow piperine zone (T2) at R_f ~ 0.25. The essential oil compounds are seen as blue zones at R_f ~ 0.6 and at the solvent front.

C **Piperis albi** (2) and **Piperis nigri fructus** (3) both contain piperine at R_f ~ 0.5 (T1) and dipiperine at R_f ~ 0.7. Piperis fructus mainly differs in the contribution of minor compounds, such as piperyline (R_f ~ 0.15), piperettine, piperine isomers (R_f ~ 0.55) and piperaesthin A at R_f ~ 0.75. Different amounts of terpenes are detectable as violet-blue zones at the solvent front.

Capsici fructus, Sinapis semen

Drug sample
1 Capsici fructus (C. frutescens)
1a Capsici fructus (C. annum)
(extracts, 10–20 µl)
4 Sinapis albae semen
5 Sinapis nigri semen

Reference compound
T1 capsaicin T4 sinalbin (R_f ~ 0.1) T5 sinigrin

Solvent system
Fig. 2A diethyl ether (100)
B n-butanol-n-propanol-glacial acetic acid-water (30:10:10:10)

Detection
A dichloroquinone chloroimide (DCC No. 10) → vis
B trichloroacetic acid-hexacyanoferrate-$FeCl_3$ reagent (TPF No.35) → vis

Fig. 2A Development of **Capsicum extracts** (1,1a) in diethyl ether and reaction with the very sensitive and specific DCC reagent reveals the characteristic blue-violet zone of capsaicin at R_f ~ 0.35.

B After treatment with the TPF reagent, the Sinapis extracts show four blue zones (vis.) in the R_f range 0.1–0.5 and one at the solvent front.
Sinapis albae semen (4) is characterized by two main zones in the R_f range 0.3–0.5 and sinalbin at R_f ~ 0.1 (T4), while **Sinapis nigri semen** (5) shows its two major zones at R_f ~ 0.1 and 0.3 and the additional zone of sinigrin at R_f ~ 0.4 (T5).

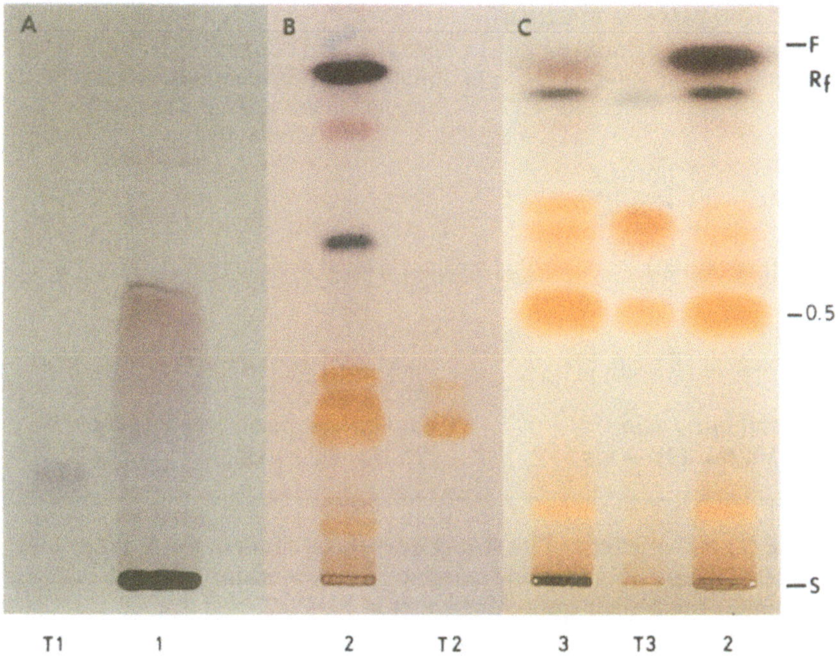

Fig. 1

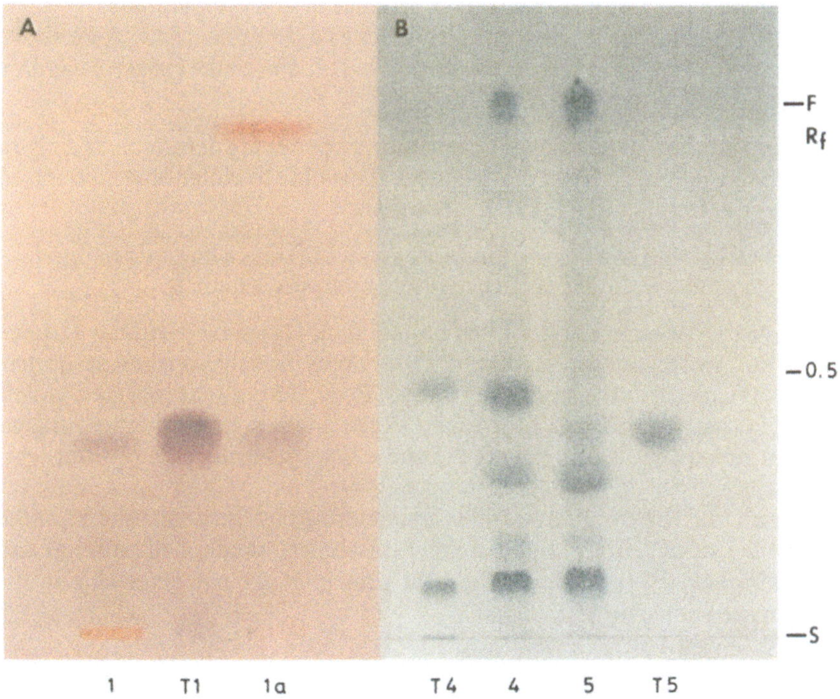

Fig. 2

Galangae and Zingiberis rhizoma

Drug sample	1 Galangae rhizoma (DCM extract) 2 Zingiberis rhizoma (DCM extract) (extracts, 20 µl)	1a Galangae aeth. (commercial oil) 2a Zingiberis aeth. (commercial oil)
Reference compound	T1 vanillin T3 borneol T2 capsaicin T4 cineol	
Solvent system	Fig. 3 n-hexane-ether (40:60)	Fig. 4 toluene-ethyl acetate (93:7)
Detection	Fig. 3A UV-254 nm B Bartons reagent (BT No. 5) → vis C Vanillin sulphuric acid reagent (VS No. 42) → vis	Fig. 4A UV-365 nm B Anisaldehyde-sulphuric acid reagent (AS No.2) → vis C Vanillin-sulphuric acid reagent (VS No. 42) → vis

Fig. 3A **Galangae rhizoma** (1) is characterized by strong quenching zones in the R_f range 0.25–0.4. The vanillin test (T1) serves as a guide compound for the major galangol. **Zingiberis rhizoma (2)** has weaker quenching zones at $R_f \sim 0.25$ and 0.5.

B With the Barton reagent **Galangae rhizoma** (1) shows two prominent dark-blue zones at $R_f \sim 0.15$ and 0.45 as well as four to five weak blue zones in between. They represent the pungent principles such as galangols, a complex mixture of diarylheptanoids. **Zingiberis rhizoma** (2) develops its pungent principles as a prominent blue zone above the start (R_f range of the capsaicin test, T2) and at $R_f \sim 0.2$, due to the gingerols and/or shogaols, the corresponding anhydro compounds.

C With VS reagent in addition to the pungent principles, terpenes are detectable mainly in the R_f range 0.6 up to the solvent front. Both classes show blue to violet-blue colours. In 1, further yellow zones in the lower R_f range are found.

Fig. 4A **Galangae rhizoma** (1,1a). In UV-365 nm, the DCM extract (1) and a commercially available essential oil (1a) show a similar sequence of blue fluorescent zones.

B With AS reagent, the DCM-extract (1) and commercial oil (1a) show a similar terpene pattern of brown and violet zones in the R_f range 0.45 up to the solvent front: sesquiterpene hydrocarbons (violet/front), ester zones (brown $R_f \sim 0.75$), 1,8-cineole (red-violet/T4). A different TLC pattern of 1 and 1a is found in the lower R_f range due to pungent principles, present only in the DCM extract (1) and e.g. terpene alcohols (T3/borneol) in (1a).

C **Zingiberis rhizoma** (2a). The commercial oil is characterized by the high amount of the blue THC zone at the solvent front (zingiberene, β-bisabolen, sesquiphellandrene) and the blue zones at R_f 0.45–0.5 (e.g. citral). Pungent principles are not detectable in the essential oil (compare with sample 2, Fig. 3C).

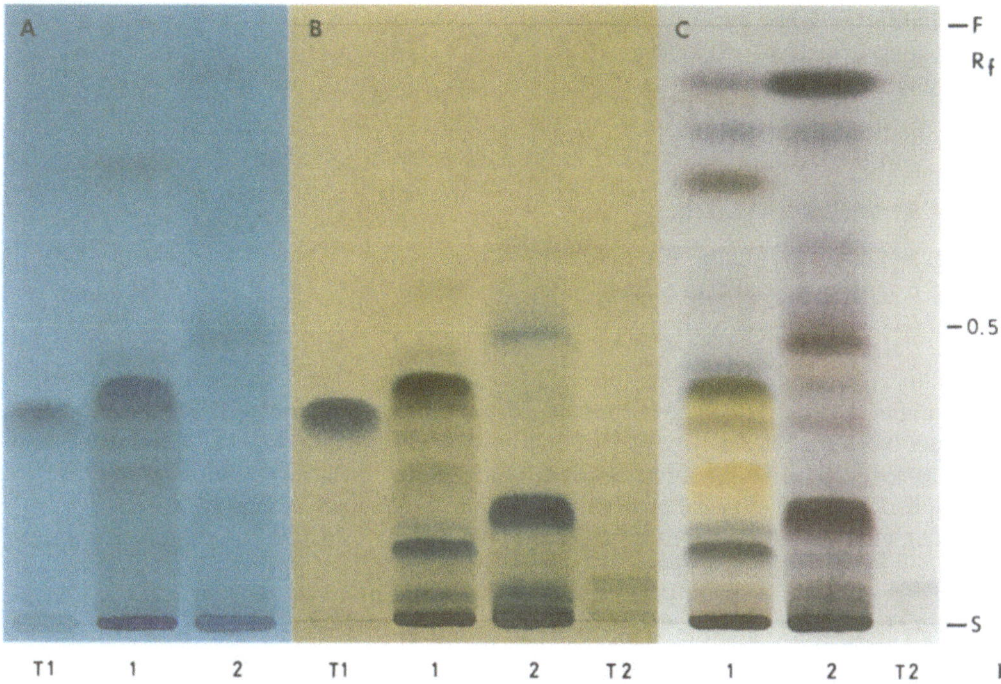

Fig. 3

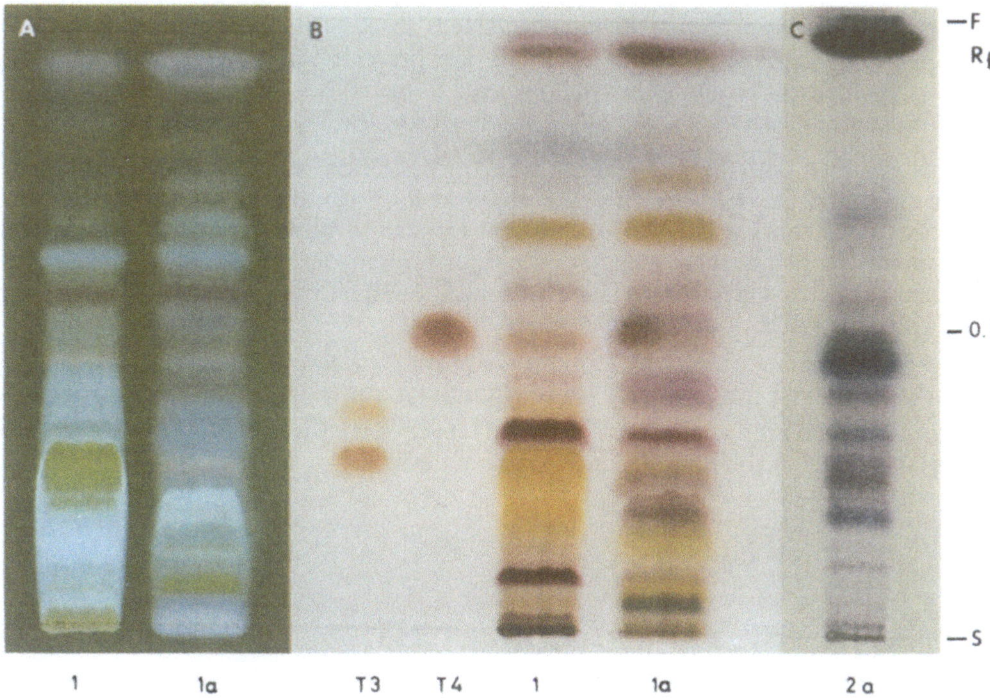

Fig. 4

Allium species

Drug sample 1 Allium sativum - garlic
2 Allium ursinum - wild garlic
3 Allium cepa -onion (dichloromethane extract, see Sect. 13.2.1)

Reference compound T1 diallylsulphide T2 allicin
T3 dipropylthiosulphinate T4 dimethylthiosulphinate

Solvent system Fig. 5,6 toluene-ethyl acetate (100:30)

Detection Fig. 5 palladium-II-chloride reagent (PC No. 32) → vis
Fig. 6 vanillin-glacial acid reagent (VGA No. 42) → vis

Fig. 5 DCM extracts of fresh bulb samples of **Allium sativum** (1) and **Allium ursinum** (2) show a similar qualitative pattern of four yellow-brown thiosulphinate (TS) zones in the R_f range 0.2–0.45. The diallylthiosulphinate allicin (T2) at $R_f \sim 0.45$ is the major compound in sample 1, while in sample 2 the zone of allicin and three yellow-brown zones at $R_f \sim$ 0.3, 0.25 and 0.20 (T4) are present in almost equal concentration.
Allylmethyl (AMTS), methylallyl (MATS) and diallylthio (DATS) sulphinates are reported as the main compounds in Allium ursinum (2).
Additional zones at the solvent front are due to sulphides, such as diallysulphide; brown zones at $R_f \sim 0.05$ are due to degradation products (see allicin test).

Freshly prepared extracts of **Allium cepa** (3) show five to seven dark-brown zones in the R_f range 0.2–0.65 with two prominent zones of thiosulphinates at $R_f \sim 0.3$ and $R_f \sim 0.45$. The dipropylthiosulphinate (T3) at $R_f \sim 0.45$ is the characteristic compound of onion extracts. Allicin with almost the same R_f value is absent. Other thiosulphinates such as dimethylthiosulphinate (T4) at $R_f \sim 0.2$ are present, which in contrast to garlic thiosulphinates (TS) show brown to brown-red colours (vis.). This is partly due to higher TS concentrations and to compounds which overlap the TS, as shown in Fig. 6.

Fig. 6 After treatment with the VGA reagent, the extract of **Allium cepa (3)** is distinguishable by the characteristic violet-brown major zones at $R_f \sim 0.3$ and at $R_f \sim 0.45$, with less concentrated zones at $R_f \sim 0.6$–0.8, whereas extracts of **Allium sativum (1)** and **Allium ursinum** (2) mainly show grey, grey-violet or brown zones in the R_f range 0.2–0.55. Allicin is seen as a grey-brown-coloured zone, the sulphides at the solvent front as blue to grey-blue zones.

The TLC pattern of various drug samples can vary according to the extraction methods. Stored powdered drug samples of Allium species can contain more degradation or transformation products such as ajoens and cepaenes, shown as yellow-brown (Fig. 5) or grey-blue (Fig. 6) zones in the low R_f range of the TLC.

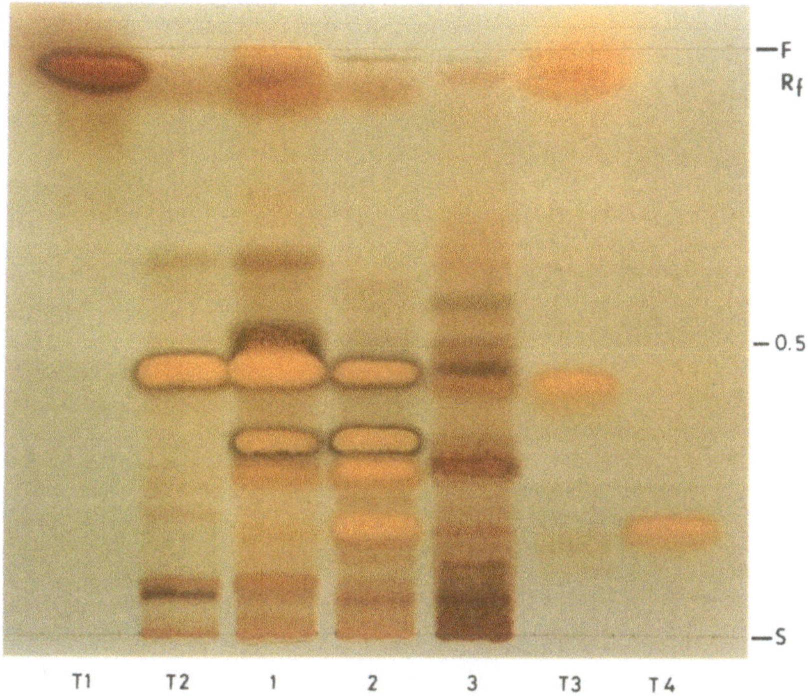

Fig. 5

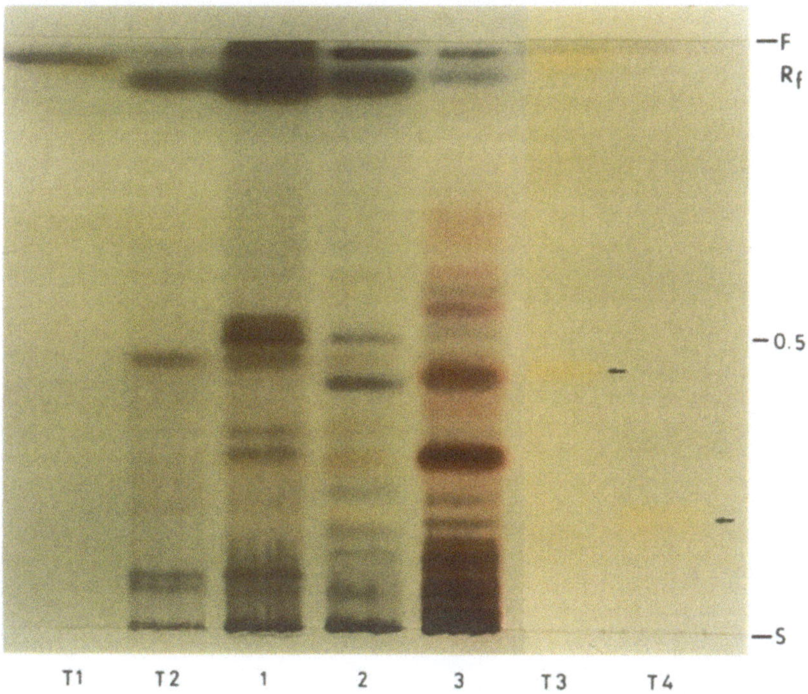

Fig. 6

14 Saponin Drugs

Most of the saponins of official saponin drugs are triterpene glycosides. Some drugs also or only contain steroid saponins.
Sugar residues may be linked via the OH group at C-3-OH of the aglycone (monodesmosidic saponins) or more rarely via two OH groups or a single OH group and a carboxyl group of the aglycone moiety (bisdesmosidic saponins).

Triterpene saponins. These saponins possess the oleanane or, more rarely, the ursane or dammarane ring system. Many have acidic properties, due to the presence of one or two carboxyl groups in the aglycone and/or sugar moiety. Other oxygen-containing groups may also be present in the sapogenin, e.g. -OH, $-CH_2OH$ or -CHO.
The carbohydrate moiety usually contains one to six monosaccharide units, the most common of these being glucose, galactose, rhamnose, arabinose, fucose, xylose, glucuronic and galacturonic acid. The horse chestnut saponins are partly esterified with aliphatic acids.
Most triterpene saponins possess hemolytic activity, which varies from strong to weak, depending on the substitution pattern.

Steroid saponins. The sapogenins of the steroid saponins are mostly spirostanols. Furostanol derivatives are usually converted into spirostanols during isolation procedures: these sapogenins do not carry carboxyl groups. Steroid saponins possess less sugar units than the triterpene saponins. In contrast to the monodesmosides, the bisdesmosidic furostanol glycosides exert no hemolytic activity.

14.1 Preparation of Extracts

General method
Powdered drug (2 g) is extracted by heating for 10 min under reflux with 10 ml 70% ethanol. The filtrate is evaporated to about 5 ml, and 20–40 µl of this solution is used for TLC.

Enrichment
A total of 3 ml of the ethanolic extract (see above) is shaken several times with 5 ml water-saturated n-butanol. The n-butanol phase is separated and concentrated to about 1 ml; 20 µl is used for TLC.

Exceptions
Ginseng radix is extracted under the same conditions (see above), but with 90% ethanol.
Liquiritiae radix: an ethanolic extract (see above) is evaporated to dryness, and the residue is dissolved in 2.0 ml chloroform-methanol (1:1); 20 µl is used for the detection of glycyrrhizin.

Hydrolysis of glycyrrhizin Powdered drug (2 g) is heated under reflux for 1 h with 30 ml 0.5 *M* sulphuric acid. The filtrate is shaken twice with 20-ml quantities of chloroform. The combined chloroform extracts are dried over anhydrous sodium sulphate, filtered and evaporated to dryness. The residue is dissolved in 2.0 ml chloroform-methanol (1:1), and 10 µl of this solution is used for the detection of glycyrrhetic acid.

14.2 Thin-Layer Chromatography

Reference solutions The commercially available reference compounds such as aescin, primula acid, glycyrrhizin and standard saponin are each prepared as a 0.1% solution in methanol; 10 µl is used for TLC.

Adsorbent Silica gel 60 F_{254}-precoated TLC plates (Merck, Darmstadt).

Chromatography solvents

- Chloroform-glacial acetic acid-methanol-water (64:32:12:8)
  This system is suitable for separation of the saponin mixtures from the listed drugs.
- Chloroform-methanol-water (70:30:4) ▶ ginsenosides (Ginseng radix)
- Ethyl acetate-ethanol-water-ammonia (65:25:9:1) ▶ glycyrrhetic acid (Liquiritiae radix).

14.3 Detection

- Without chemical treatment
  With the exception of glycyrrhizin and glycyrrhetic acid (Liquiritiae radix), no saponins are detectable by exposure to UV-254 or UV-365 nm.
- Spray reagents (see Appendix A)

- Blood reagent (BL No. 8)
  Hemolytically active saponins are detected as white zones on a reddish background. Hemolysis may occur immediately, after allowing the TLC plate to stand or after drying the plate in a warm airstream.
- Vanillin-sulphuric acid reagent (VS No. 42)
  Evaluation in vis.: saponins form mainly blue, blue-violet and sometimes red or yellow-brown zones (vis).
- Anisaldehyde-sulphuric acid reagent (AS No. 3)
  Evaluation in vis.: colours are similar to those with VS reagent; inspection under UV-365 nm light results in blue, violet and green fluorescent zones.
- Vanillin-phosphoric acid reagent (VPA No. 41)
  Ginsenosides give red-violet colours in vis. and reddish or blue fluorescence in UV-365 nm.

14.4 **Drug List** (in alphabetical order)

With the exception of Avenae sativae herba, Rusci aculeati radix, Sarsaparillae radix (steroid saponins) most saponin drugs contain a complex mixture of monodesmosidic or bidesmosidic triterpene glycosides.

Drug/plant source Family/pharmacopoeia	Main constitutents Hemolytic index (HI)	
Avenae sativae herba Avenae sativae fructus (excorticatus) Oat, Kernel Avena sativa L. Poaceae MD, BHP 83	Steroid saponins: avenacosides A and B with nuatigenin as aglycone and glucose and rhamnose as sugars Avenacoside A: 0.025% semen, 0.3% herba Avenacoside B: 0.015% semen, 1.3% herba Triterpene saponins: e.g. avenarin 3%–4% free sugars: fructose, glucose Flavonoids: vitexin-2″-O-rhamnoside, isoorientin-2″-O-arabinoside	**Fig. 9**
Centellae herba Indian pennywort Centella asiatica (L.) URBAN (syn. Hydrocotyle asiatica) Apiaceae MD, BHP 83	Ester saponins (monodesmosidic acyl-glycosides) derived from asiatic and 6-hydroxy asiatic acid, betulinic and terminolic acid Asiaticoside A and B, "madecassoside" (=mixture of A and B)	**Fig. 8**
Ginseng radix Ginseng root Panax ginseng MEY. and other Panax ssp. Araliaceae DAB 10, ÖAB, Helv. VII, MD, Jap XI, Chin PIX	2%–3% tetracyclic triterpene glycosides ginsenosides R*x* ($x = a, b_1, b_2, c, d, e, f, g_1, h$) derived from 20-S-proto-panaxadiol and 20-S-proto-panaxatriol (dammarane ring system; neutral bisdesmosides) Ginsenoside R_0: oleanolic acid as aglycone The glycosides contain glucose, arabinose, rhamnose and glucuronic acid HI (drug) < 1000	**Fig. 1,2**
Hederae folium Ivy leaves Hedera helix L. Araliaceae	4%–5% hederasaponins (A–J) oleanolic and 28-hydroxy oleanolic acid glycosides 1.7%–4.8% hederacoside C, 0.1%–0.2% hederacoside B as neutral bisdesmoside, 0.4%–0.8% hederasaponin D hederagenin-arabinoside In dried material 0.1%–0.3% α-hederin and β-hederin as acid monodesmosides HI (drug) 1000–1500, HI (β-hederin) 15000 ▶ Flavonoids: rutin, kaempferol-rhamno-glucosid ▶ Chlorogenic and isochlorogenic acid ▶ Coumarins: scopoletin-7-0-glucoside	**Fig. 5,6**

	Drug/plant source Family/pharmacopoeia	Main constitutents Hemolytic index (HI)
Fig. 5,6	**Hederae terrestris herba** Glechomae hederaceae herba Ground ivy (Nepeta hederacea) Glechoma hederacea L. Lamiaceae	0.5%–0.7% triterpene glycosides α-, β-ursolic acids, oleanolic acid ► caffeic acid derivatives, chlorogenic acid ► Flavonoids: luteolin-7-O-glucoside and 7-O-glucobioside, hyperoside, isoquercitrin
Fig. 1,2,3	**Hippocastani semen** Horse chestnut seeds Aesculus hippocastanum L. Hippocastanaceae DAB 10, MD	3%–6% pentacyclic triterpene glycosides Aescins: a complex mixture of diesters of penta- and hexahydroxy-β-amyrine, linked to glucuronic acid and glucose and esterified with angelica, tiglic, α-butyric or isobutyric acetic acids; (aglycone protoaescigenin and barringtogenol C) (<15%) β-aescin: C-21 and C-23 ester; kryptoaescin: C-21 and C-23 ester; α-aescin: mixture of β-aescin+kryptoaescin (4:6) Aescinols (e.g. aescinol-21, 22, 28-triol derivatives) are artefacts, due to hydrolysis of aescins HI (drug) < 6000; HI (aescin) 9500–12 500 ► 0.3% Flavonoids (biosides and triosides of quercetin and kaempferol)
Fig. 10	**Liquiritiae radix** Licorice root (peeled/unpeeled) Glycyrrhiza glabra L. Fabaceae DAB 10, DAC 86, Ph.Eur. II, ÖAB 90, Helv. VII, MD, USP XXI Jap XI, Chin PIX	Saponins: 8%–12% glycyrrhizin calcium salt of glycyrrhizic acid no hemolytic activity; the aglycone glycyrrhetic acid is active. ► Flavonoids: 1%–1.5% with liquiritin (4′, 7-dihydroxy-flavanone-7-O-glucoside) as main compound, the corresponding chalcone; isoflavone (glabridin) HI (drug) 250–300
Fig. 3	**Primulae radix** Primrose root Primula elatior (L.) HILL. Primula veris L. Primulaceae DAB 10, ÖAB 90, MD	5%–10% Tetra- and pentacyclic triterpene glycosides (monodesmosidic) Primula acids: 1%–6.5% primula saponin 1 (~90% primula acid A, a protoprimulagenin-A-penta glycoside); 1.9%–4.5% primula saponin 2 (a protoprimula-genin-A-tetra glycoside); Glycosides derived from priverogenin A-16- and -B-22-acetate HI 2500–5000 Phenolglycosides: primulaverin 0.2%–2.3%, primverine 0.4%–2.2%.

Drug/plant source Family/pharmacopoeia	Main constitutents Hemolytic index (HI)	
Quillajae cortex Quillaja bark Soap bark Quillaja saponaria MOLINA Rosaceae DAC 86,ÖAB, Helv. VII, MD	9%–10% "Quillaja saponin", a complex mixture derived from 16-α-hydroxygypsogenin (quillaic acid) with glucuronic or galacturonic acid as parts of the sugar moiety HI 3500–4500	**Fig. 4**
Rusci aculeati rhizoma Butcher's broom Ruscus aculeatus L. Asparagaceae MD	0.5%–1.5% Steroid saponins Aglycone neoruscogenin and ruscogenin. Neoruscogenin glycosides: monodesmosidic spirostanol type: ruscin (trioside), desglucoruscin, desglucorhamnoruscin (=neo-ruscogenin-1-arabinoside); bisdesmosidic furostanol type: ruscoside (1-OH-trioside/-26-OH-glucoside), desglucoruscoside Ruscogenin glycosides (1-β-hydroxydiosgenin): present in low concentrations only	**Fig. 7,8**
Saponariae radix S. rubrae radix Red soapwood root Saponaria officinalis L. Caryophyllaceae	3%–5% bisdesmosidic triterpene saponins Saponin mixture derived from gypsogenin with saponaside A and D (two branched sugar chains with five monoses) HI 1200–2000	**Fig. 3**
Saponariae albae radix White soapwood root Gypsophila paniculata L., and other G. species Caryophyllaceae MD	15%–20% triterpene saponins G. paniculata: gypsoside A (with nine sugars) as a main compound HI 2600–3900	
Sarsaparillae radix Sarsaparilla, Sarsa Smilax regelii KILLIP et MORTON (Honduras drug) S. aristolochiifolia MILL. (veracruz drug) Liliaceae/Smilaceae MD	1.8%–3% steroid saponins ("smilax saponins"=bisdesmosidic furostanol saponins sarsaparilloside, easily cleaved (enzymes/H^+) into parillin, a monodesmosidic spirostanol saponin Aglycone: sarsapogenin (=parigenin) and its isomer, smilagenin HI 3500–4200	**Fig. 1**

Drug/plant source Family/pharmacopoeia	Main constitutents Hemolytic index (HI)
Fig. 1 **Senegae radix** Polygalae radix Milkwort root Polygala senega L. or Polygala senega var. latifolia TORR et GREY Polygalaceae DAB 9, ÖAB, BP 88; MD, Japan	6%–10% triterpene ester saponins ("senegines"); presenegine as aglycone; senegine II as main saponin (bisdesmosidic) with glucose, galactose, rhamnose, xylose and fucose; the fucose is esterified with 3,4-dimethoxy cinnamic acid; senegine IV: fucose is esterified with 4- methoxy-cinnamic acid and contains additional rhamnose units; senegine III=desrhamnosyl senegine IV HI 2500–4500

Note:
Herniariae herba is characterized on the basis of saponins, flavonoids and coumarins; these chromatograms are shown in Chap. 5.

Betulae folium and **Verbasci flos**, which contain saponins (betulin, verbascosaponin), with little or no hemolytic activity and **Equiseti herba** ("equisetonin") are easily identified on the basis of their flavonoid compounds; the relevant chromatograms are reproduced in Chap. 7.

14.5 Formulae

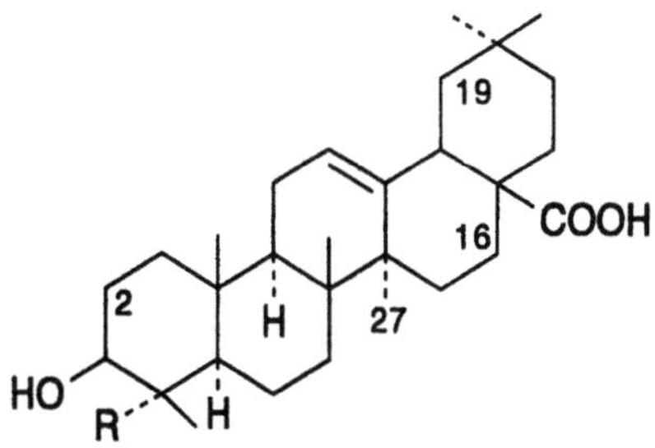

R	OH	Aglycon
CH_3	-	Oleanolic acid
CH_3	16α	Echinocystic acid
CH_2OH	-	Hederagenin
CHO	16α	Quillaic acid

Primulae radix

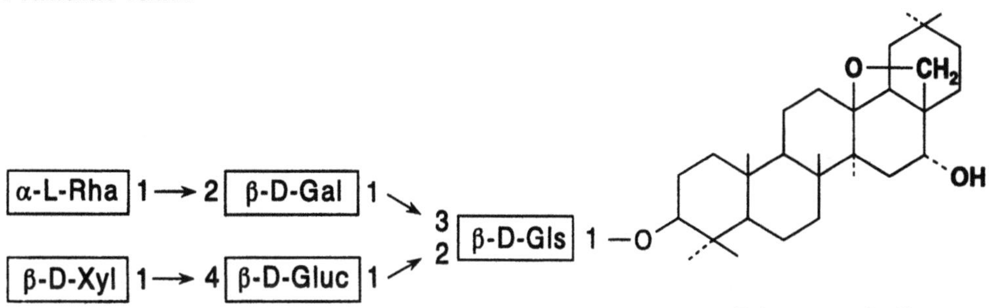

Priverogenin B

Primula acid I

Ginseng radix

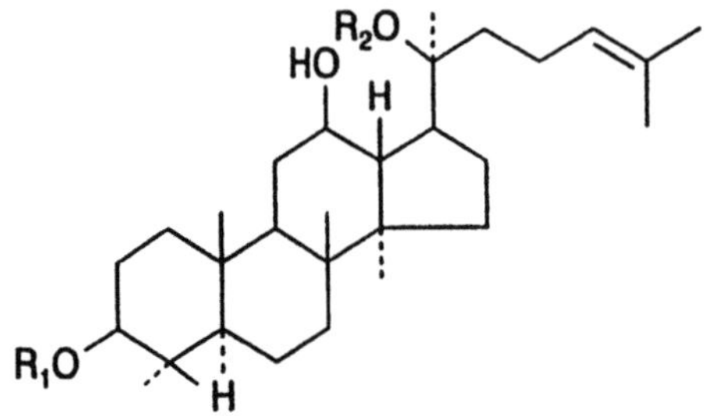

A	R_1			R_2	
(20 S – Protopanaxadiol)	H			H	
Ginsenoside Rb_1	β-D-Glu (1→2)	β-D-Glu	β-D-Glu	(1→6)	β-D-Glu
Ginsenoside Rb_2	β-D-Glu (1→2)	β-D-Glu	α-L-Ara*p*	(1→6)	β-D-Glu
Ginsenoside Rc	β-D-Glu (1→2)	β-D-Glu	α-L-Ara*f*	(1→6)	β-D-Glu
Ginsenoside Rd	β-D-Glu (1→2)	β-D-Glu	β-D-Glu		

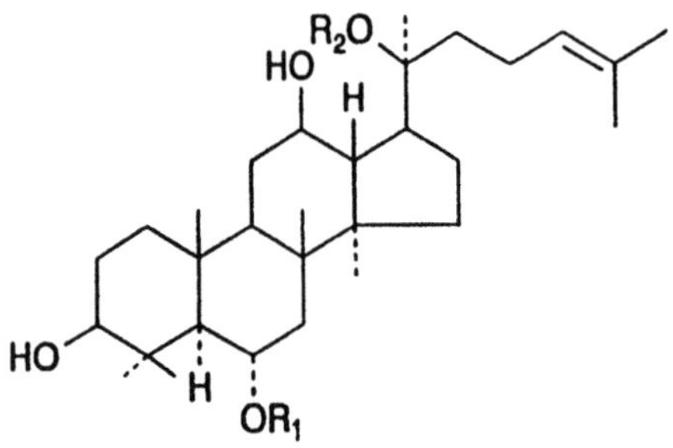

B	R_1		R_2
(20 S-Protopanaxatriol)	H		H
Ginsenoside Re	α-L-Rha (1→2)	β-D-Glu	β-D-Glu
Ginsenoside Rf	β-D-Glu (1→2)	β-D-Glu	H
Ginsenoside Rg_1	β-D-Glu		β-D-Glu
Ginsenoside Rg_2	α-L-Rha (1→2)	β-D-Glu	H
Ginsenoside Rh_1	β-D-Glu		H

Hederae folium

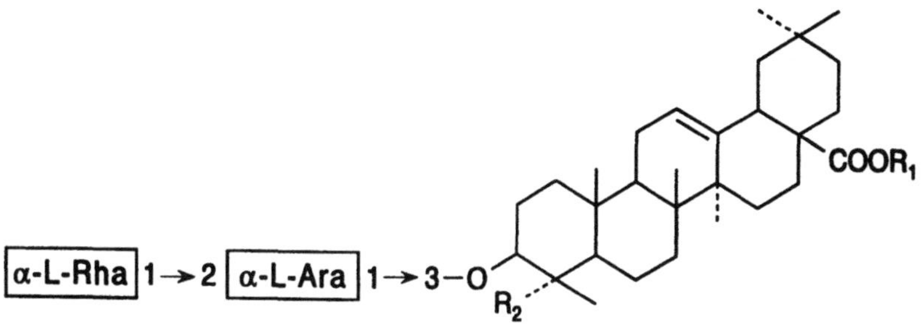

Hederacoside C (Hederasaponin C)	R_1 = ←1) β-D-Glu (6 ← 1) β-D-Glu (4 ← 1) –α-L-Rha	R_2 = $-CH_2OH$
α-Hederin	R_1 = –H	R_2 = $-CH_2OH$
Hederacoside B (Hederasaponin B)	R_1 = ←1) β-D-Glu (6 ← 1) β-D-Glu (4 ← 1) –α-L-Rha	R_2 = $-CH_3$

Hippocastani semen

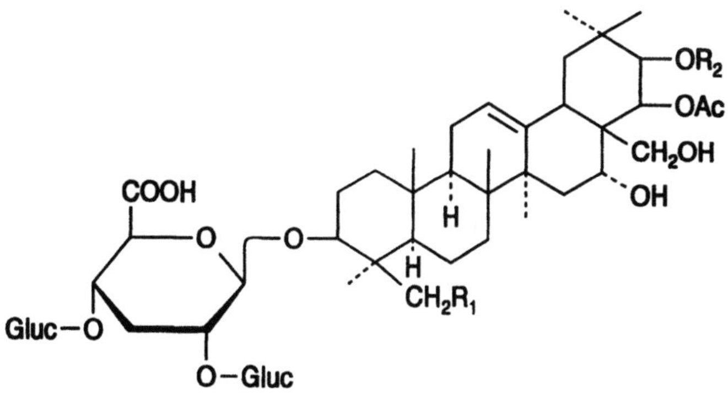

"Escin" [Aescin]

R_1 = OH Aglycone: Barringtogenol C
R_1 = H Aglycone: Protoaescigenin
R_2 = Tigloyl-, Angelicoyl-, 2-Methylbutyryl- or Isobutyryl-

Centellae herba

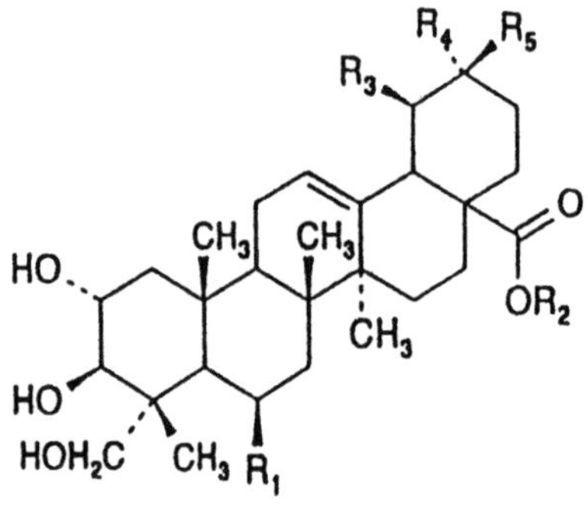

	R_1	R_2	R_3	R_4	R_5
Asiatic acid	-H	-H	$-CH_3$	$-CH_3$	-H
Madecass(ic) acid	-OH	-H	$-CH_3$	$-CH_3$	-H
Asiaticoside	-H	→1)-β-Gluc-(6→1)-β-D-Gluc-(4→1)-α-L-Rha	$-CH_3$	$-CH_3$	-H
Asiaticoside A	-OH	→1)-β-Gluc-(6→1)-β-D-Gluc-(4→1)-α-L-Rha	$-CH_3$	$-CH_3$	-H
(Terminolic acid	-OH	-H	-H	$-CH_3$	$-CH_3$)
(Asiaticoside B	-OH	→1)-β-Gluc-(6→1)-β-D-Gluc-(4→1)-α-L-Rha	-H	$-CH_3$	$-CH_3$)

Liquiritiae radix

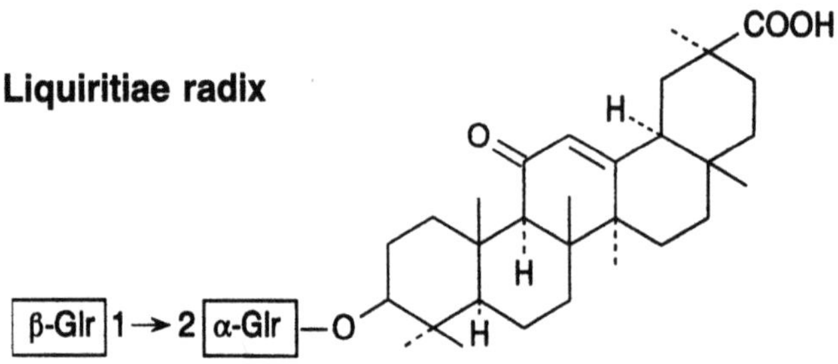

Glycyrrhizic acid, Glycyrrhizin

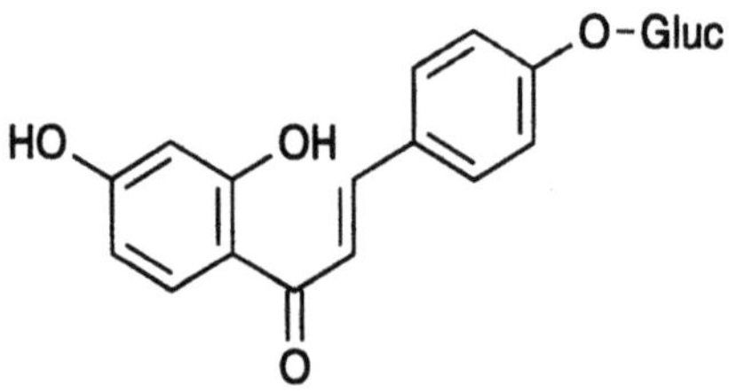

Isoliquiritin

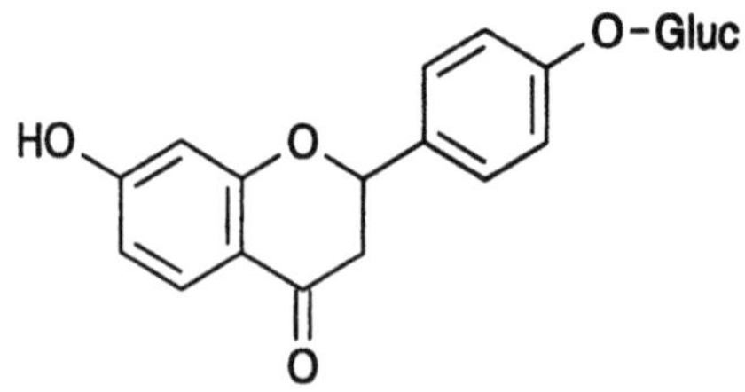

Liquiritin

Saponariae radix

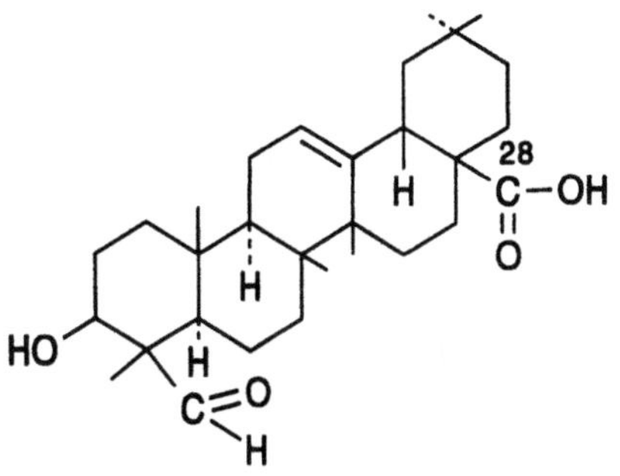

Gypsogenin

Gypsoside A

Senegae radix

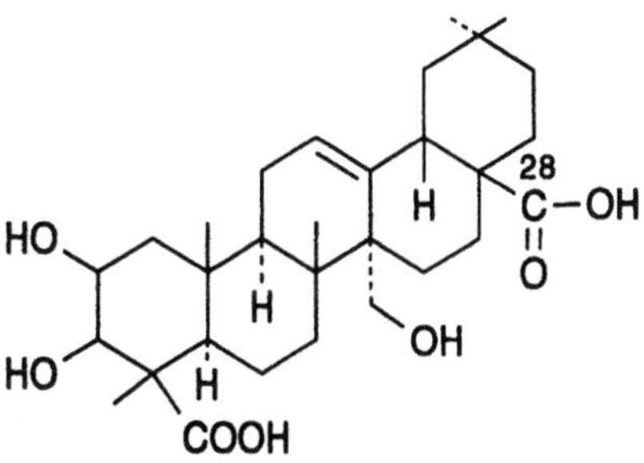

Presenegin

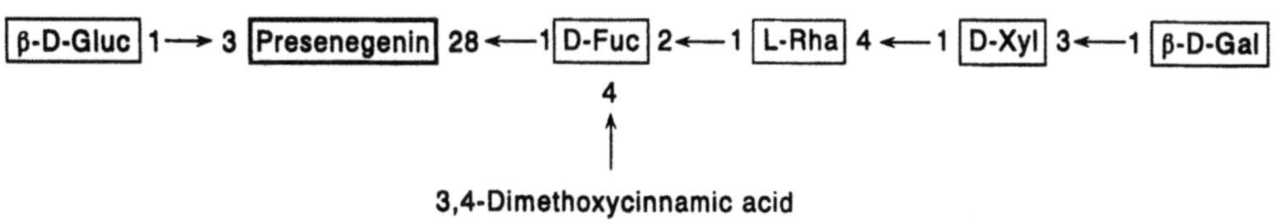

Senegin II

Rusci aculeati rhizoma

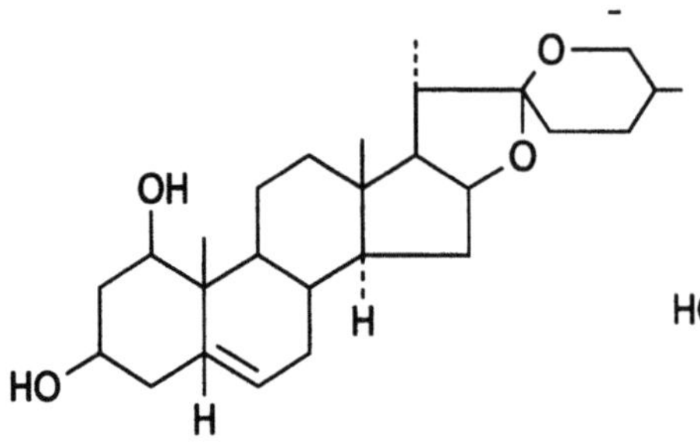

Ruscogenin

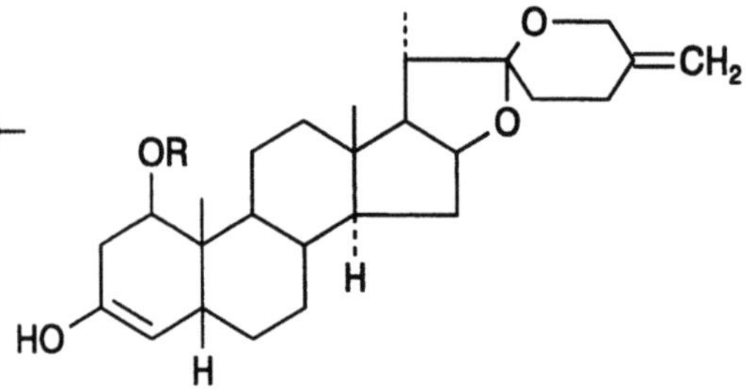

Neoruscogenin R = H

Ruscin R = Oβ-D Gluc-(1→3)-O-α-L-Rha-(1→2)-O-α-L-Ara(1→)

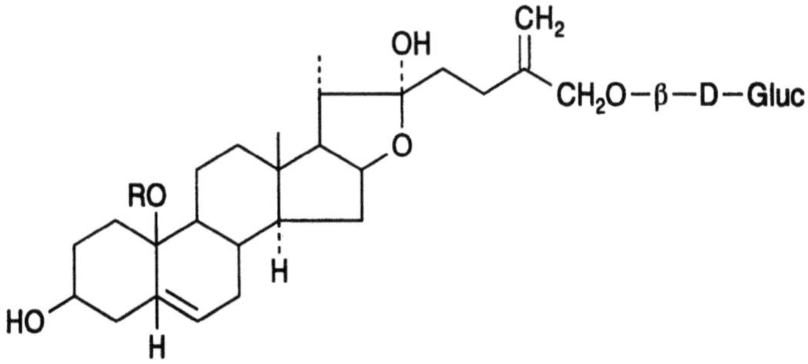

Ruscoside R = O-β-D-Gluc-(1→3)-O-α-L-Rha-(1→2)-O-α-L-Ara(1→)

Furosta -5,25 (27)-dien-26-glucopyranosyloxy-1β,3β,22α-triol R = H

Sarsaparillae radix

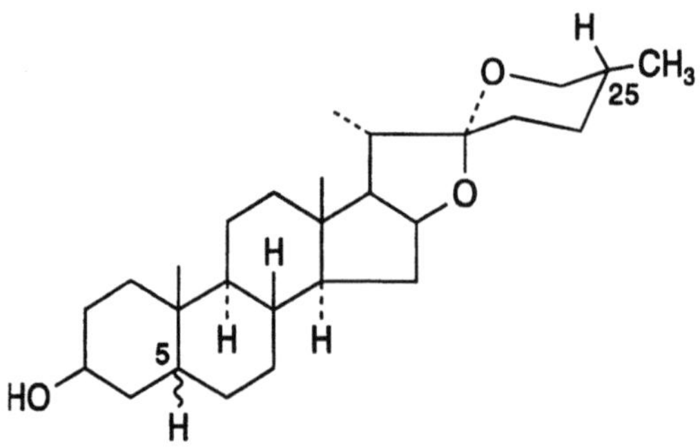

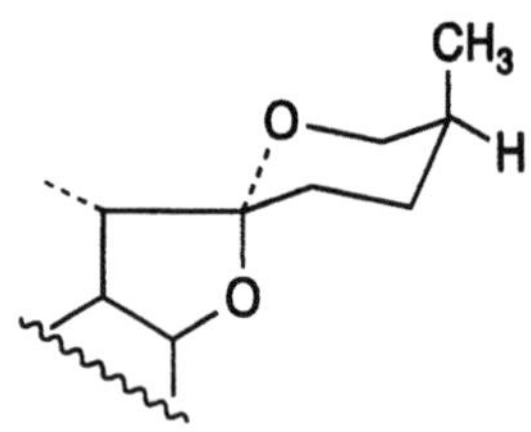

Smilagenin (5β, 25α) Sarsapogenin (5β, 25β)

Sarsaparillae radix

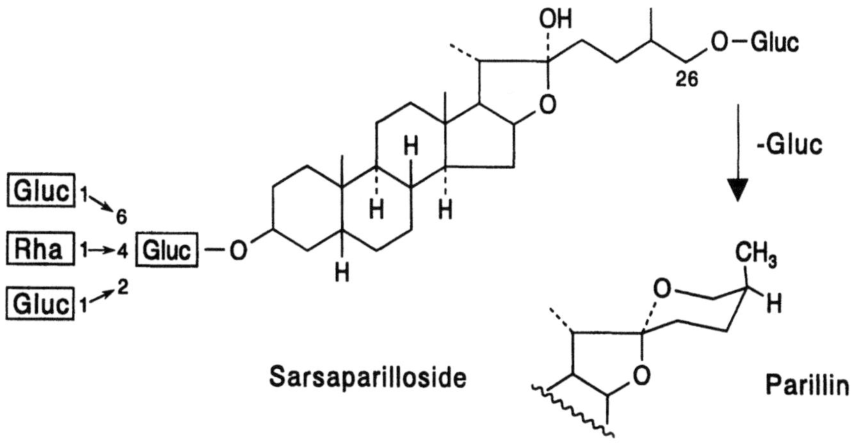

Avenae sativae herba/fructus

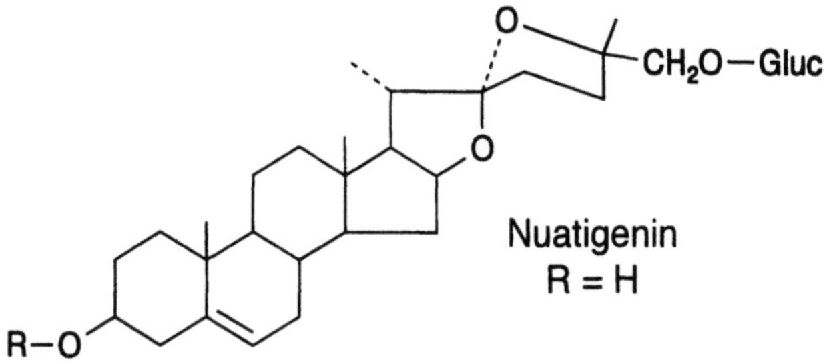

	R:
Avenacoside A	-O-β-D-gluc-(4←1) rham (2←1) gluc
Avenacoside B	-O-β-D-gluc-(4←1) rham (2←1) gluc-(3←1)-gluc

14.6 Chromatograms

TLC Synopsis of Saponin Drugs

Drug sample 1 Senegae radix 3 Ginseng radix
2 Sarsaparillae radix 4 Hippocastani semen (ethanolic extracts, 20 µl)

Reference compound T1 aescin

Solvent system Fig. 1 chloroform-glacial acetic acid-methanol-water (60:32:12:8)

Detection A Anisaldehyde-sulphuric acid reagent (AS No. 3) → vis
B Blood reagent (BL No.8) → vis.

Fig. 1 This solvent system and the AS reagent are suitable for the separation and the detection of triterpene-saponins, e.g. senegins in Senegae radix (1), as well as steroid (ester) saponins, e.g. Smilax saponins in Sarsaparillae radix (2).

A **Senegae radix** (1) is characterized by four to five red saponin zones (R_f 0.2–0.4) with senegin II as the major compound at $R_f \sim 0.2$.
Sarsaparillae radix (2) generates six yellow-brown saponin zones (R_f 0.2–0.75) such as sarsaparilloside and parillin.
Ginseng radix (3) shows eight grey-blue zones of ginseng saponins in the R_f range 0.2–0.65 (→see also Fig. 2)
Hippocastani semen (4) is characterized by the violet-black zones of aescins (T1) in the R_f range 0.4–0.5 (→see also Fig. 3).

B All saponins of the drug samples 1–4 show white hemolysis zones on a red-brown background with the blood reagent.

Ginseng radix

Drug sample G = Ginseng radix (ethanolic extract, 20 µl)

Reference compound 1 = Rc 3 = Rb_1 5 = Re 7 = Rh_1
2 = Rb_2 4 = Rd 6 = Rg_1

Solvent system Fig. 2 chloroform-methanol-water (70:30:4)

Detection Vanillin-phosphoric acid reagent (VPA No. 41A)
A → vis. B → UV-365 nm

Fig. 2A,B **Ginseng radix** (G) is characterized by the ginsenosides Rb_1, Rb_2, Rc, Rd, Re, Rg′. With VS reagent they form red zones (vis./→A) and prominent red fluorescent zones in UV-365 nm (→B). The ginsenosides Rb_1, Rb_2, Rc (1–3) with four to five sugar units appear in the lower R_f range 0.05–0.15, the less polar Rd, Re (4,5) at $R_f \sim 0.25$, Rg' (6) with two sugars is found at R_f 0.4 and Rh' (7) with one sugar at R_f 0.55.

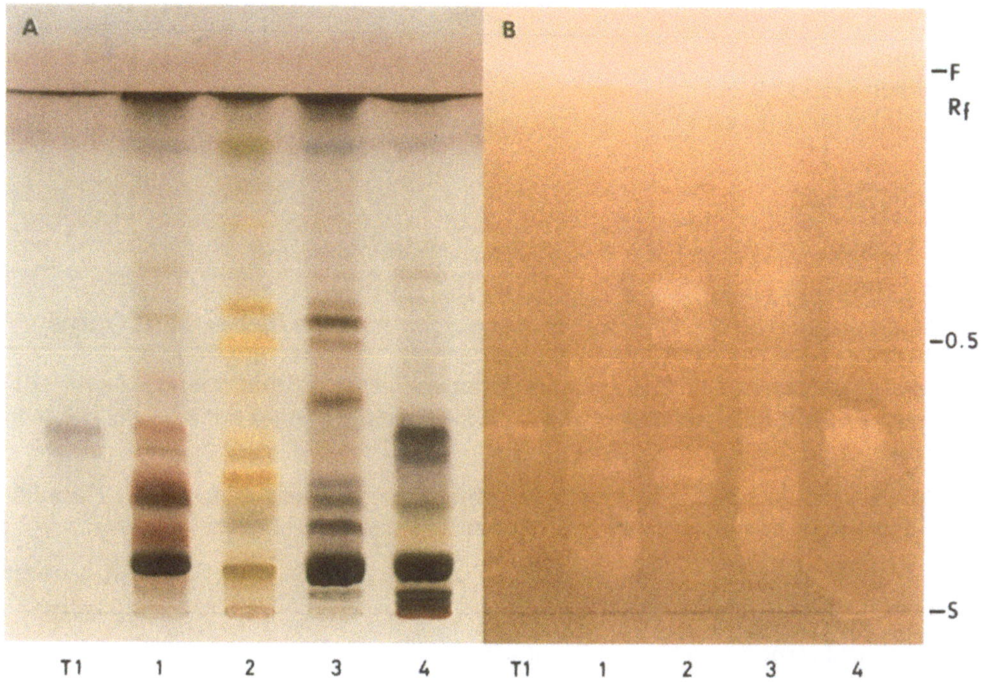

Fig. 1

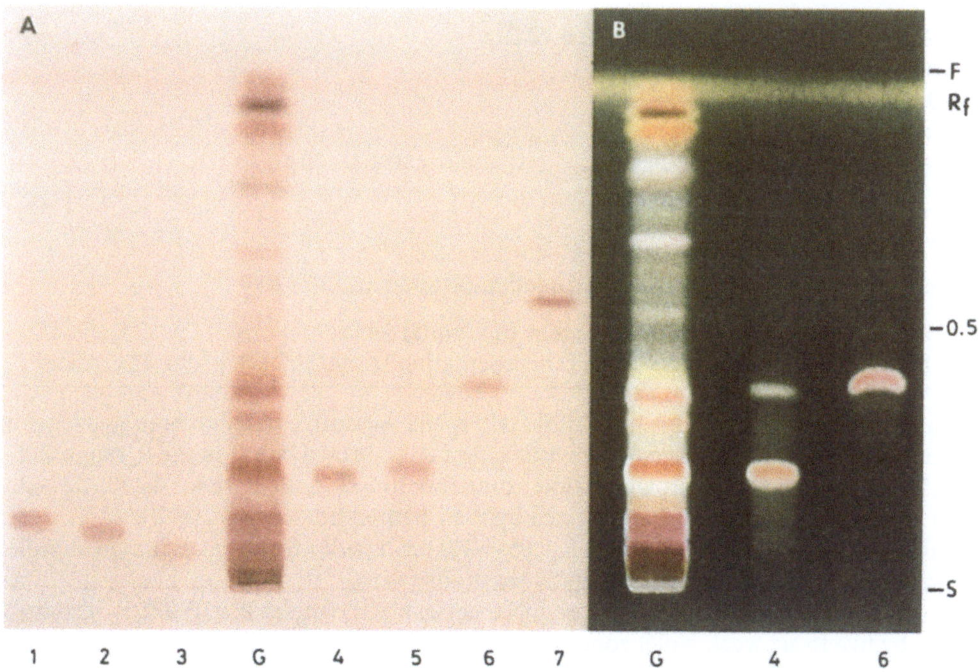

Fig. 2

Hippocastani semen, Primulae radix

Drug sample 1 Hippocastani semen
2 Primulae radix (P. elatior)
3 Primulae radix (P. veris) (ethanolic extracts, 20 µl)

Reference compound T1 aescin
T2 primula acid

Solvent system Fig. 3 chloroform-glacial acetic acid-methanol-water (60:32:12:8)

Detection A Anisaldehyde-sulphuric acid reagent (AS No. 3) → vis
B Blood reagent (BL No. 8) → vis (documentation from the back of the TLC plate)

Fig. 3

A **Hippocastani semen** (1): The complex triterpene ester saponin mixture aescin (T1) generates a main blue-violet zone at $R_f \sim 0.45$. A prominent zone at $R_f \sim 0.2$ is due to glucose.

B Treatment with blood reagent reveals the white zones of aescin at $R_f \sim 0.45$. The additional two weak hemolytic zones in test T1 at $R_f \sim 0.6$ are aescinoles (artefacts).

A **The Primula species** (2,3) show the saponin mixture primula acid (T2) as two to three red-violet zones in the R_f range 0.25–0.35.

B Primula acid responds to blood reagent with hemolysis, seen as white zones (vis.).

Quillajae cortex, Saponariae radix

Drug sample 1 Quillajae cortex
2 Saponariae radix (S. alba)
3 Saponariae radix (S. rubra) (ethanolic extracts, 30 µl)

Reference compound T1 standard saponin
(Gypsophila saponin)

Solvent system Fig. 4 chloroform-glacial acetic acid-methanol-water (60:32:12:8)

Detection A Anisaldehyde-sulphuric acid reagent (AS No. 3) → vis
B Blood reagent (BL No. 6) → vis (documentation from the back of the TLC plate)

Fig. 4A, B The complex mixture of bisdesmosidic triterpene saponins, derived from gypsogenin (quillaic acid), of the extracts 1–3 reveals major dark-brown bands in the R_f range 0.05–0.15 four to eight minor brown or violet zones are found in the R_f range 0.2–0.75 (→A). All zones are more easily characterized by their hemolytic reactions (→B).

Quillajae cortex (1). The saponins (R_f 0.05–0.45) react with AS reagent as brown or violet zones (→A) and give strong hemolytic reactions (→B).

Saponariae albae radix (2). One broad, brownish-black band at R_f 0.05–0.1 is accompanied by five to six weak violet zones between R_f 0.15 and 0.4 (→A). All react with blood reagent to give white zones (→B).

Saponariae rubrae radix (3). Besides two main brownish-black zones at R_f 0.05–0.1, some additional violet zones are seen in the R_f range 0.75–0.8 (→A). The characteristic hemolytic zones are found between R_f 0.05 and 0.1, at $R_f \sim 0.4$ and at $R_f \sim 0.7$ (→B).

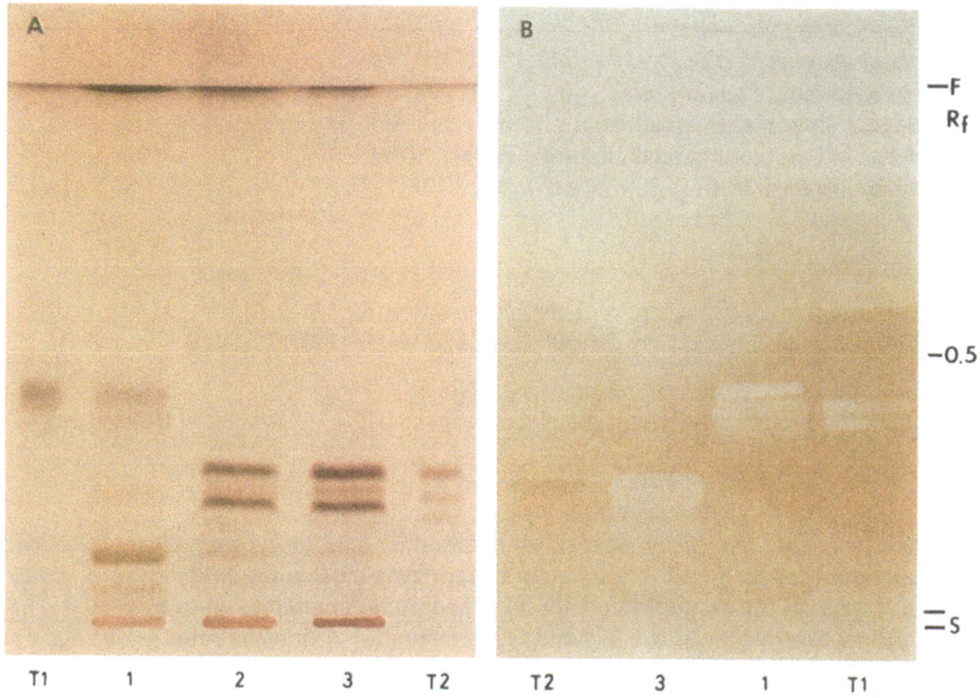

Fig. 3

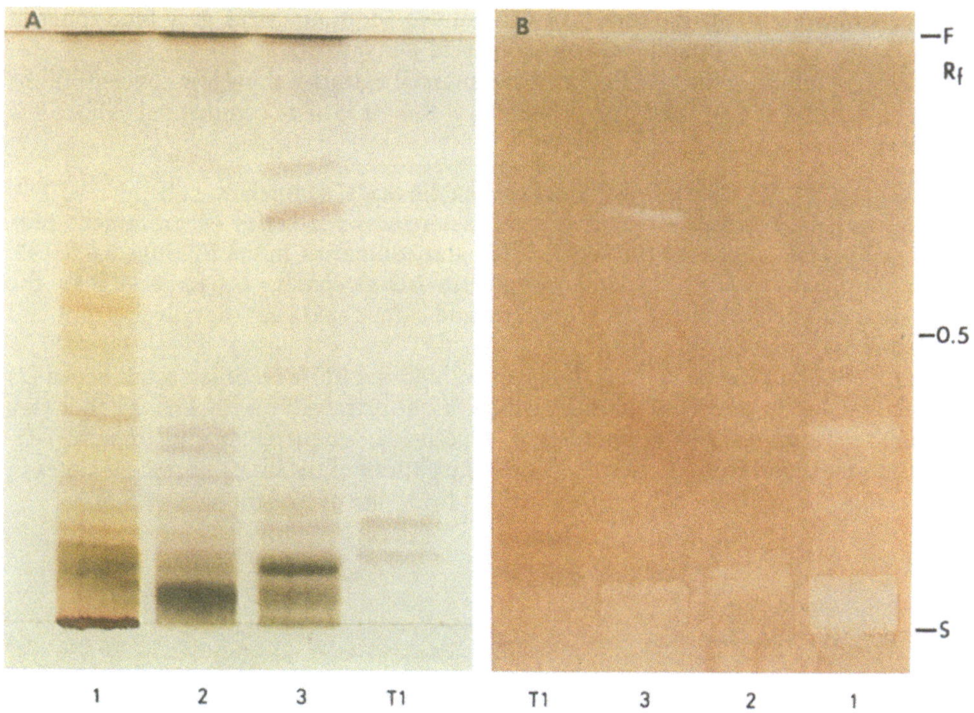

Fig. 4

Hederae folium

Drug sample
1 Hederae folium (H. helix)
2 Hederae terrestris herba (trade sample)
3 Hederae folium (trade sample)
4 Hederae folium (commercial ethanolic extract, 15%)
5 Hederae folium (commercial ethanolic extract, 50%)
(ethanolic extracts, 30 µl)

Reference compound
T1 β-hederin
T2 chlorogenic acid

Solvent system
Fig. 5 chloroform-glacial acetic acid-methanol-water (60:32:12:8)
Fig. 6 ethyl acetate-formic acid-glacial acetic acid-water (100:11:11:26)

Detection
Fig. 5 Vanillin-phosporic acid reagent (VPA No. 41) → vis.
Fig. 6 Natural products-polyethylene glycol reagent (NP/PEG No.28) → UV-365 nm

---

Fig. 5 VPA reagent vis. → **Hederasaponins**

The bisdesmosidic triterpene glycosides of oleanolic and 28-hydroxy-oleanolic acids, such as hederacoside B and C, are found as dark grey-blue zones in the lower R_f range 0.15–0.2, whereas the monodesmosides α-, β-hederin (T1) migrate up to R_f 0.7–0.8. The weak yellow-brown zone at R_f 0.2 is due to quercetin-3-0-rutinoside rutin (→see Fig. 6).
Hedera helix (ivy) drug (1) sample represents a good-quality drug with hederacosides B and C as the major compounds and smaller amounts of α-, β-hederin.
Hederae terrestris herba (ground ivy) (2) shows different and less concentrated triterpenoid zones in the R_f range 0.15–0.2 and two additional weak grey-blue zones in the R_f range 0.25–0.35 (ursolic and oleanolic acid derivatives).
The Hederae folium trade sample 3 and commercial extract 5 show low concentrations of saponin zones at R_f 0.2–0.3 and β-hederin at R_f ~ 0.75. The commercial extract 4 is prepared from Hederae terrestris herba.

Fig. 6 NP/PEG reagent, UV-365 nm → **Phenol carboxylic acids, coumarins, rutin**

The **Hedera helix** samples (1,3,5) are characterized by a series of prominent blue fluorescent zones of phenol carboxylic acids and coumarins in the R_f range 0.45–0.95: e.g. chlorogenic acid (T2) and scopoletin-7-O-glucoside at R_f 0.45–0.5, the isochlorogenic acids at R_f ~ 0.75, scopoletin and caffeic acid near the solvent front and the yellow rutin zone at R_f ~ 0.4.
A prominent orange zone in the R_f range of rutin found in **Hederae terrestris herba** (2) is probably due to luteolin-7-bioside, which is not extractable with 15% alcohol (see sample 4). Traces of yellow fluorescent hyperoside, isoquercetin and luteolin-7-O-glucoside are found in the R_f range 0.6–0.7. The pattern of the blue fluorescent zones in the R_f range 0.6 up to the solvent front differs from that of Hedera helix (1,3,5).

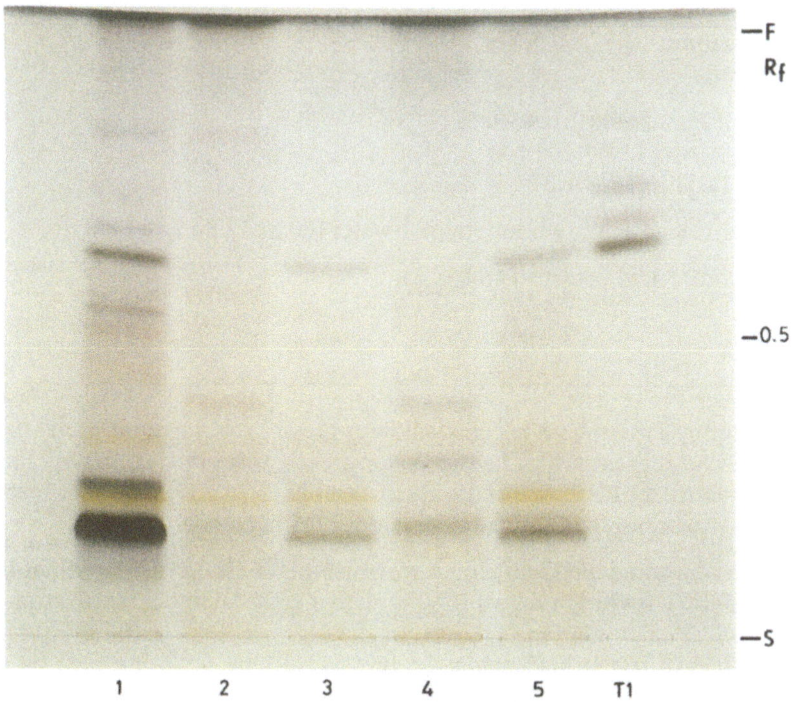

Fig. 5

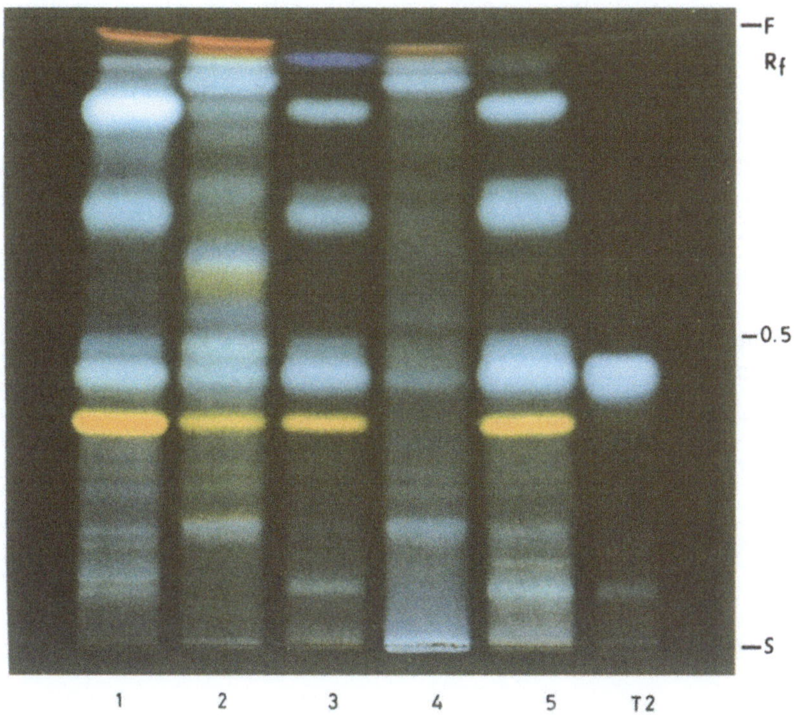

Fig. 6

Rusci rhizoma, Centellae herba

Drug sample 1 Rusci aculeati rhizoma
2 Centellae herba (ethanolic extracts, 20 µl)

Reference compound T1 aescin ($R_f \sim 0.35$)/aescinol (R_f 0.45–0.5)
T2 asiaticoside
T3 madecassoside (asiaticoside A,B)

Solvent system Fig. 7,8 chloroform-glacial acetic acid-methanol-water (60:32:12:8)

Detection Anisaldehyde-sulphuric acid reagent (AS No. 3)
A → vis
B → UV-365 nm

Fig. 7A **Rusci aculeati rhizoma** (1) shows six to eight yellow or green-brown saponin zones in the R_f range 0.1–0.7 (→ AS reagent vis.). The two major green-brown (vis.) zones are found directly below and above the blue zone of the aescin test (T1, $R_f \sim 0.35$). The brown zones in the R_f range 0.4–0.6 are in the R_f range of aescinols (T1, $R_f \sim$ 0.45–0.5).

B In UV-365 nm the main zones of (1) develop a brownish-black (R_f 0.1–0.2) or violet-blue fluorescence (R_f 0.3–0.7). In the R_f range 0.75–0.9, further zones of low concentration are detectable.
According to the literature the bisdesmosidic, furostanol-type saponins such as ruscoside and desglucoruscoside are found the R_f range 0.1–0.3. The monodesmosidic, spirostanol-type steroid saponins such as ruscin, desglucoruscin and desglucodesrhamnoruscin migrate into the R_f range 0.35–0.6, and the aglycones neoruscogenine and ruscogenine into the R_f range 0.8–0.9.

Note: The zones in the R_f range 0.05–0.75 show strong hemolytic activity with blood reagent. (BL No. 8)

Fig. 8A **Centellae herba** (2) is characterized by the ester saponins asiaticoside (T2) and "madecassoside", a mixture of asiaticoside A and B (T3) seen as brown-violet to violet zones in the R_f range 0.2–0.35 and the blue aglycone zone at $R_f \sim 0.85$.

B In UV-365 nm, the saponins appear with violet-blue (T1/T2) or red-violet (aglycone) fluorescence.
Depending on the drug origin (e.g. India or Africa), asiaticoside and/or madecassoside (asiaticoside A and B) are present in various concentrations.

Note: The ester saponins show only very weak hemolytic activity.

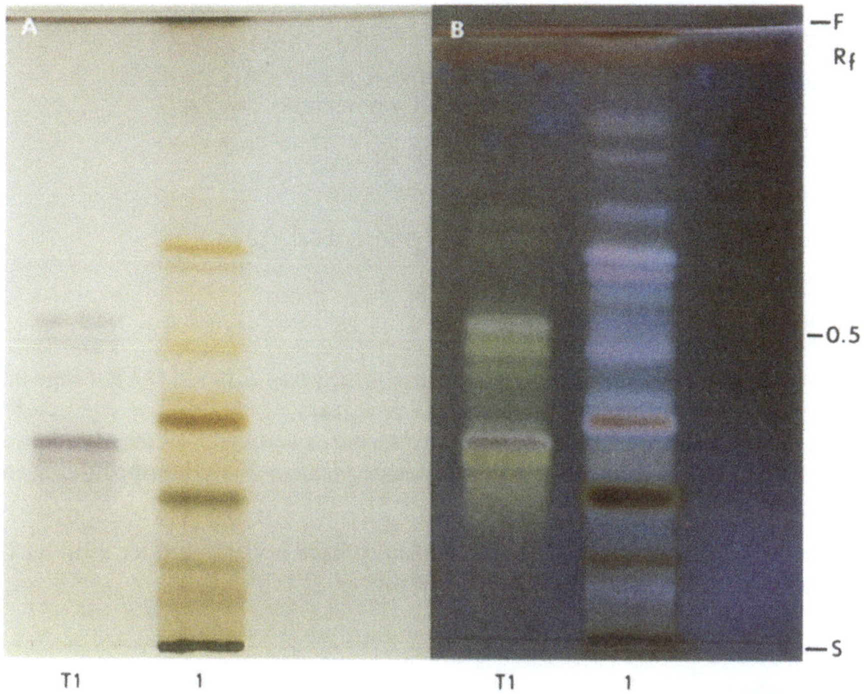

Fig. 7

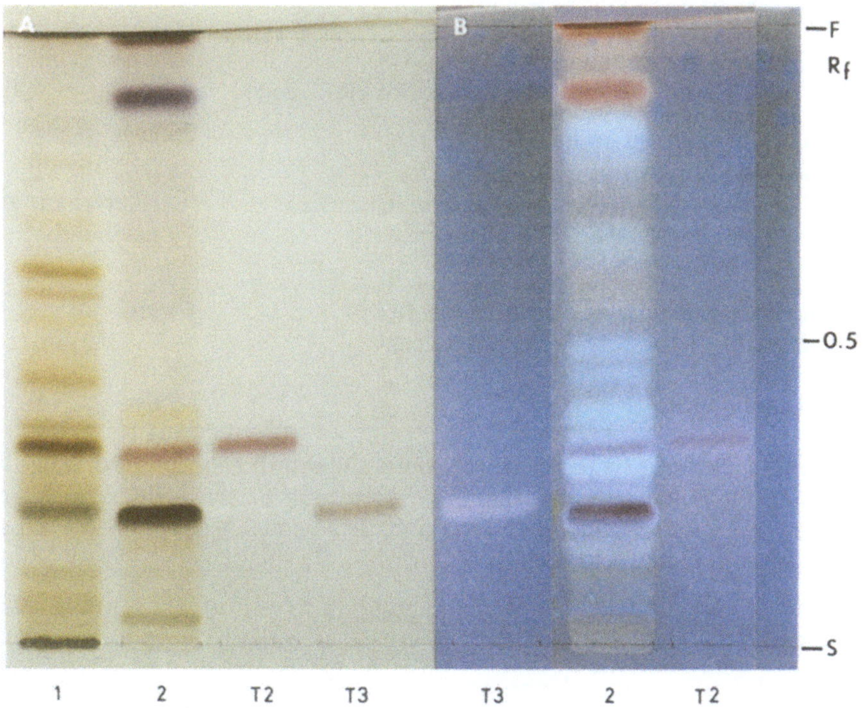

Fig. 8

Avenae sativae

Drug sample 1 Avenae sativae fructus
2 Avenae sativae herba (n-BuOH enrichment, 30 µl)

Reference compound T1 avenacoside B
T2 avenacoside A
T3 vanillin glucoside

Solvent system Fig. 9 chloroform-glacial acetic acid-methanol-water (60:32:12:8)

Detection A Anisaldehyde-sulphuric acid reagent (AS No. 3) → vis
B UV-254 nm (without chemical treatment)

Fig. 9A **Avena sativa samples** (1,2) reveal eight to ten grey-blue (vis.) zones in the R_f range 0.15–0.9. Avenacoside A and B are found in the lower R_f range (T1/T2). The characteristic saponin avenacoside B is more concentrated in the herba sample 2, which also contains flavon glycosides (e.g. isoorientin-2″-O-arabinoside, vitexin-2″-O-rhamnoside) seen as one yellow zone at R_f ~ 0.25.

B Vanillin glucoside (T3, grey zone at R ~ 0.5 →A) reported in the literature as well as the flavonoids show a strong quenching in UV-254 nm.

Liquiritiae radix

Drug sample 1 Liquiritiae radix (ethanolic extract, 20 µl)

Reference compound T1 glycyrrhizin T3 glycyrrhetic acid
T2 aescin T4 rutin (R_f ~ 0.3) ▶ hyperoside (R_f ~ 0.55)

Solvent system

Fig. 10A	chloroform-glacial acetic acid-methanol-water (60:32:12:8)	→ saponins
B	ethyl acetate-ethanol-water-ammonia (65:25:9:1)	→ glycyrrhetic acid
C + D	ethyl acetate-glacial acetic acid-formic acid-water (100:11:11:26)	→ flavonoids

Detection

A,B	Anisaldehyde-sulphuric acid reagent (AS No. 3)	→ vis.
C	Natural products-polyethylene glycol reagent (NP/PEG No. 28)	→UV-365 nm.
D	50% ethanolic H_2SO_4 (No. 37)	→vis.

Fig. 10A **Liquiritiae radix** (1) shows six to seven blue, violet and brown zones in the R_f range 0.1–0.65 in solvent system A. The main saponin glycyrrhizin is detected with AS reagent as a violet zone in the R_f range 0.35–0.4 (T1, R_f similar to aescin/T2), directly below a major brown zone (flavonoids and chalcones).

B The aglycone glycyrrhetic acid (T3), which migrates in solvent system A to the solvent front, is found in solvent B at R_f ~ 0.45.

C The flavanon glycosides and chalcones are separated in solvent C. They fluoresce with NP/PEG reagent yellow-white (R_f 0.15–0.3) and dark green (R_f ~ 0.4 and R_f ~ 0.75) in UV-365 nm.

D With sulphuric acid the flavanon glycosides (e.g. liquiritin, liquiritoside) and the corresponding chalcones appear as characteristically orange-yellowish brown zones (vis).

Note: For the detection of glycyrrhizin, see also Chap. 15.

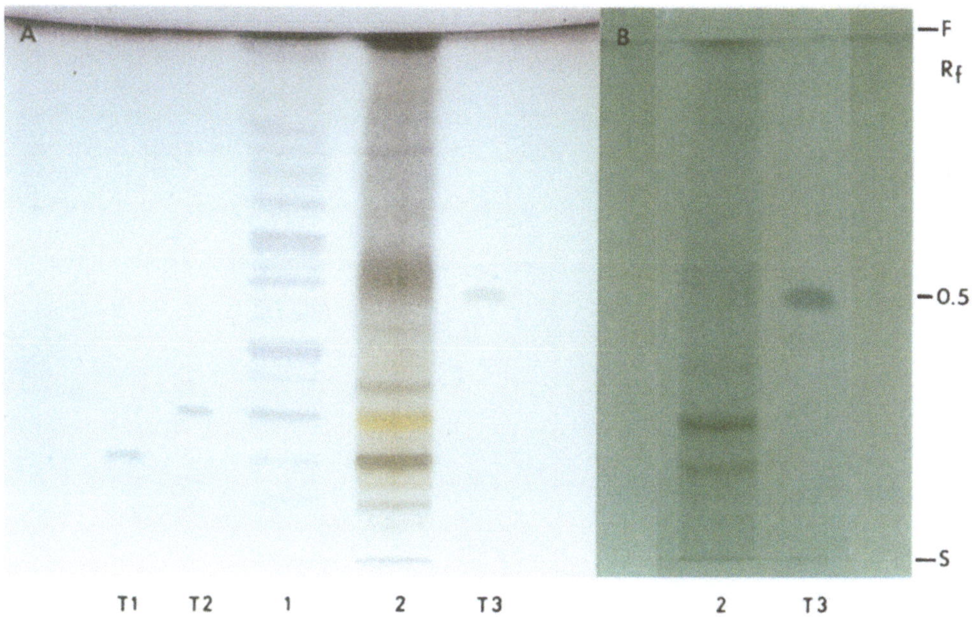

Fig. 9

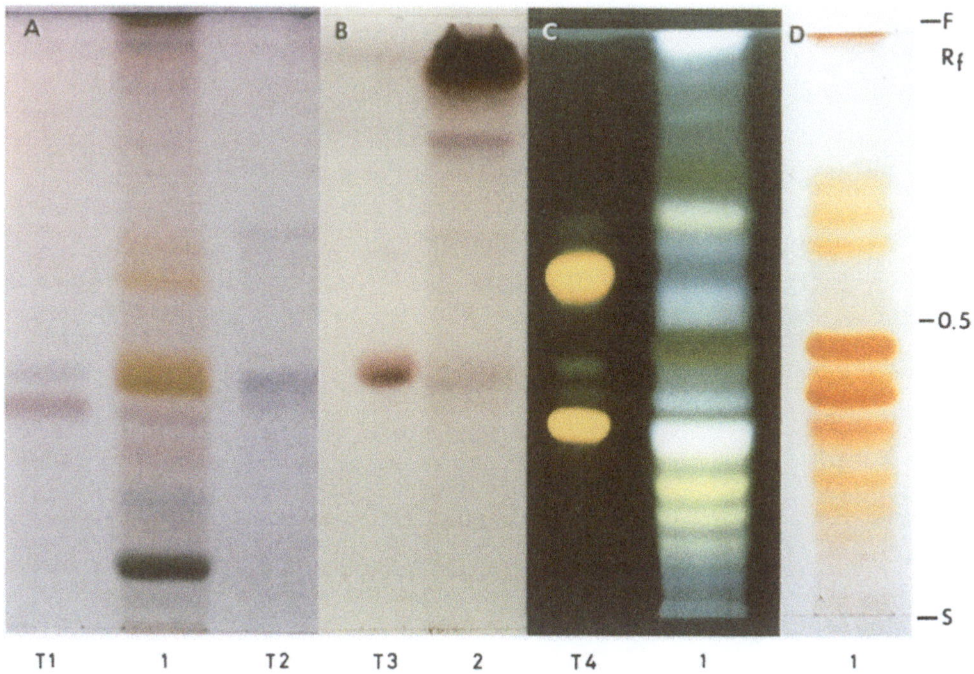

Fig. 10

15 Drugs Containing Sweet-Tasting Terpene Glycosides

(Steviae folium – Diterpene glycosides; Liquiritiae radix – Triterpene glycoside)

15.1 Preparation of Extracts

Steviae folium

Powdered drug (1 g) is extracted for 15 min with 15 ml methanol under reflux. The filtrate is evaporated to 3 ml and 30 µl is used for TLC.

Enrichment

Powdered drug (1 g) is extracted for 2 h with 20 ml water, with occasional shaking. For further enrichment of the diterpene glycosides, the extract is shaken with 20 ml n-butanol and the n-butanol-phase separated. The filtrate is evaporated to 3 ml and 20 µl is used for TLC.

Liquiritiae radix

Powered drug (1 g) is extracted with 10 ml methanol (50%) for 1 h under reflux; 20 µl of the filtrate is used for TLC.

15.2 Thin-Layer Chromatography

Reference solutions

1 mg stevioside and rebaudioside A are dissolved in 1 ml methanol → Steviae fol.
2 mg glycyrrhizin is dissolved in 1 ml methanol (50%) → Liquiritiae radix

Adsorbent

Silica gel 60 F_{254}-precoated plates (Merck, Germany)

Chromatography solvents

Chloroform-methanol-water (65:25:4) → Steviae fol.
Chloroform-methanol-water (64:50:10) → Liquiritiae rad.

15.3 Detection

- UV-254 nm → glycyrrhizin shows quenching
- Liebermann-Burchard reagent (LB No. 25) → Steviae folium
  The plate is heated for 5–10 min at 110°C; evaluation in vis: grey to red-brown zones

15.4 Drug List

	Drug/plant source Family/pharmacopoeia	Main constituents Sweetening agents
Fig. 1	**Liquiritiae radix** Licorice root (peeled/unpeeled) Glycyrrhiza glabra L. var. typica var. glandulifera Fabaceae DAB 10, DAC 86, ÖAB, Helv. VII, MD	6%–14% pentacyclic triterpene glycosides glycyrrhizinic acid (diglucuronide of 18-β-glycyrrhetic acid), glycyrrhizin as Na^+, K^+ or Ca^{2+} salt In addition, flavonoids (liquiritigenine glycoside) → Spanish licorice → Russian licorice (see also Chap. 14, Fig. 10)
Fig. 2	**Steviae folium** Yerba dulce, Azucá Stevia rebaudiana (BERT) HEMSL. Asteraceae MD	5%–14% diterpene glycosides stevioside (steviosin, phyllodulcin 5%–10%), rebaudiosides A (2%–4%), B, C, D, E, dulcosides A (0.3%–0.7%) and B Aglycone steviol (13-hydroxy-kaur-16-en-18-oic acid)

15.5 Formulae

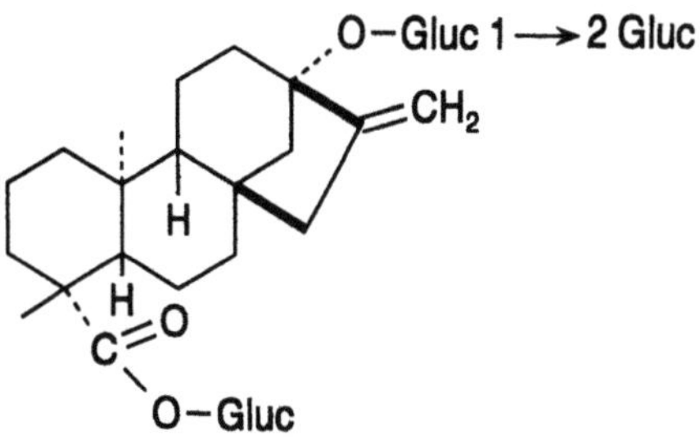

Stevioside
Gluc = glucose
Gluc 1 → 2 Gluc = sophorose

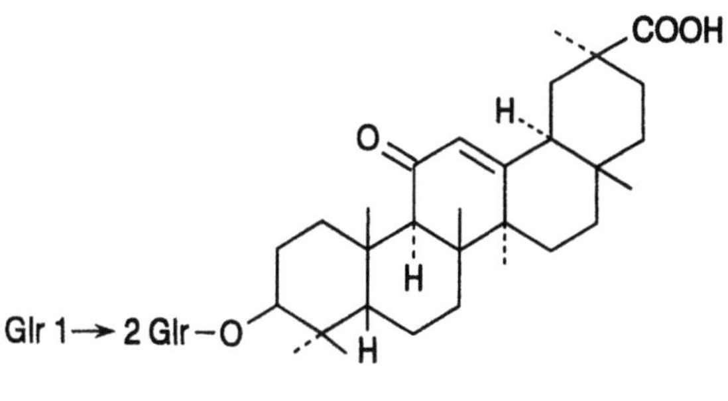

Glycyrrhizinic acid, Glycyrrhizin
Glr = galacturonic acid

15.6 Chromatograms

Liquiritiae radix

Drug sample 1 Liquiritiae radix (methanolic extract, 20 µl)

Reference compound T1 glycyrrhizin (K^+ salt)

Solvent system Fig. 1 chloroform-methanol-water (64:50:10)

Detection A UV-254 nm
B Anisaldehyde-sulphuric acid reagent (AS No. 3) → vis

Fig. 1A **Liquiritiae radix** (1) shows glycyrrhizin (T1) as a quenching band at R_f 0.25–0.3, three quenching zones of flavonoid glycosides and chalcones in the R_f range 0.65–0.75 and the aglycones at the solvent front.

B With AS reagent, glycyrrhizin (T1) develops a pink-violet colour. *Glucose*, as a non-quenching compound (UV 254 nm), reacts as a prominent black-grey zone at $R_f \sim 0.25$, partly overlapping the broad band of glycyrrhizin. The chalcones and flavanone glycosides, e.g. liquiritin, are seen as prominent yellow bands in the R_f range 0.6–0.8. Glycyrrhetic acid runs with the solvent front (→ identification, see Chap. 14, Fig. 10)

Steviae folium

Drug sample 1 Steviae folium (n-BuOH extract, 20 µl)
2 Steviae folium (water extract, 30 µl)
3 Steviae folium (methanol extract, 20 µl)

Reference compound T1 stevioside
T2 rebaudioside A

Solvent system Fig. 2 chloroform-methanol-water (65:25:4)
A with chamber saturation B without chamber saturation

Detection Liebermann-Burchard reagent (LB No. 25) → vis
A 5 min/110°C B 8 min/110°C

Fig. 2A In a **Steviae folium** BuOH extract (1) the sweet-tasting diterpene glycosides are found as four grey zones in the R_f range 0.1–0.3 with rebaudioside A (T2) at $R_f \sim 0.2$ and stevioside (T1) at $R_f \sim 0.3$. The three weak grey zones in the upper R_f range 0.75 up to the solvent front are due to less polar diterpene glycosides and aglycones.

B Stevioside (T1) and rebaudioside A (T2) are easily soluble and detectable in the water extract (2).
The methanolic extract (3) also contains flavonoids (yellow-brown zones at R_f 0.25–0.5), lipophilic compounds (R_f 0.8–0.9) and chlorophyll at the solvent front.

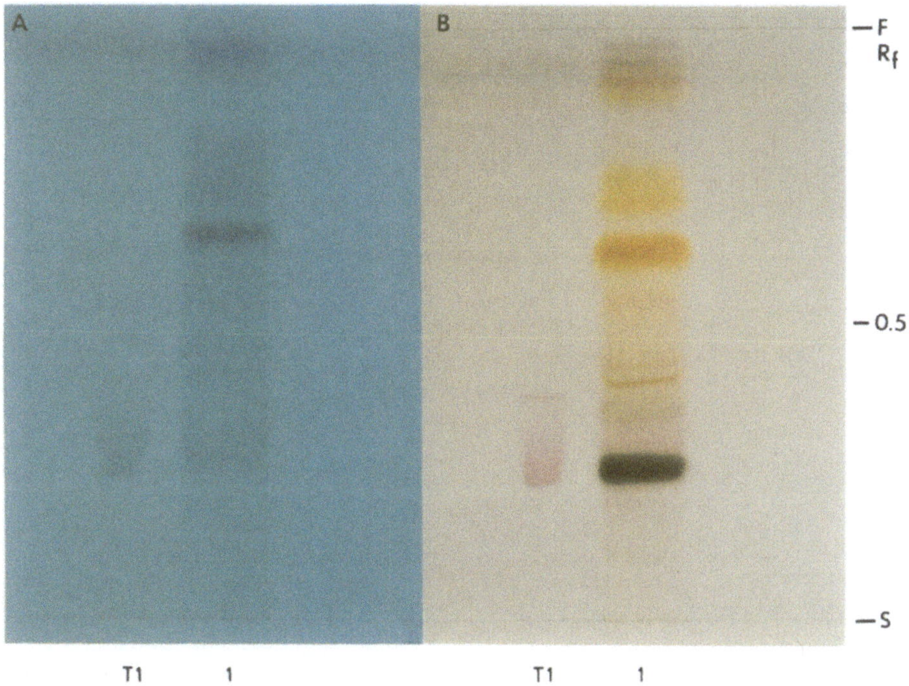

Fig. 1

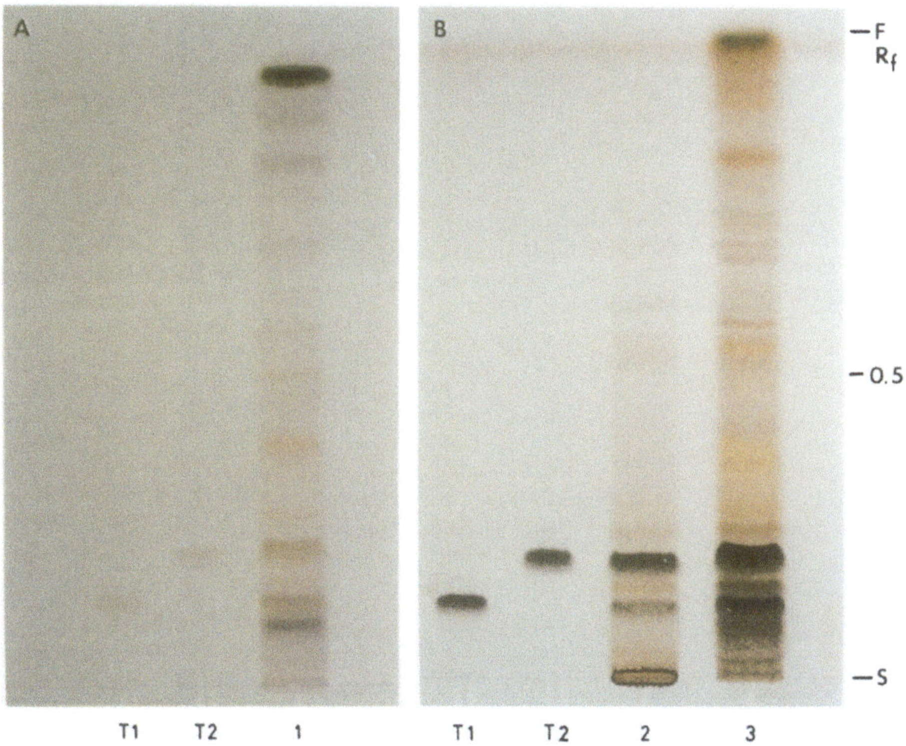

Fig. 2

16 Drugs Containing Triterpenes

This drug group includes Cimicifugae rhizoma (Tetracyclic triterpene glycosides, actein, cimifugoside) and Ononidis radix (Triterpene α-onocerin = onocol).

16.1 Preparation of Extracts

Cimicifugae rhizoma

Powdered drug (1 g) is extracted with 10 ml methanol for 15 min on a water bath; 30 µl of the filtrate is used for TLC.

Ononidis radix

Powdered drug (1 g) is extracted with chloroform for 1 h under reflux; 20 µl of the filtrate is used for TLC of terpenes.
Powdered drug (1 g) is extracted with 10 ml methanol for 15 min on a water bath; 20 µl is used for TLC of flavonoid glycosides and terpenes.

16.2 Thin-Layer Chromatography

Reference solutions

Caffeic and ferulic acid; ononin, rutin, chlorogenic acid, hyperoside as a 0.1% solution in methanol; 10 µl is used for TLC.

Adsorbent

Silica gel 60 F_{254}-precoated plates (Merck, Germany)

Chromatography solvents

Ethyl formiate-toluene-formic acid (50:50:15) → Cimicifugae rhizoms
Toluene-chloroform-ethanol (40:40:10) → Ononidis radix
Ethyl acetate-glacial acetic acid-formic acid-water (100:11:11:26) → polar compounds → ononin (isoflavone)

16.3 Detection

- UV-254 nm — caffeic acid, its derivatives and isoflavones show quenching
- UV-365 nm — caffeic acid, its derivatives and isoflavones fluoresce blue
- Anisaldehyde-sulphuric acid reagent (AS No. 3) (see Appendix A) — The sprayed TLC is heated for 6 min at 100°C; evaluation in vis.: triterpenes blue-violet (Cimicifugae rhizoma) and red to red-violet (Ononidis radix)

16.4 Drug List

Fig. 1	**Cimicifugae rhizoma** Cimicifuga, black cohosh Cimicifuga racemosa (L.) NUTT (syn. Actaea racemosa L.) Ranunculaceae MD, Japan, China (other Cimicifuga species)	▸ Tetracyclic triterpene glycosides: actein (acetyl-acteol xyloside), cimicifugoside (cimicigenol xyloside) ▸ Isoflavone: formononetin ▸ Caffeic and isoferulic acid ▸ 15%–20% resins (cimicifugin)
Fig. 2	**Ononidis radix** Restharrow root Ononis spinosa L. Fabaceae ÖAB, MD	▸ Triterpenes: α-onocerin (= onocol) ▸ Isoflavones: ononin (= formononetin-7-glucoside), formononetin-7-(6″-O-maloyl)-glucoside, biochanin A-7-glucoside ▸ 0.02%–1% essential oil: anethole, carvon, menthol

16.5 Formulae

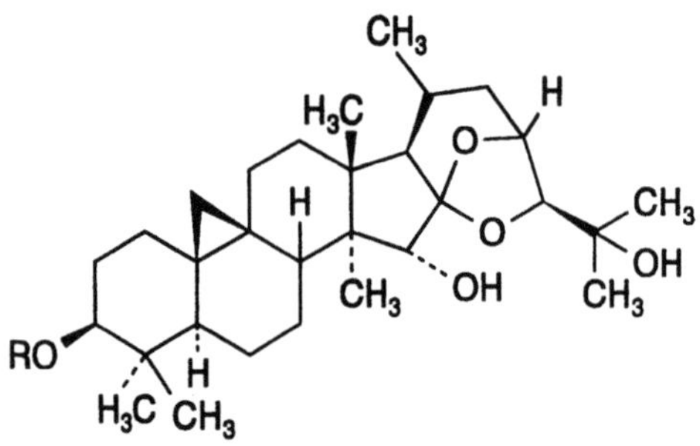

Cimifugoside R = Xylose
Cimicigenol R = H

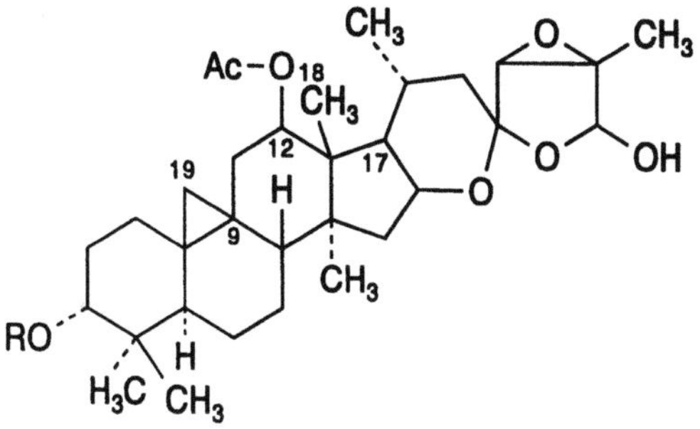

Actein R = Xylose
Acetyl-acteol R = H

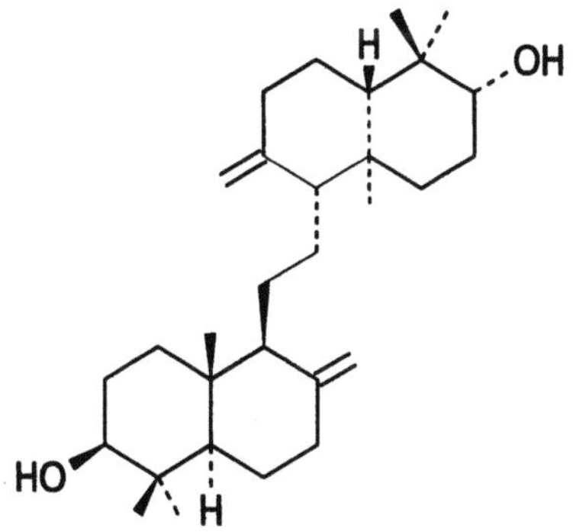

α-Onocerin (=Onocol)

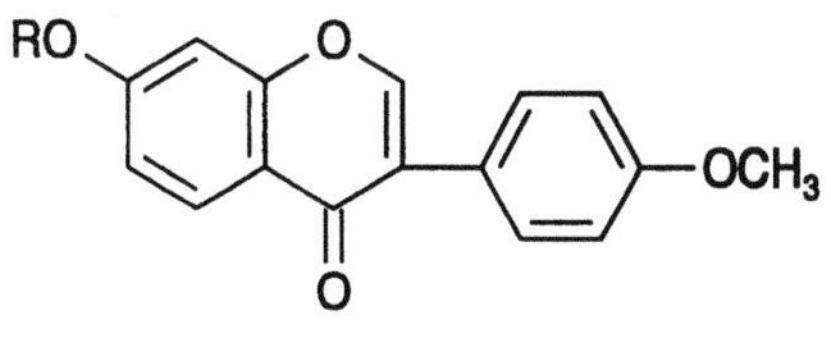

Ononin R = Glucose
Formononetin R = H

16.6 Chromatograms

Cimicifugae rhizoma

Drug sample 1 Cimicifugae rhizoma (methanolic extract, 10 µl)

Reference compound T1 caffeic acid (+ methylester of caffeic acid)
T2 formononetin

Solvent system Fig. 1 ethyl formiate-toluene-formic acid (50:50:15)

Detection A UV-254 nm
B Natural product polyethylene reagent (NST/PEG No. 28) → UV-365 nm
C Anisaldehyde-sulphuric acid reagent (AS No. 3) → vis.

Fig. 1A In UV-254 nm **Cimicifugae rhizoma** (1) shows two prominent quenching zones at R_f 0.55–0.65 due to caffeic acid (T1) and the isoflavone formononetin (T2). The zones at R_f 0.05–0.1 are phenol carboxylic acids.

B After treatment with the NST/PEG reagent, the main zones develop a bright blue fluorescence in UV-365 nm.

C Treatment with AS reagent reveals the violet zones of the triterpene glycosides in the R_f range 0.3–0.55. The prominent zones are due to actein and cimifugoside.

Ononidis radix

Drug sample 1 Ononidis radix (methanolic extract, 20 µl)

Reference compound T1 ononin
T2 rutin (R_f ~ 0.4) ► chlorogenic acid (R_f ~ 0.5) ► hyperoside (R_f ~ 0.6)

Solvent system Fig. 2 A ethyl acetate-glacial acetic acid-formic acid-water (100:11:11:26) → system 1
B+C toluene-chloroform-ethanol (40:40:10) → system 2

Detection A UV-254 nm
B UV-365 nm
C Anisaldehyde-sulphuric acid reagent (AS No. 3) → vis

Fig. 2A **Ononidis radix** (1). The characteristic isoflavone formononetin-7-O-glucoside (ononin/T1) and its -6″-malonate are found as quenching zones in the R_f range 0.65–0.75 in system 1. The terpenes move with the solvent front.

B In the lipophilic solvent system 2 the green-blue fluorescent ononin (T1) remains close to the start; additional blue fluorescent zones are found in the R_f range 0.2–0.45. The terpenes migrate to the lower R_f range, detectable after treatment with the AS reagent.

C Treatment with AS reagent reveals five violet-red zones (R_f 0.05–0.55) with onocerin as the major terpene zone at R_f ~ 0.4 (vis).

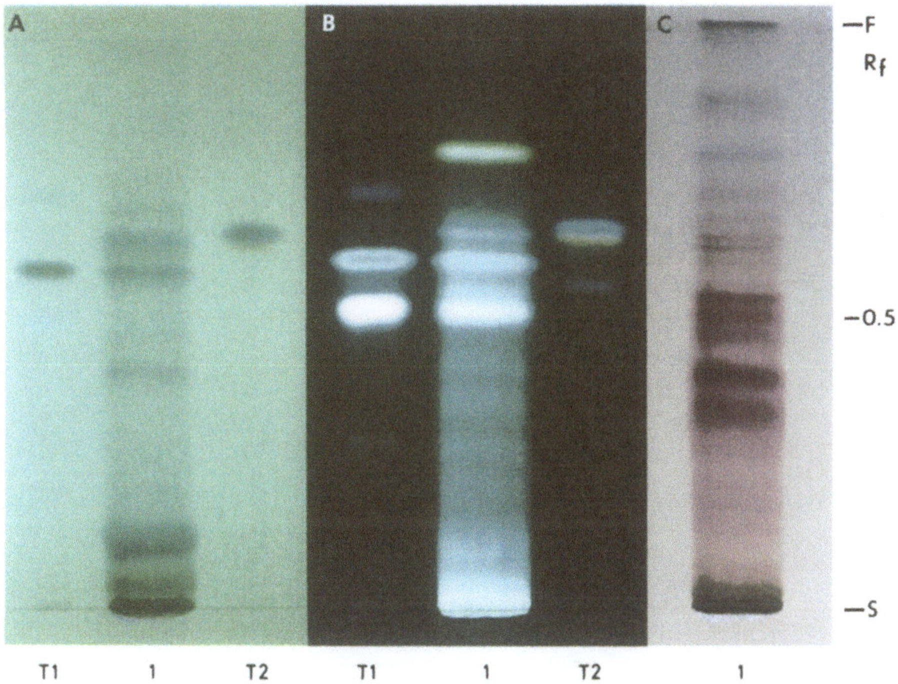

Fig. 1

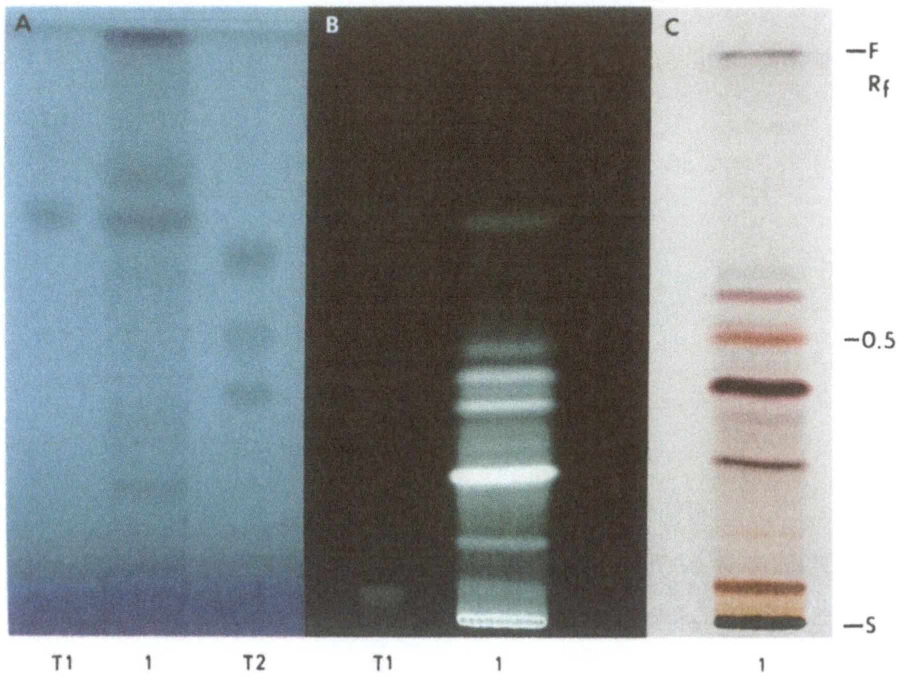

Fig. 2

17 Drugs Containing Valepotriates (Valerianae radix)

The major compounds of valerian, the valepotriates (e.g. valtrate, acevaltrate, didrovaltrate), are triesters of a monoterpene alcohol, with the structure of an iridoid cyclopenta-(c)-pyran and an attached epoxide ring. These iridoid esters are instable and easily form degradation products (e.g. polymers, baldrinals).
The essential oil contains esters of isovalerianic acid, α-hydroxyisovalerianic acid, eugenol, bornyl acetate, terpene hydrocarbons and sesquiterpenes (e.g. valeranone). The sesquiterpene carboxylic acids (e.g. valerenic and acetoxy valerenic acids) are stable, non-volatile compounds. Both classes of compounds contribute to the sedative effect of valerian root.

17.1 Preparation of Extract

DCM extract

Powdered drug (0.5 g) is extracted with 5 ml dichloromethane for 5 min at about 40°C. After filtration, the drug powder is washed with a further 2 ml dichloromethane. The combined extracts are evaporated to dryness and the residue dissolved in 0.2 ml ethyl acetate; 10 µl is used for TLC investigation.

Pharmaceutical preparations

Two to three powdered tablets or dragées are extracted by shaking them with 5 ml dichloromethane for 10 min; 20 µl of the filtrate is used for chromatography.

17.2 Thin-Layer Chromatography

Reference solutions

Valtrate, isovaltrate, didrovaltrate, acevaltrate and valerenic acids:
5 mg is dissolved in 5 ml methanol; 10 µl of each solution is used for TLC.

Adsorbent

Silica gel 60 F_{254}-precoated TLC plates (Merck, Germany).

Sample concentrations

A total of 10 µl valerian extract of good-quality drugs is sufficent for the chromatographic detection of the major constituents. Commercial valerian roots, however, often have a low valtrate content and need 50 µl extract for TLC investigation.

Chromatography solvents

Toluene-ethyl acetate (75:25) → valepotriates
n-Hexane-ethyl acetate-glacial acetic acid (65:35:0.5) → sesquiterpene carboxylic acids

17.3 Detection

- UV-254 nm Valepotriates show quenching
- UV-365 nm Valepotriates do not fluoresce; yellow fluorescent zones are due to degradation products, e.g. baldrinal and homobaldrinal.
- Spray reagents (see Appendix A)
- Hydrochloric acid-acetic acid reagent (No.17)
  Specific detection of valtrate, IVHD valtrate and acevaltrate. They form blue zones (halazuchrome reaction). Didrovaltrate turns brown (vis).
- Anisaldehyde-H_2SO_4 reagent (AS No.3)
  General detection of terpenes (essential oil) and valerenic acids.
  After 10 min at 100°C, violet or blue zones (vis) develop.

17.4 Drug List

	Drug/plant source Family/pharmacopoeia	Main constituents Valeriana-epoxytriester
Fig. 1,2	**Valerianae radix** Valerian, allheal Valeriana officinalis L. ssp./var. geographical race oecotype diploid→octaploid	0.5%–2% iridoids (valepotriates) Valtrate, isovaltrate, IVHD valtrate, didrovaltrate, acevaltrate (for amounts, see Table 1) 0.35%–1% essential oil Eugenol, isoeugenol, (−) borneol, bornyl acetate, α-, β-pinene, (−)-limonene, (−)-camphene, caryophyllene, β-bisabolene, valerenal (fresh root), esters of valerianic acid, α-hydroxy-valerianic acid, isovalerianic acid
	Valerianaceae DAB 10, ÖAB, Helv VI, MD	0.1%–0.3% sesquiterpene carboxylic acids Valerenic, acetoxy and α-hydroxyisovalerenic acid (non-volatile)
	Valerianae indicae radix Valeriana jatamansi JONES (V. wallichii DC)	3%–6% iridoids Indian or Pakistan valerian see Table 1
	Valerianae mex. radix Valeriana edulis NUTT. ex TORR et GRAY ssp. procera MEY. (V. mexicana DC.)	3%–8% iridoids Mexican valerian see Table 1

Table 1

Constituents	Valeriana officinalis ~0.95%	Valeriana edulis 5–8%	Valeriana jatamansi Valtrate/Acevaltrate-race (3%)	Didrovaltrate race (4%)
Valtrate/isovaltrate	0.4%–2%	3%–4%	2%–3%	0.3–0.7%
Didrovaltrate	0.1%	1.5–4%	0.1%	1.5–4%
Acevaltrate	0.1%	0.1%	0.2%	0.2%
Other valepotriates	0.1%	1.0% (IVHD)	0.1%	0.2%

17.5 Formulae

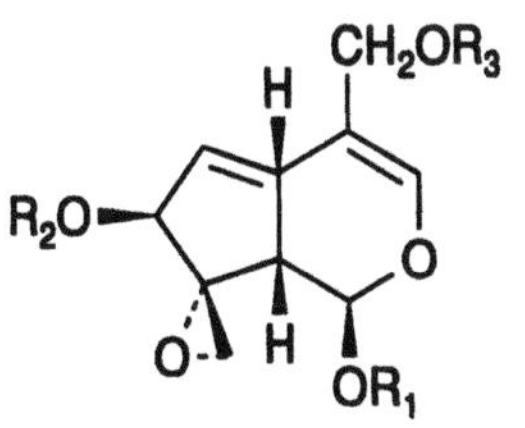

Dien	R_1	R_2	R_3
Valtrate	Isovaleryl-	Isovaleryl-	Acetyl-
Isovaltrate	Isovaleryl-	Acetyl-	Isovaleryl-
Acevaltrate	Isovaleryl-	β-Acetoxy-isovaleryl	Acetyl-

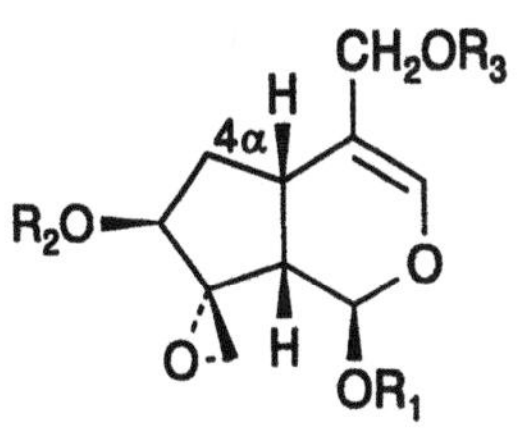

Monoen	R_1	R_2	R_3	4α
Didrovaltrate	Isovaleryl-	Acetyl-	Isovaleryl-	H
IVHD-valtrate	Isovaleryl-	Acetyl-	2-(Isovaleryl-oxy)-isovaleryl-	OH

(IVHD = Isovaleroxyhydroxy didrovaltrate)

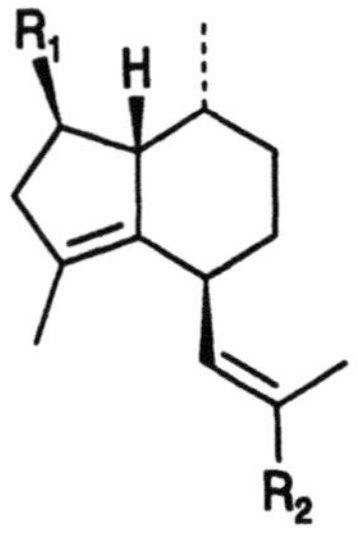

Cyclopentan sesquiterpene	R_1	R_2
Valerenic acid	H	COOH
Acetoxy valerenic acid	$OCOCH_3$	COOH
Valerenal	H	CHO

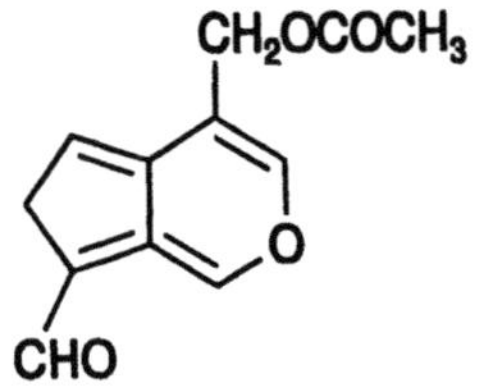

Baldrinal

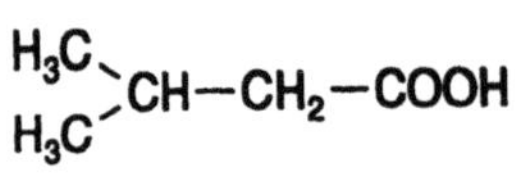

Isovalerianic acid

17.6 Chromatograms

Valerianae radix

Valerian samples

1	Mexican valerian	6	Official valerian (fresh root)
2	Indian valerian	7	Valerian tincture (from a dispensary)
3–5	Official valerian (powdered roots) (DCM extracts, 10–50 µl)	8–11	Valerian extracts (from different pharmaceutical preparations)

Reference compound

T1 valtrate	T3 didrovaltrate	T5 valerenic acids	T7 anisaldehyde
T2 acevaltrate	T4 IVHD valtrate	T6 vanillic acid	

Solvent system Fig. 1,2A,B,C toluene-ethyl acetate (75:25)
D,E hexane-ethyl acetate-glacial acetic acid (65:35:0.5)

Detection A,B,C HCl-glacial acetic acid (2:8) (No.17) → vis
D without chemical treatment → UV-254 nm
E Anisaldehyde-sulphuric acid reagent (AS No. 3) → vis

Fig. 1A,B TLC Comparison of **valerian samples** (1–11)
Composition and concentration of valepotriates (T_1, T_2) varies in the valerian samples. They are detected with HCl/glacial acetic acid (**) as various blue (e.g. valtrate) and brown zones (e.g. didrovaltrate) in the R_f range 0.05–0.75:

0.7–0.75	Isovaltrate/valtrate	T1 (*)	blue (**)	►(*): quenching in UV-254 nm
0.65	Didrovaltrate	T3	brown (**)	►(**): colours (vis): blue to black-blue,
0.55	Acevaltrate	T2 (*)	blue	light to dark brown depending on
0.4	IVHD valtrate	T4	blue	the heating time with reagent No.19
↑	Valtrathydrins	(*)	blue	► sample 9: baldrinal, homobaldrinal
Start	Degradation products			yellow R_f 0.4–0.5 (vis./UV-365 nm)

Characteristic of a good-quality Valeriana officinalis root is the high amount of valtrate (>80%, sample 6). Powdered and stored valerian roots show mainly degradation/polymerization products (sample 3–5). For pharmaceutical preparations, Mexican (1) or Indian (2) valerian are often used.

(1) high content of valtrate/isolvaltrate; medium content of didro- and acevaltrate
(2) predominantly didrovaltate, low concentration of valtrate and IVHD valtrate
(3,4,5) mainly degradation products of valepotriates, traces of valtrate (5)
(6) freshly harvested valerian root with valtrate as main compound
(7) traces of valtrate, medium concentration of ace-, IVHD valtrate, valtrathydrins
(8,10) good valerian quality with high content of valtrate
(9,11) minor quality (degradation products), yellow zone of baldrinal (9)

Fig. 2C TLC comparison of a good quality (sample 10) and minor quality (sample 11) of official valerian, concerning the amount of valtrate/iso-, ace- and IVHD valtrates.

D Valerenic acids are present in official valerians only. They show weak quenching in UV-254 nm (T5: hydroxy, acetoxy and valerenic acid). The commercially available vanillic acid (T6) and anisaldehyde (T7) can be used as guide substances.

E Treatment with AS reagent reveals valerenic acids, valepotriates and essential oil compounds as partly overlapping violet zones (vis).

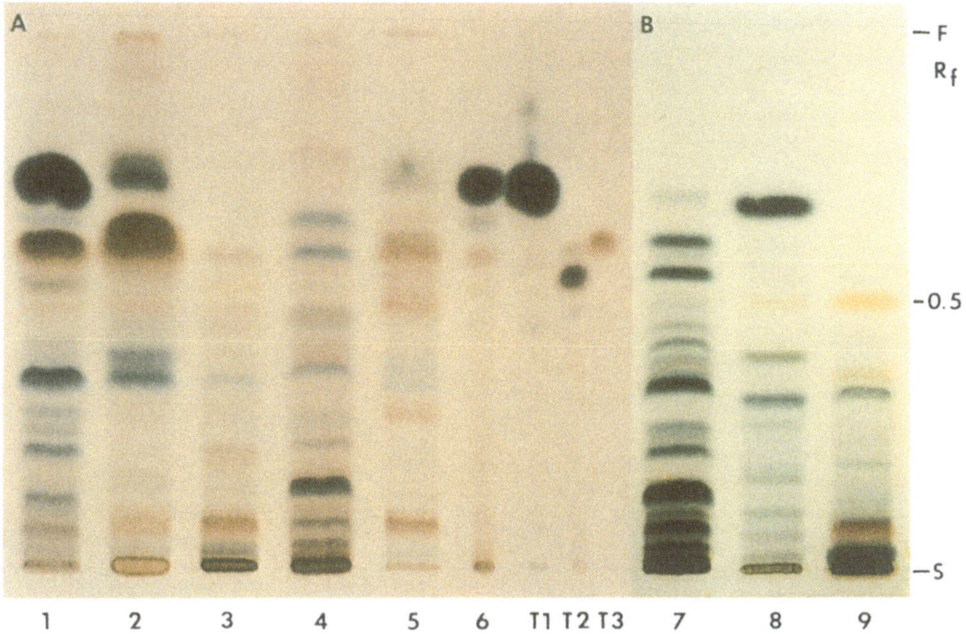

Fig. 1

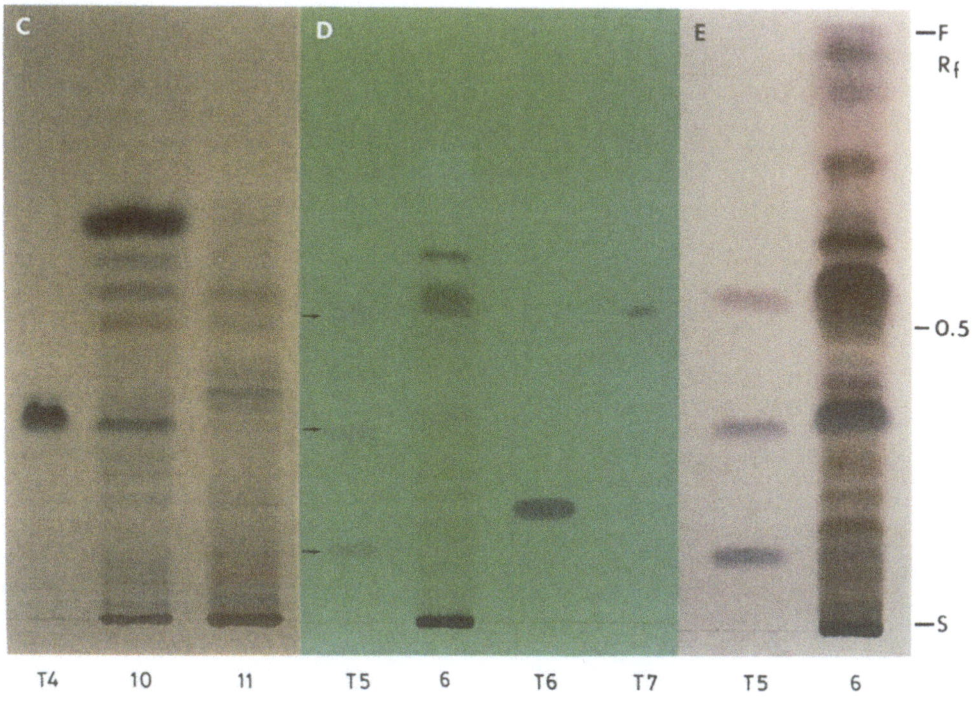

Fig. 2

18 Screening of Unknown Commercial Drugs

With the following analytical procedure, a drug can be assigned to a group of plant constituents or identified on the basis of its constituents. Analyses are performed for the following main active constituents:

- alkaloids
- anthraglycosides
- arbutin
- cardiac glycosides
- bitter principles
- flavonoids, phenolcarboxylic acids
- saponins
- essential oils
- coumarins and phenol carboxylic acids
- valepotriates

18.1 Preparation of Drug Extracts for Analysis

Anthraglycosides, bitter principles, flavonoids and **arbutin**

Powdered drug (1 g) is extracted by heating on a water bath for 10 min with 5 ml methanol; 20 µl of the filtrate is used for TLC investigation.

Alkaloids

Powdered drug (1 g) is moistened with about 1 ml 10% ammonia solution; 5 ml methanol is added and the drug is then extracted for 10 min on a water bath; 20 µl and 100 µl of the filtrate are used for TLC analyses.

Saponins

A methanolic extract is prepared according to method described for anthraglycosides. The extract is evaporated to about 1 ml, mixed with 0.5 ml water and then extracted with 3 ml n-butanol (saturated with water); 20 µl and 100 µl of the butanol phase are used for TLC investigation.

Cardiac glycosides

Powdered drug (1 g) is mixed with 5 ml 50% methanol and 10 ml 10% lead (II) acetate solution and then heated for 10 min on the water bath. The filtrate is cooled to room temperature and then extracted twice with 10-ml quantities of dichloromethane. The combined DCM extracts are evaporated and the residue is dissolved in DCM-methanol (1:1); 100 µl of this solution is used for TLC investigation.

Terpenes, coumarins, phenol carboxylic acids, valepotriates

- Dichloromethane extract (DCM extract) – for lipophilic compounds
  Powdered drug (1 g) is extracted by heating under reflux for 15 min with 10 ml DCM. The filtrate is evaporated to dryness, and the residue is dissolved in 0.5 ml toluene; 20–40 µl is used for TLC investigation.

Essential oils

- Microdistillation of essential oils (LUCKNER or TAS method (see Chap. 6).
  Using the TAS method, all those compounds that are volatile at about 200°C can be obtained, such as terpenes or propylphenols in essential oils, naphthoquinones and coumarins.

18.2 Thin-Layer Chromatography

Absorbent From each extract, prepared according to the methods described above, 20 µl and, if indicated, 100 µl are applied to a TLC silica gel plate (60 F_{254}, 10 cm × 10 cm).

Ten such plates are prepared to cover each of the main classes of constituents. A different selection of standard substances belonging to the class of constituents to be analyzed is also applied to each plate (see separation scheme Sect. 18.4).

Screening system Chromatography is performed in two solvent systems and both solvents are allowed to run for a distance of 8 cm.

System A
- Ethyl acetate-methanol-water (100:13.5:10) for the analysis of polar compounds (glycosides)
- ▶ anthraglycosides, arbutin, alkaloids, cardiac glycosides, bitter principles, flavonoids, saponins.

System B
- Toluene-ethyl acetate (93:7) for the analysis of lipophilic compounds (aglycones)
- ▶ essential oils, terpenes, coumarins, naphthoquinons, valepotriates, liphophilic plant acids.

18.3 Detection and Classification of Compounds

Detection The developed chromatograms are first inspected under UV-254 nm and UV-365 nm light.

UV-254 nm Quenching zones are detected
▶ quenching is caused by all compounds with conjugated double bonds
e.g. anthraglycosides, arbutin, coumarins, flavonoids, propylphenols in essential oils, some alkaloid types such as indole, isoquinoline and quinoline alkaloids

UV-365 nm Fluorescent zones are detected
▶ all anthraglycosides, coumarins, flavonoids, phenolcarboxylic acids
▶ some alkaloid types (e.g. China, Rauwolfia, Ipecacuanha alkaloids)

No fluorescence is observed
▶ cardiac glycosides, bitter principles, saponins, terpenoids in essential oils, valepotriates

Spray reagents After preliminary inspection in UV-254 and UV-365 nm light, each chromatogram is analyzed for the presence of drug constituents by spraying with an appropriate group reagent (see separation scheme, section 18.4). The following reactions and spray reagents (see Appendix A) can be used to determine the types of compounds present.

10% ethanolic KOH reagent (No.35) "Bornträger reaction"
▶ red zones (vis); red fluorescence (UV-365 nm) → anthraquinones (e.g. frangulin A, B; glucofrangulin A, B; emodin, rhein)

▶ yellow zones (vis); yellow fluorescence (UV-365 nm) → anthrones (e.g. aloin, cascarosides; sennosides do not react and need specific treatment)
• For further identification, see Chap. 2, Figs. 1–10.

10% ethanolic KOH reagent (No.35)

▶ bright-blue fluorescent zones in UV-365 nm → coumarins (e.g. scopoletin, umbelliferone)
▶ green-blue, yellow, yellow-brown in UV-365 nm → furano- and pyranocoumarins
Remark: In the polar screening system A, coumarins aglycones migrate unresolved with the solvent front; in the lipophilic screening system B they are separated in the lower and middle R_f range.
• For further identification, see Chap. 5, Figs. 1–16.

Kedde reagent (No.23)

▶ pink and blue-violet (vis) zones → very specific for cardenolides
Remarks: Bufadienolides do not react.
Detection with antimony-(III)-chloride reagent ($SbCl_3$, No. 4) → blue (vis) zones (e.g. proscillaridin)
Detection with anisaldehyde sulphuric acid reagent (AS No. 3) → blue (vis) zones (e.g. hellebrin)
• For further identification, see Chap. 4, Figs. 1–12.

Dragendorff reagent (DRG No.13)

▶ red-brown (vis) zones; the colour may be unstable → alkaloids.
Remark: Some of the strongly basic alkaloids do not migrate in the screening system A; if at the start-line of the chromatogram a positive DRG reaction is shown, a second chromatogram should be run in solvent system toluene-ethyl acetate-diethylamine (70:20:10)
• For further identification, see Chap. 1, Figs. 1–32.

Natural products-polyethylene glycol reagent (NP/PEG No.28)

▶ intense yellow, orange and green fluorescent zones in UV-365 nm → flavonoids.
Without chemical treatment, flavonoids show a distinct quenching of fluorescence in UV-254 nm, and yellow, green or weak blue fluorecence in UV-365 nm.
The screening system A does not produce sharply separated zones of flavonoid glycosides. For positive identification, chromatography should be repeated in the specific flavonoid solvent system ethyl acetate-formic acid-glacial acetic acid-water (100:11:11:26).
Chlorogenic acid, which is frequently present in flavonoid-containing extracts, remains at the start in the screening system A and migrates at $R_f \sim 0.5$ in the flavonoid separation system.
• For further identification, see Chap. 7, Figs. 1–23.

Vanillin-sulphuric acid (VS No.42)
Anisaldehyde-sulphuric acid reagent (AS No.3)

Most plant constituents react with VS and AS reagent with coloured zones in vis. Both reagents are sufficent to detect bitter principles, saponins and essential oil compounds.

▶ **Bitter principles**
If the extract tastes distinctly bitter and the screening system A shows red-brown, yellow-brown or dark green (vis.) zones in the R_f range 0.3–0.6, the drug may be one of the known bitter principle drugs.

Very lipophilic bitter principles, such as quassin, absinthin and cnicin migrate unresolved up to the solvent front in screening system A; a lipophilic solvent system is appropriate.

- For further identification, see Chap. 3, Figs. 1–14.

Remarks: Extracts containing alkaloids or cardiac glycosides also taste bitter.

▶ **Saponins**

Saponins also form coloured (vis.) zones with VS or AS reagent. In the screening system, however, the known saponins (e.g. aescin, primulaic acids and the saponin test mixture) do not migrate and remain at the start.

For a precise differentiation, chromatography must be performed in the saponin solvent system chloroform-glacial acetic acid-methanol-water (64:32:12:8).

- For further identification, see Chap. 14, Figs. 1–10.

▶ **Essential oils**

Blue, brown or red zones in vis. In the polar screening system A, essential oils migrate unresolved at the solvent front.

Classification is possible after chromatography in the lipophilic solvent system toluene-ethyl acetate (93:7)

- For further identification, see Chap. 6, Figs. 1–28.

Specific reactions Berlin blue reaction (No. 7) → arbutin, blue in vis.; (see Chap. 8, Figs. 1, 2) Halazuchrom reaction (HCl/AA No. 17) → valepotriates, blue (in vis) (see Chap. 17, Figs. 1, 2).

18.4 Scheme of Separation and Identification

Ethyl acetate-methanol-water (100:13.5:10) TLC 1–TLC 7

TLC 1	**Anthraglycosides** Extract 20 µl Tests aloin 5 µl frangulin 10 µl Detection KOH reagent No.35	→ red (vis): anthraquinones yellow (vis): anthrones	→ Identification: see Anthraquinone Drugs, Chap.2
TLC 2	**Arbutin** Extract 20 µl Tests arbutin 10 µl hydroquinone 10 µl Detection Berlin blue No.7	→ blue (vis)	→ Identification: see Arbutin Drugs, Chap. 8

	Sample	Result	Next step
TLC 3	**Cardiac glycosides** Extract 20 µl/100 µl Tests lanatosides A–C 10 µl k-strophanthin 10 µl proscillaridin 10 µl Detection Kedde reagent No.23 ($SbCl_3$ reagent No.4)	→ pink/violet (vis) → blue (vis) $SbCl_3$ reagent only	→ Identification: see Cardiac Glycoside Drugs, Chap. 4 → Identification: see Bufadienolides, Chap. 4
TLC 4	**Bitter principles** Extract 20 µl/100 µl Tests naringin 10 µl (rutin 10 µl) Detection VS reagent No.42	→ red/yellow-brown/blue-green	→ Identification: see Bitter Principle Drugs, Chap. 3
TLC 5	**Alkaloids** Extract 20 µl/100 µl Tests atropine 10 µl reserpine 10 µl papaverine 10 µl Detection Dragendorff reagent No.13	→ orange-brown (vis)	Solvent system → toluene-ethyl acetate-diethylamine (70:20:10) ↓ Identification: see Alkaloid Drugs, Chap. 1
TLC 6	**Flavonoids** Extract 20 µl/100 µl Tests rutin 10 µl chlorogenic acid 10 µl hyperoside 10 µl Detection NP/PEG reagent No.28	→ yellow/green/orange (UV-365 nm)	Solvent system → ethyl acetate-formic acid-glacial acetic acid-water (100:11:11:26) ↓ Identification: see Flavonoid Drugs, Chap. 7

TLC 7	**Saponins** Extract 20 µl/100 µl Tests aescin 10 µl primula acid 10 µl Detection VS reagent No. 42 AS reagent No. 3	→ blue (vis)	Solvent system → chloroform-glacial acetic acid-methanol-water (64:32:12:8) ↓ Identification: see Saponin Drugs, Chap. 14

Toluene-ethyl acetate, 93:7 TLC 8–TLC 10

TLC 8	**Essential oils** Extract 20 µl/100 µl Test linalool, thymol, linalyl acetate, anethole 5 µl each Detection VS reagent No. 42	→ red/yellow/blue/brown (vis)	→ Identification: see Drugs with essential Oils, Chap. 6
TLC 9	**Valepotriates** Extract 20 µl/100 µl Test Valtrate or standard pharmaceuticals Detection Hydrochloric acid-acetic acid reagent No.17	→ blue/brown (vis)	→ Identification: see Valerianae radix, Chap. 17
TLC 10	**Coumarins** Extract 20 µl Tests scopoletin 5 µl umbelliferone 5 µl Detection UV-365 nm without chemical treatment; ethanol 10% KOH	→ light blue/brown (UV-365 nm)	Solvent system → diethyl ether-toluene (1:1; saturated with 10% acetic acid) ↓ Identification: see Coumarin Drugs, Chap. 5

19 Thin-Layer Chromatography Analysis of Herbal Drug Mixtures

Many phytopreparations contain mixtures of drug extracts. Therefore, chromatograms display a large number of more or less overlapping zones (UV and vis.) making the identification or classification of the compounds present difficult or only partly successful. In such cases it is necessary to submit the preparation to column chromatographic fractionation or other special procedures for the separation of individual classes of compounds.

If the various drugs of the herbal formulation contain the same classes of compounds and active principles, identification of the characteristic components is usually possible. Salviathymol, a mixture of essential oil components (Fig. 1) and a laxative preparation containing various anthraglycosides (Fig. 2) are chosen as an example for the TLC of mixed herbal preparations.

Salviathymol®

1 g of the composition contains:

2 mg Salviae aeth. (standardized at not less than 40% thujone); 2 mg Eucalypti aeth. (not less than 75% cineole); 23 mg Menthae pip. aeth. (not less than 50% menthol); 2 mg Cinnamomi aeth. (not less than 75% cinnamaldehyde); 5 mg Caryophylli aeth. (not less than 80% eugenol); 10 mg Foeniculi aeth. (not less than 60% anethole and 10% fenchone); 5 mg Anisi aeth. (not less than 90% anethole); 10 mg Myrrhae tincture (DAB 10); 4 mg Rathanhiae tincture (DAB 10); 20 mg Alchemillae tincture (1:5); 20 mg menthol; 1 mg thymol; 6 mg phenylsalicylate; 0.4 mg guajazulene. 5 µl are applied for TLC investigation.

Chromatography and detection

Adsorbent	Silica gel 60 F_{254} precoated TLC plates (Merck, Germany)
Solvent system	toluene-ethyl acetate (93 : 7)
Detection	Vanillin-sulphuric acid reagent (VS No.42) or phosphomolybdic acid reagent (PMA No.34)

Commercial laxative phytopreparations

Mixed herbal preparation with anthraglycosides as the main components

Preparation of extracts

Three finely powdered dragées are extracted by heating on the water bath for 5 min with 6 ml methanol; 10 µl of the clear filtrate is used for chromatography.

Chromatography and detection

Adsorbent	Silica gel 60 F_{254}-precoated TLC plates (Merck, Germany)
Solvent system	Ethyl acetate-methanol-water (100 : 13.5 : 10) → 10 cm
Detection	UV-365 nm

Fig. 1 **Salviathymol®**

Interpretation of chromatograms

Identifiable terpenoids	~R_f value	VS (vis)	PMA (vis)
Azulene and terpene hydrocarbons	0.98	violet-blue	blue
Anethole (► T1)	0.9	violet-blue	blue
Thujone (after PMA)	0.7	–	red-violet
Thymol (► T2)	0.5	red-violet	blue
Cinnamaldehyde/eugenol	0.45	brown-orange	blue
cineole/piperitone	0.4	blue-orange	blue
Menthol (► T3)	0.2	blue	blue

Fig. 2 **Laxans**

Nine commercial laxative formulations are used for TLC (samples 1–9). They all represent mixtures of two to five anthraglycoside-containing drug extracts. In some cases, extracts of other drugs are also present (e.g. Gentianae radix, Bryoniae radix or Curcumae rhizoma).

The identifiable components are labelled I–VI

	I–VI	~R_f value	Samples
I	Anthraquinone aglycones	Solvent front	1, 2, 3, 4, 5, 6, 7, 8, 9
II	A-monoglycosides Frangulins A and B	0.8–0.85	2, 4, 5, 6, (7), 9
III	Deoxyaloin	0.6	1, 2, (3), (4), (5), (6), (7), 8
IV	Aloin Rhein	0.5	1, 2, 3, 4, 6, 7, 8, 9
V	Glucofrangulins Aloinosides	0.35–0.4	2, 3, 4, 5, 8
VI	Cascarosides A, B, C, D Sennosides	0.05–0.2 Start (UV-254 nm)	(1), 2, 3, 4, (5), (6), (7), 8 3, 4, 8

For further differentiation, the TLC plates are sprayed with the KOH reagent and NP/PEG reagent (see Chap. 2).
For analysis of sennosides in Sennae folium or fructus, the solvent system and detection method descriped in Fig. 7, 8, chap. 2) should be used.

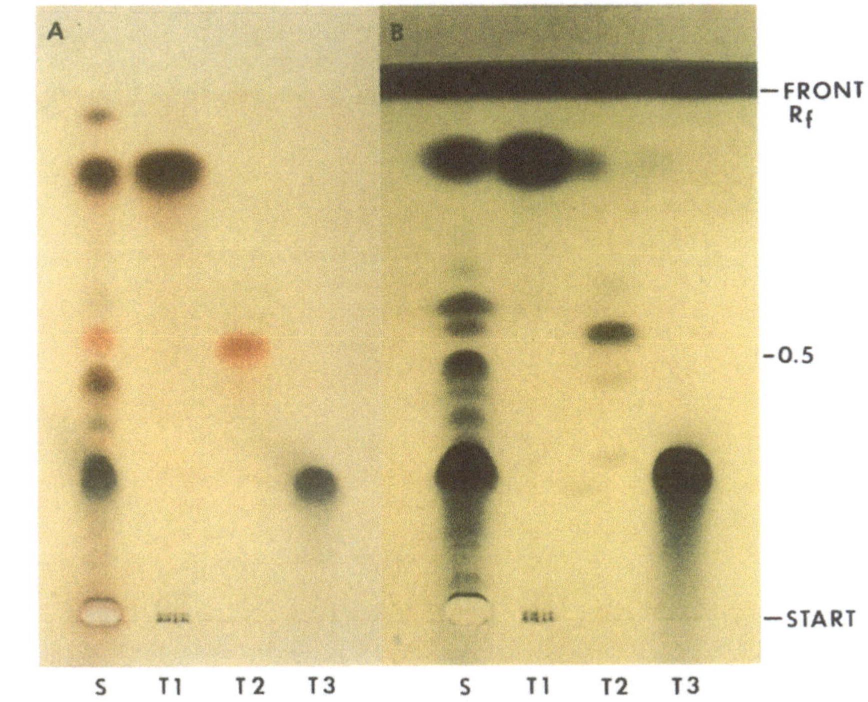
A
B
FRONT
Rf
0.5
START
S
T1
T2
T3
S
T1
T2
T3

Fig. 1

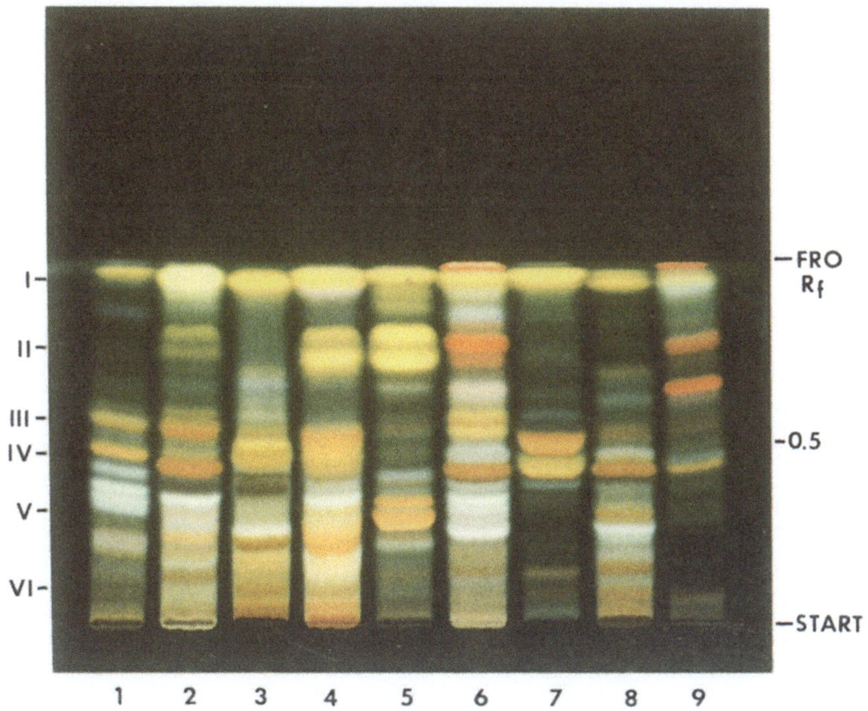
I
II
III
IV
V
VI
FRO
Rf
0.5
START
1
2
3
4
5
6
7
8
9

Fig. 2

Appendix A: Spray Reagents

Acetic anhydride reagent (AN) **No. 1**

The TLC plate is sprayed with 10 ml acetic anhydride, heated at 150°C for about 30 min and then inspected in UV-365 nm.

▶ Detection of ginkgolides.

Anisaldehyde-acetic acid reagent (AA) **No. 2**

0.5 ml anisaldehyde is mixed with 10 ml glacial acetic acid.
The plate is sprayed with 5–10 ml and then heated at 120°C for 7–10 min.

▶ Detection of petasin/isopetasin.

Anisaldehyde-sulphuric acid reagent (AS) **No. 3**

0.5 ml anisaldehyde is mixed with 10 ml glacial acetic acid, followed by 85 ml methanol and 5 ml concentrated sulphuric acid, in that order.
The TLC plate is sprayed with about 10 ml, heated at 100°C for 5–10 min, then evaluated in vis. or UV-365 nm.
The reagent has only limited stability and is no longer useable when the colour has turned to red-violet.

▶ Detection of terpenoids, propylpropanoids, pungent and bitter principles, saponins.

Antimony-III-chloride reagent ($SbCl_3$) **No. 4**

20% solution of antimony-III-chloride in chloroform (or ethanol).
The TLC plate must be sprayed with 15–20 ml of the reagent and then heated for 5–6 min at 110°C. Evaluation in vis. or UV-365 nm.

▶ Detection of cardiac glycosides, saponins.

Bartons reagent **No. 5**

(a) 1 g potassium hexacyanoferrate (III) in 100 ml water.
(b) 2 g iron-III-chloride in 100 ml water.
The TLC plate is sprayed with a.1:1 mixture of (a) and (b). Evaluation in vis.

▶ Detection of gingeroles (Zingiberis rhizoma).

Benzidine reagent (BZ) **No. 6**

0.5 g benzidine is dissolved in 10 ml glacial acetic acid and the volume adjusted to 100 ml with ethanol. Evaluation in vis.

▶ Detection of aucubin (Plantaginis folium).

Berlin blue reagent (BB) **No. 7**

A freshly prepared solution of 10 g iron-III-chloride and 0.5 g potassium hexacyanoferrate in 100 ml water. The plate is sprayed with 5–8 ml. Evaluation in vis.

▶ Detection of arbutin.

No. 8 **Blood reagent (BL)**

10 ml of 3.6% sodium citrate solution is added to 90 ml fresh bovine blood; 2 ml blood is mixed with 30 ml phosphate buffer pH 7.4. The plate is sprayed in a horizontal position. Phosphate buffer pH 7.4: 20.00 ml potassium dihydrogen phosphate solution (27.281 g potassium dihydrogen phosphate dissolved in double-distilled, CO_2-free water and volume adjusted to 10.00 ml) mixed with 39.34 ml 0.1 *M* sodium hydroxide, and volume made up to 100 ml with CO_2-free, double-distilled water.

▶ Detection of saponins: white zones are formed against the reddish background of the plate. Hemolysis may be immediate or may occur when the plate has been dried under slight warming.

No. 9 **Chloramine-trichloroacetic acid reagent (CTA)**

10 ml freshly prepared 3% aqueous chloramine T solution (syn. sodium sulphamide chloride or sodium tosylchloramide) is mixed with 40 ml 25% ethanolic trichloroacetic acid.

The plate is sprayed with 10 ml, heated at 100°C for 5–10 min; evaluated in UV-365 nm.

▶ Detection of cardiac glycosides.

No. 10 **Dichloroquinone chloroimide = Gibb's reagent (DCC)**

0.5% methanolic solution of 2,6-dichloroquinone chloroimide.

The plate is sprayed with 10 ml, then immediately exposed to ammonia vapour.

▶ Detection of arbutin, capsaicin.

No. 11 **Dinitrophenylhydrazine reagent (DNPH)**

0.1 g 2,4-dinitrophenylhydrazine is dissolved in 100 ml methanol, followed by the addition of 1 ml of 36% hydrochloric acid.

After spraying with about 10 ml, the plate is evaluated immediately in vis.

▶ Detection of ketones and aldehydes.

No. 12 **DNPH-acetic acid-hydrochloric acid reagent**

0.2 g 2,4-dinitrophenylhydrazine in a solvent mixture consisting of 40 ml glacial acetic acid (98%), 40 ml hydrochloric acid (25%) and 20 ml methanol.

The plate is sprayed with 10 ml and evaluated in vis. It is then heated at 100°C for 5–10 min and evaluated again in vis.

▶ Detection of valepotriates (Valeriana). Chromogenic dienes react without warming. Dienes can also be detected with HCl-AA reagent (No. 17).

No. 13 **Dragendorff reagent (DRG; MUNIER and MACHEBOEUF)**

Solution (a): Dissolve 0.85 g basic bismuth nitrate in 10 ml glacial acetic acid and 40 ml water under heating. If necessary, filter.

Solution (b): Dissolve 8 g potassium iodide in 30 ml water.

Stock solution: (a) + (b) are mixed 1:1.

Spray reagent: 1 ml stock solution is mixed with 2 ml glacial acetic acid and 10 ml water.

▶ Detection of alkaloids, heterocyclic nitrogen compounds.

Dragendorff reagent, followed by sodium nitrite or H_2SO_4

After treatment with Dragendorff reagent, the plate may be additionally sprayed with 10% aqueous sodium nitrite or with 10% ethanolic sulphuric acid, thereby intensifying the coloured zones (→ $NaNO_2$, dark brown; → H_2SO_4, bright orange).

EP reagent (EP) No. 14

0.25 g 4-dimethylamino benzaldehyde is dissolved in a mixture of 45 ml 98% acetic acid, 5 ml 85% o-phosphoric acid and 45 ml water, followed by 50 ml concentrated sulphuric acid (under cooling with ice).
The sprayed plate is evaluated in vis.
▶ Detection of proazulene (Matricariae flos); After heating at 100°C for 5–10 min, proazulene gives a blue-green colour (vis.) The blue colour of azulene is intensified by EP reagent.

Fast blue salt reagent (FBS) No. 15

0.5 g fast blue salt B is dissolved in 100 ml water. (Fast blue B = 3,3′-dimethoxybiphenyl-4,4′-bis(diazonium)-dichloride).
The plate is sprayed with 6–8 ml, dried and inspected in vis. Spraying may be repeated, using 10% ethanolic NaOH, followed again by inspection in vis.
▶ Detection of phenolic compounds.

Fast red salt reagent (FRS) No. 16

0.5% aqueous solution of fast red salt B (= diazotized 5-nitro-2-aminoanisole).
The plate is spraed with 10 ml, followed immediately by either 10% ethanolic NaOH or exposure to ammonia vapour.
▶ Detection of amarogentin.

Hydrochloric acid – glacial acetic acid reagent (HCl/AA) No. 17

Eight parts of concentrated hydrochloric acid and two parts of glacial acetic acid are mixed.
After spraying, the plate is heated at 110°C for 10 min. Evaluation in vis. or in UV-365 nm.
▶ Detection of valepotriates with diene structure (halazuchrome reaction).

Iodine reagent No. 18

About 10 g solid iodine are spread on the bottom of a chromatograph tank; the developed TLC plate is placed into the tank and exposed to iodine vapour.
▶ all compounds containing conjugated double bonds give yellow-brown (vis.) zones.

Iodine-chloroform reagent (I/$CHCl_3$) No. 19

0.5% Iodine in chloroform.
The sprayed plate is warmed at 60°C for about 5 min. The plate is evaluated after 20 min at room temperature in vis or in UV-365 nm.
▶ Detection of Ipecacuanha alkaloids.

Iodine-hydrochloric acid reagent (I/HCl) No. 20

(a) 1 g potassium iodide and 1 g iodine are dissolved in 100 ml ethanol.
(b) 25 ml 25% HCl are mixed with 25 ml 96% ethanol.
The plate is first sprayed with 5 ml of (a) followed by 5 ml of (b).
▶ Detection of the purine derivatives (caffeine, theophylline, theobromine)

Iodoplatinate reagent (IP) No. 21

0.3 g hydrogen hexachloroplatinate (IV) hydrate is dissolved in 100 ml water and mixed with 100 ml 6% potassium iodide solution.

The plate is sprayed with 10 ml and evaluated in vis.
▶ Detection of nitrogen-containing compounds, e.g. alkaloids (blue-violet).
▶ Detection of Cinchona alkaloids: the plate is first sprayed with 10% ethanolic H_2SO_4 and then with IP reagent.

No. 22 **Iron-III-chloride reagent ($FeCl_3$)**
10% aqueous solution.
The plate is sprayed with 5–10 ml and evaluated in vis.
▶ Detection of oleuropeine and hop bitter principles.

No. 23 **Kedde reagent (Kedde)**
5 ml freshly prepared 3% ethanolic 3,5-dinitrobenzoic acid is mixed with 5 ml 2 *M* NaOH.
The plate is sprayed with 5–8 ml and evaluated in vis.
▶ Detection of cardenolides.

No. 24 **Komarowsky reagent (KOM)**
1 ml 50% ethanolic sulphuric acid and 10 ml 2% methanolic 4-hydroxybenzaldehyde are mixed shortly before use.
The sprayed plate is heated at 100°C for 5–10 min. Evaluation in vis.
▶ Detection of essential oils, pungent principles, bitter principles, saponins, etc.

No. 25 **Liebermann-Burchard reagent (LB)**
5 ml acetic anhydride and 5 ml concentrated sulphuric acid are added carefully to 50 ml absolute ethanol, while cooling in ice. The reagent must be freshly prepared.
The sprayed plate is warmed at 100°C for 5–10 min and them inspected in UV-365 nm.
▶ Detection of triterpenes, steroids (saponins, bitter principles).

No. 26 **Marquis reagent**
3 ml formaldehyde is diluted to 100 ml with concentrated sulphuric acid. The plate is evaluated in vis, immediately after spraying.
▶ Detection of morphine, codeine, thebaine.

No. 27 **Millons reagent (ML)**
3 ml mercury is dissolved in 27 ml fuming nitric acid and the solution diluted with an equal volume of water.
▶ Detection of arbutin and phenolglycosides.

No. 28 **Natural products-polyethylene glycol reagent (NP/PEG) (=NEU-reagent)**
The plate is sprayed with 1% methanolic diphenylboric acid-β-ethylamino ester (= diphenylboryloxyethylamine, NP), followed by 5% ethanolic polyethylene glycol-4000 (PEG) (10 ml and 8 ml, respectively).
▶ Detection of flavonoids, aloin. Intense fluorescence is produced in UV-365 nm. PEG increases the sensitivity (from 10 μg to 2.5 μg). The fluorescence behaviour is structure dependent.

No. 29 **Ninhydrin reagent (NIH)**
30 mg ninhydrin is dissolved in 10 ml n-butanol, followed by 0.3 ml 98% acetic acid. After spraying (8–10 ml), the plate is heated for 5–10 min under observation and evaluated in vis.
▶ Detection of amino acids, biogenic amines, ephedrine.

Nitric acid (HNO_3 concentrate) **No. 30**

The TLC plate is inspected immediately after spraying.

► Detection of ajmaline and brucine, red in vis.

► Detection of sennosides: after spraying with HNO_3 concentrated the plate is heated for 15 min at 120°C; the plate then is sprayed with 10% ethanolic KOH reagent. Red-brown (vis) or yellow-brown fluorescent (UV-365 nm) zones are formed.

Nitrosodimethylaniline reagent (NDA) **No. 31**

10 mg nitrosodimethylaniline is dissolved in 10 ml pyridine and used immediately to spray the TLC plate.

► Detection of anthrone derivatives (grey-blue zones, vis).

Palladium-II-chloride reagent (PC) **No. 32**

Solution of 0.5% palladium-II-chloride in water, with 1 ml concentrated HCl.

► Detection of Allium species (yellow-brown zones in vis.).

Phenylenediamine reagent (PD) **No. 33**

0.5% ethanolic solution.

Evaluation in vis. or in UV-365 nm.

► Detection of constituents of Lichen islandicus (e.g. fumarprotocetraric acid).

Phosphomolybdic acid reagent (PMA) **No. 34**

20% ethanolic solution of phosphomolybdic acid. The plate is sprayed with 10 ml and then heated at 100°C for 5 min under observation.

► Detection of constituents of essential oils.

► Detection of rhaponticosides: 4 g phosphomolybdic acid is dissolved in 40 ml hot water; 60 ml concentrated sulphuric acid is carefully added to the cooled solution. Rhaponticoside and deoxyrhaponticoside form strong, blue (vis.) zones.

Potassium hydroxide reagent (KOH) **No. 35**

5% or 10% ethanolic potassium hydroxide (Bornträger reaction).

The plate is sprayed with 10 ml and evaluated in vis. or in UV-365 nm, with or without warming.

► Detection of anthraquinones (red), anthrones (yellow, UV-365 nm); ► Detection of coumarins (blue, UV-365 nm).

Potassium permanganate-sulphuric acid reagent (PPM) **No. 36**

0.5 g potassium permanganate is dissolved carefully in 15 ml concentrated sulphuric acid, while cooling in ice (warning: explosive manganese heptoxide is formed).

► Detection of fenchone; the plate is sprayed first with phosphomolybdic acid reagent (PMA No.34/10 min/110°C), followed by PPM reagent (5 min/110°C; blue, vis.).

Sulphuric acid (H_2SO_4) **No. 37**

(a) (5%) 10% ethanolic H_2SO_4

(b) 50% ethanolic H_2SO_4

(c) concentrated H_2SO_4

(a,b) The plate is heated at 100°C for 3–5 min, evaluation in vis.

(c) coloured (vis.) zones appear immediately.

► Detection of e.g. cardiac glycosides, lignans.

No. 38 Trichloroacetic acid-potassium hexacyanoferrate-iron-III-chloride reagent (TPF)
(a) 25% trichloroacetic acid in chloroform.
(b) 1% aqueous potassium hexacyanoferrate mixed with an equal volume of 5% aqueous iron-III-chloride.
The plate is sprayed with solution (a) and heated at 110°C for 10 min. It is then sprayed with solution (b) and evaluated in vis.
► Detection of sinalbin and sinigrin.

No. 39 Vanillin-glacial acetic acid reagent (VGA)
0.8 g vanillin are dissolved in 40 ml glacial acetic acid, 2 ml concentrated H_2SO_4 are added.
The plate is sprayed with 10 ml solution and heated for 3–5 min (110°C), evaluation in vis.
► Detection of salicin and derivatives.

No. 40 Vanillin-hydrochloric acid reagent (VHCl)
The plate is sprayed with 5 ml of 1% ethanolic vanillin solution, followed by 3 ml concentrated HCl, then evaluated in vis. Colours are intensified by heating for 5 min at 100°C.
► Detection of myrrh constituents.

No. 41 Vanillin-phosphoric acid reagent (VP)
(a) Dissolve 1 g vanillin in 100 ml of 50% phosphoric acid.
(b) Two parts 24% phosphoric acid and eight parts 2% ethanolic vanillic acid.
After spraying with either (a) or (b), the plate is heated for 10 min at 100°C, and evaluated in vis. or in UV-365 nm.
► Detection of e.g. terpenoids, lignanes and cucurbitacins.

No. 42 Vanillin-sulphuric acid reagent (VS)
1% ethanolic vanillin (solution I).
10% ethanolic sulphuric acid (solution II).
The plate is sprayed with 10 ml solution I, followed immediately by 10 ml solution II. After heating at 110°C for 5–10 min under observation, the plate is evaluated in vis.
► Detection of e.g. components of essential oils (terpenoids, phenylpropanoids).

No. 43 Van Urk reagent
0.2 g of 4-dimethylaminobenzaldehyde is dissolved in 100 ml 25% HCl with the addition of one drop of 10% iron-II-chloride solution.
► Detection of Secale alkaloids.

No. 44 Zimmermann reagent (ZM)
(a) 10 g dinitro benzene + 90 ml of toluene.
(b) 6 g NaOH + 25 ml water + 45 ml methanol.
The TLC is first sprayed with (a), followed by (b).
► Detection of sesquiterpenes (e.g. Arnica species).

Appendix B: Definitions

TLC: thin layer chromatography
Silica gel: specific surface area 500 cm^3/g; pore volume 0.75 cm^3/g; pore diameter 60 Å. The TLC is performed on siliacagel 60 F_{254} glas-coated TLC-plates from Merck (Germany), a company with world wide representation.
The use of silicagel-coated plates on aluminium or plastic or the use of material of other companies can give slight variations in the Rf-values, or may alter the TLC-fingerprint of plant extracts. This is due to different binding agents, which can influence the phenolic compounds, for example.

UV-254 nm: shorter wavelength ultraviolet light, used to detect substances that quench fluorescence. On TLC plates marked "60 F_{254}", the compounds with C=C double-bonds in conjugation, appeaer as dark zones against a yellow-green fluorescent background.
UV-365 nm: for the detection of substances that fluoresce in long wave ultraviolet light.
UV-lamps: commercially available e.g. 8 W low pressure mercury vapour tubes with selected filters (e.g. SCHOTT) or 125 W high pressure mercury discharge lamp, 365 nm.
vis.: visible light or daylight.

General concepts without chamber saturation: the chromatography solvent is poured into the chromatography tank, and swirled around for a few seconds. The TLC plate is then placed in position, and chromatography allowed to proceed.
With chamber saturation: the solvent is allowed to remain in the closed tank for 0.5–1 h before chromatography. The inside of the tank should be lined with filter paper.
Volume of chromatography solvent: about 100 ml is normally used. Chormatography tank dimensions: 20 × 9 × 20 cm.

Extraction procedures
Powdered drugs are used for extraction e.g. "medium-fine powder" corresponds to mesh size 300. Sample weights quoted for drug extraction refer to the dried drug.

Sample volume
The volumes quoted are recommended averages. Depending on the quality of the drug, larger and smaller volumes should also be used. Exact volumes can be applied with the aid of commercially available, standardized capillaries and application pipettes. If melting point capillaries are used, it can be assumed that 1 cm is roughly equivalent to 4–5 µl. As a rule, the sample should be applied to the start as a line about 0.5–1 cm wide. Small sample volumes (1–3 µl), however, are applied as a spot.

Standard Literature

Literature on medicinal plants, plant constituents and their pharmacological activities has piled up enormously in the last ten years and it is not possible to sum up all the available literature and to cite references on 200 herbal drugs and more. Instead we have tried to summarize all the relevant chemical constituents of plant parts which can be detected by the TLC methods in drug lists for each chapter.

Baerheim Svendsen A, Verpoorte R (1983) Chromatography of Alkaloids. Journal of Chromatography Library, Volume 23A. Elsevier Amsterdam Oxford New York

Budavari S (ed) (1989) The Merck Index: an encyclopedia of chemicals, drugs and biologicals, 11th edn. ISBN 911910- 28-X. Merck & Co., Inc, Rahway, NJ (or earlier editions where stated)

Goodman, Gilman A, Rall TW, Nies AS, Taylor P (eds) (1990) Goodman and Gilman's: the pharmacological basis of therapeutics, 8th edn. ISBN 0-08-040296-8. Pergamon Press, New York

Hänsel R, Keller K, Rimpler H, Schneider G (eds) (1992, 1993, 1994) Hagers Handbuch der Pharmazeutischen Praxis, Band 4, (Drogen A–D; Band 5, Drogen E–O; Band 6, Drogen P–Z). Springer, Berlin Heidelberg New York

Hartke K, Mutschler E (eds) (1991) DAB 10-Kommentar: Deutsches Arzneibuch, 10. Aufl. mit wissenschaftlichen Erklärungen. Bände 1–4; Band 1: Allgemeiner Teil (Methoden und Reagenzien); Band 2: Monographien A–F; Band 3: Monographien G–O; Band 4: Monographien P–Z. Wissenschaftliche Verlagsgesellschaft mbH. Stuttgart, or Govi-Verlag GmbH, Frankfurt

Heftmann E, (1982) Chromatography, fundamentals and applications of chromatographic and electophoretic methods (Part A, B). Journal of Chromatography Library Volume 22A and 22B, Elsevier Amsterdam Oxford New York

Kraus L, Koch A, Hoffstetter-Kuhn S (1995) Dünnschichtchromatographie. Springer Berlin Heidelberg New York Tokyo

Leung AY (1980) Encyclopedia of common natural ingredients used in food, drugs and cosmetics. ISBN 0-471-04954-9. John Wiley & Sons, New York

Pachaly P (1995) DC-Atlas. Wissenschaftliche Verlagsgesellschaft mbH, Stuttgart

Popl M, Fähnrich J, Tatar V (1990) Chromatographic analysis of alkaloids. In: Cazes (ed) Chromatographic science Volume 53 Marcel Dekker New York Basel

Reynolds JEF (ed) (1989) Martindale – the extra pharmacopoeia, 29th edn. The Pharmaceutical Press, London (or earlier editions where stated)

Steinegger E, Hänsel R (1988) Lehrbuch der Pharmacognosie und Phytopharmazie. ISBN 3-540-17830-9. Springer, Berlin Heidelberg New York

Wren RC (revised by Williamson EM, Evans FJ) (1988) Potter's new cyclopaedia of botanical drugs and preparations. ISBN 0-85207-197-3. The C.W. Daniel company Ltd, Saffron Walden

Wagner H (ed) (1993) Pharmazeutische Biologie, 5. Aufl. ISBN 3-437-20498-X. Gustav Fischer Verlag, Stuttgart

Wagner H (ed) (1995) Phytotherapie. ISBN 3-437-00775-0. Gustav Fischer Verlag, Stuttgart

Weiss RF (1989) Herbal medicine. Translated by AR Meuss from the 6th German edition of Lehrbuch der Phytotherapie. ISBN 3-8047-1009-3. Wissenschaftliche Verlagsgesellschaft mbH, Stuttgart

Wichtl M (ed) (1989) Teedrogen: Ein Handbuch für die Praxis auf wissenschaftlicher Grundlage, 2. Aufl. ISBN 3-8047-1009. Wissenschaftliche Verlagsgesellschaft mbH, Stuttgart

Pharmacopoeias

DAB 10	Deutsches Arzneibuch. 10. Ausgabe 1991. ISBN 3-7692-1461-7. Deutscher Apotheker Verlag, Stuttgart, or Govi-Verlag GmbH, Frankfurt
DAC 86	Deutscher Arzneimittelcodex, 1986 mit bis zu 3 Ergänzungen 1991. Issued by the Bundesvereinigung Deutscher Apothekerverbände. Deutscher Apotheker Verlag, Stuttgart, or Govi-Verlag GmbH, Frankfurt
Ph.Eur.2	European Pharmacopoeia, 2nd edn. Part I, 1980; Part II, 1st fascicle (1980) to 15th fascicle (1991). Published by Maisonneuve SA, 57160 Saint-Ruffine, France. Available in the UK from The Pharmaceutical Press, London
BP 88	British Pharmacopoeia 1988, with subsequent addenda up to 1992. Her Majesty's Stationery Office, London
BPC	British Pharmaceutical Codex. The Pharmaceutical Press, London. Prepared by the Pharmaceutical Society of Great Britain. Publication has ceased; 29th edn, 1989
BHP 90	British Herbal Pharmacopoeia, vol 1 (1990) of the revised edition. ISBN 0-903032-08-2. Published by the British Herbal Medicine Association. Obtainable by mail order from: BHMA Publications, P.O. Box 304, Bournemouth, Dorset BH7 6JZ
ÖAB 90	Österreichisches Arzneibuch (1990) und 1. Nachtrag, Verlag der Österreichischen Staatsdruckerei, Wien (or earlier edition ÖAB 81 (1981))
Ph.Helv.VII	Pharmacopoea Helvetica, Editio Septima, 1987 with later supplements. Distributed by Office Central Fédéral des Imprimés et du Matériel, 3000 Bern
USP XXII	The United States Pharmacopeia - The National Formulary (USP XXII-NF XVII). US Pharmacopeial Convention Inc, 1990 (or earlier editions when cited)
FCC III	Food Chemicals Codex, 3rd edn. National Academy Press, Washington, DC, 1981
USSR X	State Pharmacopoeia of the (former) Union of Soviet Socialist Republics, 10th edn (English version). Ministry of Health of the USSR, Moscow (ca. 1973)
Japan Jap XI	The Pharmacopoeia of Japan, 11th edn, 1986 (English version), The Society of Japanese Pharmacopoeia
China	Arzneibuch der chinesischen Medizin Monographien des Arzneibuchs der Volksrepublik China 1985 und nachfolgender Ausgaben 2. überarbeitete Aufl. 1991. Deutscher Apotheker Verlag, Stuttgart (3 Ergänzungslieferung 1994)

Subject Index

Page numbers referring to the description of chromatograms are printed in **bold** letters

MIX
Papier aus verantwortungsvollen Quellen
Paper from responsible sources
FSC® C105338

If you have any concerns about our products,
you can contact us on
ProductSafety@springernature.com

In case Publisher is established outside the EU,
the EU authorized representative is:
Springer Nature Customer Service Center GmbH
Europaplatz 3, 69115 Heidelberg, Germany

Printed by Libri Plureos GmbH
in Hamburg, Germany